ORGANIC PHOSPHORUS COMPOUNDS

ORGANIC PHOSPHORUS COMPOUNDS

Volume 5

G. M. KOSOLAPOFF
Auburn University

and

L. MAIER
Monsanto Research S. A.

WILEY-INTERSCIENCE, a Division of John Wiley & Sons, Inc.

New York · London · Sydney · Toronto

Library of Congress Cataloging in Publication Data:

Kosolapoff, Gennady M
Organic phosphorus compounds.

1950 ed. published under title: Organophosphorus compounds.
Includes bibliographies.
1. Organophosphorus compounds. I. Maier, L., joint author. II. Title.

QD421.P1K55 1973 547'.07 72-1359
ISBN 0-471-50444-0 (v. 5)

Printed in the United States of America

10 9 8 7 6 5 4 3 2 1

Contents

Chapter 12. Organic Derivatives of Hypophosphorous, Hypodiphosphorous, and Hypophosphoric Acid

M. BAUDLER

Institut für Anorganische Chemie der Universität, 5-Köln 1, Germany

The compounds of hypophosphorous acid H_3PO_2, hypodiphosphorous acid $H_4P_2O_4$, and hypophosphoric acid $H_4P_2O_6$ all contain phosphorus in a lower oxidation state than +5. The characteristic structural element of the derivatives of both diphosphorus acids is a P-P framework whose other valences are taken up by ester or amide groups together with oxygen and sulphur. The following classes of compounds are treated in detail:

1. Derivatives of hypophosphorous acid: $H_2P(O)OR$; corresponding thiocompounds are not yet known.
2. Derivatives of hypodiphosphorous acid: $(RO)_2P\text{-}P(OR)_2$, $(R_2N)_2P\text{-}P(NR_2)_2$; corresponding thioesters are not yet known.
3. Derivatives of hypophosphoric acid:

$$(RO)_2\overset{O}{\overset{\|}{P}}-\overset{O}{\overset{\|}{P}}(OR)_2,\quad (RO)_2\overset{S}{\overset{\|}{P}}-\overset{O}{\overset{\|}{P}}(OR)_2,\quad (RO)_2\overset{S}{\overset{\|}{P}}-\overset{S}{\overset{\|}{P}}(OR)_2,$$

$$(R_2N)_2\overset{O}{\overset{\|}{P}}-\overset{O}{\overset{\|}{P}}(NR_2)_2,\quad (R_2N)_2\overset{S}{\overset{\|}{P}}-\overset{S}{\overset{\|}{P}}(NR_2)_2.$$

A. METHODS OF SYNTHESIS

A.1. Derivatives of Hypophosphorous Acid

Esters of hypophosphorous acid $H_2P(O)OR$ were not synthesized until 1960. The corresponding thiocompounds $H_2P(S)OR$ and $H_2P(S)SR$ have not yet been prepared.

I. REACTION OF HYPOPHOSPHOROUS ACID WITH DIAZOALKANES

The treatment of hypophosphorous acid with diazoalkanes[19] has proved to be a suitable method for the preparation of monoesters of monoalcohols of low molecular weight (Kabachnik et al.). The reaction of the acid with diazomethane or diazoethane proceeds smoothly, even in the cold.

$$H_2\overset{O}{\overset{\|}{P}}\text{-}OH + R'N_2 \longrightarrow H_2\overset{O}{\overset{\|}{P}}\text{-}OR + N_2;\ R' + H = R\ (\text{Me-, Et-})$$

The monoester results even in the presence of a large excess of diazoalkane.

II. REACTION OF HYPOPHOSPHOROUS ACID WITH ALCOHOLS

The reaction of the acid with alcohols has been used for the preparation of certain monoalkyl esters of hypophosphorous acid, the resulting water being azeotropically distilled off with benzene. At 110°C in an atmosphere of inert gas, the reaction is complete within a few hours and can be catalyzed by aluminum chloride, zinc chloride, or sodium acetate. The yields of the hypophosphites formed lie between 80% and 87%.[34,35]

The esterification of the acid with high polymer alcohols such as polyvinyl alcohol and cellulose[32-34,41] has been carried out using the same method. Reacting 3 mole of hypophosphorous acid per monomer unit of polyvinyl alcohol

$$(-CH_2-\underset{\displaystyle OH}{\underset{|}{CH}}-)$$

results in a product which contains 7.5% phosphorus.[32] A paper chromatography investigation of the hydrolysis product shows the expected hypophosphorous acid and also some phosphorous acid, so that under the conditions of the ester synthesis, a partial disproportionation to the phosphorous acid and phosphine evidently occurred.[32]

For the preparation of the cellulose ester of hypophosphorous acid, the fibers are soaked for some time in a methanolic H_3PO_2- solution, dried, and then heated in an inert gas atmosphere. Using this method, cellulose hypophosphites are prepared with a phosphorus content up to 14%, corresponding to a degree of substitution of 0.95. The hydrolysis of the ester forms H_3PO_2, which can be proven unequivocally by paper chromatography. If the temperature is allowed to rise above 120°C during the reaction, cellulose hypophosphite is formed as well as the corresponding phosphite. Under the conditions of the esterification, the cellulose evidently does not undergo any noticeable cleavage to components of lower molecular weight (glucose or oligosaccharides).[41] Instead of the acid, ammonium hypophosphite can also be used for the preparation of cellulose hypophosphite. However, the salt is less reactive than the acid. Moreover the esterification products contain varying proportions of nitrogen which is interpreted by partial disproportionation of ammonium hypophosphite, thus enabling the formation of ammonium cellulose phosphite.[41]

III. REACTION OF HYPOPHOSPHOROUS ACID WITH ORTHO-CARBONYL COMPOUNDS

A particularly easy preparation of alkyl hypophosphites is the reaction of crystalline hypophosphorous acid with orthocarbonyl compounds[12] (Fitch)

$$H_2\overset{\overset{O}{\|}}{P}\text{-}OH + R_nC(OR')_{4-n} \longrightarrow H_2\overset{\overset{O}{\|}}{P}OR' + R'OH + R_n\overset{\overset{O}{\|}}{C}(OR')_{2-n}$$

(orthocarbonate n = 0, orthocarboxylate n = 1). The starting materials are mixed at +2 to +5°C in an inert gas atmosphere and subsequently warmed to room temperature while being thoroughly mixed. With increasing esterification, a one phase system is formed, and 1 to 30 min is therefore necessary. This time depends on the orthocarbonyl compound used, the water contents of the acid, and the intensity of mixing of the reaction components. In most cases, the esterification is almost quantitative. The yields of the alkyl hypophosphates, however, are reduced by disproportionating reactions during the distillation of the products.

When using ketales or acetales (n = 2) for this type of reaction, the alkyl hypophosphite reacts with the ketone or aldehyde simultaneously formed, giving phosphinic acid esters. In this case, the preparation of hypophosphorous acid esters is impossible.

A.2. Derivatives of Hypodiphosphorous Acid

Several methods have been attempted to prepare the tetraalkyl and tetraarylesters of hypodiphosphorous acid $(RO)_2P\text{-}P(OR)_2$ without obtaining pure compounds, however.

It has been reported[31] that small amounts of esters of hypodiphosphorous acid have been detected UV spectroscopically after alcoholysis of diphosphorus tetraiodide solutions in hydrocarbons. However, other phosphorus compounds absorb in the same range of wavelengths, thus necessitating further confirmation to be made.

In regard to the known properties of hypodiphosphorous acid $H_4P_2O_4$--easily oxidisable in air and having a marked tendency to disproportionate to P(III) compounds and compounds of lower oxidation number of phosphorus[7,9]--it would be expected that the esters could only be prepared under mild temperature conditions and complete exclusion of air.

IV. REACTION OF $(RX)_2PY$ WITH SODIUM

According to a patent,[11] corresponding esters are said

to be prepared by the reaction of the chlorides of the diesters of phosphorous acid with sodium in xylene at about 100°C, separation of sodium chloride, followed by washing the solution with water until neutral and then removing the solvent. However, these results could not be repeated by other workers.[10] Under the reaction conditions a mixture of substances was obtained which could not be separated by distillation and which contained a considerable proportion of phosphorous acid triester.

In the same patent,[11] the class of compounds, thioesters of hypodiphosphorous acids $(RS)_2P-P(SR)_2$ is mentioned, but even to the present time no member has been prepared in the pure state.

However, pure hypodiphosphorous acid tetrakis-[dimethylamide] (tetrakis-[dimethylamino]-biphosphine) can be prepared in 25% yield from the reaction of the chloride of phosphorous acid bisdimethylamide with sodium in petrolether or hexane under dried inert gas (Nöth et al.).[36,37]

$$2\ [Me_2N]_2P\text{-}Cl + 2\ Na \longrightarrow [Me_2N]_2P\text{-}P[NMe_2]_2 + 2\ NaCl$$

Pentakis-[dimethylamino]-triphosphine $P_3(NMe_2)_5$, tris-[dimethylamino]-phosphine $P(NMe_2)_3$ (phosphorous acid-tris-[dimethylamide]) and polymeric dimethylaminopolyphosphines $P_x(NMe_2)_y$ are formed in lower yield,[37] (see Vol. 1, p. 346).

A.3. Derivatives of Hypophosphoric Acid

Several methods may be used for the preparation of tetraesters of hypophosphoric acid $(RO)_2(O)P-P(O)(OR)_2$. One can either start with derivatives of various monophosphoric acids and link them by a coupling reaction to form a P-P bond or with compounds already containing P-P groups.

V. REACTION OF SODIUM DIALKYLPHOSPHITES WITH CHLORIDES OF DIALKYLPHOSPHORIC ACID

One of the best and simplest methods for the preparation of symmetrical and unsymmetrical tetraalkylesters of hypodiphosphoric acid is the reaction of sodium dialkylphosphite with chlorides of dialkylphosphoric acids (Michalski, Zwierszak[28,29]).

$$(RO)_2P\text{-}ONa + Cl\text{-}P(O)(OR)_2 \longrightarrow (RO)_2(O)P\text{-}P(O)(OR)_2 + NaCl$$

The reaction proceeds at 0-5°C in benzene with an average yield of 50% w.r.t. tetraalkylhypophosphates. P(III)-O-P(V) acid esters $(RO)_2P-O-P(O)(OR)_2$, tetraalkyl pyrophosphate $(RO)_2(O)P-O-P(O)(OR)_2$, and tetraalkyl pyrophosphite $(RO)_2P-O-P(OR)_2$ are formed in lower yield. In earlier investigations, these compounds were not separated.[30] Using a large excess (25-50%) of sodium dialkylphosphite, the formation of pyrophosphoric acid ester is almost completely repressed.[28,29,50] Because of its lower boiling point and particularly its substantially faster rate of hydrolysis at room temperature, the P(III)-O-P(V) acid ester can be separated from the hypophosphoric acid ester.

The first step which is proposed for the reaction is the following, in which the ambidentic character of the phosphite anion plays an important factor:[29,50]

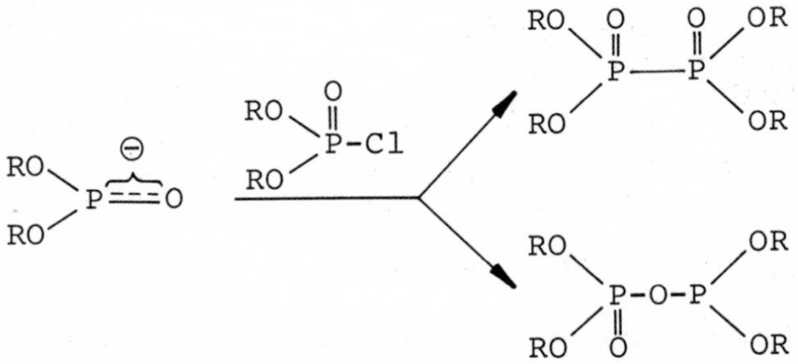

Monothiohypophosphate $(RO)_2(O)P-P(S)(OR)_2$ and dithiohypophosphate $(RO)_2(S)P-P(S)(OR)_2$ may also be prepared by the same type of synthesis.

In principle, the preparation of monothiohypophosphate may be carried out in two ways,[22] i.e., the reaction of sodium dialkylphosphites with dialkyl monothiophosphoric acid chlorides

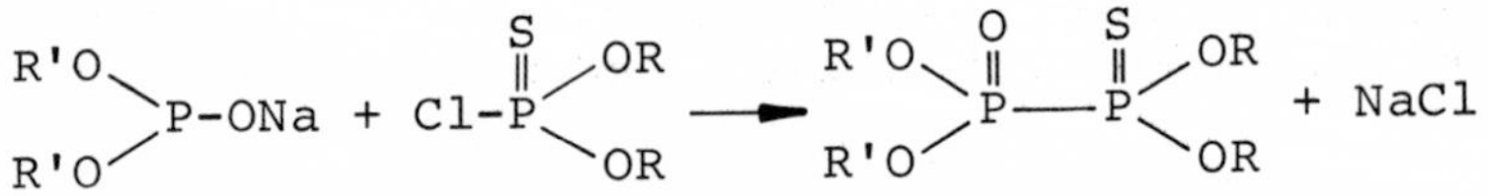

or the reaction of sodium dialkylthiophosphites with dialkylphosphoric acid chlorides

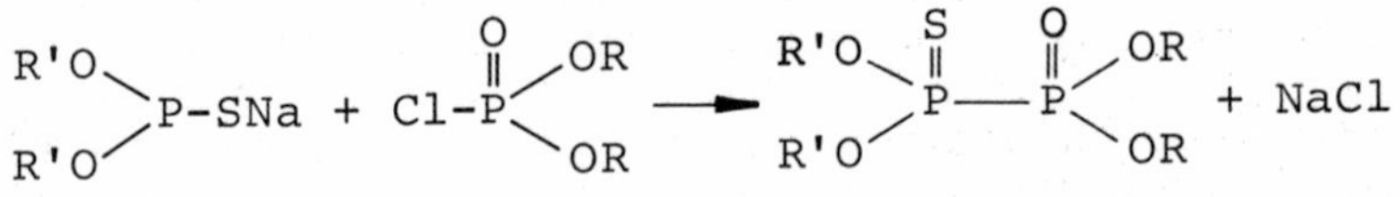

(Michalski, Stec, Zwierzak).

However, only the second method of synthesis is of practical importance due to the fact that in the first reaction simultaneous formation of dithiohypophosphate and further products takes place, some of which are difficult to remove by distillation. The reaction of the dialkylphosphoric acid chlorides with sodium dialkylphosphites is carried out in benzene at 40 to 50°C. With potassium dialkyl thiophosphites, room temperature is sufficient. The working up of the reaction products with water, followed by high vacuum distillation yields 40-50% of the pure tetraalkyl monothiohypophosphates.

The reaction of sodium dialkyl thiophosphite with dialkyl monothiophosphoric acid chloride in boiling benzene gives tetraalkyldithiohypophosphate[1] (Almasi, Paskucz).

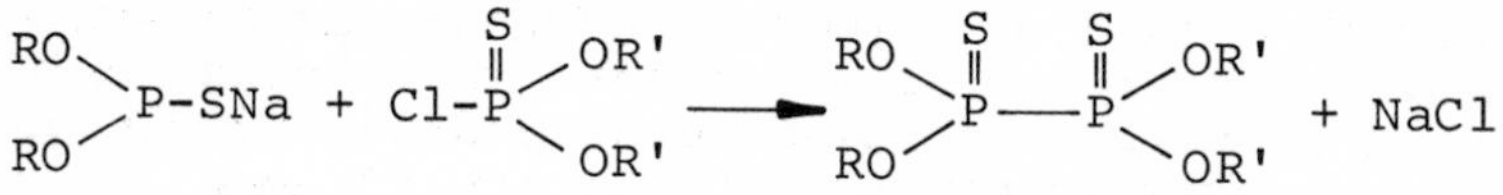

After removal of the precipitate formed in the reaction and high vacuum fractional distillation, pure tetraalkyldithiohypophosphates are obtained in 45-50% yields. The same reaction path also makes possible the synthesis of cyclic hypophosphoric acid esters together with their monothio and dithio derivatives. Thus the reaction of the sodium salt of 2-hydroxy-5,5-dimethyl-1,3,2-dioxaphosphorinane with 2-chloro-2-oxo-5,5-dimethyl-1,3,2-dioxaphosphorinane yields the corresponding hypophosphoric acid ester [bis-(5,5-dimethyl-2-oxo-1,3,2-dioxaphosphorinanyle]:[48]

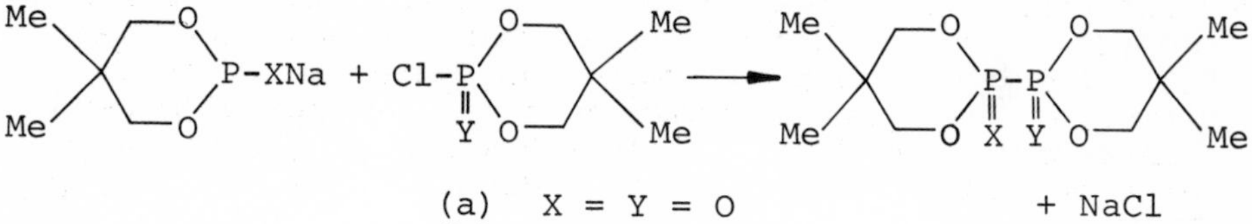

(a) X = Y = O
(b) X = S, Y = O
(c) X = Y = S

VI. REACTION OF DIALKYLPHOSPHORIC ACID CHLORIDE WITH SODIUM

The reaction of sodium on dialkylphosphoric acid chlorides leads preferably to the formation of a P-P bond, concomitant with the formation of tetraalkylesters of hypophosphoric acid[5,6] (Baudler).

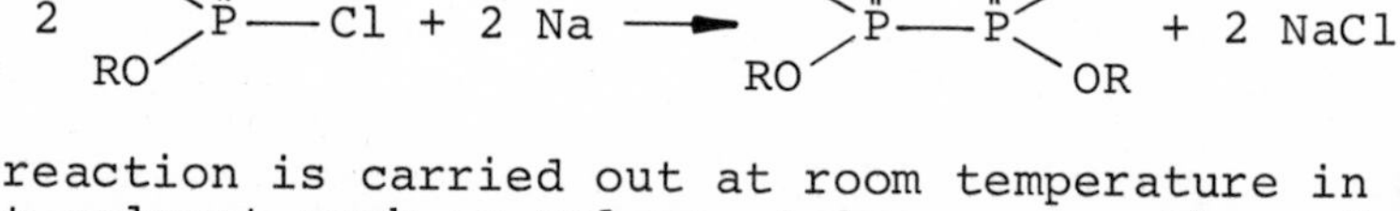

The reaction is carried out at room temperature in an inert solvent such as xylene, toluene, or ether, followed by gentle warming. After separation of the precipitate formed during the reaction and high vacuum distillation, there results a mixture of products consisting, in the main, of hypophosphoric acid ester together with considerable amounts of pyrophosphoric acid ester. Further separation can be achieved by high vacuum fractional distillation in a special apparatus.

Dialkylphosphoric acid bromides and iodides react similarly with sodium. Difficulties are experienced however, during the distillation of the reaction mixture, due to the thermal decomposition of unreacted starting material.[6] According to a patent, the corresponding esters of dithiohypophosphoric acid are said to be obtained from thiophosphoric acid 0,0-diesterchlorides with sodium.[11] As the obtained products were not thoroughly characterized, further investigations are necessary.

VII. FURTHER ATTEMPTS FOR THE PREPARATION USING MONOPHOSPHORIC ACID DERIVATIVES

1. Tetraalkylesters of hypophosphoric acid are said to be obtained by heating dialkylphosphoric acid chlorides with trialkylphosphites.[40] As investigations from other workers have demonstrated,[8,29,51] the reaction mechanism is complex and leads essentially to isomeric compounds in which both phosphorus atoms are linked via an oxygen bridge such as is found in anhydrides. In the reaction mixture, tetraalkylhypophosphites were only detected in small quantities by thin layer chromatography[51] (compare, however, the reaction of $PhPCl_2$ + $(MeO)_3P$ which has been reported to give $(MeO)_2(O)P-P(Ph)-P(O)(OMe)_2$,[12a] (see Vol. 1, p. 338).

2. Tetraethyl hypophosphate was formed by the reaction of sodium diethylphosphite with 1,2-dibromoderivatives such as 1,2-dibromocyclohexane or 1,2-dibromoethane as well as bromine.[3] Simultaneously with the formation of the resulting unsaturated hydrocarbons are produced esters of various other mono- and diphosphoric acids which do not allow a satisfactory preparative separation.

3. As was earlier reported,[15,20,21] the reaction of sulfuryl chloride on sodium dialkylphosphate does not lead, under the formation of a P-P bond, to pure tetraalkylhypophosphates. In fact, a very complex reaction mixture results, containing mainly esters of pyrophosphoric and thiopyrophosphoric acid.[3,29,49,51]

4. Tetraalkyl dithiohypophosphates are said to be prepared in satisfactory yield by reacting nitrosylchloride with thiophosphorous acid 0,0-dialkylesters in the presence of pyridine.[16,26,27] Other workers have reported tetraalkyl monothiopyrophosphate to be the main product.[1]

VIII.   REACTION OF HYPOPHOSPHORIC ACID WITH DIAZOALKANES

The direct esterification of the anhydrous acid with diazoalkanes[4,6,42] (Baudler) is a very clean reaction for the synthesis of pure tetraalkylhypophosphates.

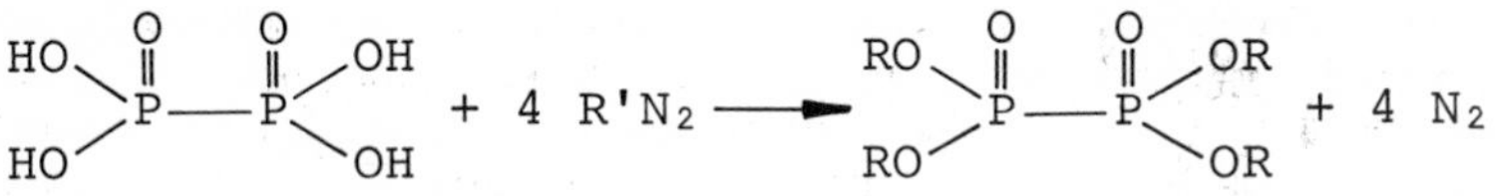

$$
\underset{HO}{\overset{HO}{>}}\!\!P\!\!\overset{\overset{O}{\|}}{-}\!\!P\!\!\underset{OH}{\overset{OH}{<}} \;+\; 4\ R'N_2 \;\longrightarrow\; \underset{RO}{\overset{RO}{>}}\!\!P\!\!\overset{\overset{O}{\|}}{-}\!\!P\!\!\underset{OR}{\overset{OR}{<}} \;+\; 4\ N_2
$$

R' + H = R (Me-, Et-)

Suitably, the reaction is carried out at 0°C by adding carefully dried acid in small portions to an excess of ethereal diazoalkane solution.

After removal of the solvent and high vacuum fractionation, the pure tetraalkylhypophosphates are obtained in 40-50% yield.

IX.   ATTEMPTED PREPARATION OF TETRAALKYLESTERS USING SILVER HYPOPHOSPHATE

Contrary to earlier reports,[43-46] the reaction of silver hypophosphate with alkyl halides does not only lead to hypophosphoric acid esters. Instead it is found that a mixture of esters of various mono- and diphosphoric acids is formed. Under mild conditions in boiling ether, however, noticeable amounts of hypophosphoric acid ester are produced.[42] Simultaneously formed pyrophosphoric acid ester can only be separated with difficulty.

X.   REACTION OF HYPODIPHOSPHOROUS ACID-TETRAKIS-(DI-METHYLAMIDE) WITH OXYGEN AND SULPHUR

By controlled oxidation of hypodiphosphorous acid-tetrakis-[dimethylamide] (tetrakis-[dimethylamino]-biphosphine) with aerobic oxygen, hypophosphoric acid-tetrakis-[dimethylamide] (tetrakis-[dimethylamino]-biphosphinedioxide) is obtained[37] (Nöth, Vetter).

$$
(Me_2N)_2P\text{-}P(NMe_2)_2 \;+\; O_2 \;\longrightarrow\; (Me_2N)_2\overset{\overset{O}{\|}}{P}\text{-}\overset{\overset{O}{\|}}{P}(NMe_2)_2
$$

Sulphur reacts analogously giving dithiohypophosphoric acid-tetrakis-[dimethylamide][36,37] (Nöth, Vetter).

B. CHEMICAL PROPERTIES

B.1. Hypophosphorous Acid Derivatives

The monomethyl and monoethylesters of hypophosphorous acid decompose easily at room temperature but are stable at $-60°C$ to $-70°C$.[12,19] They are oxidized by air.[19] On the other hand, esters of long chain alcohols are noticeably more stable.[17,18,35] Water causes a hydrolytic reaction, forming the alcohol and hypophosphorous acid. Even under very mild conditions this reaction proceeds at a considerable rate with polyvinyl and cellulose ester.[32,41] Cellulosehypophosphite hydrolyzes faster than the corresponding phosphite.

All the esters of the hypophosphorous acid react with aldehydes or ketones even under relative mild conditions forming the corresponding phosphinic acid esters.[12]

$$H_2\overset{\displaystyle O}{\overset{\displaystyle \|}{P}}\text{-OR} + R_2CO \longrightarrow RO-\underset{H}{\overset{\displaystyle O}{\overset{\displaystyle \|}{P}}}\underset{R}{\overset{R}{-C}}\text{-OH}$$

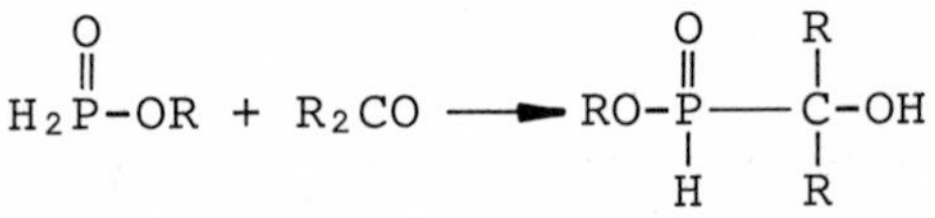

Using chloral, one obtains bis-(α-hydroxy-β,β,β-trichlorethyl)-phosphinic acid ester.[18]

Alkyl hypophosphites react with bis-(diethylamino)-methand $CH_2(MEt_2)_2$ or diethylamino methyl ethyl ether $EtOCH_2NEt_2$ forming a mixture of esters of bis-(diethyl-amino-methyl)-phosphinic acid $(Et_2NCH_2)_2PO_2R$ and diethyl-amino-methyl-phosphonic acid $Et_2NCH_2P(O)(OR)_2$.[17] They also add to activated olefins to give phosphinic acid derivatives.[19a]

B.2. Hypodiphosphorous Acid Derivatives

From results of previous experiments[10] in preparing tetra-alkyl and tetraarylesters of hypodiphosphorous acid, these compounds with a phosphorus oxidation number of +2 have a marked tendency to disproportionate into P(III) compounds and substances of lower phosphorus oxidation number. Even at $0°C$ this rearrangement has a noticeable reaction rate. Together with a great susceptibility to oxidation, these properties have not yet permitted their preparation in the pure state.

Principally analogous behavior is shown by hypodiphosphorous acid-tetrakis-[dimethylamide] (tetrakis-[dimethylamido]-biphosphine) which has been isolated in pure form and whose chemical properties are much better known.[36,37] The colorless crystalline compound even slowly decomposes at room temperature, giving a yellow coloration and disproportionating to $P(NMe_2)_3$, $P_3(NMe_2)_5$, and $P_x(NMe_2)_y$. With hydrogen chloride at $-90°C$, a diphosphonium salt is formed

$$(Me_2N)_2P\text{-}P(NMe_2)_2 + 2\ HCl \longrightarrow (Me_2N)_2P\text{-}P(NMe_2)_2 \cdot 2\ HCl$$

which disproportionates above -80°C.

Similarly, methyl iodide reacts at -70°C forming $[(Me_2N)_3PMe]I$ and dimethylaminopolyphosphine. Generally, Lewis acids such as BCl_3, PCl_3, and $(Me_2N)_2PCl$ promote the disproportionation of hypodiphosphorous acid-tetrakis-[dimethylamide]; also simultaneous ligand exchange reactions take place. Hypodiphosphorous acid-tetrakis-[dimethylamide] behaves toward oxygen, sulphur, diborane, and carbon disulphide like a tetraorganylbiphosphine. This extremely oxygen sensitive and spontaneously inflammable compound gives hypophosphoric acid-tetrakis-[dimethylamide] under controlled oxidation with air. Sulphur reacts in an analogous way to form the corresponding dithiophosphoric acid derivative. These compounds are thermally and hydrolytically considerably more stable than the starting material. Using diborane in a molar ratio 1:1, a reaction product is obtained in which every phosphorus atom is bonded to a BH_3 group. It is relatively stable to hydrolysis but decomposes on heating above 130°C. Carbon disulphide also forms an oily, stable, deep brown-red colored adduct of the composition $P_2(NMe_2)_4 \cdot CS_2$. Hypodiphosphorous acid-tetrakis-[dimethylamide] hydrolyzes relatively slowly, but fast in the presence of an acid. As with tetraorganyl biphosphines, the reaction with bromine causes cleavage of the P-P bond. The first reaction step forms $(Me_2N)_2PBr$ which reacts further to $(Me_2N)_2PBr_3$.

B.3. Hypophosphoric Acid Derivatives

Only since it has been possible to synthesize a number of corresponding compounds during the last few years, the chemical behavior of esters of hypophosphoric acid, monothio-, and dithiohypophosphoric acid has become known in more detail.

Tetraalkylhypophosphates are hydrolyzed by water to monophosphoric acid derivatives:[6,28,29,51]

$$(RO)_2\overset{O}{\overset{\|}{P}}\text{-}\overset{O}{\overset{\|}{P}}(OR)_2 \xrightarrow[16\ h/60°]{H_2O} (RO)_2\overset{O}{\overset{\|}{P}}\text{-}OH + (RO)_2\overset{O}{\overset{\|}{P}}\text{-}H$$

Hydrolytic cleavage occurs with about the same rate as in tetraalkylpyrophosphates but is noticeably slower as with the isomeric esters of P(III)-O-P(V) acid. This may be made use of in the analytical determination in mixtures of both classes of compounds and for their preparative separation.[51] Cyclic hypophosphoric acid esters such as

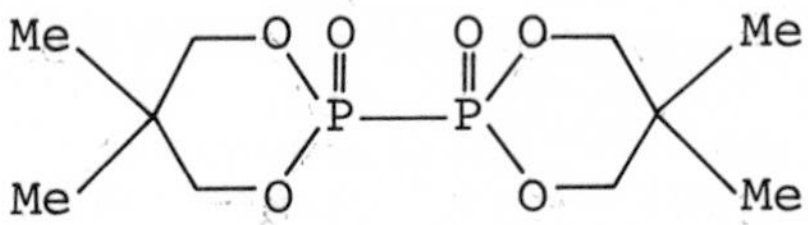

are considerably more stable towards hydrolysis and show almost no change after 20 hr of refluxing with water in tetrahydrofuran.[48]

Tetraalkylmonothiohypophosphates are even more resistant toward hydrolytic decomposition. In contrast to the isomeric

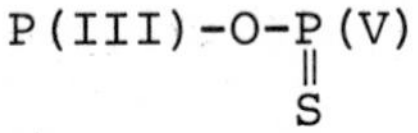

esters, no reaction with excess water in dioxane takes place on heating at 60°C for 16 hr. There is no reaction even in a weak alkaline medium ($NaHCO_3$).[22]

Chlorolysis of hypophosphoric acid esters gives, under homolytic cleavage of the P-P bond, the corresponding phosphoric acid ester dichlorides.[51] This can be done using sulphuryl chloride or elemental chlorine. Tetraethylhypophosphate reacts with sulphuryl chloride even at 50 to 60°C in 55% yield, while with chlorine only at 100°C or above in 40% yield, forming the phosphoric acid diester chloride.[28,29,51] On the other hand, cyclic hypophosphates react only with chlorine at still higher temperatures, while they are cleaved with bromine even at room temperature.[48]

At low temperature (0 to 5°C) in benzene solution, tetraalkyl monothiohypophosphates react with sulphuryl chloride forming an equimolecular mixture of the diester-chlorides of phosphoric acid and the monothiophosphoric acid (total yield approximately 50%).[22] Under the same conditions, dithiohypophosphates react quantitatively to the monothiophosphoric acid diester chlorides.[1]

No reaction occurs between hypophosphoric acid esters and copper(I)-halides. They are relatively stable toward air and other oxidizing agents. No reaction occurs between the pure compounds and iodine solution; only after a certain period of standing in air does any discoloration become apparent due to beginning hydrolysis.[4,6] There is also no reaction when heated for several hours with ethanol or acetic acid or with hydrogen sulphide in the presence of pyridine at room temperature.[51]

At 140°C in boiling xylene, tetraalkylhypophosphates as well as tetraalkylmonothiophosphates react smoothly with mercury oxide to give pyrophosphoric acid esters.[22,51] Obviously, this reaction is characteristic for compounds containing a P-P bond. The mechanism is still unknown. Dinitrogen tetroxide can also convert hypophosphoric acid esters into the pyrophosphate even at 0°C.[47]

Using sodium iodide or ammonium thiocyanate, tetraalkyl hypophosphates can be converted by anionic dealkylation into salts of symmetrical dialkyldihydrogenhypophosphates.[24,25,51]

$$RO\diagdown\overset{\overset{O}{\|}}{P}\text{---}\overset{\overset{O}{\|}}{P}\diagup OR \quad \underset{\text{------------------------}}{\xrightarrow{\text{NaI resp. NH}_4\text{SCN}}} \quad RO\diagdown P\text{---}P\diagup OR$$

$$RO\diagup \qquad \diagdown OR \qquad\qquad\qquad (NH_4)NaO\diagup \qquad \diagdown ONa(NH_4)$$

On carrying out the reaction in boiling acetone or cyclohexanone, the salts are immediately precipitated out. When treated with cation exchangers (e.g., "Dowex 50") the corresponding free acids are obtained and can be characterized as their anilinium or cyclohexylammonium salts.[24,25] On stronger heating, the tetraalkylhypophosphates rearrange to the isomeric esters of the P(III)-O-P(V) acid.[23,51] At temperatures of 190 to 200°C for 3 to 4 hr, 60% rearrangement occurs. This reaction is of theoretical interest because the hypophosphates are thermodynamically more stable and a 3-valent phosphorus atom is formed from a 5-valent phosphorus atom. A satisfactory explanation of the mechanism has not yet been proposed.

C. GENERAL DISCUSSION OF THE PHYSICAL PROPERTIES

C.1, Hypophosphorous Acid Derivatives

Alkylesters of hypophosphorous acid are--according to the size of the alkyl group--colorless liquids or solids which, in the case of methyl and ethyl compounds, decompose at room temperature.[19,35]

Esters of hypophosphorous acid derived from high polymer alcohols such as polyvinyl alcohol or cellulose are solids. Polyvinylhypophosphite has a lower combustibility, but poorer thermal stability than polyvinyl alcohol and is a suitable plastic stabilizer.[32,33] The cellulose ester also shows considerable heat resistance which increases with increasing phosphorus content.[41]

The hypophosphites have a ^{31}P NMR spectrum similar to the hypophosphorous acid; however, one observes a shift of 6 to 12 ppm for the triplet to lower field.

In the spectrum of methyl hypophosphite, every single signal is therefore split into a quartet with a coupling constant of 13 cps. In the ethyl and butyl ester only the hydrogen atoms closest to the phosphorus atom give a perceptible coupling.[12]

C.2. Hypodiphosphorous Acid Derivatives

Because tetraalkyl or tetraaryl esters of hypodiphosphorous acid have not yet been prepared in a pure state, their physical properties are not known in detail. It has been suspected that their ultraviolet absorption lies between λ_{max} = 220 mμ (R = Me) and 244 mμ (R = Bu),[31] although this needs further confirmation.

Hypodiphosphorous acid tetrakis-[dimethylamide] (tetrakis-[dimethylamino]-biphosphine) forms colorless crystals which even at room temperature decompose slowly with yellow coloration and disproportionation. The compound dissolves well in ethers, aliphatic and aromatic hydrocarbons, and acetonitrile. The IR spectrum has only been investigated in the NaCl region and, as was expected, showed great similarity with that of tris(dimethylamino)-phosphine.[36,37]

C.3. Hypophosphoric Acid Derivatives

Tetraalkylesters of hypophosphoric acid are colorless, low viscous liquids which are distillable in high vacuum and generally easily soluble in organic solvents.[4,6,29,51] The tetramethyl and tetraethyl esters have a fruity, aromatic odor. Their solubility in ether noticeably increases with increasing size of the alkyl group.[6]

All hypophosphoric acid esters have considerably higher refractive indices ($n_D \simeq 1.4400$) than the corresponding isomeric P(III)-O-P(V) esters and the pyrophosphates. This is a relatively simple criterium, therefore, for the purity of these compounds.[4,6,29,51] The P-P bond refraction of the group

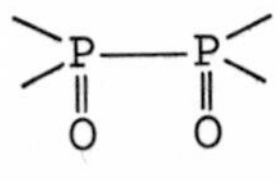

has an average value of R_D = 6.68 ccm[29] as derived from the molar refraction of several alkyl esters.

The Raman spectra of tetramethyl and tetraethyl hypophosphates show an intensive line at 260 and 245 cm^{-1}, respectively, attributable to the P-P stretching vibration.[6] Several tetraalkyl hypophosphates have been investigated using ^{31}P and ^{1}H NMR spectroscopy.[14] The chemical shift of the ^{31}P signal of the ethyl ester has a value of -7.2 ppm and is only slightly changed when proceeding to the n-propyl or n-butyl ester. However, the influence of the i-propyl group is greater.[14]

The symmetrical dialkyl dihydrogen hypophosphates are faint yellow, sirupy liquids which are difficult to purify. In contrast, using a suitable solvent mixture, the colorless dianilinium and dicyclohexylammonium salts can be relatively easily purified by crystallization and characterized by their melting points. The disodium salts are white, crystalline, high melting powders, easily soluble in water and unsoluble in the usual organic solvents.[24,25,51]

The tetraalkyl monothio and dithio hypophosphates with the exception of i-butyl are also liquids which can be high vacuum distilled. They are insoluble in water but dissolve easily in organic solvents. The refractive indices lie even higher than those of the pure oxygen compounds; they increase from monothio to dithio hypophosphates according to increasing sulphur content.[1,22] The P-P bond refraction of the group

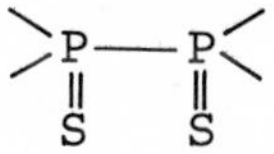

has an average value of R_D = 7.95 ccm[1] as derived from the molar refraction of several alkyl esters; a somewhat higher value than that of the hypophosphates. The IR spectra of tetraalkyl dithio hypophosphates show among others an intensive band at 630 to 648 cm^{-1} (P=S group) and a weaker absorption between 430 and 470 cm^{-1} attributable to the P-P bond.[1]

The ^{31}P NMR spectrum of the tetraethyl ester of monothio phosphoric acid shows two signals, having chemical shifts of -6.3 (P=O) and -73.5 ppm (P=S). The coupling constant is 583 cps. For the corresponding dithiophosphoric acid ester, the chemical shift of the ^{31}P signal is -75.8 ppm.[14] Cyclic hypophosphoric acid esters with the structure

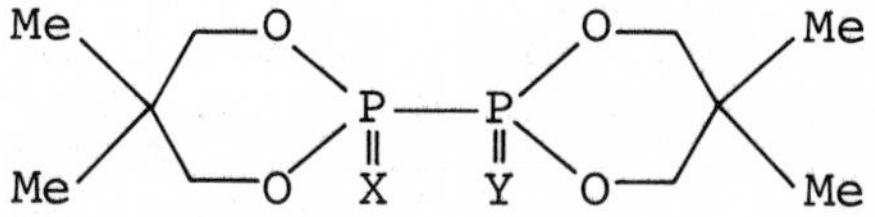

X = Y = 0; X = Y = S; X = 0, Y = S

are colorless, crystalline, high melting solids which are insoluble in water and only slightly soluble in organic solvents.[48] Single crystals of these compounds have been investigated morphologically and by x-ray diffraction.[13] Hypophosphoric acid-tetrakis-[dimethylamide] (tetrakis-[dimethylamino]-biphosphine dioxide) is a colorless liquid which is distillable in high vacuum, slightly soluble in water but, however, well soluble in alcohol, ether, and benzene.[37] Dithiohypophosphoric acid tetrakis-[dimethylamide] (tetrakis-[dimethylamino]-biphosphine-disulphide) forms high melting, laminaform crystals, which are difficultly soluble in water but dissolve well in benzene, ether, and alcohol.[36,37]

D. LIST OF COMPOUNDS

D.1. Hypophosphorous Derivatives

$MeOP(O)H_2$. I.[19] III.[12] Colorless liquid, $b_{2.5}$ 25-5.5°,[19] $b_{0.7}$ 22°,[12] n_D^{20} 1.4275,[19] d_4^{20} 1.2177,[19] ^{31}P-43.1/-18.9/+4.6 ppm,[12] ^{1}H NMR.[12]
$EtOP(O)H_2$. I.[19] III.[12] Colorless liquid, b_2 31-2°,[19] $b_{1.3}$ 33-5°,[12] n_D^{20} 1.4250,[19] d_4^{20} 1.1120,[19] ^{31}P-38.7/-15.0/+8.4 ppm,[12] ^{1}H NMR.[12]
$PrOP(O)H_2$. II.[17] Liquid, b_{10} 90-100°,[17] n_D^{20} 1.4222,[17] d_4^{20} 1.0447.[17]
$BuOP(O)H_2$. II.[17] III.[12] Liquid, b_{10} 120-5°,[17] n_D^{20} 1.4288,[17] d_4^{20} 1.0045.[17]
i-$BuOP(O)H_2$. II.[17] Liquid, b_{10} 110-5°,[17] n_D^{20} 1.475,[17] d_4^{20} 0.9760.[17]
$AmOP(O)H_2$. II.[17] Liquid, $b_{0.2}$ 70-80°,[17] n_D^{20} 1.4332,[17] d_4^{20} 1.0126.[17]
$HexOP(O)H_2$. II.[17,34,35] Liquid, $b_{0.2}$ 105-10°,[17] n_D^{20} 1.4530,[35] n_D^{20} 1.4355,[17] d_4^{20} 0.9614.[17]
$C_7H_{15}OP(O)H_2$. II.[17] Liquid, $b_{0.004}$ 100-10°,[17] n_D^{20} 1.4390,[17] d_4^{20} 0.9619.[17]
$C_8H_{17}OP(O)H_2$. II.[17] Liquid, $b_{0.006}$ 120-30°,[17] n_D^{20} 1.4430,[17] d_4^{20} 0.9709.
$C_9H_{19}OP(O)H_2$. II.[17] Liquid, $b_{0.006}$ 160-5°,[17] n_D^{20} 1.4460,[17] d_4^{20} 0.9524.[17]
$C_{10}H_{21}OP(O)H_2$. II.[17] Liquid, $b_{0.002}$ 130-40°,[17] n_D^{20} 1.4465.[17]
$C_6H_5CH_2OP(O)H_2$. II.[17] Liquid, $b_{0.05}$ 160-70°,[17] n_D^{20} 1.5415,[17] d_4^{20} 1.1496.[17]
$HOCH_2OCH_2OP(O)H_2$. II.[17] n_D^{20} 1.4505,[17] d_4^{20} 1.4259.[17]
$(CH_2)_{10}O_2[P(O)H_2]_2$. II.[34] Solid, m. 76-8°.[34]
Lauryl-$OP(O)H_2$. II.[34,35] Solid, m. 63-4°.[34,35]
Polyvinyl-$OP(O)H_2$. II.[32,33,34] Solid, decomp. 170-8°.[34]
Cell-$OP(O)H_2$. II.[41] Solid.[41]

D.2. Hypodiphosphorous Derivatives

$(Me_2N)_4P_2$. IV.[36,37] Colorless needles, m. 48°,[36,37] $b_{0.01}$ 50°.[36,37]

D.3. Hypophosphoric Derivatives

$(MeO)_4P_2O_2$. VIII.[4,6,42] Colorless liquid, $b_{10^{-4}}$ 85-7°,[4,6] n_D^{20} 1.4421,[4,6] d_4^{20} 1.331,[4,6] Raman (P-P-valencyvibration 260 cm^{-1}).[6,42]
$(EtO)_4P_2O_2$. V.[29] VI.[5,6] VIII.[4,6] Colorless liquid, $b_{10^{-4}}$ 90-3°,[6] $b_{0.01}$ 88-9°,[29] n_D^{20} 1.4400,[6] n_D^{20} 1.4394,[29]

d_4^{20} 1.172,[29] IR,[3] Raman (P-P-valencyvibration 245 cm^{-1}),[3,6] ^{31}P-7.2 ppm,[14] ^1H NMR,[14] MR_D.[29]

$(EtO)_2(MO)_2P_2O_2$. 2C. M = Na. High-melting solid.[24,25]
M = NH_4. Solid.[24,25]
M = $PhNH_3$. Colorless leaflets, m. 171-3° (from MeOH/Me_2CO).[24,25]
M = c - $C_6H_{11}NH_3$. Small, colorless needles, m. 218-20° (from EtOH/C_6H_6).[24,25]

$(PrO)_4P_2O_2$. V.[29] Colorless liquid, $b_{0.03}$ 100-2°,[29] n_D^{20} 1.4394,[29] d_4^{20} 1.087,[29] MR_D.[29]

$(PrO)_2(MO)_2P_2O_2$. 2C. M = Na. High-melting solid.[24,25]
M = $PhNH_3$. Small, colorless leaflets, m. 172-3° (from i-PrOH/H_2O).[24,25]
M = c - $C_6H_{11}NH_3$. Small, colorless leaflets, m. 209-13° (from i-PrOH/H_2O).[24,25]

$(iso\text{-}PrO)_4P_2O_2$. V.[29] Colorless liquid, $b_{0.001}$ 87-8°,[29] n_D^{20} 1.4313,[29] d_4^{20} 1.067,[29] MR_D.[29]

$(BuO)_4P_2O_2$. V.[29] Colorless liquid, $b_{0.05}$ 120-2° [29] n_D^{20} 1.4434,[29] d_4^{20} 1.041,[29] MR_D.[29]

$(BuO)_2(MO)_2P_2O_2$. 2C. M = Na. High-melting solid.[24,25]
M = $PhNH_3$. Small, colorless leaflets, m. 166-8° (from i-PrOH/H_2O).[24,25]

M = c - $C_6H_{11}NH_3$. White, crystalline powder, m. 206-8°.[24,25]

$(EtO)_2(PrO)_2P_2O_2$. V.[29] Colorless liquid, $b_{0.03}$ 99-101°,[29] n_D^{20} 1.4389,[29] d_4^{20} 1.119,[29] MR_D.[29]

$(EtO)_2(BuO)_2P_2O_2$. V.[29] Colorless liquid, $b_{0.01}$ 110-1°,[29] n_D^{20} 1.4409,[29] d_4^{20} 1.085,[29] MR_D.[29]

$(PrO)_2(BuO)_2P_2O_2$. V.[29] Colorless liquid, $b_{0.001}$ 119-20°,[29] n_D^{20} 1.4406,[29] d_4^{20} 1.063,[29] MR_D.[29]

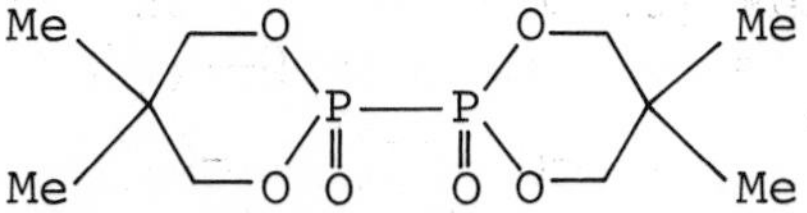

. V.[48] Colorless needles, m. 232-5°,[48] IR,[48] x-ray structure P-P bond length 2.2 Å.[13]

$(Me_2N)_4P_2O_2$. X.[37] Colorless liquid, $b_{1.5}$ 145-50°.[37]

$(EtO)_4P_2(S)(O)$. V.[22] Colorless liquid, $b_{0.01}$ 99°,[22] n_D^{20} 1.4724,[22] d_4^{20} 1.167,[22] R_f value,[22] ^{31}P-6.3/-73.5 ppm,[14] ^1H NMR.[14]

$(PrO)_4P_2(S)(O)$. V.[22] Colorless liquid, $b_{0.01}$ 110°,[22] n_D^{20} 1.4696,[22] R_f value.[22]

$(BuO)_4P_2(S)(O)$. V.[22] Colorless liquid, $b_{0.03}$ 128°,[22] n_D^{20} 1.4685,[22] d_4^{20} 1.045,[22] R_f value.[22]

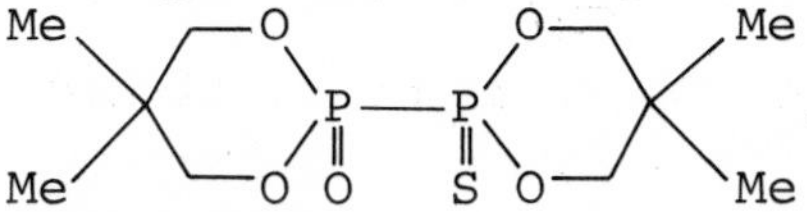

. V.[48] Colorless needles, m. 243-5°,[48] IR,[48] x-ray structure.[13]

$(EtO)_4P_2S_2$. V.[1] Liquid, m. 0-1°,[1] $b_{0.1}$ 109°,[1] n_D^{20} 1.5080,[1] d_4^{20} 1.1639,[1] IR,[1] MR_D,[1] ^{31}P -75.8 ppm,[14] ^1H NMR.[14]

$(EtO)_2(PrO)_2P_2S_2$. V.[1] Liquid, $b_{0.15}$ 117°,[1] n_D^{20} 1.5015,[1]

d_4^{20} 1.1225,[1] MR_D.[1]
$(EtO)_2(iso\text{-}PrO)_2P_2S_2$. V.[1] Liquid, $b_{0.2}$ 105°,[1] n_D^{20} 1.4963,[1] d_4^{20} 1.1124,[1] MR_D.[1]
$(EtO)_2(BuO)_2P_2S_2$. V.[1] Liquid, $b_{0.3}$ 136°,[1] n_D^{20} 1.4970,[1] d_4^{20} 1.0919,[1] MR_D.[1]
$(EtO)_2(iso\text{-}BuO)_2P_2S_2$. V.[1] Liquid, $b_{0.15}$ 119°,[1] n_D^{20} 1.4940,[1] d_4^{20} 1.0840,[1] MR_D.[1]
$(EtO)_2(AmO)_2P_2S_2$. V.[1] Liquid, $b_{0.1}$ 142°,[1] n_D^{20} 1.4938,[1] d_4^{20} 1.0656,[1] MR_D.[1]
$(EtO)_2(iso\text{-}AmO)_2P_2S_2$. V.[1] Liquid, $b_{0.3}$ 144°,[1] n_D^{20} 1.4920,[1] d_4^{20} 1.0646,[1] MR_D.[1]
$(PrO)_4P_2S_2$. V.[1] Liquid, $b_{0.07}$ 123°,[1] n_D^{20} 1.4970,[1] d_4^{20} 1.0913,[1] MR_D.[1]
$(iso\text{-}PrO)_4P_2S_2$. V.[1] Solid, m. 43-4°,[1] $b_{0.1}$ 113°,[1] n_D^{20} 1.4865,[1] d_4^{20} 1.0695,[1] MR_D.[1]
$(BuO)_4P_2S_2$. V.[1] Liquid, $b_{0.1}$ 152°,[1] n_D^{20} 1.4910,[1] d_4^{20} 1.0480,[1] MR_D.[1]
$(iso\text{-}BuO)_4P_2S_2$. V.[1] Liquid, $b_{0.1}$ 136°,[1] n_D^{20} 1.4855,[1] d_4^{20} 1.0349,[1] MR_D.[1]

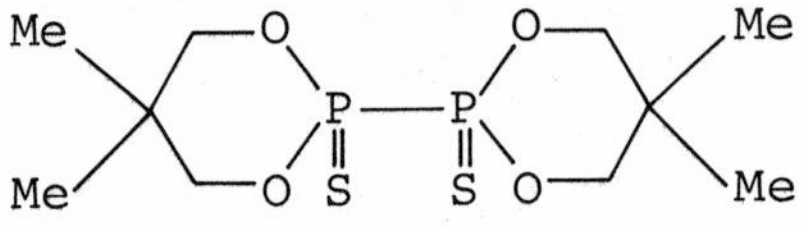

. V.[48] Colorless needles, m. 233-5°,[48] IR,[48] x-ray structure P-P bond length 2.18 Å.[13]

$(Me_2N)_4P_2S_2$. X.[36,37] Leaflets, m. 227°.[36,37]

(received November 9, 1970)

REFERENCES

1. Almasi, L., and L. Paskucz, Chem. Ber., 96, 2024 (1963).
2. Arbuzov, A. E., and B. A. Arbuzov, J. Prakt. Chem. [N.F.], 130, 103 (1931).
3. Arbuzov, B. A., E. N. Dianova, V. S. Vinogradova, and A. K. Shamsutdinova, Dokl. Acad. Nauk SSSR, 158, 137 (1964); C.A. 61, 16088h (1964).
4. Baudler, M., Z. Naturforsch., 8b, 326 (1953).
5. Baudler, M., Z. Naturforsch., 9b, 447 (1954).
6. Baudler, M., Z. Anorg. Allg. Chem., 288, 171 (1956).
7. Baudler, M., Z. Naturforsch., 14b, 464 (1959).
8. Baudler, M., and W. Giese, Z. Anorg. Allg. Chem., 290, 258 (1957).
9. Baudler, M., and M. Mengel, Z. Anorg. Allg. Chem., 373, 285 (1970).
10. Baudler, M., G. Sadri, and A. Moog, to be published.
11. Engelke, E. F., U.S. 2 403 792 (1946); C.A. 40, 60997 (1946).
12. Fitch, S. J., J. Am. Chem. Soc., 86, 61 (1964).
12a. Fluck, E., and H. Binder, Inorg. Nucl. Chem. Lett., 3, 307 (1967).
13. Galdeki, Z., and J. Karolak-Wojciechowska, to be published.

14. Harris, R. K., A. R. Katritzky, S. Musierowicz, and B. Ternai, J. Chem. Soc. A, 1967, 37.
15. Houben-Weyl, Methoden der organischen Chemie, Stuttgart, 1964, Vol. 12/II, p. 2.
16. Houben-Weyl, Methoden der organischen Chemie, Stuttgart, 1964, Vol. 12/II, p. 3.
17. Ivanov, B. E., and L. A. Kudryavtseva, Izv. Akad. Nauk SSSR, Ser. Khim., 1967, 1498; C.A. 68, 78359x (1968).
18. Ivanov, B. E., and L. A. Kudryavtseva, Izv. Akad. Nauk SSSR, Ser. Khim., 1968, 1633; C.A. 69, 87114m (1968).
19. Kabachnik, M. J., A. E. Shipov, and T. A. Mastrjukova, Izv. Akad. Nauk SSSR, Ser. Khim., 1960, 146; C.A. 54, 20838a (1960).
19a. Maier, L., Helv. Chim. Acta, 56, 489 (1973).
20. Michalski, J., and T. Modro, Chem. Ind. (London), 1960, 1570.
21. Michalski, J., and T. Modro, Rocz. Chem., 36, 483 (1962); C.A. 59, 1468d (1963).
22. Michalski, J., W. Stec, and A. Zwierzak, Bull. Acad. Polon. Sci., Ser. Sci. Chim., 13, 677 (1965).
23. Michalski, J., W. Stec, and A. Zwierzak, Chem. Ind. (London), 1965, 347.
24. Michalski, J., W. Stec, and A. Zwierzak, Bull. Acad. Polon. Sci., Ser. Sci. Chim., 14, 843 (1966).
25. Michalski, J., W. Stec, and A. Zwierzak, Chem. Ind. (London), 1966, 856.
26. Michalski, J., and A. Zwierzak, Angew. Chem., 73, 142 (1961).
27. Michalski, J., and A. Zwierzak, Rocz. Chem., 36, 489 (1962); C.A. 59, 1468f (1963).
28. Michalski, J., and A. Zwierzak, Proc. Chem. Soc., 1964, 80.
29. Michalski, J., and A. Zwierzak, Bull. Acad. Polon. Sci., Ser. Sci. Chim., 13, 253 (1965).
30. Milobedzki, T., and J. Walczynska, Rocz. Chem., 8, 486 (1928).
31. Moeller, T., and J. E. Huheey, J. Inorg. Nuclear Chem., 24, 315 (1962).
32. Nifant'ev, E. E., I. V. Fursenko, and V. L. L'vov, Vysokomolekul. Soedin., Ser. B, 9, 18 (1967); C.A. 66, 86481e (1967).
33. Nifant'ev, E. E., and L. P. Levitan, U.S.S.R 166694 (1964); C.A. 62, 7895h (1965).
34. Nifant'ev, E. E., and L. P. Levitan, Zh. Obshch. Khim., 35, 758 (1965); C.A. 63, 4152a (1965).
35. Nifant'ev, E. E., L. P. Levitan, and Z. V. Vil'danov, U.S.S.R. 166028 (1964); C.A. 62, 10338b (1965).
36. Nöth, H., and H. -J. Vetter, Naturwiss., 47, 204 (1960).
37. Nöth, H., and H. -J. Vetter, Chem. Ber., 94, 1505 (1961).

38. Nylen, P., Studien über organische Phosphorverbindungen (Diss.), Upsala, 1930.
39. Nylen, P., and O. Stelling, Z. Anorg. Allg. Chem., 212, 169 (1933).
40. Petrov, K. A., N. K. Bliznyuk, and V. E. Burigin, Zh. Obshch. Khim., 29, 1486 (1959); C.A. 54, 8606a (1960).
41. Predvoditelev, D. A., E. E. Nifant'ev, and Z. A. Rogovin, Vysokomolekul. Soedin., 7, 791 (1965); C.A. 63, 18443d (1965).
42. Remy, H., and H. Falius, Z. Anorg. Allg. Chem., 282, 217 (1955).
43. Rosenheim, A., and J. Pinsker, Chem. Ber., 43, 2003 (1910).
44. Rosenheim, A., and M. Pritzke, Chem. Ber., 41, 2708 (1908).
45. Rosenheim, A., W. Stadler, and F. Jakobsohn, Chem. Ber., 39, 2837 (1906).
46. Sänger, A., Ann. Chem., 232, 1 (1886).
47. Samuel, D., and B. Silver, Chem. Ind. (London), 1961, 556.
48. Stec, W., and A. Zwierzak, Can. J. Chem., 45, 2513 (1967).
49. Stec, W., A. Zwierzak, and J. Michalski, Tetrahedron Letters, 56, 5873 (1968).
50. Stec, W., A. Zwierzak, and J. Michalski, Bull. Acad. Polon. Sci., Ser. Sci. Chim., 18, 23 (1970).
51. Zwierzak, A., Lecture delivered at the conference on organophosphorus compounds, Moscow, October 1965.

Chapter 13. Organic Derivatives of Phosphorous Acid and Thiophosphorous Acid

W. GERRARD AND H. R. HUDSON

Department of Chemistry, Polytechnic of North London, Holloway Road, London N7 8DB, England

A. SYNTHESIS OF PHOSPHITES AND HALOGENOPHOSPHITES

Ia. INTERACTIONS OF HYDROXY COMPOUNDS AND PHOSPHORUS TRIHALIDES OR HALIDITES IN THE ABSENCE OF A TERTIARY BASE

The esters of phosphorous acid in the main may be deemed to come directly or indirectly from phosphorus trichloride, or analogously from the tribromide or triiodide; and a remarkable variety of reaction systems and factors can be involved. Two aspects of alcohol-phosphorus trihalide systems developed almost independently up to about 1940; one related to the formation of phosphorus esters and the other to the preparation of alkyl halides, early publications having shown that both types of product were obtainable. (See Refs. 358, 415, 755, 793, 854, 1013, 1156, 1215, 1283, 1333, 1367, 1368, 1384, 1390, and 1398.)

By 1917 the pattern of reactions for phosphorus trichloride began to emerge[898] when it was declared that the action of the trichloride on an alcohol results in the formation of the triester which is then stepwise dealkylated to phosphorous acid:

$$PCl_3 + 3\ ROH \longrightarrow (RO)_3P + 3\ HCl$$

$$(RO)_3P + HCl \longrightarrow (RO)_2POH + RCl$$

$$(RO)_2POH + HCl \longrightarrow ROP(OH)_2 + RCl$$

$$ROP(OH)_2 + HCl \longrightarrow P(OH)_3 + RCl$$

Despite this revelation, the formation of alkyl halides by the action of phosphorus trihalides in the experiments on the Walden inversion (1925-1937)[573,683,684] was deemed to be due to an ionic decomposition of the compound $ROPX_2$. By analogy with chlorosulphites, the dichlorides were supposed to undergo essentially two types of reaction.

$$ROPCl_2 \longrightarrow Cl^- + ROP^+Cl \longrightarrow RCl + (OPCl)$$

$$ROPCl_2 \longrightarrow R^+ + O^-PCl_2 \longrightarrow R^+ + Cl^- + (OPCl) \longrightarrow RCl + (OPCl)$$

Dichlorophosphites of aliphatic primary or secondary alcohols are in fact remarkably stable, both thermally and to dealkylation by hydrogen chloride.

In 1940 detailed work was reported[449] on the initial stages of a comprehensive study of alcohol-phosphorus trihalide systems in the attempt to elucidate the nature and sequences of reactions from as many aspects as could then be conceived, especially with reference to steric course and consequent changes in the magnitude and sign of optical rotatory power. The essential factors involved are now seen to be the electron density on oxygen, and the reactivities of the C-O and O-H bonds. These are determined by the nature of the atoms or groups attached to C. Since the interaction of a hydroxy-compound and a phosphorus trihalide involves the formation of a hydrogen halide, which intervenes in definite ways, the electron density on oxygen is most appropriately measured in terms of hydrogen bonding with hydrogen halide as revealed by the solubility of a hydrogen halide (HCl, HBr, HI) in the compound containing it. Solubility data have vital significance relating to the constitution of reaction systems, and to operational factors.[27,452,454-456,472,473]

Phosphorus trichloride systems will first be described. Interaction of the trichloride with alcohols involves a complex system of consecutive and concurrent reactions which should never be referred to as "the reaction." Although reports enable one to arrive at a rough general description, there is still much to be done to assess the various operational factors, especially in relation to the degree of reactivity of the C-O bond, and the O-H bond, and to steric requirements. It is necessary to distinguish

between compelling and permissive factors in an attempt
to discern optimum procedure. It is essential to appre-
ciate that much can depend upon the constitution of the
particular alcohol. For alcohols of "normal reactivity"
(i.e., most primary and secondary aliphatic alcohols) and
for phenols, reaction may be depicted in the following
tentative diagrams:

$$\begin{array}{ccc} R & Cl\ Cl \\ \backslash & \backslash / \\ O\!-\!-\!\blacktriangleright P \\ | & | \\ H\!\blacktriangleleft\!-\!-\!Cl \end{array} \quad \begin{array}{c} (a) \\ \xrightarrow{\ quick\ } \end{array} \quad ROPCl_2 + HCl \quad \begin{array}{c} partly \\ held\ as \\ Cl_2P\!-\!\underset{R}{O}\!\cdots\!HCl \end{array}$$

ROH added to PCl_3 alkyl (or aryl) phosphorodichloridite

n-Bu or Ph   alkyl (or aryl) dichlorophosphite

            alkyl (or aryl) oxydichlorophosphine

$$\begin{array}{ccc} R & Cl\ OR \\ \backslash & \backslash / \\ O\!-\!-\!\blacktriangleright P \\ | & | \\ H\!\blacktriangleleft\!-\!-\!Cl \end{array} \quad \begin{array}{c} (b) \\ \xrightarrow{\ quick\ } \\ but\ slower \\ than\ (a) \end{array} \quad (RO)_2PCl + HCl \quad \begin{array}{c} more \\ effectively \\ held\ as \\ Cl\!-\!P\overset{OR}{\underset{R}{\diagdown}O\!-\!-\!-\!H\!-\!Cl} \end{array}$$

            dialkyl(diaryl)phosphorochloridite

            dialkyl(diaryl)chlorophosphite

            dialkyl(diaryl)oxychlorophosphine

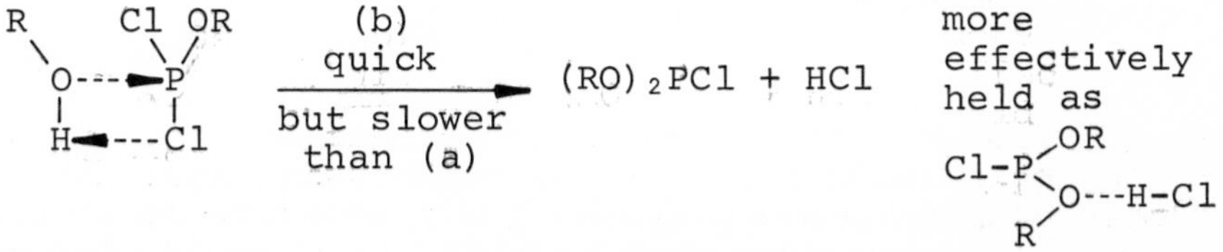

$$\begin{array}{ccc} R & RO\ OR \\ \backslash & \backslash / \\ O\!-\!-\!\blacktriangleright P \\ | & | \\ H\!\blacktriangleleft\!-\!-\!Cl \end{array} \quad \begin{array}{c} (c) \\ \xrightarrow{\quad\quad} \\ slower\ than \\ (b) \end{array} \quad (RO)_3P + HCl \quad \begin{array}{c} no\ HCl \\ evolved \\ from\ this \\ step \end{array}$$

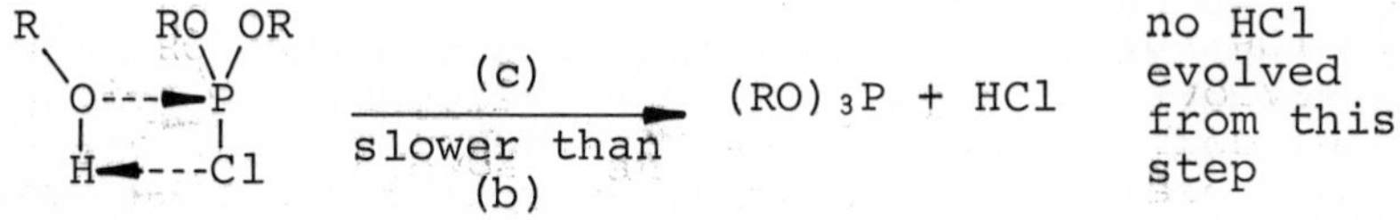

Rapid dealkylation of the trialkyl phosphite by hydrogen
halide may then ensue so that the principal products are
alkyl chloride and dialkyl hydrogen phosphite.

$$(RO)_2P\!-\!O\underset{HCl}{\overset{R}{\diagup}} \longrightarrow Cl^- + R\!-\!\overset{+}{\underset{H}{O}}\!-\!P(OR)_2 \longrightarrow$$

$$RCl + (RO)_2POH \longrightarrow (RO)_2P\underset{O}{\overset{H}{\diagup}}$$

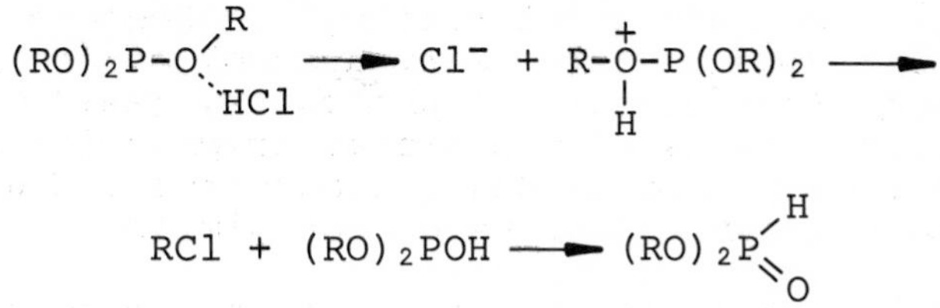

or

$$(RO)_3P\cdots HCl \longrightarrow Cl^- + R-O-\overset{+}{\underset{\underset{H}{|}}{P}}(OR)_2 \longrightarrow RCl + (RO)_2P\overset{H}{\underset{O}{\diagdown}}$$

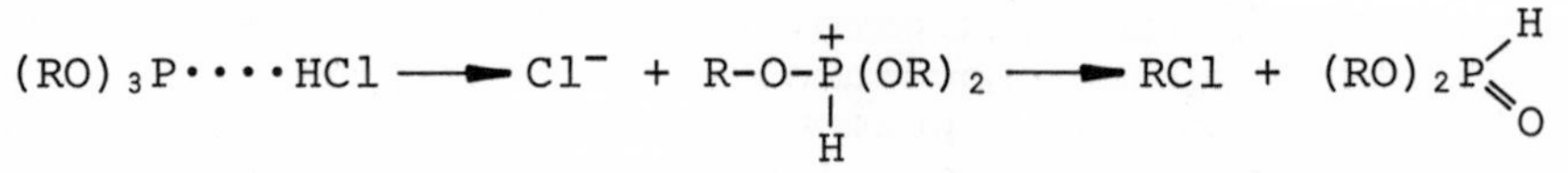

In the reverse order of addition, trichloride to alcohol, hydrogen chloride would tend to be held as

$$R-\underset{\underset{H}{|}}{O}\cdots HCl$$

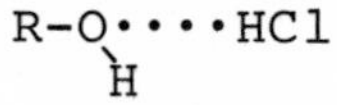

and, although the three alkyloxylation steps can occur from the beginning of the dropwise addition, there could come a stage where the third alkyloxylation is slow, because the electron density on the oxygen of the alcohol has been reduced.

$$\underset{\underset{\underset{HCl}{|}}{R-O-H}}{(RO)_2P-Cl} \quad \xrightarrow{\text{slow}} \quad (RO)_3P + 2\ HCl$$

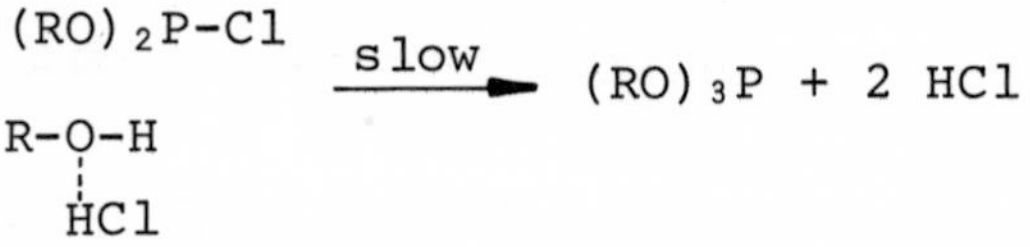

However, from the beginning, the hydrogen phosphite and alkyl chloride will form. It is conceivable that the hydrogen bonding would increase the reactivity of the O-H bond and give some compensation. In the example of phenol, this compensation is probably more effective because the electron density on the oxygen atom of phenol is much lower than for n-butanol. Aryloxylation will proceed through the three stages from the beginning of the procedure and will not be complicated by the process corresponding to dealkylation.

In the examples of tertiary alcohols such as t-BuOH and secondary alcohols such as Ph(Me)CHOH, the main products obtained in the absence of a tertiary base are alkyl chloride and phosphorous acid, no matter what the order of addition. These alcohols quickly give alkyl chloride with hydrogen chloride itself. An early generalization[572,1399] that primary alcohols give the dichloridite, secondary give primarily olefins, and tertiary give alkyl chloride is untenable. The tertiary alcohol, $Me_2(Cl_3C)COH$, is slow to form even the dichloridite.[481] Acetone cyanhydrin has been mentioned as a tertiary alcohol;[19] the formation of so-called α-cyanoisopropyl phosphite, $[Me_2C(CN)O]_3P$, in 74% yield by means of phosphorus trichloride alone in boiling benzene[290] seems odd, although the C≡N group is supposed to have an inductive effect similar to that of chlorine.[27]

Steric hindrance is an effect to be looked for in certain systems; and there are two entirely different approaches to be considered. Neopentyl alcohol, for example, shows no ordinarily perceptible hindrance to the "broadside" mechanism of alkoxylation, but shows a decided one to the three-center, end-on mechanism, referred to as the S_N2, which is involved in the dealkylation process.[474,475] Trineopentyl phosphite is thus obtained from neopentyl alcohol and phosphorus trichloride at room temperature in the absence of a tertiary base, although considerable dealkylation may occur on distillation unless the hydrogen chloride is removed by the addition of base.[278,474,475,575]

The interplay of steric hindrance and the drastic inductive effect of chlorine with reference to the end-on (3-center) or broadside (4-center) reactions was demonstrated for the model alcohols Me_3CCH_2OH (1), Cl_3CCH_2OH (2), $Me_2(CCl_3)COH$ (3), and $Cl_3CCHOHCCl_3$ (4).[19,459,464,474,475,481] Whereas (1) readily forms both the dichloridite and the triester, (2) reacted more slowly, and gave the two chloridites and the triester, according to the proportions of reagents; (3) required heating at 100° (16 hr) to form the two chloridites; and (4) did not react in boiling n-hexane (72 hr).

As indicated above, the addition of phosphorus trichloride to an alcohol of normal reactivity leads to the formation of dialkyl hydrogen phosphite. An outstanding feature, which is still ill-defined in much of the literature, is the function of product hydrogen chloride. It intervenes in several definite ways and its removal at low pressure, before the temperature is raised for distillation, is in general an essential operation for optimum results. The tenacity with which it can be held has been attributed to the formation of a coordination compound; and this aspect can now be given a much clearer definition in terms of hydrogen bonding as indicated by the solubility value x_{HCl} (mole of HCl/mole of liquid compound).[27,452,454-456,472] A study of the x_{HCl} p_{HCl} plots (p_{HCl} being the pressure of HCl above the liquid system) shows that as p_{HCl} is reduced from 760 mm the value of x_{HCl} can fall rather steeply at first, but much more slowly at the lower range of p_{HCl}. At 0°, e.g., for n-octanol, x_{HCl} drops from 1 mole/mole at 760 mm to about 0.3 mole/mole at 10 mm. The removal of HCl by continual pumping at 10 mm is very slow indeed, and a considerable amount of HCl is held even at 1 mm.

In earlier work[837,838] yields of $(RO)_2PHO$ of 90% were obtained by cooling during mixing, and making a special effort to remove HCl before distillation. When ether was used as solvent, ammonia was found necessary to remove all the hydrogen halide; but when CCl_4 was the solvent, the passage of ammonia could be omitted. Such observations add weight to the significance of the x_{HCl} values.

It appears unlikely that there could have been any triester, and more than a small amount of diester present in the distilled material obtained from the reaction of phosphorus trichloride and alkyl lactates,[50] cyclohexanol,[50] or dodecyl alcohol,[574] by the described procedures.

For the economic preparation of dialkyl hydrogen phosphites, phosphorus trichloride (1 mole) was added to a mixture of methanol (1 mole) and a higher alcohol (2 mole) at -5° to -10°.[817,1043] Methyl chloride and hydrogen chloride were then removed in vacuo, previous to heating at low pressure on a steam bath. The yields were 70-85%.[817]

$$2\ ROH + MeOH + PCl_3 \longrightarrow (RO)_2PHO + MeCl + 2\ HCl$$

Technical details for the manufacture of dimethyl hydrogen phosphite[539] include the precooling of methanol and phosphorus trichloride mixed with liquid methyl chloride to assist in the rapid removal of hydrogen chloride from the reaction mixture. Interaction is said to take place spontaneously, being exothermic to the extent of about 32 kcal/mole of PCl_3. "At least one-half of the hydrogen chloride remains in the liquid product, at least partially in chemical combination with the dimethyl hydrogen phosphite." Liquid butane was recommended[207] for heat dissipation and rapid removal of hydrogen chloride, for at temperatures "much above 0° (the hydrogen phosphite) reacts with hydrogen chloride to form methyl chloride."

In another technical process[264] for the production of hydrogen phosphites, the alcohol and phosphorus trichloride were separately mixed at a nozzle where the temperature of the reaction system was about 70° to 90° (or more) according to the alcohol. The collection zone was cooled at about -20°, and most of the gaseous by-product was rapidly separated under reduced pressure. Apparently up to 1 mole of water for 1 mole of phosphorus trichloride could be mixed with the alcohol to reduce the amount of by-product alkyl chloride.

$$PCl_3 + 2\ ROH + H_2O \longrightarrow (RO)_2PHO + 3\ HCl$$

It was stated that phosphorus tribromide could be used; however, no examples were quoted.

Mixed dialkyl phosphites are obtainable by the addition of phosphorus trichloride to a mixture of the two alcohols and water.[590,816] For the preparation of dialkyl phosphites, the use of benzene and temperature control to < 10° were considered advisable.[845] An alcohol-water mixture in benzene has also been recommended.[1322] Other technical details have been published.[157,162,198,199,203,206,266,294,295,672,726,732,768,805,812,814,834,841,1093,1252,1262,1290,1374]

In an omnibus coverage (1969) without details it was

stated[574] that phosphite esters, $(R^1O)(R^2O)(R^3O)P$, are readily prepared by the interaction of phosphorus trihalide (PCl_3 or PBr_3) (1 mole) with 1, 2, or 3 moles of ROH, according to whether one requires the mono-, di-, or triester. It is stated that "in those cases where mono- or di-esters are formed, it is sometimes desirable . . . to treat the reacted mixture with water, dilute aqueous caustic, or dilute aqueous mineral acid in order to hydrolyze off the residual chlorine or bromine atoms."

The influence of the phenyl and the naphthyl group[682, 846] in simple alcohols is clearly shown by preparing the trialkyl phosphites $(RO)_3P$ by Method Ib and passing in hydrogen chloride at -10°.[478] When the aryl group is attached directly to the alcoholic carbon atom, hydrogen chloride easily causes the fission of the C-O bond. In such examples, the formation of alkyl chloride in the absence of tertiary base is the main process, and may involve more than one route. Benzyl alcohol is said to yield a crude dichloridite, which explodes on attempted distillation.[1200] On the other hand, halogen at the β-carbon has a marked effect in the opposite sense.[459,464,481] In one report,[302] 1,3-dichloroisopropanol was converted into the hydrogen phosphite by phosphorus trichloride; but in another,[776] addition of 1 mole of the alcohol to 1 mole of the trichloride gave the dichloridite, the chloridite, and the trialkyl phosphite, which was converted into the phosphonate on distillation.

When the fluoroalkyl compounds, F_3CCH_2OH and $C_3F_7CH_2OH$ (3 mole), were mixed with phosphorus trichloride, hydrogen chloride was evolved from the start, and after the system had been heated at 80 to 90° for 4 hr, the triesters, $(F_3CCH_2O)_3P$ (78% yield) and $(C_3F_7CH_2O)_3P$ (83% yield), were isolated by direct distillation.[749] The ready evolution of hydrogen chloride is in accordance with its low solubility in such halogenated alcohols and esters.[27,452] Whereas 1,1,1-trifluoropropan-2-ol and phosphorus tribromide were stated[1305] to give the dibromidite, $CF_3CH(Me)OPBr_2$, heptafluorobutanol gave the phosphite (unstated yield).[749]

The phosphorus tribromide systems are much more difficult to handle, because of the greater polarizability effects of bromine, and the greater nucleophilic function of the bromine anion. The solubility of hydrogen bromide is also decidedly greater than that of hydrogen chloride, in the same compound.[27] Detailed work with optically active 2-octanol[451] and with the isomeric butanols[463] enables the essential pattern of factors and results to be fairly clearly discerned for alcohols of "ordinary" reactivity. By carefully removing hydrogen bromide before distillation, n-butyl phosphorodibromidite was prepared from butan-1-ol (1 mole) and the tribromide (1 mole), but alkyl bromide, the tribromide, and undistillable material were also isolated, for further alkoxylation occurred concurrently, and dealkylation of the esters produced gives alkyl bromide. The dibromidite was stable at 100° (2 hr) but slowly decomposed at 160°, giving alkyl bromide, undistillable products, and, unexpectedly, phosphorus tri-

$$ROH + PBr_3 \longrightarrow PBr_2{\cdot}OR + HBr$$

$$ROH + PBr_2{\cdot}OR \longrightarrow PBr(OR)_2 + HBr;$$

$$ROH + PBr(OR)_2 \longrightarrow P(OR)_3 + HBr$$

bromide. This decomposition was to some extent facilitated by pyridine hydrobromide. An alkyl bromide and phosphorus tribromide were also slowly formed when the dibromidite reacted with hydrogen bromide at 22°.

Di-n-butyl phosphorobromidite could not be isolated by using more alcohol than the 1:1 proportion but was prepared by the addition of a mixture of alcohol and pyridine to the dibromidite. It decomposed on being stored at 15°, and at 100° afforded alkyl bromide in 90% yield in 2 hr. When the ester was dealkylated by hydrogen bromide, alkyl bromide (1 mole) was formed in 4.5 hr at 25°.

The phosphorus triiodide systems are even more difficult to analyze. In what appears to be the most detailed study reported so far, the interactions of phosphorus triiodide in carbon disulfide and a number of alcohols were examined.[185] The reagents were mixed at -10°, and the mixtures were allowed to remain for about 24 hr at room temperature prior to aqueous treatment before distillation. The main product isolated in a pure condition was alkyl iodide, the anlaytical evidence pointing clearly to "phosphite" esters as the other main products. No direct evidence for the intervention of iodophosphites has yet been obtained although diethyl phosphoroiodidite $(EtO)_2PI$ is stated to be formed from the chloridite and lithium iodide.[1349]

There are two outstanding points about the phenolic systems, the low electron density on oxygen, and the resistance of the C-O bond to nucleophilic attack. Although there was a general impression that phenol and its homologs reacted readily with phosphorus trichloride at about 20° to 60° when mixed in equimolar proportions, and in the absence of catalysts or hydrogen chloride acceptors,[63,219,274,299,982,1299] there appears to be some confusion over the matter of yields of the dichloridites. Yields of 25-50% were reported more recently;[190,274] although a temperature of 150° was deemed necessary for a yield of 47% of phenyl phosphorodichloridite in earlier work.[299] With 2 mole of PhOH, the yield of $(PhO)_2PCl$ was 18%. Even with phosphorus trichloride in large excess, only a small yield of p-chlorophenyl phosphorodichloridite was obtained.[1299] In a detailed study[1326] of the results of adding the phenol dropwise to the trichloride in excess (about 6 mole/mole of the phenol) at about 20°, followed by heating (> 10 hr) to about 80°,[923] a reduction of the concentration of hydrogen chloride was deemed an essential factor. High yields for the halogenated phenols were achieved by adding magnesium chloride (0.01 mole/mole of the phenol) when about one-third of the calculated amount of hydrogen chloride was found in the scrubber. "The reaction rate" was deemed to decrease with the increasing dissociation constants of the phenols; but on careful consideration this aspect can be shown to be less satisfying than a recognition that the essential factor is the electron density on the oxygen atom of the phenol, i.e., the basic function. This is much smaller than for such as n-butanol, because of the mesomeric interaction (delocalization) between the oxygen lone pair and the π system. Alkyl substituents on the ring will oppose this and increase the rate of interaction, as was found, but progressive substitution by chlorine will considerably decrease the electron density on oxygen. Of course this last effect will tend to increase the reactivity of the O-H bond; however, the driving force appears to be the nucleophilic function of oxygen. o- or p-Substituted chloro- or nitrophenols gave good yields of the triaryl phosphites when heated with phosphorus trichloride.[666]

Considerable interest attaches to the preparation and uses of ring phosphites containing O-P-O linkages in the ring, and although a considerable number of examples have been named, by no means all of them have been characterized by analytical and/or physical data. One method of preparation of the triesters is from the ring chloridite, prepared from phosphorus trichloride and a dihydroxy compound. 1,2- and 1,3-Glycols,[438,799] pentaerythritol,[721,900,1819] and catechol[59,520] give the following ring systems.

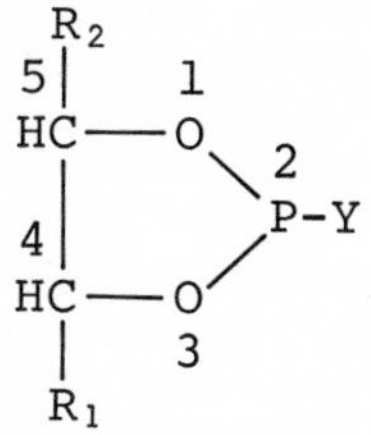

1,3,2-dioxaphospholane

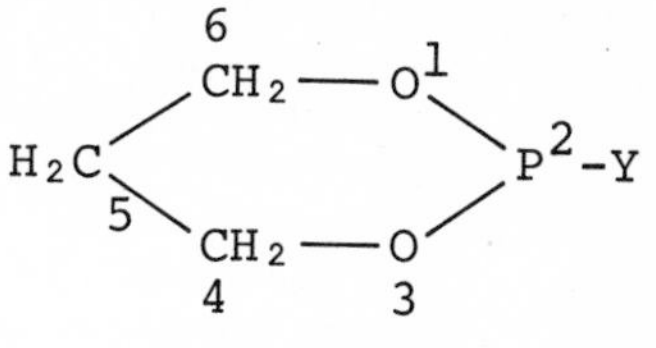

1,3,2-dioxaphosphorinane

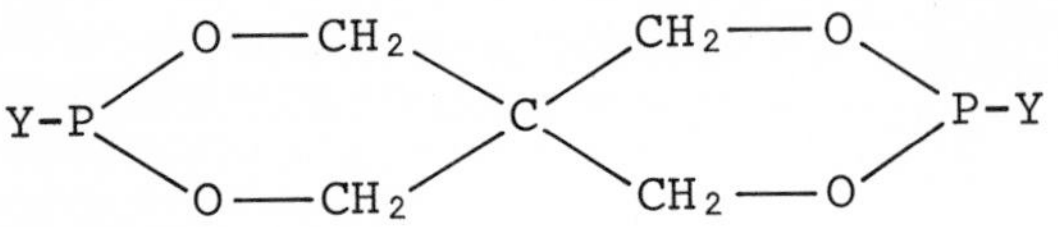

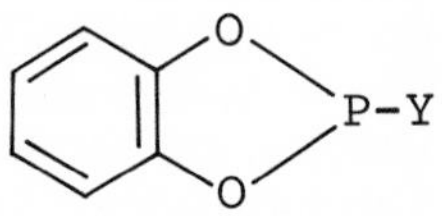

4,5-benzo-1,3,2-dioxaphospholane

Earlier reports[54-56,61,62,82,96,97,105,269,301,629,714,969,1226,1244] show the need for much more detailed analysis of operational factors. The pattern of reactions is complicated by the possibility of the formation of such arrangements as shown.

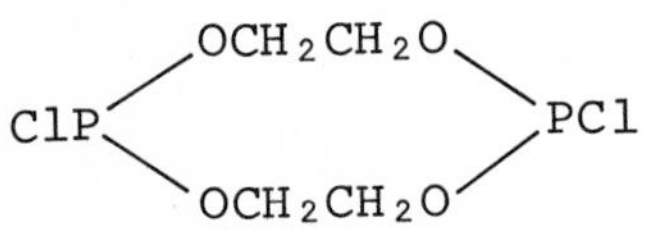

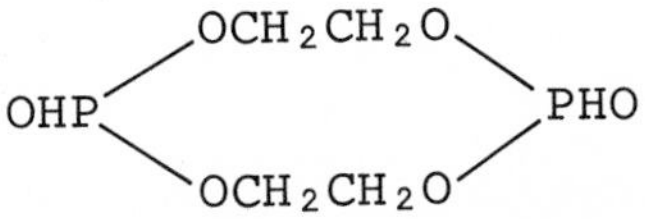

A substituted 1,3-propandiol (2 mole) and phosphorus trichloride (1 mole) gave a bicyclic diphosphite in 72% yield, probably due to the interaction of the intermediate chloridite.[840]

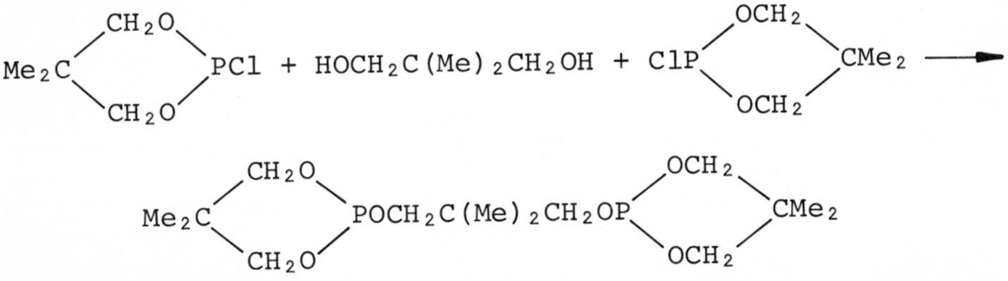

It is not clear that all the factors have been separated. According to a recent authority the previous methods of synthesis of cyclic hydrogen phosphites were not properly organized, and "much of the data reported, especially concerning the physical properties, are contradictory and controversial."[1415]

Ib. INTERACTION OF HYDROXY COMPOUNDS AND PHOSPHORUS TRIHALIDES OR HALIDITES IN THE PRESENCE OF A TERTIARY BASE OR AMMONIA

It is the intervention of hydrogen chloride which makes it necessary to use what is commonly referred to as a hydrogen halide acceptor, usually a tertiary base, when trialkyl phosphites are to be prepared. For the isolation of the triester, the trichloride (1 mole) was added to a solution of the alcohol (3 mole) and pyridine (3 mole) in ether at 0°. The hydrogen phosphites were obtained by using 2 mole of pyridine instead of 3.[898] Although the complications attending the functions of hydrogen halide are thus mostly eliminated, it cannot be said that the operational factors have been properly worked out. Certain earlier workers[449,898] happened to use pyridine, others, dimethylaniline or diethylaniline,[404,448] and ammonia is also recommended in certain technical processes. Triethylamine has also been mentioned.[138,763,764,766,1023,1024,1334] A solvent is used, and the choice is usually amongst diethyl ether, pentane, benzene. The best procedure would seem to be the dropwise addition of the trihalide or halidite in the solvent to a well stirred (or shaken) and cooled solution of the hydroxy compound. Subsequent treatment may involve storage at room temperature for a time, or even some measure of heating, before filtration from the base hydrohalide, followed by distillation, after or without aqueous treatment.

The preparations of simple or mixed phosphites and of ring phosphites are illustrated as follows. The methods are applicable to both alkyl and aryl systems.[581,661,945,1196,1197,1297,1301,1334]

$$PCl_3 + 3\ ROH + base \longrightarrow (RO)_3P + 3\ base\ HCl$$

$$PCl_3 \xrightarrow{2\ ROH\ +\ 2\ base} \underset{\text{in situ}}{(RO)_2PCl} \xrightarrow{R'OH\ +\ base} (RO)_2POR'$$

$$PCl_3 \xrightarrow{ROH} ROPCl_2 \xrightarrow{2\ R'OH\ +\ base} ROP(OR')_2$$

Dichloridites and 1 mole of a hydroxy compound in the presence of a tertiary base can lead to the formation of the chloridite.[646,1193]

Even when the C-O bond is too reactive for the permanent formation of the C-O-P link by phosphorus trichloride alone, in the presence of a tertiary base, and by properly controlled operations, all the halogen can be precipitated as base hydrochloride, and the trialkyl phosphite remains as a final residue, which might not be distillable.[449,450,467,478,494] Alcohols with phenyl on a carbon at least one from the alcoholic one afford the triester in about 85% yield; even such as 1-phenylethanol gave the undistillable phosphite, and base hydrochloride in calculated amounts, and tribenzyl phosphite was likewise prepared.[478,848]

Care must be exercised in resorting to general statements about tertiary carbon compounds; much depends on the groups attached as well as on operational procedures. Thus, although tri-t-butyl phosphite was obtained in 54%

yield by the careful addition of phosphorus trichloride (1 mole) to t-butanol (3 mole) and dimethylaniline (3 mole) in ligroin (3 liter/mole) at 0-5°,[733] the triester could not be isolated when pyridine was the base used. It was stated in another report that the triester had not been obtainable by an "analogous" procedure; instead, di-t-butyl hydrogen phosphite was isolated in 51% yield.[1401] The conditions were, however, not strictly "analogous," for triethylamine (3 mole) was the base, and the proportions of PCl_3 (1 mole) and t-butanol (6 mole) showed the latter to be in considerable excess, about 6 liter of "petroleum ether" being used. The formation of both the diester and the triester from t-butanol, phosphorus trichloride, and pyridine has been demonstrated by NMR,[827] although decomposition to diisobutylene and triisobutylene occurred on attempted distillation.[898]

The use of 2 mole of base has been advocated for the preparation of hydrogen phosphites, thus relying on the production of enough hydrogen chloride to effect one dealkylation.[535,536,898]

$$3\ ROH + 2\ C_5H_5N + PCl_3 \longrightarrow (RO)_2PHO + 2\ C_5H_5NHCl + RCl$$

Mixed dialkyl phosphites have been prepared in an analogous way from the dichloridites but there was no evidence presented to show that the order of mixing is immaterial, nor that only one alkyl chloride was produced.

$$2\ ROH + C_5H_5N + R'OPCl_2 \longrightarrow R'O(RO)PHO + C_5H_5NHCl + RCl$$

Hydrogen phosphites from hydroxy-carboxylic esters have been prepared by the in situ hydrolysis of the chloridite.[50]

For the preparation of dialkyl hydrogen phosphites, ammonia was passed into the reaction mixture after the addition of the trichloride in ether to the alcohol in ether through which dry air was passed to effect the removal of as much hydrogen chloride as possible.[302,303,837] A similar procedure was adopted for the preparation of trialkyl phosphites. Thus it was claimed[388] that trialkyl phosphites are stable to hydrogen halides at below -20°, preferably below -30°, and that the alkoxylation process between a phosphorus trihalide (Cl, Br, I) and an alcohol at these low temperatures is complete. The details were restricted to the example of ethanol (3 mole) to which was added phosphorus trichloride (1 mole) at -25°, followed by ammonia to the "neutralization" condition as seen by methyl orange. After aqueous treatment the triester was isolated in 80% yield. In the absence of more data, the explanation can be other than the one given; for the condition at the end of the pre-ammonia stage could be:

$$\text{ClP} \begin{array}{c} \diagup \text{O-Et} \\ \diagdown \underset{\underset{\text{HCl}}{\vdots}}{\text{O-Et}} \end{array} \quad + \quad \text{R-O} \begin{array}{c} \diagup \text{H} \\ \diagdown \underset{\text{HCl}}{\vdots} \end{array}$$

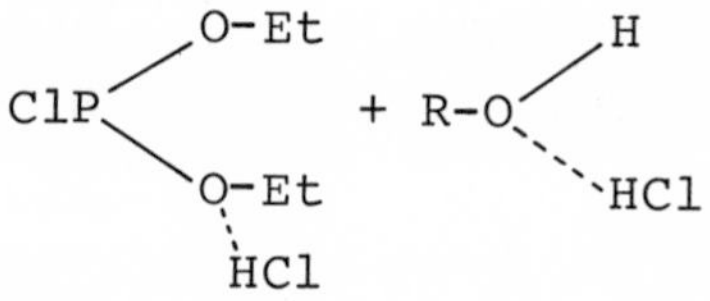

with little if any triester present. Addition of ammonia
will remove the hydrogen-bonded hydrogen chloride and enable the triester to form in the absence of hydrogen chloride. Provided the ammonia is added simultaneously with
the trichloride to maintain the pH at 7.0 to 8.5 (methyl
red) (in a hydrocarbon solvent), the system may be held
at +30° for the preparation of a range of triesters in
yields of about 70-80% after aqueous treatment. Production
of trimethyl phosphite required the low temperature.[1353]
   A considerable number of heterocycles containing
O-P-O links have been prepared by means of tertiary bases,
involving the prior formation of the cyclic chloridite as
described under Ia.[82,83,99,100,105,270,546,840,1226,1268]
In this way mixed phosphites containing 5-, 6-, 7-, and 8-
membered rings were prepared. A series of 1,3,2-dioxa-
phospholanes and phosphorinanes, RO_2POR' (where R' = an
alkyl group) were prepared from the ring chloridite and
the alcohol, R'OH, in a ligroin solution of N-ethylmorphol-
ine, which was preferred because of the easier removal of
its hydrochloride from the product.[799]   There appear to
be no examples of other than the 1,2- positions in the
benzene ring being involved in ring formation. A number
of mixed phosphites from ethyl tartrate have been prepared
in good yield.[104]

$$\begin{array}{c} CO_2Et \\ | \\ CHOH \\ | \\ CHOH \\ | \\ CO_2Et \end{array} \quad \xrightarrow[\text{NEt}_3 \,+\, \text{Et}_2\text{O}]{\text{ROPCl}_2} \quad \begin{array}{c} CO_2Et \\ | \\ CHO \\ \diagdown \\ P\text{-OR} \\ \diagup \\ CHO \\ | \\ CO_2Et \end{array}$$

   Ammonium carbamate and a substituted carbamate have
been used as hydrogen chloride acceptors.[580]   Ethylene
oxide has also been used to hold the liberated hydrogen
chloride,[199,201] and heterocyclic bases have been used as
catalysts in the preparation of triaryl phosphites.[206]
Other technical details have been given.[170,202,798,1210,
1211,1300,1407]
   The interaction of the chloridite of ethylene glycol
with ethanolamine in ether has been described as resulting
in the exclusive formation of the P-O instead of the P-N
link.[693]

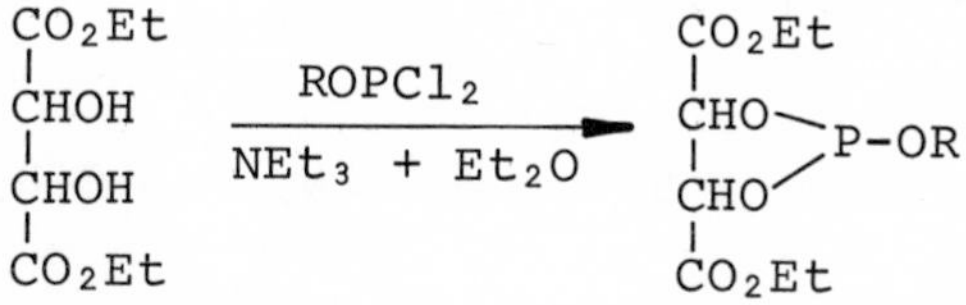

$$3 \quad \overset{CH_2-O}{\underset{CH_2-O}{|}} PCl \; + \; 3 \; HOCH_2CH_2NH_2 \longrightarrow$$

$$\overset{CH_2-O}{\underset{CH_2-O}{|}} P-OCH_2CH_2NH_3{}^+Cl^- \; + \; ClCH_2CH_2OPH(O)OCH_2CH_2NH_3^+Cl^- \; +$$

$$\overset{CH_2-O}{\underset{CH_2-O}{|}} P-OCH_2CH_2NH_2$$

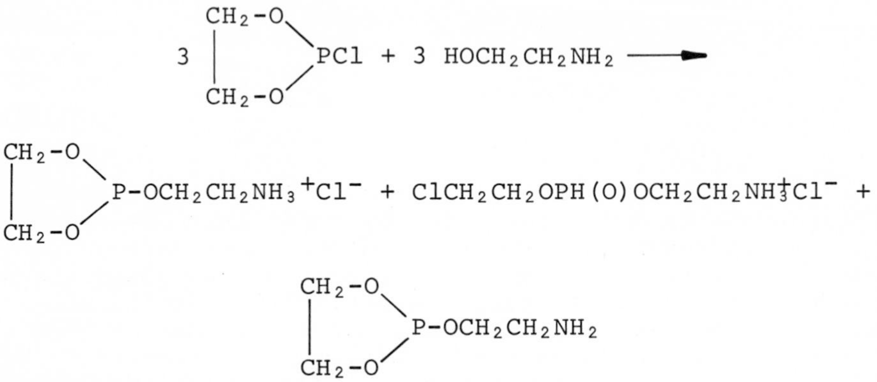

In the presence of triethylamine, however, there appeared to be exclusive formation of the P-N link.

$$\overset{CH_2O}{\underset{CH_2O}{|}} PCl \; + \; HOCH_2CH_2NH_2 \; + \; Et_3N \longrightarrow$$

$$\overset{CH_2O}{\underset{CH_2O}{|}} P-NHCH_2CH_2OH \; + \; Et_3NHCl$$

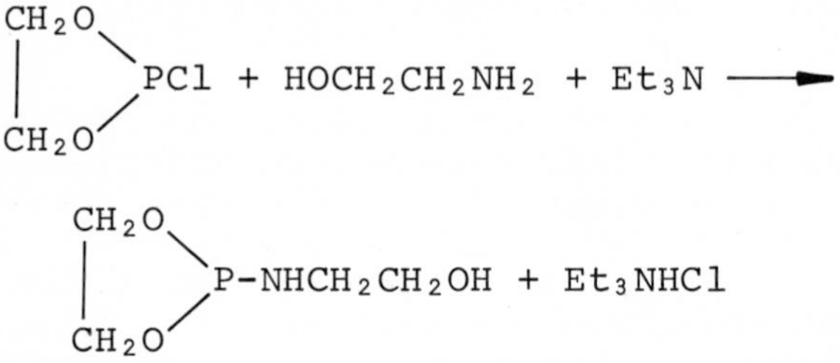

The implication in the original paper appears to be that the P-N compound is formed even in the absence of added tertiary base, but is converted into the P-O one by means of hydrogen chloride.

$$\overset{CH_2O}{\underset{CH_2O}{|}} P-NHCH_2CH_2OH \; + \; 2 \; HCl \longrightarrow ClCH_2CH_2O\overset{\overset{H}{|}}{\underset{\underset{O}{\parallel}}{P}}OCH_2CH_2NH_3^+Cl^-$$

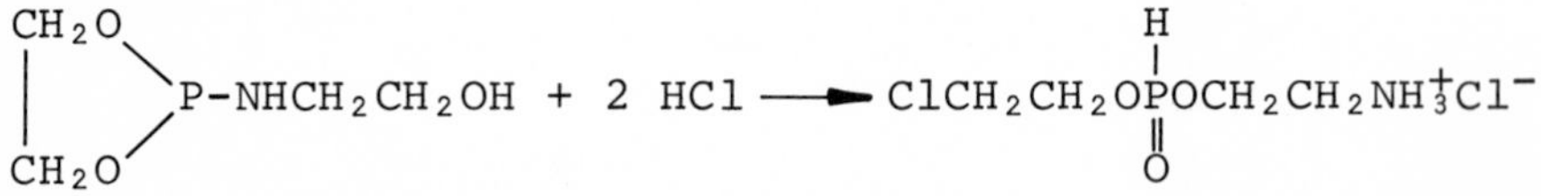

It is not clear that the evidence shows this. Other work on the interaction of $(CH_2O)_2PNMe_2$ with ethanolamine (cf. Section D) shows a spirophosphorane to be formed and the possibility of a tautomeric equilibrium between this and the open-chain compound above is suggested.[1238]
    In the presence of triethylamine, diethyl phosphorochloridite and an oxime readily gave the corresponding amide having a P-N bond.[752] ^{31}P NMR showed the presence of the intermediate P-O-N compound in the reaction system at <0°, even when the tertiary base was omitted, and isomerization to the P(O)N above 0°.[751]

$$(EtO)_2PCl + HON:C(X)(Y) \xrightarrow{<0°} (EtO)_2PON:C(X)(Y) \xrightarrow{>0°}$$

$$(EtO)_2P(O)N:C(X)(Y)$$

$$(X = Me, Y = Et; \text{ or } X = Me, Y = Et)$$

Ic. INTERACTION OF PHOSPHORUS TRIHALIDES OR HALO-
PHOSPHITES WITH CYCLIC OXIDES

This procedure is illustrated by the following scheme,
the essential point being that no hydrogen halide is formed,
although no tertiary base is used.[582,629,643-646,1227,
1280,1312]

$$(ClCH_2CH_2O)_2PCl \xrightarrow{(CH_2)_2O} (ClCH_2CH_2O)_3P$$

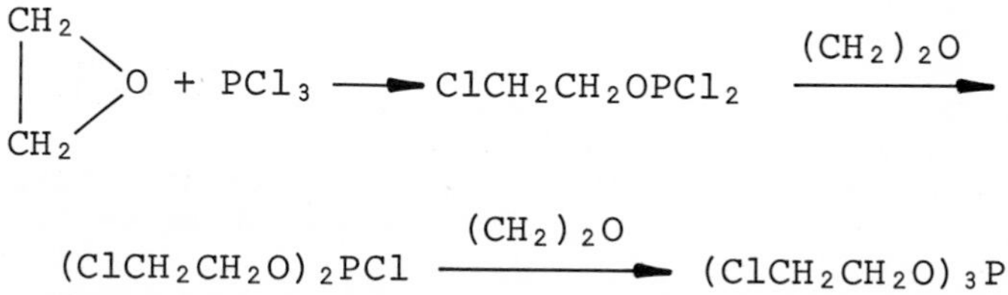

The interaction is highly exothermic, and cooling is desirable. In earlier work it appears that satisfactory
yields of the dichloridite were obtainable, although the
chloridites underwent disproportionation and the triesters
underwent isomerization, during distillation. Examples
have been given of preparations from various alkylene
oxides,[1280] from tetrahydrofuran or tetrahydrosylvan,[1276]
and from cyclohexane epoxides.[444] Preferential opening
of the epoxide ring attached to the cyclohexane ring was
observed in the following:[933]

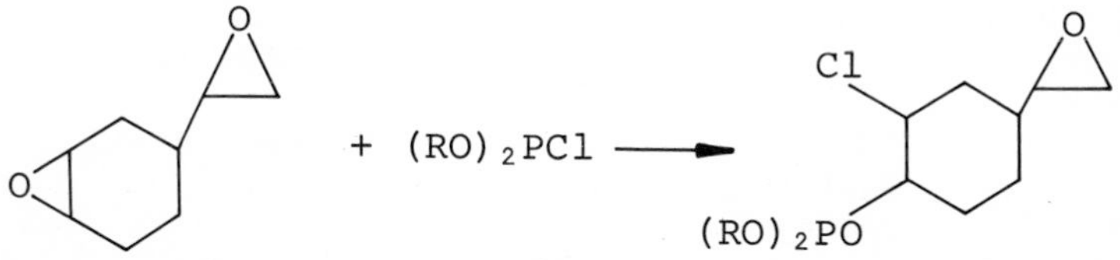

The unsymmetrical epichlorohydrin opens to give the primary
alkyl ester.[526,1234]

$$(ClCH_2CHXO)_2PX \longrightarrow (ClCH_2CHXO)_3P$$

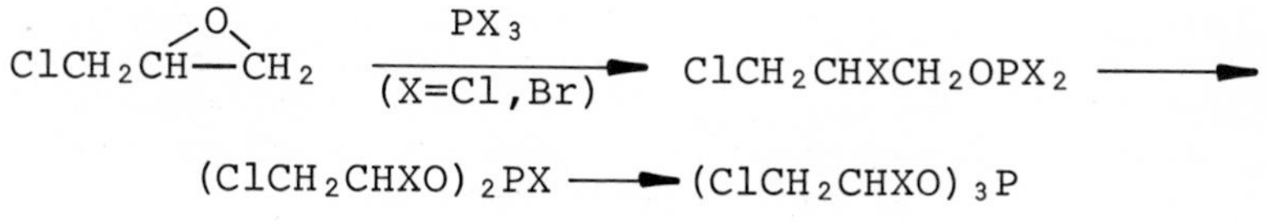

Id. MUTUAL EXCHANGE (DISPROPORTIONATION) IN PHOSPHITE-PHOSPHORUS TRIHALIDE-HALOPHOSPHITE SYSTEMS

That alkoxyl and halogen groups can undergo mutual exchange under certain conditions was recognized in the earlier work.[54,55,299,449] The 4-center mechanism is the most helpful.

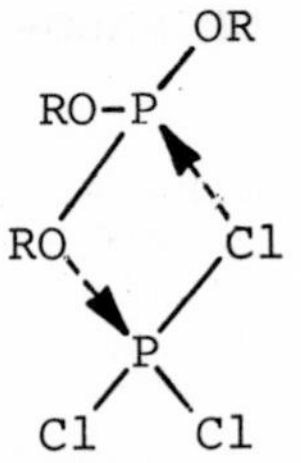

$$\longrightarrow ROPCl_2 + (RO)_2PCl \xrightarrow{PCl_3} 2\ ROPCl_2$$

Various alkyl dichloridites or dialkyl chloridites have thus been obtained from the corresponding trialkyl phosphites, readjustment of equilibrium tending to be quick during distillation,[449,1314] e.g., 1 mole of tributyl phosphite and 2 mole of trichloride afforded the dichloridite in 80% yield after the system had been heated at 50° for 1 hr;[449] aryl systems may require higher temperatures. Mutual exchange may be an undesirable side reaction in other preparative procedures, e.g., during the alkoxylation or aryloxylation of phosphorus trichloride by alcohols or phenols.

As fluoridites do not appear to be obtainable from phosphorus trifluoride and alcohols, the formation of fluoridites under this heading is significant.[221]

$$(MeO)_3P + PF_5 \longrightarrow MeOPF_4 + (MeO)_2PF$$

$$(EtO)_3P + PF_5 \longrightarrow EtOPF_4 + (EtO)_2PF$$

$$(MeO)_3P + PF_3 \longrightarrow MeOPF_2 + \text{other compounds}$$

As a variant of this procedure is the interaction of antimony trifluoride and the corresponding chloridite, giving yields which can be as high as 95%.[1201,1249] The reactions are conducted at about 60° under nitrogen. In one example NaF was used as well as SbF_3, and in another KSO_2F in benzene.[1249]

Gas chromatography has been used to study the random exchange of alkoxyl groups between trimethyl and triethyl phosphites[914] and the formation of methyl n-propyl hydrogen phosphite from the two simple esters at room temperature.[343] Mixed dialkyl phosphites have been prepared by heating two simple dialkyl phosphites under reflux.[389]

Ie. TRANSESTERIFICATION OF PHOSPHITES

This is a secondary process and requires the primary formation of a trialkyl, or triaryl phosphite (usually ethyl or phenyl) by a direct procedure. In principle it is a matter of heating one of these phosphites with an alcoholic or phenolic compound to effect the distillation of ethanol or phenol and leave the new-made phosphite, which might be simple or mixed. Tris(2-chloroethyl) phosphite has also been used.[281,1027] The 4-center broadside mechanism is the most satisfying.

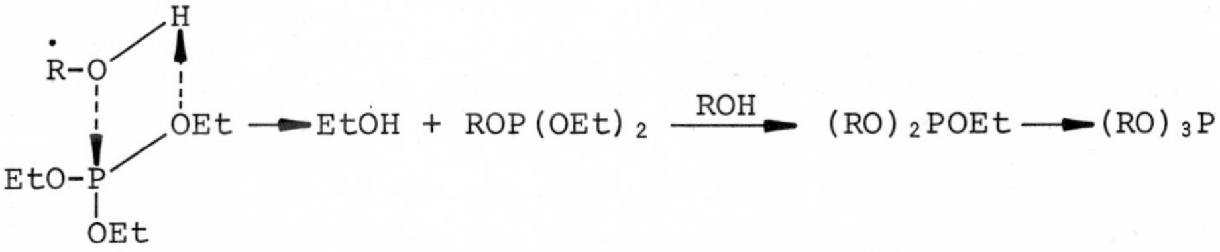

The exchange is catalyzed by sodium alkoxides,[144,1228] sodium phenate,[545] and dialkyl or diaryl hydrogen phosphites.[521,522,782,783] Several trialkyl phosphites and mixed hydrogen phosphites have also been obtained by using phosphoric acid as a catalyst,[100,128,132] but continuous disproportionation during distillation was a difficulty, and certain products could not be distilled.[132]

Some details have been reported on the rates of hydrolysis, transesterification, and mutual exchange of trialkyl phosphites.[587-589] Thus, pure triethyl phosphite was not transesterified by i-amyl alcohol even at 150° in 4 hr. When the phosphite had not been purified by treatment with sodium to remove dialkyl hydrogen phosphite, traces of the latter catalyzed the transesterification. The mutual exchange of alkoxy groups in the presence of alkoxide or hydrogen phosphite was described in some detail.

In earlier work[899] successive replacement of aryl by alkyl was effected by stoichiometric amounts of sodium alkoxide.

$$(ArO)_3P + 3\ AlkONa \longrightarrow (AlkO)_3P + 3\ ArONa$$

Tricresyl phosphite was stated to react with alcohols having sufficiently high boiling points in the absence of catalyst giving successive replacement of aryl by alkyl,[523,1214] although propanol and triphenyl phosphite, on being heated above 150°, gave dipropyl propanephosphonate.[899] Phosphites having free hydroxyl groups were obtained by interaction of a simple triester with a substituted pyranol or substituted cyclohexanol; reaction was catalyzed by diaryl or dialkyl hydrogen phosphites, or sodium alcoholates.[521,522]

Other complex mixed phosphites and ring phosphites have been obtained by transesterification with diethylene glycol or hexane-1,6-diol,[281] polyethylene glycol,[782,783] neopentylene glycol,[144] dipropylene glycol,[144,422] hydroxy-oxetanes,[144] pentaerythritol,[161,545] mannitol,[1360,1361] etc.

1,2- and 1,3- glycols afford 1,3,2-dioxaphospholanes and phosphorinanes, respectively.[546] Trimethyl phosphite and 2-chloromethyl-2-methylpropane-1,3-diol at 90° gave a mixture of two conformers.[1383]

The alcoholysis of acyl phosphites has also been described for the preparation of trialkyl phosphites and appears to have a future.[430,958,961,962] Phosphites such as tris(2-hydroxyethyl) phosphite are deemed unstable above −20°, giving rise to compounds which could be regarded as the result of intramolecular transesterification.

$$(HOCH_2CH_2O)_3P \longrightarrow \begin{array}{c} CH_2-O \\ | \qquad\qquad\quad P \cdot OCH_2CH_2OH \\ CH_2-O \end{array}$$

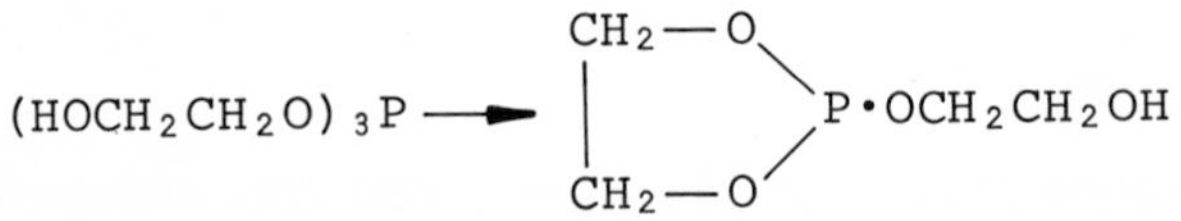

This, and related compounds, resulted from the transesterification of triethyl phosphite or ethyl salicyl phosphite by ethylene glycol.[954]

Transesterification of dialkyl hydrogen phosphites occurs readily in the presence of catalytic amounts of sodium alkoxide. Diethyl hydrogen phosphite reacted with higher alcohols to give up to 85% yield of the higher dialkyl ester, the course of reaction being unchanged by the use of molar equivalents of alkoxide.[770] The previous claim that diethyl ether and sodium ethyl hydrogen phosphite are also formed was contradicted.[770]

The transesterification of dialkyl hydrogen phosphites with ethylene glycol[1053] appears to give the cyclic biphosphite shown, rather than the linear biphosphite ROP(H)(O)OCH2CH2OP(O)(OR)H.

$$2 \begin{array}{c} CH_2OH \\ | \\ CH_2OH \end{array} + (RO)_2PHO \longrightarrow \begin{array}{c} H\quad\quad OCH_2CH_2O\quad\quad H \\ P \qquad\qquad\qquad P \\ O \quad OCH_2CH_2O \quad\quad O \end{array}$$

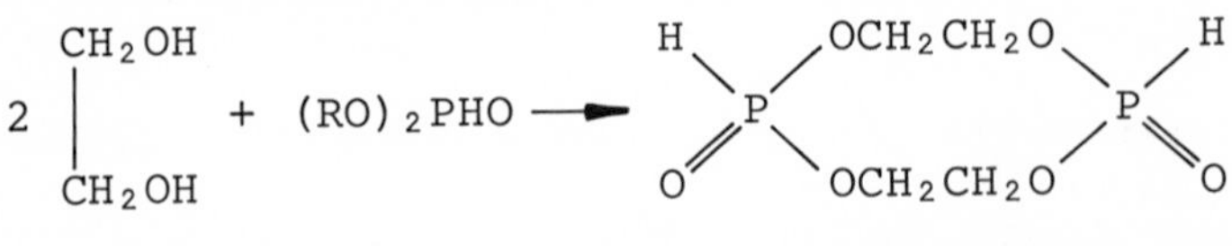

The formation of hydrogen phosphites containing the phospholane or phosphorinane ring systems has also been reported, by reaction of diethyl hydrogen phosphite with ethylene glycol or butane-1,3-diol.[1027] Many other examples of transesterification have been studied.[427,540, 542,544,552,915,1055,1105,1110,1257] The study of catalytic effects, and of interactions with cellulose are topics in active development.[1007,1051,1052,1233] Sulfur-containing polyphosphites have likewise been obtained.[700]

If. HYDROLYSIS OF PHOSPHITES

Although dialkyl or diaryl hydrogen phosphites are obtainable from the trialkyl esters by hydrolysis with aqueous acids or alkalis under mild conditions,[73,85,893] they are usually much more conveniently prepared (see Ia, Ib). It is stated that trialkyl phosphites are hydrolyzed very rapidly to the dialkyl hydrogen phosphites in the presence of acids, but slowly in alkaline media.[73] Trimethyl phosphite is rapidly hydrolyzed to the diester and methanol in water; but triethyl and higher trialkyl phosphites are said to be stable in pure water even at 100°.[73,105]

However, relative rates of hydrolysis for a series of trialkyl phosphites in pure water (exothermic) have been determined dilatometrically.[587-589] Autocatalysis was indicated by the increase in rate caused by the addition of diethyl hydrogen phosphite to the triethyl phosphite-water system. Pyridine (and more effectively triethylamine), or sodium or potassium hydroxide lowered the rate. Diethyl hydrogen phosphite and triethylamine were stated to form a "salt" (an expression in urgent need of redefinition); but the two initial compounds were separated on distillation. Changes in density and refractive index were taken to indicate the formation of a "salt" in the $(MeO)_2PHO$-C_5H_5N system; but in this example separation into the original constituents by distillation was much less effective, and there was a viscous residue.

The ring phosphites may undergo either ring opening or side chain hydrolysis. Although ring fission has been deemed a propensity, there are examples to the contrary. The calculated amount of water containing a "little HCl" gave the ring hydrogen phosphites,[83,99] e.g., as follows:

$$\overline{OCHMeCH_2CH_2OP}OCH_2Ph \xrightarrow{H_2O} \overline{OCHMeCH_2CH_2OP}HO + PhCH_2OH$$

On the other hand, ethyl ethylene phosphite undergoes ring fission:[100]

$$(CH_2O)_2POEt \xrightarrow{H_2O} HOCH_2CH_2OPHO(OEt)$$

The hydrolysis of dialkyl hydrogen phosphites to the monoalkyl phosphites is a significant way of producing the latter, and is catalyzed both by acids and alkalis.[771,772] Esters such as the diethyl ester lose one ester group too rapidly in aqueous alkali for accurate rate measurement.[996]

The sodium salt, (RO)(NaO)PHO, can be isolated by careful evaporation at low pressure of the water-alcohol mixture (1 mole NaOH per mole of diester).[997]   The treatment of the sodium derivative of the hydrogen phosphite, $(RO)_2PONa$, with 1 mole of water in dry alcohol leads to the same result.[622,996,997]   Limited hydrolysis of the cyclic biphosphite derived from ethylene glycol (see Ie) afforded the dihydrogen biphosphite as follows:[1053]

$$\underset{O}{\overset{H}{\diagup}}P\underset{OCH_2CH_2O}{\overset{OCH_2CH_2O}{<}}P\underset{O}{\overset{H}{\diagdown}} \quad \xrightarrow{H_2O} \quad (HO)_2POCH_2CH_2OP(OH)_2$$

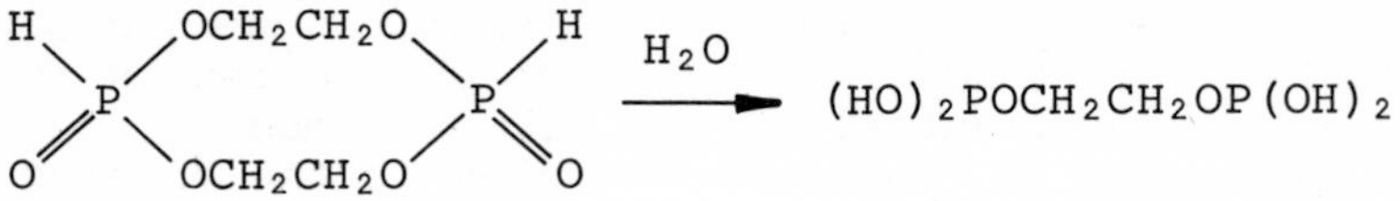

Kinetic and other studies have been described.[177,412] Hydrogen phosphites also result from the hydrolysis of an anhydride such as $(RO)_2POP(O)(OR)_2$ by aqueous sodium bicarbonate.[876]

Ig. HYDROLYSIS OF HALOPHITES

Chloridites and dichloridites are readily hydrolyzed; but the usefulness of the procedure depends on the purpose and experimental details. In earlier work reaction mixtures were treated with ice water, and sodium carbonate solution.[449,450,769]

$$ROPCl_2 \quad or \quad (RO)_2PCl \quad \xrightarrow[\text{(ii)}Na_2CO_3]{\text{(i)}H_2O}$$

$$\begin{cases} ROPHO(ONa) \text{ (dissolved)} \xrightarrow{HCl} ROPHO(OH) \\ (RO)_2PHO \text{ (liquid, separated)} \end{cases}$$

Aryl phosphorodichloridites and a calculated amount of water have been made to yield the monoester,[769] although further hydrolysis occurs if water is added in excess. Hydrolysis of the cyclic chloridite, $(CH_2O)_2PCl$, gave the cyclic hydrogen phosphite.[351]

Ih. INTERACTION OF ALKALI ALKOXIDES WITH PHOSPHORUS
     TRIHALIDES OR HALOPHOSPHITES

Although the interaction of phosphorus trichloride and an alkali (usually sodium) alkoxide was featured in earlier work,[69-71,269,620,1156,1246,1384,1412] there appears to be little to commend it as a general rule. An operational difficulty is the separation of finely divided sodium chloride, although centrifuge techniques are now much more available. The process has been used to prepare chloridites, especially the mixed esters, by addition of alkoxide (suspended in Et_2O or petroleum) to the dichloridite.[80,94,1193]

$$ROPCl_2 + NaOR' \longrightarrow (RO)(R'O)PCl + NaCl$$

Phosphites of diacetoneglucose have been made from the potassium derivative,[940] and for examples of this kind there may still be occasions for such practice; but for general application the procedure may now be deemed outmoded.

Magnesium alcoholates are worth considering. Trialkyl phosphites were obtained in moderate yields by adding phosphorus trichloride to the magnesium alkoxide in diethyl ether, the temperature being allowed to rise to the boiling point of the ether, the system then being heated under reflux for 1.5 hr.

$$2\ PCl_3 + (RO)_2Mg \longrightarrow 2\ ROPCl_2 + MgCl_2 \xrightarrow{\text{etc.}}$$

$$(RO)_2PCl \text{ and } (RO)_3P$$

Although no data were given it was stated that the dichloridite and chloridite could be isolated readily.[849]

Ii. INTERACTION INVOLVING PHOSPHOROUS ACID OR PHOSPHOROUS OXIDE

Earlier work on the reaction of phosphorous acid with alcohols[73,1232] or glycols[269,802] leads to the view that the process is not attractive. The reaction has nevertheless been used more recently for the preparation of monoalkyl phosphites from menthol, various unsaturated alcohols, and hydroxynitriles, the products being isolated as sodium, calcium, or barium salts.[283] The preparations of dialkyl phosphites from phosphorous acid and alcohols in the presence of zinc chloride, sulphuric acid, or potassium acetate as catalysts, have also been reported.[1054] A variation consists in the interaction of trialkyl phosphites with molten phosphorous acid at 25° to 60° to yield both mono- and dialkyl phosphites.[275,967]

High yields of dialkyl hydrogen phosphites are claimed from the interaction of alkyl halides and phosphorous acid in the presence of a tertiary base at 80-5°[624] and nucleoside phosphites are prepared from nucleosides, phosphorous acid, and di-(p-tolyl)-carbodiimide or dicyclohexylcarbodiimide in pyridine.[1251] By heating a mixture of dipotassium tartrate and phosphorous acid, followed by treatment with acetic acid, the compound designated was precipitated.[330]

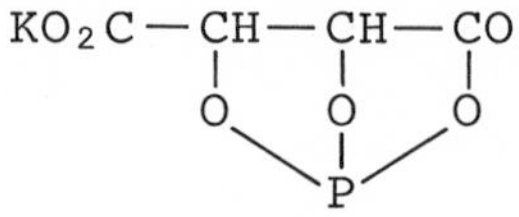

Reactions of phosphorous acid with epoxides have been described.[781] In early work[1323] diethyl hydrogen phosphite was obtained by a very vigorous interaction between ethanol and phosphorous trioxide but the method is not of practical significance.

Ij. MISCELLANEOUS REACTIONS

A variety of reactions in which phosphites or their derivatives are formed is given in Table 1.

Table 1. Miscellaneous Reactions in which Trialkyl or Dialkyl Phosphites or Their Derivatives Are Formed

Reactants	Products	Ref.
$(EtO)_2PONa + Cl_2$	$(EtO)_2PCl$	80
Pb phosphite + RI	$(RO)_2PHO$	858
$(RO)_2POAg + Ar_3CCl$	$(RO)_2POCAr_3$	75
$(RO)_3PS + Na$	$(RO)_3P$	1071
$Na_2HPO_3 + R_2SO_4 + C_5H_5N$	$(RO)_2PHO$	484
$(RO)_2PONa + (R'O)_2P(O)Cl$	$(RO)_2POP(O)(OR')_2$	887
$(RO)_3P + (R'O)_2P(S)OH$	$(RO)_2PHO + (R'O)_2P(S)OR$	1136
$MeCHCH_2OC(O)O + PCl_3$ (cyclic)	$(MeCHClCH_2O)_3P$	829
$ROP(OH)ONa + R'OCOCl$	$(RO)(R'O)PHO$	297
$EtOCO_2Na + PCl_3$	$(EtO)_3P$	737
$(RO)_2PCl + Hg(CH_2CHO)_2$	$(RO)_2POCH{=}CH_2$	948
$ROPCl_2 + Ag_2O$(in C_6H_6)	$ROPOP(OR)OP(OR)O$ (cyclic)	1259
$(RO)_2PNHPh + PhCHO$	$(RO)_2PHO$	901
$(RO)_2PCH{=}C(OR')R'' + H_2O$	$(RO)_2PHO$	1083
$(RO)_2PH + Et_3N$(in CCl_4)	$(RO)_2PHO$	968
$(EtO)_3P + (HSCH_2)_2PO_2H$	$(EtO)_2PHO$ (byproduct)	614
$(EtO)_2POP(O)(OEt)_2 + H_2S$	$(EtO)_2PHO$	871
ROH + P(white) + O_2	$(RO)_2PHO$	372
Olefin + P(white) + O_2	"Phosphorate polymer" which with ROH gives $(RO)_2PHO$	1373
$(RO)_2POCOMe + MeCO_2H$	$(RO)_2PHO + (MeCO)_2O$	959

Table 1 (Continued)

Reactants	Products	Ref.
$(EtO)_2POP(OEt)_2$ + $AcNH_2$	$(EtO)_2PHO$ + $(EtO)_2PNHAc$	633
$RCO_2P(OR)_2$ + $(RO)_2PSSH$	$(RO)_2PHO$ + $(RO)_2P(S)SCOR$	1139

B. CARBOXYLIC ACID SYSTEMS

II. REACTIONS WITH CARBOXYLIC ACID

Up to about 1950 reports on the phosphorus halide-carboxylic acid systems were considered almost exclusively from the aspect of production of the acyl chloride.[167,169,212,220,293,433,447,1321,1324,1328,1341] There was a difference of opinion relating to the stoichiometric equation which may be used, although the one adopted by writers of standard textbooks was $3\ RCOOH + PCl_3 \rightarrow P(OH)_3 + 3\ RCOCl$.[220,800,1216,1233] There was an early mention of acetyl dihydrogen phosphite, and diacetyl hydrogen phosphite, relating to the investigation into the phosphorus trichloride system;[220] but there was no definite isolation; the monoacetyl compound was stated to be formed by the interaction of acetyl chloride and phosphorous acid in acetic anhydride at 50°, and isolated by filtration.[220] Phosphorous acid and acetic anhydride were stated to give a similar product.[1365]

$$AcCl + HOP(O)(OH)H \longrightarrow HCl + AcOP(O)(OH)H$$

More recently it was said that hydroxy-ethylidenediphosphonic acid, $H_2O_3PC(OH)(CH_3)PO_3H_2$, is formed in these systems[811a] (see also Chapter 12).

The formation of a mixed alkyl-acyl phosphite as a reaction intermediate was postulated for the following scheme.[52]

$$R\cdot CO_2H + ClP(OR')(OR'') \xrightarrow{Et_3N} [RCO_2P(OR')(OR'')] \xrightarrow{R'''NH_2}$$

$$RCONHR''' + HOP(OR')(OR'')$$

In a review on acyl phosphorus compounds it is concluded that on the derivatives of phosphorous acid "reports often disagree."[677]

In the phosphorus trichloride-aliphatic acid system

there was no evidence of R•CO•O-P bond formation, either in the absence or in the presence of base.[242-247] It was believed that two of the chlorine atoms readily cause direct formation of acyl chloride (possibly by a 4-center mechanism); but there can be complications over the third

$$R\cdot CO_2H + PCl_3 \longrightarrow R\cdot COCl + PCl_2\cdot OH$$

$$R\cdot CO_2H + PCl_2\cdot OH \longrightarrow R\cdot COCl + PCl(OH)_2$$

atom. From the yields of acyl chloride it appeared that the compound $PCl(OH)_2$ can continue to form acyl chloride, but some of it is degraded to hydrogen chloride and metaphosphorous acid. Addition of phosphorus trichloride (1 mole) to benzoic acid (3 mole) and pyridine (3 mole) in benzene at 0°, however, gave base hydrochloride (98%) and tribenzoyl phosphite (96%).

The observation that potassium chloride (1 g per mole) increased the rate of conversion of CCl_3COOH (14-fold), and $ClCH_2COOH$ (4-fold) into the acyl chloride by phosphorus trichloride, and a study of the kinetic curves for the evolution of hydrogen chloride were deemed to point to the intermediate formation of the acyl phosphorodichloridite, $RCOOPCl_2$.[743] Other workers[434] found that potassium chloride caused "no significant gains in conversion over the 2h period," and disregarded the foregoing inference[743] on the grounds that the rate of evolution of hydrogen chloride cannot be taken as the rate of reaction. This is a valid point, and relates to the solubility of hydrogen chloride in the compounds mixed initially, and those formed in changing amounts.[27,452] Not more than 70% of the calculated amount of acyl chloride was isolated.[743] In the presence of pyridine (1 mole), in pentane, an aliphatic carboxylic acid reacted with di-n-butyl phosphorochloridite (1 mole) to yield base hydrochloride, acid anhydride, and dibutyl hydrogen phosphite.[245-247]

$$2\ RCOOH + (n\text{-}BuO)_2PCl + C_5H_5N \longrightarrow$$

$$(n\text{-}BuO)_2P(O)H + (RCO)_2O + C_5H_5NHCl$$

Benzoic acid behaved differently, for no hydrogen phosphite could be isolated, and there were definite indications of the formation of benzoyl dibutyl phosphite, which could not be isolated in a pure state.

Addition of n-butyl phosphorodichloridite (1 mole) to an aliphatic carboxylic acid and pyridine in pentane at -10° gave the base hydrochloride and, amongst other products, diacetyl butyl phosphite. There were indications that the dibenzoyl phosphite was similarly formed.

Hydrogen chloride rapidly deacylates the acyl alkyl

phosphites, giving acyl chloride; but, remarkably enough, tribenzoyl phosphite gives phosphorus trichloride and benzoic acid, with a small amount of benzoyl chloride,[247] the latter being attributed to the interaction of phosphorus trichloride and the directly formed acid.

Polyhydric alcohols are stated to react more satisfactorily with acetyl phosphites than with phosphorochloridites, for the preparations of cyclic phosphites.[431]

Phosphorus trichloride and salicyclic acid at 70° yield a cyclic phosphite I(X = Cl),[698,949,952,953] although an alternative metaphosphite structure has been considered.[51]

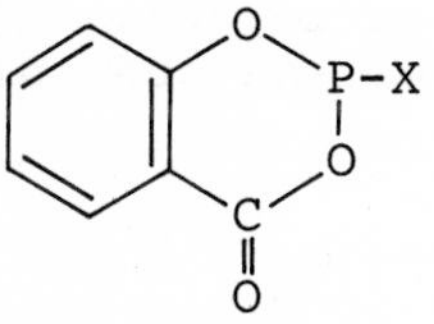

I

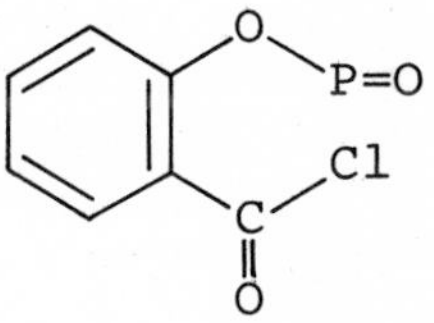

II

An improved yield (85%) is obtained by the use of pyridine at -10°.[245,247] The reaction of I(X = Cl) with butan-1-ol and pyridine, or of n-butyl phosphorodichloridite with salicyclic acid and pyridine, yields I(X = BuO). With hydrogen chloride, I(X = Cl or OBu) reverted to salicyclic acid and phosphorus trichloride or the dichloridite; with acetic acid, acetyl chloride and salicyclic acid were formed. In the presence of pyridine, an aliphatic carboxylic acid afforded the hydrogen phosphite I(X = OH) and the acid anhydride, although benzoic acid gave the benzoyl phosphite I(R = OCOPh) in what appeared to be almost quantitative yield.

Triethyl phosphite and carboxylic acids at 110-70° give diethyl hydrogen phosphite and the ethyl carboxylate,[103] $[(EtO)_3P + RCOOH \rightarrow (EtO)_2P(O)H + RCOOEt]$ thereby affording a resemblance to dealkylation by hydrogen halides, although the process is very much slower. Dealkylation by acetic acid, so-called acetolysis, was examined to reveal the behavior of mixed phosphites with reference to the preference for removal of a particular group.

The speculation on mechanism[581] would seem to go beyond the limited extent of the data presented. That 2-octyl acetate was produced from the (+)-2-octyl phosphite, $(EtO)_2POC_8H_{17}$ with "92-94% inversion of configuration" led to the conclusion that an S_N2 displacement on the alkyl group is the predominant course in the "simpler systems," although such a result is equally consistent with an S_N1 process.[440]

The greater ease of removal of the secondary alkyl

group over that of the n-alkyl group was recognized as contrary to the S_N2 requirement, but was attributed to the smaller steric energy of diethyl hydrogen phosphite compared with that of, e.g., (EtO)PH(O)(OPr-i).[581]

Dialkyl hydrogen phosphites (so-called dialkyl phosphonates) were considered as reagents for the preparation of alkyl carboxylates, R'COOR in excellent yields.[554] The hydrogen phosphite $(RO)_2P(O)H$ (0.5 to 1.0 mole) was heated under reflux with 1 mole of the carboxylic acid at a temperature (100° to 200°) and for a time (usually 3 or more hours) according to the system, e.g., acetic acid (1 mole) and diethyl hydrogen phosphite (1 mole) gave the ethyl acetate in 95% yield in 3.5 hr at 153°, and for this particular pair, the dealkylation rate ratio was estimated as 1st : 2nd = 3:1; deemed inadequate for the convenient isolation of monoalkyl phosphites.

$$R'COOH + (RO)_2P(O)H \longrightarrow R'COOR + ROP(O)H(OH) \xrightarrow{R'COOH} R'COOR + H_3PO_3$$

Further reactions of trialkyl phosphites with organic acids or anhydrides have been studied.[667]

C. PROPERTIES AND INTERACTIONS OF PHOSPHITES AND HALOGENOPHOSPHITES

C.1. Thermal Stability

Primary or secondary alkyl dichloridites or chloridites yield alkyl chloride remarkably slowly at elevated temperatures by comparison with, for example, the chlorosulphites.[467] Rearrangement of alkyl groups other than p-n-alkyl occurs and olefins may be formed.[278] Mutual exchange of alkoxyl and Cl has a finite probability at all temperatures. Under reduced pressure, the formation of the more volatile phosphorus trichloride from the dichloridite is encouraged.

As the reactivity of the alkyl-O bond increases so will the formation of alkyl chloride, or more deepseated decomposition take precedence over mutual exchange; and indeed it is far from clear in most cases to what extent the chloridite or dichloridite are stable in the examples of t-alkyl or 1-phenylethyl groups. Di-t-butyl phosphorochloridite has however been identified by NMR spectroscopy.[827]

The bromophosphites yield alkyl halide more readily than do the chloro compounds although both p- and s-alkyl derivatives have been isolated by distillation under reduced pressure.[463] Neopentyl dibromophosphite is particularly stable.[575] Only one iodophosphite has been reported.[1349]

In the examples of much less reactive R-O bonds, as in aryl or 2-chloroethyl systems, and more especially 2,2,2-trichloroethyl, then mutual exchange will be a significant process. Symmetrical trialkyl phosphites have some tendency to isomerize to the phosphonate structure but the extent of this during distillation will depend on several factors. Mixed phosphites will have a finite probability of mutual exchange in favor of the more symmetrical compounds.

Geometrical isomerism[31] and configurational stability of phosphite rings[403] are topics of current activity.

Trialkyl phosphites containing the simple alkyl groups of few carbon atoms (groups of "ordinary reactivity") have generally been described as having considerable thermal stability.[1144] The stability of tributyl phosphite has been compared with that of the phosphate at 223°. It has been pointed out that conversion of the P-O-C linkage into P(O)C involves a gain in total bond stability of at least 32 and possibly 56 kcal/mole; but what is needed is an accessible path for the necessary redistribution of electron density.[825]

An examination of a series of trialkyl phosphites[825] showed that there was no observed conversion at 120° (16 hr); but at 200° (17.5 hr) the methyl ester underwent 100% conversion into methyl methanephosphonate, the ethyl ester showed no change, the isopropyl ester underwent 25% conversion into the hydrogen phosphite and propylene, and the allyl ester polymerized. At 300° (28 hr) triethyl phosphite gave ethyl ethylphosphonate as the major product, together with diethyl hydrogen phosphite, triethyl phosphate, and ethylene. The difference between the methyl and the ethyl systems affords another example of the need for caution in accepting the methyl pattern as a general one.

Ready isomerization of trimethyl phosphite affords a very satisfactory method of preparing dimethyl methylphosphonate.[317] Rate studies by NMR indicate an autocatalytic effect in which the first formed phosphonate enters into bimolecular reaction with further phosphite, e.g., as follows:[825]

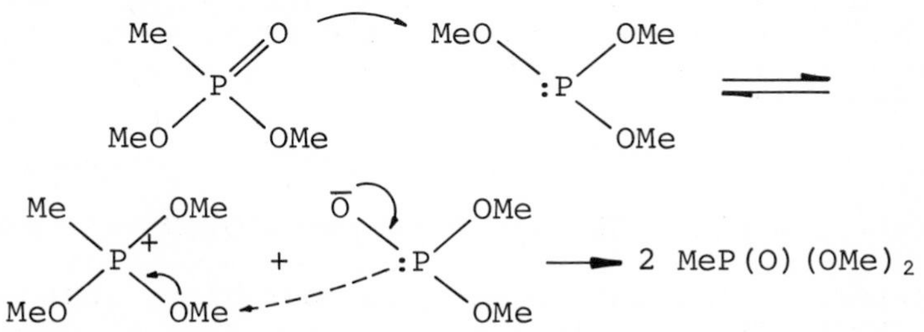

Other phosphoryl, thiophosphoryl, or carbonyl compounds may react similarly with phosphites.[823,825]

It appears probable that a trialkyl phosphite cannot undergo isomerization to the phosphonate form, unless the necessary phosphoryl compound is already present in small amounts as an impurity.[1144,1229,1278] Diethyl hydrogen phosphite is a likely impurity in triethyl phosphite, and deliberately added has a marked accelerating effect on the formation of diethyl ethylphosphonate. Oxidation of some phosphite to the phosphate can likewise provide the necessary functional material.[1144] It has been claimed that saturated trialkyl phosphites are not isomerized merely by heating.[1144]

The isomerization of tris(2-chloroethyl) phosphite is stated to be "improved" by the use of a solvent such as xylene at 150°, polymerization being thus mainly avoided.[443]

The so-called mixed anhydrides of dialkyl phsophorous acids and acrylic acids,[1127] or carboxylic acids,[1118,1119] undergo thermal isomerization at higher temperatures.

When diethyl acetyl phosphite is heated at 160° to 170° for 5 to 6 hr the following sequence of reactions occurs:

$$(EtO)_2POCOMe \xrightarrow{160-170°} (EtO)_2\overset{\overset{O}{\|}}{P}-\overset{\overset{O}{\|}}{C}Me \xrightarrow[\text{(even at room temperature)}]{(EtO)_2POCOMe}$$

$$\text{50\% yield}$$

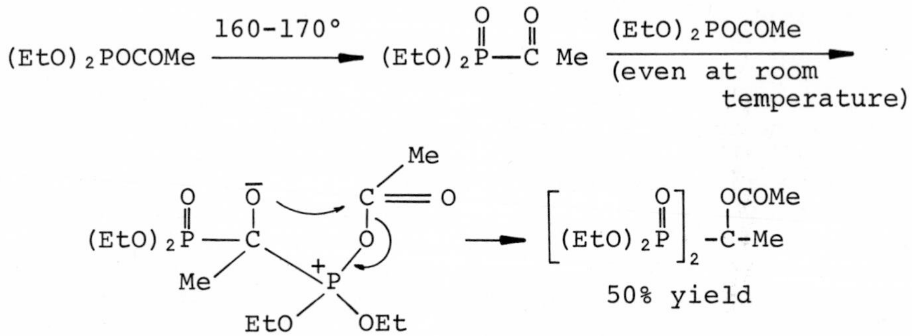

Small amounts of diethyl hydrogen phosphite and diethyl α-acetoxyvinylphosphonate were also isolated.

Further synthetic aspects of the second step involve the independent interaction of dialkyl acetyl phosphites with α-ketophosphonic esters. At elevated temperatures the diethyl acetophosphonate gives diethyl hydrogen phosphite and ketene; the latter undergoes an exothermic reaction with diethyl acetyl phosphite, giving the vinylphosphonate.[1118]

$$(EtO)_2POCOMe + CH_2 = C = O \longrightarrow (EtO)_2\overset{\overset{O}{\|}}{P}-\overset{\overset{OCOMe}{|}}{C} = CH_2$$

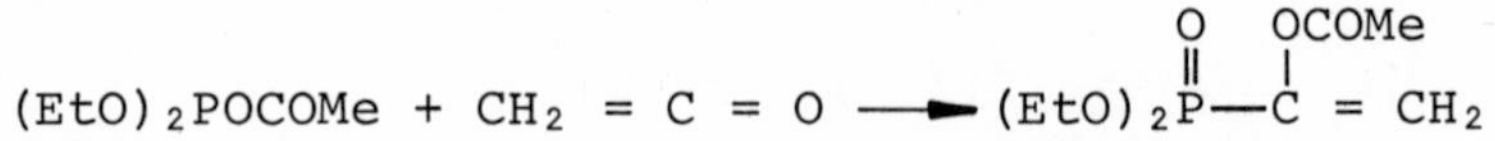

The thermal isomerization of ring phosphites such as $(R_2CO)_2POCH_2(CH_2)_3Cl$ has also been reported.[1115] While triallyl phosphite and tris(2-methylallyl) phosphite at 200° gave viscous or glasslike products, of an apparent polymeric nature,[1097] diethyl 1,1-di-methylallyl phosphite rearranged completely in 1 hr at 200°, and in 5 hr at 120°, to the 3,3-dimethylallyl ester.[825] The process was depicted as involving an S_Ni' route.

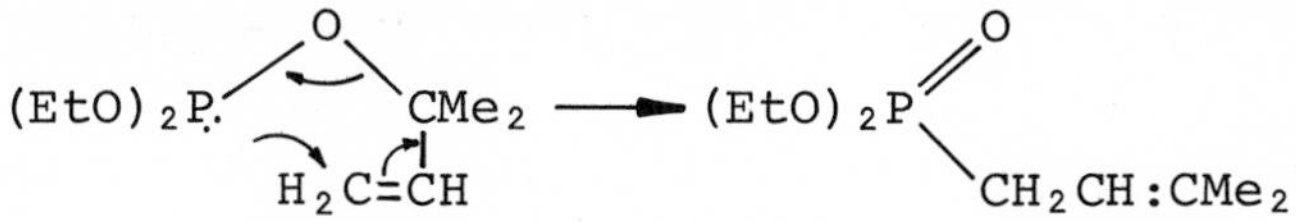

The concurrent formation of the hydrogen phosphite and isoprene was attributed to an intramolecular elimination, similar to that proposed for tri-t-butyl phosphite.[827]

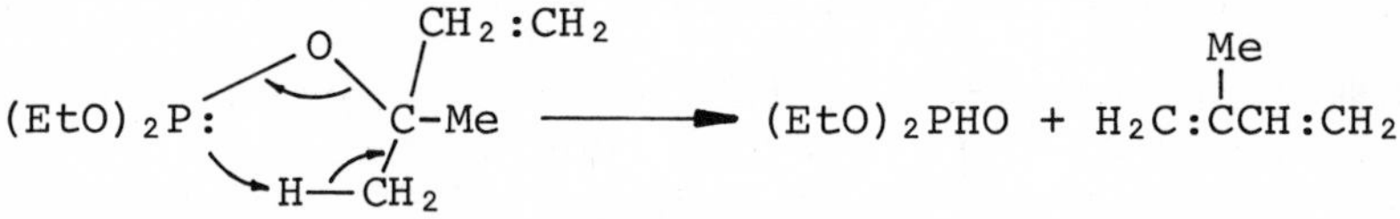

Other allyl phosphites have also been examined,[785,1097] the first order rearrangement of α-methylallyl phosphite being in accord with the S_Ni' mechanism.[784]

Tri-2-alkynyl phosphites exhibit a pronounced tendency to undergo rearrangement to allenic compounds.[586,825,1034,1095,1113] Data in the earlier literature on the supposed esters of propargyl alcohol and its homologs are thus deemed in error. A consideration of bond energies shows that the rearrangement can entail a drop in energy of some 37 to 73 kcal/mole; an intramolecular mechanism involving a five-membered cyclic transition state is envisaged.

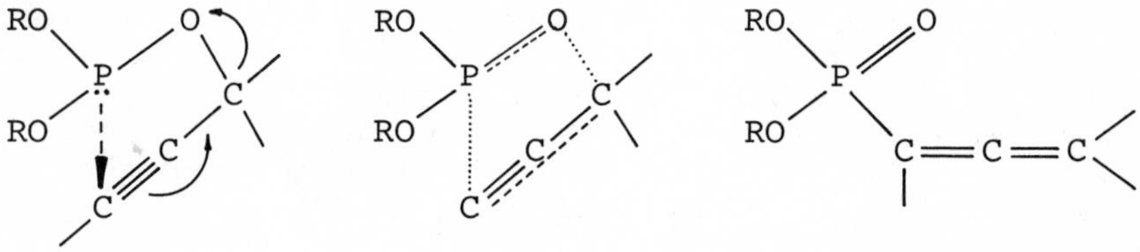

Diethyl 2-butynyl phosphite, $MeC{\equiv}CCH_2OP(OEt)_2$ appeared to be the most stable member examined, being isolable at or below room temperature, and undergoing smooth isomerization to the allenic phosphonate at about 100°. Diethyl propargyl phosphite was distillable at low temperature and pressure but rearranged too rapidly between room temperature and 50° to react with ethyl iodide by the Michaelis-Arbuzov reaction. Hexachlorocyclopentadiene did, however, react below room temperature with this and other phosphites to give alkylated pentachlorocyclopentadienes.

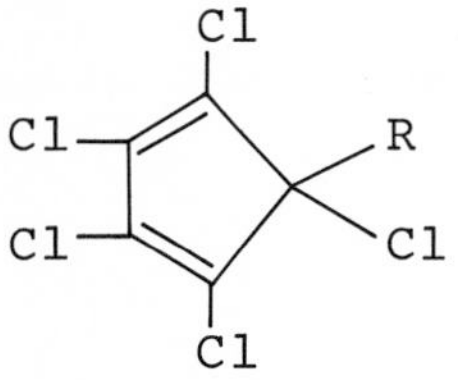

$R = -CH_2C{\vdots}CH$ $R = -CH_2CH_2C{\vdots}CH$

$R = -CH_2C{\vdots}CMe$ $R = -CH(Me)C{\vdots}CH$

When R was $-C(Me)_2C{\vdots}CH$ isomerization occurred too quickly for interception or detection by spectroscopy. Interaction with sulfur and chloral are other examples of interception.

$$(EtO)_2P(OCH_2C{\vdots}CH) + S \longrightarrow (EtO)_2P(S)(OCH_2C{\vdots}CH)$$

$$(EtO)_2P(OCH_2C{\vdots}CH) + Cl_3CCHO \longrightarrow (EtO)_2P(O)(OCH_2{:}CCl_2)$$

Alkyl substitution at the 1-position appears to reduce the stability of the propargyl phosphites; e.g., propargyl 1,2-phenylene phosphite is stable at room temperature and isomerized in boiling benzene;[598] but after 2-phenylene phosphorochloridite was mixed with the tertiary alcohol $MeEtC(OH)C{\equiv}CH$ in the presence of trimethylamine in ether at -50°, only the rearranged allenic phosphonate was isolated on evaporating the solvent at room temperature.[724] Similar observations have been made for the propargyl chloridites and dichloridites.[584,585,615,1033,1035] The intramolecular nature of the rearrangement is confirmed since the deuterated alcohol, $DC{\vdots}CCH_2OD$, gave the dichloridate, $CH_2{:}C{:}CDOP(O)Cl_2$, in the presence of triethylamine.[725] 2,3-Butadienyl phosphites can be distilled below 70° under reduced pressure and are stable enough at room temperature to react with oxygen, sulfur, selenium and hexachlorocyclopentadiene. Isomerization to butadienephosphonic esters occurs at 112-14° at a rate which could be followed by IR and NMR.

Much of the primary product, however, underwent the Diels-Alder dimerization. The rearrangement mechanism was deemed to be analogous to that suggested for the acetylenic esters. Substitution by methyl at the 1-position is rate accelerating.

C.2. Structure of Hydrogen Phosphites

From the beginning of the purposeful study of dialkyl and diaryl hydrogen phosphites, two tautomeric forms, having the phosphite or phosphonate structure, were recognized as possibilities.

$$(RO)_2P - OH \qquad or \qquad (RO)_2P\overset{\displaystyle O}{\underset{\displaystyle H}{\diagup}}$$

Dialkyl phosphite   Dialkyl phosphonate

Spectroscopic evidence is now overwhelmingly in favor of the phosphonate structure. The infrared spectrum clearly shows the presence of P=O and P-H, no hydroxyl group being detectable,[174,175,181,237,323,408,571,807,844,890,1317] whilst the magnitudes of the ^{31}P NMR chemical shift and P-H coupling constant[826] are entirely consistent with the phosphonate form.[263,396,935,1342] There is no evidence for intermolecular hydrogen bonding. Other data (parachor, viscosity, molecular weight, boiling point, dipole moment, molecular refraction, magnetic rotation, etc.) have been reviewed;[354] they are less conclusive but in general are in line with the above conclusions. Certain reactions, e.g., the addition of sulphur in the presence of triethylamine,[107] are more readily explained on the basis of the phosphite structure and the possibility of an equilibrium remains.

$$(EtO)_2PHO \rightleftharpoons (EtO)_2\overset{..}{P}-OH \xrightarrow{\ S\ } (EtO)_2P(S)OH$$

It has been pointed out that the lower limit for the detection of the phosphite form by the methods used is about 1%, and an estimated figure of 10^{-4}% has been given.[354]

The dialkyl hydrogen phosphites show both basic and acidic function. The former is demonstrated by the solubility of hydrogen chloride which presumably forms hydrogen-bonded species[480] and the latter by the replacement of hydrogen by metals.[354]

Sodium causes the vigorous evolution of hydrogen when added to a solution of the usual simple hydrogen phosphites in ether, hexane, benzene, or toluene. The sodium derivative of a lower alkyl ester is moderately soluble in these solvents, and freely soluble even in petroleum when R is Bu or higher. The sodium alkoxides may also be used. Certain operational factors have been revealed by using droplets of sodium in hot toluene, and by using sodium dust at about 0°.[734] The magnesium derivatives are prepared in the presence of pyridine. The silver derivative is formed in aqueous systems, by the addition of the calculated amount of ammoniacal silver nitrate to an aqueous suspension of the hydrogen phosphite followed by neutralization with nitric acid. There is more recent work on the compound, $(EtO)_2POAg$.[94,524] The sparingly soluble cuprous compounds have been obtained by means of hydrated cuprous oxide.[354]

The ^{31}P NMR spectra of the sodium or potassium salts in aqueous solution show chemical shifts which exclude the possibility of a metal-phosphorus bond, the ionic structure $M^+[OP(OR)_2]^-$ being assumed.[913] Infrared evidence for the lithium, sodium, potassium, and silver salts indicates the absence of a phosphoryl group,[323] a structure $MOP(OR)_2$ having considerable covalent character being favored because of the high solubility in petroleum ether.[354] X-ray absorption spectroscopy of the solid sodium derivative had earlier indicated the structure $(RO)_2P(O)Na$; but $(RO)_2POAg$ for the silver salt.[1292]

Examples of mercuric "salts," $(RO)_2P(O)HgX$ (R = Et, Pr, etc.; X = AcO, Cl, Br, I, SCN) and bis(dialkoxy-phosphoryl)-mercury, $[(RO)_2P(O)]_2Hg$ (R = Me, Et, to i-Bu), have been prepared,[410,1041,1346] interaction between diethyl hydrogen phosphite (named as "diethyl phosphonate") and mercuric acetate being exothermic. The halides could be obtained from the acetate by means of an alkali metal halide, but were better prepared directly in benzene, product water being azeotropically removed.

$$2\ (RO)_2PHO + HgX_2 + HgO \longrightarrow 2\ (RO)_2P(O)HgX + H_2O$$

$$2\ (RO)_2PHO + HgO \xrightarrow{\text{slow}} [(RO)_2P(O)]_2Hg + H_2O \xrightarrow[\text{fast}]{HgX_2} 2\ (RO)_2P(O)HgX$$

Heating the hydrogen phosphite and mercuric oxide in benzene (or certain other solvents) at 200 mm < 50°, the intermediate bis-ester could be isolated, and then converted into the -HgX compounds by treatment with HgX_2.[410,1346] Solubility and stability characteristics were outlined. The hydrogen phosphites were thus deemed to function as "phosphonates," possessing "a reactive hydrogen atom somewhat similar in properties to an α-hydrogen atom in a ketone, or the hydrogen atoms in a reactive methylene group." The characteristic P-H infrared band is lacking in the mercuric acetate salts. The P-Hg bond was deemed to be "relatively weak," free mercury being eliminated by a trace of alkali or on heating to about 70° in the presence of solvents. The increase in melting point in passing from n-Pr to i-Pr and n-Bu to i-Bu is remarkable.

C.3.IIIc. DEALKYLATION BY HYDROGEN HALIDES

This is a special feature of phosphite chemistry and was outlined in Section Ia. The rates of dealkylation are in the order $(RO)_3P \gg (RO)_2PHO > ROPHO(OH)$, and HI > HBr >> HCl, for the same R group, but the rate of each dealkylation varies considerably according to the nature of R, being overall very quick for such as t-Bu, or Ph(Me)CH, and very slow for such as 2,2,2-trichloroethyl.[185,304,449,450,463,467,474,476-478,1092] The removal of the first alkyl group from a trialkyl phosphite is very rapid and no kinetic data are at present available; for HCl or HBr it is complete in a time during which removal of the second alkyl group is imperceptible.[480] In the absence of solvent, the hydrogen phosphite immediately absorbs 1 mole of hydrogen chloride with the evolution of approximately 4260 cal/mole at 25°; but dealkylation proceeds slowly, 0.5 mole of alkyl group being removed during 4 days, and 0.75 mole during 30 days.[480] A kinetic study of the dealkylation of diisopropyl hydrogen phosphite by hydrogen chloride showed the process to be second order in hydrogen halide.[265] Work with hydrogen bromide at 25° shows that the removal of the first alkyl group from the diester proceeds about 9 times as fast as the dealkylation of the monoester.[480]

The detailed mechanism is probably far from simple. There is most likely only a very small concentration of "free" hydrogen chloride in dynamic equilibrium with a variety of hydrogen-bonded species.

Stereochemical studies[172,185,218,449,450,459-462,470,476,477] especially with 2-butyl,[497] 2-pentyl,[279] or 2-octyl phosphites[300,311] point to an S_N2 mechanism for the first rapid dealkylation of these triesters, e.g.,

$$Cl^- \curvearrowright R \text{—} O \curvearrowright \text{—} P^+H(OR)_2 \longrightarrow RCl + (RO)_2PHO$$

this being supported by the absence of rearrangement[300,311] (< 0.1% by GLC[280] in the preparation of 3-bromopentane[577]) which would be expected for a carbonium ion process involving such groups. This reaction provides one of the few authenticated methods for the preparation of s-n-alkyl halides without rearrangement and (for chlorides and bromides) in an optically pure form.[576] Although some overlap of the second and third dealkylation steps may occur, the stereochemical and analytical (GLC) evidence indicates that the second dealkylation is also an S_N2 process for such groups while the third involves considerable loss in rotatory power and rearrangement.[311] Primary n-alkyl groups yield isomerically pure halide in all three stages of the dealkylation.[311] The more branched groups (neopentyl, 3-methyl-2-butyl) rearrange in the first dealkylation step, except for neopentyl and hydrogen iodide.[67,278,575] Olefinic by-products are formed, more especially from the branched alkyl esters, and addition of hydrogen halide could account for some of the rearrangement products observed.

Negligible reaction occurs between p- or s-alkyl dichloridites and hydrogen chloride and indeed the solubility of the latter in these compounds is low.[278,467] The bromoesters are dealkylated by hydrogen bromide somewhat more readily although it is unlikely that this is a major route to alkyl bromide formation in alcohol-phosphorus tribromide systems.[463]

C.4.IIId. INTERACTION WITH HALOGENS AND INORGANIC NONMETAL HALIDES

Parts of the chemistry coming under this heading have been reviewed up to about March 1964 under the title "reactions between trivalent phosphorus derivatives and positive halogen sources."[888]

The work will now be summarized under the following headings.

1. INTERACTION WITH HALOGENS. Trialkyl phosphites react readily with chlorine,[449] bromine,[469] or iodine (solution in ether)[469] at low temperatures (e.g., -10° to 0°) to form the corresponding halidates and alkyl chloride.

$$(RO)_3P + X_2 \longrightarrow (RO)_2P(O)X + RX$$

The iodidate was immediately converted into the anilide. Reaction of chlorine or bromine with tri-2-octyl phosphite proceeds with inversion of configuration, in accordance with the following mechanism:[469,476]

$$(RO)_3P{:}\ Cl{-}Cl \longrightarrow Cl^-\ R{-}\overset{+}{O}P(OR)_2Cl \longrightarrow$$

$$RCl + O{=}P(OR)_2Cl$$

The dialkyl hydrogen phosphites also afford chloridates by reaction with chlorine.[150,200,459,468,475,476,813,837,838,1366]

$$(RO)_2PHO + Cl_2 \longrightarrow (RO)_2P(O)Cl + HCl$$

The process was initially depicted as a 4-center reaction[476] not necessarily requiring the tricovalent phosphite form as suggested later.[888]

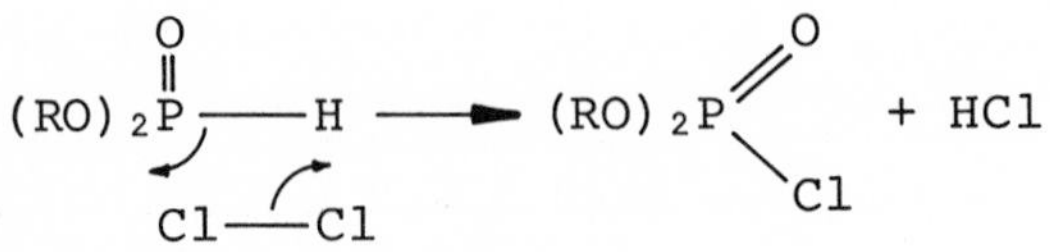

or

$$(RO)_2(HO)P{:}\ \frown\ Cl-\frown Cl \longrightarrow Cl^- + H\text{-}OP^+(OR)_2Cl \longrightarrow$$

$$HCl + O{=}P(OR)_2Cl$$

With bromine, the dibromidate is likewise formed from the neopentyl ester,[475] but from esters such as di-n-butyl hydrogen phosphite the bromidate undergoes further dealkylation by the hydrogen bromide produced to give alkyl bromide.[469]

$$(RO)_2PHO + Br_2 \longrightarrow (RO)_2P(O)Br + HBr \longrightarrow$$

$$(RO)(HO)P(O)Br + RBr$$

Although it was stated,[838] and later reiterated,[888] that iodine does not react with diethyl hydrogen phosphite because of the unfavorable position of the following equilibrium, no definite experimental evidence was given.

$$(EtO)_2P(O)I + HI \rightleftharpoons (EtO)_2PHO + I_2$$

It was shown, however,[469] that iodine was decolorized by the hydrogen phosphite in pentane, slowly at 0°, more quickly at 15°, affording ethyl iodide (50%, based on 2 alkyl groups), a result more in accordance with the bromine system, and the dealkylating propensity of hydrogen iodide. Titration by a solution of iodine in benzene was considered as a means of determining triester in the presence of diester; but the method was not reliable.[838] In contrast it was stated[275] that the rate of transfer of alkyl groups (R = Me, Et, i-Pr) from the diester to phosphorous acid could be followed by titrating with iodine in a sodium bicarbonate buffered solution, for "both the dialkyl hydrogen phosphite and phosphorous acid react with iodine whereas the monoalkyl phosphite does not." See Refs. 994-1000. Kinetic studies on the "bromination and iodination" of dialkyl hydrogen phosphites (R = Me, Et, Pr, i-Pr, Bu) in buffered solutions in the presence of potassium bromide, or iodide seem to indicate the existence of a phosphite-phosphonate equilibrium $POH \rightleftharpoons P(O)H$.

The results for iodine and dibutyl hydrogen phosphite in butanol (nonaqueous) are open to several interpretations.[408,409]

Thiocyanogen and a dialkyl hydrogen phosphite afford the compound $(RO)_2P(O)NCS$.[879] 2-Octyl phosphorodichloridite reacted readily with chlorine to give alkyl chloride and phosphorus oxychloride, or with bromine to give alkyl bromide and $POCl_2Br$. Ethyl bromide and $POCl_2Br$ were similarly obtained in early work.[854] Other studies[534] include the butyl phosphorodifluoridite-bromine system.[356]

Early work on the interaction of halogens with triaryl

phosphites entailed the formation of what were deemed to be dihalides as reactive oils.[53] The pattern of reaction sequences and products are complex.[296,1230] A conductometric examination of the triphenyl phosphite-bromine system in acetonitrile is relevant.[529] From triphenyl phosphite (2 mole) and bromine (1 mole) in chlorobenzene, the bromidite $(PhO)_2PBr$ and the crystalline compound $(PhO)_4PBr$ were obtained.[1230] The iodine system again shows points of difference.[407] However, a recommended procedure for the preparation of alkyl iodides is by the reaction of iodine in a suitable solvent with the mixed alkyl diphenyl phosphite or more conveniently the alkyl o-phenylene phosphite.[319] Cyclohexyl iodide was thus obtained in 83% yield with the minimum of elimination.[307]

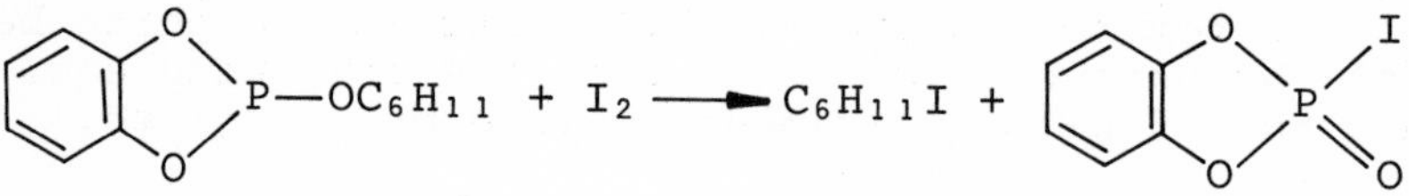

2. BORON HALIDES. Tributyl phosphite and boron trichloride react immediately and irreversibly, even at -80°, apparently by the mutual exchange of butoxy and chlorine.

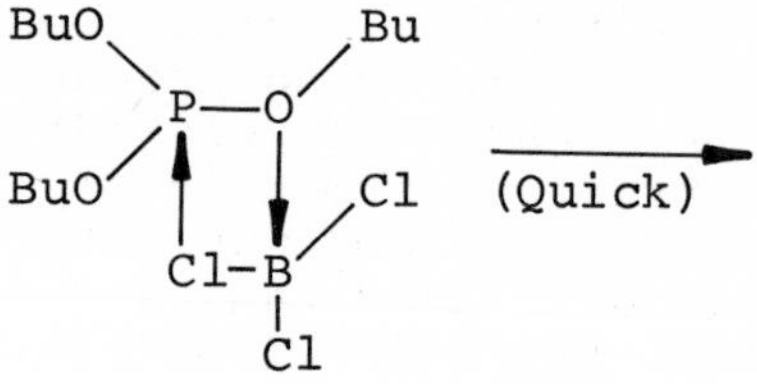

$$BuOBCl_2 + (BuO)_2PCl \xrightarrow[\text{(much slower)}]{BCl_3} 2\ BuOPCl_2$$

The third replacement to yield phosphorus trichloride was not clearly observed, because of stringent temperature control to avoid subsidiary decomposition of the boron compounds.[471] Below about -10°, boron trichloride and triphenyl phosphite could be quantitatively separated from their admixture, showing at the most, a weak complexing. Replacement reactions occurred readily at 25°.[416-418] An explanation is to be found in the suggestion that the driving force is the nucleophilic function of oxygen, the electron density on the phenoxy-oxygen being much lower than on the butoxy-oxygen because of mesomeric interaction between the oxygen lone-pairs and the π-system.

Stepwise replacement occurs in one direction only, according to the relative amounts of reagents, and the time-temperature factors. Phosphorus trichloride makes no impression on triphenyl borate even at 80°. Mutual replacement likewise occurred in the triphenyl phosphite-boron tribromide system, there being a general drift in one direction, resulting in the ultimate formation of phosphorus tribromide.

$$(PhO)_3P + BX_3 \longrightarrow (PhO)_2PX \xrightarrow{etc.} PX_3 + (PhO)_3B$$

(X = Cl or Br)

By significant contrast, trialkyl phosphates form complexes $(RO)_3P = O, BCl_3$ at temperatures below 0°; elimination of alkyl chloride occurs on heating, thus forming a network of -P-O-B-O- links.[460-462] Dialkyl hydrogen phosphites similarly yield 1:1 complexes which eliminate alkyl chloride gradually from 20 to 300°C to leave a polymeric solid.[172] Hydrogen chloride was evolved more slowly, and even at 300° the solid residue contained easily hydrolyzed chlorine. Reactions of dialkyl phosphites with phenylboron dichloride or diphenylboron chloride yield alkyl chloride at 20°, while boron trichloride and diphenyl hydrogen phosphite gave hydrogen chloride (0.93 mole) and the compound $(PhO)_2POBCl_2$.

Diphenyl hydrogen phosphite and diphenyl chloroboronate react as follows:

$$(PhO)_2PHO + (PhO)_2BCl \longrightarrow (PhO)_2P(O)B(OPh)_2 + HCl$$

Reaction of triethyl phosphite with $PrBCl_2$ gave ethyl chloride, ethoxydi-n-propylborane and a polymer designated $(EtPO_2)_n$.[1349]

3. INTERACTION WITH ORGANIC HYPOCHLORITES. The first published work is summarized as follows:[1066]

$$(EtO)_3P + EtOCl \longrightarrow [(EtO)_3P(OEt)Cl] \longrightarrow (EtO)_3PO + EtCl$$

$$(MeO)_3P + EtOCl \longrightarrow (MeO)_2P(O)OEt + MeCl$$

$$(EtO)_2PHO + EtOCl \longrightarrow (EtO)_3PO + HCl$$

Tetrahydrolinalyl and methylethylisobutylcarbinyl hypochlorites were said to react rapidly with triphenyl phosphite at about -78°, giving triphenyl phosphate, t-alkyl chlorides, olefins, and the parent alcohols. The best yields, < 50%, of RCl were obtained with pentane as solvent at -75°.[331] A bicyclic phosphite similarly gave the corresponding phosphate on treatment with t-butyl hypochlorite.[334]

4. SULFUR CHLORIDES. Sulfur monochloride reacts with

triethyl phosphite to yield the thiophosphate, together with diethyl phosphorochloridate, triethyl phosphite dichloride, and ethyl chloride.[920]

$$S_2Cl_2 + 3\ (EtO)_3P \longrightarrow 2\ (EtO)_3PS + (EtO)_3PCl_2 + (EtO)_2P(O)Cl + EtCl$$

Pinacolyl or s-butyl dichlorophosphites also give the corresponding thiophosphate $ROP(S)Cl_2$,[1253] while dialkyl hydrogen phosphites are reported to yield disulphides.[373-375,1409]

$$2\ (RO)_2PHO + S_2Cl_2 \xrightarrow[-5°]{Et_2O} (RO)_2P(O)SSP(O)(OR)_2 + 2\ HCl$$

In the initial study of the triethyl phosphite-sulfur dichloride system no products were isolated,[81] but later,[1080] the thiophosphate (82% yield) and the chloridate (64% yield, converted into anilide) were obtained in petroleum at 0°.

$$2\ (EtO)_3P + SCl_2 \longrightarrow (EtO)_2P(O)Cl + (EtO)_3P{=}S + EtCl$$

Other workers[874] found the initial reaction mixture to contain tetraethyl thiopyrophosphate in addition. From equimolar amounts of reagents, the sulphenyl chloride was identified and deemed an intermediate in the reaction system for 2 mole of phosphite.[868]

$$(EtO)_3P \xrightarrow{SCl_2} (EtO)_2P(O)SCl \xrightarrow{(EtO)_3P}$$

$$EtCl + (EtO)_2P(O)Cl + (EtO)_3PS + (EtO)_2P(S)OP(O)(OEt)_2$$

Triphenyl phosphite gives two main products.[1082]

$$2\ (PhO)_3P + SCl_2 \longrightarrow (PhO)_3P{=}S + (PhO)_3PCl_2$$

Free-radical and ionic mechanisms have both been discussed.[888,1081]

Diethyl hydrogen phosphite is stated to give a compound,[796] obtained also by dialkylaminosulfenyl chloride (see below), containing a P-S-P link, although the scheme later given was for the unsymmetrical P-O-P compound.[874]

$$(EtO)_2POH \xrightarrow{SCl_2} (EtO)_2P(O)SCl \xrightarrow{(EtO)_2POH}$$

$$(EtO)_2P(S)OP(O)(OEt)_2$$

Distillation of a crude mixture of sulphur dichloride and sulphur monochloride with di- or trialkyl phosphites has been used for the purification of the dichloride, which is then said to be stabilized by the presence of additional phosphite.[1248]

Sulfenyl chlorides (R' = Ph, $ClCH_2CH_2$, Me, Et) react readily at -60° to -10° with trialkyl phosphites (R = Et, Pr, Bu) in toluene or chloroform, the mechanism being depicted as a nucleophilic replacement of chlorine by phosphorus, followed by nucleophilic attack of chloride ion on the alkoxy group,

$$(RO)_3P + R'SCl \longrightarrow [(RO)_3PSR']^+ Cl^- \longrightarrow R'SP(O)(OR)_2 + RCl$$

although no intermediate could be isolated.[920,1068] The halidites react similarly.[1269]

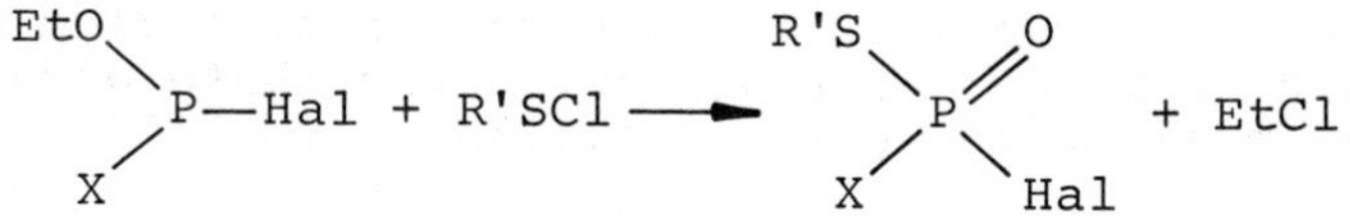

(X = OEt or Cl; Hal = Cl or F)

The formation of O,O,O-triphenyl phosphorothioate in the [$(PhO)_3P$ + EtSCl] system was called a rare example of the cleavage of a C-S bond in reactions of the "Arbuzov type"[888]

$$(PhO)_3P + EtSCl \longrightarrow (PhO)_3P{=}S + EtCl$$

but emerges from the inability of the phenyl group to accommodate the usual nucleophilic form of attack. The acetyl group is removed from the sulfur in the following example.[868]

$$MeC(=O)SCl + (EtO)_3P \longrightarrow (EtO)_3P{=}S + MeCOCl$$

Triphenyl phosphite and phenyl sulfenyl chloride give the disulphide without cleavage of either the aryl-oxygen or aryl-sulfur bonds.[1078]

$$(PhO)_3P + 2\ PhSCl \longrightarrow (PhO)_3PCl_2 + PhSSPh$$

Sulfenyl chlorides containing the P-S-Cl link react with triethyl phosphite to give products resulting from alkyl-oxygen fission or phosphorus-sulfur fission; but with triphenyl phosphite only the latter occurs.[1068,1078]

$$
\begin{array}{c}
\text{EtO} \quad \text{O} \\
\quad\backslash \;\; \parallel \\
\quad\quad \text{P-SCl} \;+\; (\text{EtO})_3\text{P} \longrightarrow \\
\quad / \\
\text{Et}
\end{array}
$$

$$
\text{EtCl} + \left[\;
\begin{array}{c}
\text{EtO} \;\; \text{O} \;\; \text{O} \\
\quad\backslash \;\; \parallel \;\; \parallel \\
\quad\quad \text{P-S-P(OEt)}_2 \longrightarrow \\
\quad / \\
\text{Et}
\end{array}
\quad
\begin{array}{c}
\text{EtO} \;\; \text{S} \;\; \text{O} \\
\quad\backslash \;\; \parallel \;\; \parallel \\
\quad\quad \text{P-O-P(OEt)}_2 \\
\quad / \\
\text{Et}
\end{array}
\;\right]
$$

$$
\begin{array}{c}
\text{EtO} \;\; \text{O} \\
\quad\backslash \;\; \parallel \\
\quad\quad \text{PSCl} \;+\; (\text{RO})_3\text{P} \longrightarrow (\text{RO})_3\text{P=S} \;+\;
\end{array}
\quad
\begin{array}{c}
\text{EtO} \;\; \text{O} \\
\quad\backslash \;\; \parallel \\
\quad\quad \text{PCl} \\
\quad / \\
\text{Et}
\end{array}
$$

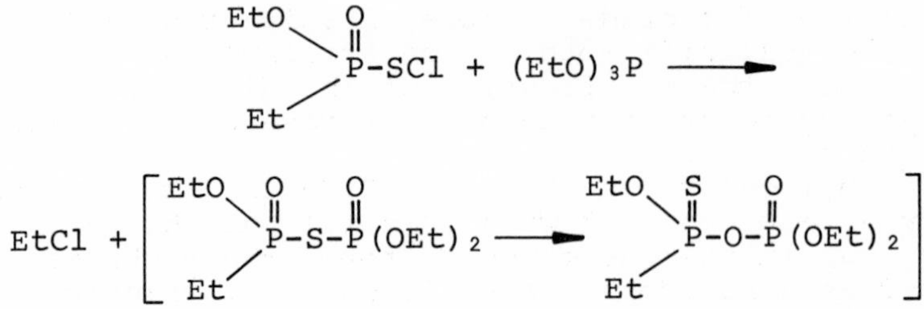

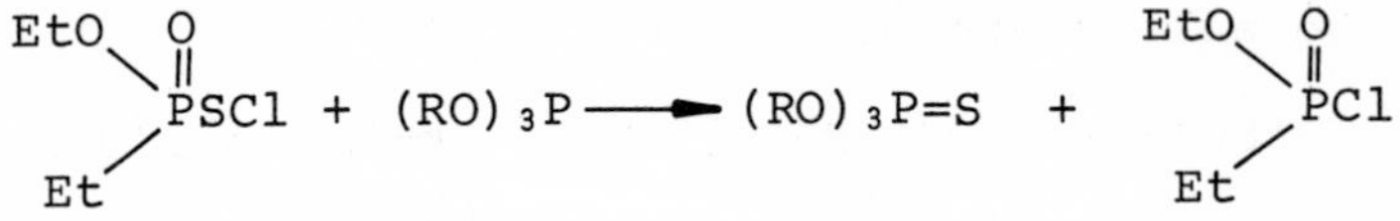

$$
R = Et, Ph
$$

There was a quantitative separation of sodium chloride when phenylsulfenyl chloride and a phosphite were mixed with sodium alkoxide in acetonitrile at 0°.[929]

$$
\text{PhSCl} + (\text{RO})_3\text{P} + \text{NaOR} \longrightarrow \text{PhSR} + (\text{RO})_3\text{P=O} + \text{NaCl}
$$

Other work with trialkyl phosphites and sulfenyl chlorides has been reported.[875]

Dialkyl phosphites interact with sulfenyl chlorides to give products[920] which were explained by invoking ter-covalent phosphorus,[888] although the

$$
(\text{RO})_2(\text{HO})\text{P} : \frown \text{S(R)Cl} \longrightarrow [(\text{RO})_2(\text{HO})\text{PSR}]^+ \text{Cl}^- \longrightarrow
$$

$$
\text{HCl} + (\text{RO})_2\text{P(O)SR}
$$

4-center mechanism is also plausible.

$$
\begin{array}{c}
\quad\quad \text{O} \\
\quad\quad \parallel \\
(\text{EtO})_2\text{P}\!-\!\text{H} \longrightarrow (\text{EtO})_2(\text{EtS})\text{PO} + \text{HCl} \\
\quad\; \text{Et}\!-\!\text{S}\!-\!\text{Cl}
\end{array}
$$

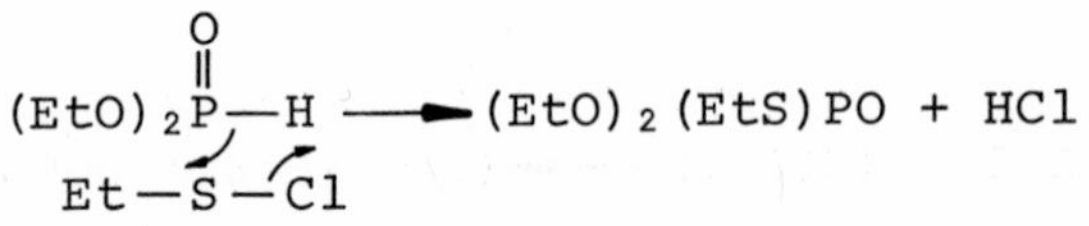

Trichloromethyl sulfenyl chloride yielded dialkyl phosphorochloridate and thiocarbonyl chloride in addition.[373-375,1409] The reactions of dialkyl phosphites with dimethyl-amino sulfenyl chloride[796] and with $(\text{EtO})_2\text{P(O)SCl}$[868] have also been reported, both yielding products containing a P-O-P link.

$$
(\text{EtO})_2\text{PHO} + \text{Me}_2\text{NSCl} \longrightarrow [(\text{EtO})_2\text{P(O)SNHMe}_2]^+ \text{Cl}^- \xrightarrow{(\text{EtO})_2\text{PHO}}
$$

$$
(\text{EtO})_2\text{P(O)SP(O)(OEt)}_2 \longrightarrow (\text{EtO})_2\text{P(O)OP(S)(OEt)}_2
$$

$$(EtO)_2P(O)SCl + (EtO)_2POH \longrightarrow$$

$$(EtO)_2P(S)OP(O)(OEt)_2(60\%) + HCl +$$

$$(EtO)_2P(O)Cl(25\%) + (EtO)_2P(S)OH(10\%)$$

When thionyl chloride in petroleum was added dropwise to triethyl phosphite in petroleum at 0°,[1081] followed by heating, there was no sign of a previously reported[173] stable intermediate $[(EtO)_3PCl]_2SO$. The following products were obtained.

$$(EtO)_3P + SOCl_2 \longrightarrow (EtO)_3P{=}O + (EtO)_3P{=}S +$$

$$(EtO)_2P(O)Cl + EtCl + SO_2$$

From triphenyl phosphite the products are explicable on the basis of the intermediate formation of sulfur dichloride.[1082]

$$(PhO)_3P + SOCl_2 \longrightarrow (PhO)_3PO + SCl_2$$

$$2\ (PhO)_3P + SCl_2 \longrightarrow (PhO)_3PS + (PhO)_3PCl_2$$

Hydrogen phosphites (Et, i-Pr) are said to react as follows:[149]

$$(RO)_2PHO + SOCl_2 \longrightarrow (RO)_2P(O)Cl$$

Dropwise addition of triethyl phosphite in petroleum to sulfuryl chloride in petroleum at 0°, under nitrogen, gave a copious evolution of sulfur dioxide and diethyl phosphorochloridate was isolated in 87% yield.[1079]

An earlier claim[173] that sulfuryl chloride forms stable intermediates with trialkyl phosphites was refuted,[1079] although a transient intermediate was depicted as follows:

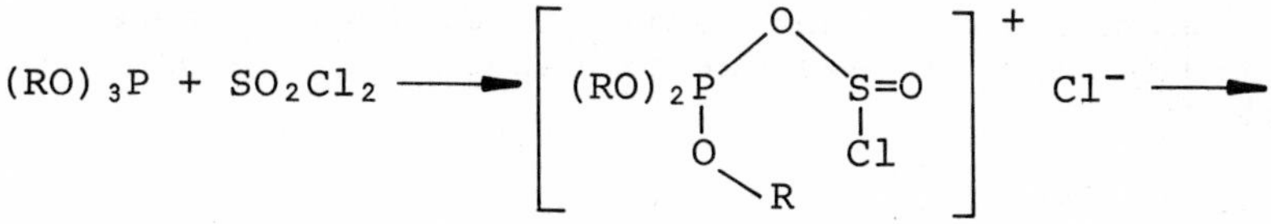

$$(RO)_2P(O)Cl + SO_2 + RCl$$

Nucleophilic displacement on chlorine has also been suggested[888] and there are other possibilities. A free radical mechanism was rejected on the grounds that addition of sulfuryl chloride to the phosphite in styrene (in excess) at 0° gave 93% yield of sulfur dioxide, but little polymerization.[1079]

Interaction with triphenyl phosphite proceeds rapidly even in the absence of light at < 10°, and again the essential distinction from the alkyl phosphite behavior is illustrated.[1082]

The chloridite behaved similarly.[148]

$$(PhO)_3P + SO_2Cl_2 \longrightarrow (PhO)_3P{=}O + SOCl_2$$

$$(PhO)_2PCl + SO_2Cl_2 \longrightarrow (PhO)_2P(O)Cl + SOCl_2$$

Mixed alkyl aryl phosphites react by removal of an alkyl group.

Hydrogen phosphites give up the P-H hydrogen,[149,200,402] and the sodium derivative gave a tetraalkyl hypophosphate, presumably via the further reaction of the first-formed dialkyl phosphorochloridate with the sodium derivative.[200,869,886] Trialkyl phosphites react with an alkyl or arylsulfonyl chloride to give the products shown, the sulfonylphosphonate not being indicated.[483,553]

$$3\ (EtO)_3P + RSO_2Cl \longrightarrow (EtO)_2P(O)SR + EtCl + 2\ (EtO)_3P{=}O$$

A possible route involves the reduction of the sulfonyl chloride to sulfenyl chloride, by phosphite and then the interaction of a third mole of phosphite, as already discussed. Aromatic sulfonyl fluorides cannot be so reduced, and did not react.

Triphenyl phosphite and benzenesulfonyl chloride in boiling benzene afforded 4 main products, the phosphate and disulfide almost quantitatively, according to the following equation.[1078]

$$6\ (PhO)_3P + 2\ PhSO_2Cl \longrightarrow$$

$$5\ (PhO)_3P{=}O + (PhS)_2 + PhCl + (PhO)_2PCl(?)$$

Dropwise addition of 3-NO_2-4-Me-benzenesulphonyl chloride to the sodium derivative of diethyl hydrogen phosphite in benzene gave the disulphone as the only compound isolated.[1289]

C.5. Disulfides

When diethyl disulfide (1 mole) and triethyl phosphite (2.5 mole, excess) were heated, a quantitative yield of diethyl sulfide was distilled, and the monothiophosphate was isolated in 77% yield.[619]

$$(EtO)_3P + EtSSEt \longrightarrow EtS\overset{+}{P}(OEt)_3(EtS^-) \longrightarrow EtSPO(OEt)_2 + Et_2S$$

Diphenyl disulfide behaves similarly and causes elimination

of an ethyl group from the phosphite; but triphenyl phosphite does not react with diphenyl disulfide.[1078]

$$(EtO)_3P + PhSSPh \longrightarrow (EtO)_2P(O)SPh + PhSEt$$

Acetyl disulfide[880] or benzoyl disulfide[533] gave the O,O,O-triethyl phosphorothioate as if the intermediate anion[880] had cleaved the acyl-sulfur bond instead of the Et-O bond.

$$(EtO)_3P + AcSSAc \longrightarrow (EtO)_3P{=}S + Ac_2S$$

Radical mechanisms have been discussed as possibilities[1371, 1372] but were not favored since the reaction was unaffected by the presence of quinol.[533] See, however, Ref. 325.

In a study of the alkylation of some heterocyclic compounds by derivatives of phosphorous acid, the formation of 2-ethylthiobenzoxazole from 2-benzoxazole disulfide (1 mole) and triethyl phosphite (2 mole) in xylene was described.[140] The same product was obtained from the corresponding sulfenyl chloride. From 2-benzothiazole disulfide and triethyl phosphite, or $EtOP(OCH_2)_2$, 2-ethylthiobenzothiazole was obtained. The sodium derivatives from dialkyl hydrogen phosphites react rapidly with disulfides in tetrahydrofuran at 0° to form O,O,S-trialkyl phosphorothioates and sodium alkyl mercaptan[882] but, when the initial mixture is allowed to stand, a further reaction occurs.[532]

$$RSSR' + NaP(O)(OEt)_2 \xrightarrow{\text{fast}} RSP(O)(OEt)_2 + NaSR' \xrightarrow{\text{slow}} RSP(O)(OEt)(ONa) + EtSR'$$

Other similar systems have been described.[1045,1361] Triaryl phosphites are converted to the thiophosphates by interaction with disulfides in refluxing benzene as follows:[852]

$$[(RO)_2P(S)]_2S_2 + (ArO)_3P \longrightarrow [(RO)_2P(S)]_2S + (ArO)_3PS$$

Surprisingly the sodium derivative of diphenyl hydrogen phosphite also gave the thiophosphate by reaction with the same reagent.

C.6. Interaction with Sulfur and Selenium

Addition of sulfur to a trialkyl or triaryl phosphite, to yield the thiophosphate, $(RO)_3P{=}S$, has long been known, and is frequently referred to in the papers and patent specifications (e.g., Ref. 362). Certain phosphites do not add sulfur for one reason or another, as is exemplified by o-phenylene 2-hydroxyphenyl phosphite.[61] Trialkyl phosphites react readily (the lower members vigorously) with selenium.[992]

$$(RO)_3P + Se \longrightarrow (RO)_3P{=}Se$$

Although the hydrogen phosphites are not reactive with sulfur, the sodium derivatives readily give the compounds $(RO)_2P(S)ONa$;[405,406] the corresponding addition of selenium to the potassium derivative, yielding $(RO)_2P(Se)OK$ has also been reported.[830]

Sulfur and selenium also add to the phosphorohalidites.[64,1071]

C.7. N-Haloamines, N-Haloimines and N-Haloamides

In what appears to be the first report on the N-haloamine systems, trialkyl phosphites were shown to react with N-chlorodiethylamine to yield amidates and (presumably) alkyl chloride, although the latter was not isolated.[1067]

$$(RO)_3P + Et_2NCl \longrightarrow Et_2NP(O)(OR)_2 + RCl$$

$$(R = Me,\ Et)$$

Triphenyl phosphite afforded a crystalline complex which lost ethyl chloride on heating, the three aryloxy groups remaining attached to phosphorus.

$$(PhO)_3P + Et_2NCl \longrightarrow [(PhO)_3P(NEt_2)Cl] \xrightarrow{\text{heat}}$$

$$(PhO)_3P{=}NEt + EtCl$$

Dialkyl phosphites gave both the amidate and the chloridate.

$$2\ (EtO)_2PHO + 2\ Et_2NCl \longrightarrow$$

$$(EtO)_2P(O)NEt_2 + (EtO)_2P(O)Cl + Et_2NH_2Cl$$

The mechanism might involve the "phosphite" form, $(EtO)_2POH$,[888] but much needs to be done before speculations on mechanisms can be properly assessed.

Although there is a quick loss of active chlorine when triethyl phosphite is mixed with ethyl N-chloroacetimide at room temperature, evolution of the alkyl chloride is much slower, and may require a higher temperature, thus indicating the formation of a stable enough intermediate.[1050] The mechanism is still obscure. Carbonic ester chloroimides

$$(EtO)_3P + ClN{=}C(Me)(OEt) \longrightarrow (EtO)_2P(O){-}N{=}C(Me)(OEt) + EtCl$$

react more vigorously, but the system still appears to require subsequent heating to decompose the "phosphonium compound." A variety of related systems have been reported, in which alkyl or aryl phosphites, cyanophosphites, or halidites, interact in an analogous manner with N-chloroimines R'R"C=NCl (R' = alkyl or alkoxyl; R = alkoxyl).[339-342,721] In interactions involving aryl phosphites, alkyl chloride is formed from the alkoxyl group originally present in the imine, e.g.,

$$(ArO)_3P + ClN{=}C(OR)_2 \longrightarrow (ArO)_3P{=}NCOOR + RCl$$

N-Chlorosuccinimide and N-2,4-trichloroacetanilide were used to convert dialkyl hydrogen phosphites into phosphorochloridates.[681]

$$(PhCH_2O)_2PHO + (CH_2CO)_2NCl \longrightarrow (PhCH_2O)_2P(O)Cl + (CH_2CO)_2NH$$

N-Bromosuccinimide and N-bromoacetamide were used in aqueous acid solution to oxidize riboflavin monophosphite to the phosphate, and other examples were given.[1311]

Sodium derivatives of N-chloroarenesulfonamides (e.g., chloramine-T) convert triaryl or trialkyl phosphites into N-arylsulfonylphosphoimidates, $ArSO_2N = P(OR)_3$, if the conditions are anhydrous (e.g., in carbon tetrachloride),[258,709] but in boiling aqueous ethanol or water the arylsulphonamide and phosphate are obtained in good yield, the formation of the latter being attributed to hydrolysis of an intermediate cation, $[(RO)_3PCl]^+$.[258] Analogous reactions occur in carbon tetrachloride between sodio-N-chlorobenzenesulphonamide and alkyl or aryl chloridites and dichloridites.[710,711]

$$ArSO_2N(Na)Cl + (RO)_nPCl_{3-n} \longrightarrow ArSO_2N{=}P(OR)_nCl_{3-n}$$

N,N-Dichloroarene sulfonamides react very readily with the dichloridites (2 mole). The reaction sequence was believed to be as follows:[606]

$$ArSO_2NCl_2 + Cl_2POR \longrightarrow Cl_2 + ArSO_2N{=}PCl_2(OR)$$

$$Cl_2 + ROPCl_2 \longrightarrow ROPCl_4 \longrightarrow RCl + POCl_3$$

In accordance with the principles discussed previously, phenyl phosphorodichloridite reacts only in the 1:1 mole ratio; but the chlorine formed is believed to become involved in the reaction sequence, entailing the formation of $PhOPCl_4$, before finally emerging as free chlorine. Alkyl and phenyl phosphorofluoridites behave similarly, giving $ArSO_2N{=}PF_2(OR)$, R = Ph or Bu, or Pr. The conversion of dialkyl phosphites to the corresponding phosphorochloridates by treatment with N,N-dichlorobenzenesulphonamide has also been described.[291]

Freshly prepared N-chloro-N-alkylsulfonamides react vigorously and exothermically with trialkyl or alkyl diaryl phosphites in benzene.[1274]

$$ArSO_2N(Me)Cl + (RO)_2P(OAlk) \longrightarrow$$

$$[ArSO_2N(Me)P(OR)_2(OAlk)]^+Cl^- \longrightarrow AlkCl + ArSO_2N(Me)P(O)(OR)_2$$

Trends are indicated by the use of N-chloro-N-alkylamides of methylphosphonic acid alkyl esters.[1272]

$$(RO)_3P + MeP(O)(OR'')NR'Cl \longrightarrow$$

$$(\text{in } C_6H_6\text{, exothermic})$$

$$(RO)_3P{=}O \quad + \quad MeP(O)(OR'')NR'P(O)(OR)_2$$

(R, R', R" are alkyl groups such as Me, Et, Pr, i-Pr, Bu)

Other examples are reported.[906]

C.8. Interaction with Organometallic Reagents

The first reported example appears to be that of diethyl zinc (1868), showing a preferential replacement of halogen rather than alkoxyl.[1384]

$$EtOPCl_2 + Et_2Zn \longrightarrow EtOPEt_2$$

Grignard reagents and organolithium compounds have been widely used to effect the formation of a P-C bond and some aspects of their applications have recently been reviewed.[186]

In earlier work,[486,487] triethyl phosphite and phenylmagnesium bromide afforded a 10% yield of triphenylphosphine oxide, while trimethyl phosphite gave a 40% yield of methyldiphenylphosphine oxide. With phenyl phosphite, triphenyl phosphine was obtained (see also Chapter 1).

$$(PhO)_3P + PhMgBr\ (\text{excess}) \longrightarrow Ph_3P$$

The nature of the phosphite and Grignard reagent[188,810,1239] and the rate and order of mixing[187,188,917] are factors in determining the extent of hydrocarbonation of phosphorus. Thus, for example, under nitrogen, trimethyl phosphite (1 mole) gave a quantitative yield of triphenylphosphine at 55° when 4 mole of the phenyl Grignard reagent was used; and the compounds, $(MeO)_2PPh$, $(MeO)PPh_2$, and Ph_3P were obtained when 3 (or less) mole of the reagent was used. Reaction of the phosphite is highly exothermic, cooling is essential, and a white solid, containing a predominance of triphenylphosphine oxide over other phosphorus compounds, immediately forms.[187,188]

Addition of 3 mole of arylmagnesium reagent to 1 mole of trimethyl phosphite in ether gave tri-p-tolylphosphine in 41.7% yield, whereas the reverse order of mixing gave a 37.4% yield.[917] (See also Chapter 1.)

Preferential reaction at the chlorine site in chloridites and dichloridites is shown by several examples.[659] Addition of the alkyl Grignard reagent (1.1 mole) [R = Me, Et, Pr, Bu, $PhCH_2$, $(CH{:}CH)_2CH$] in ether to an ethereal solution of the chloridite, $(R'O)_2PCl$ (R' = Bu, Pr) at -60 to -65°, followed by direct distillation after decantation from the solid, gave the phosphonite $RP(OR')_2$ in as much as 70% yield. The phenyl Grignard reagent also gave between 50 and 60% yields of the phosphonite, $PhP(OR')_2$.[648]

Reactions of phosphorochloridites with ethynyl[631] and p-vinylphenyl magnesium halides[650] have similarly been reported, work up being by addition of pyridine, without aqueous treatment. Reaction in ethylpyridine has also been described.[1056] Bifunctional Grignard reagents also react with chloridites and phosphites.[1240]

$$2\ (EtO)_2PCl + BrMg(CH_2)_6MgBr \longrightarrow (EtO)_2P(CH_2)_6P(OEt)_2$$

There does not appear to be any report on the functionality of mixed phosphites, $R'OP(OR)_2$, $ArOP(OR)_2$, etc. Of significance, too, is the replacement of alkoxyl groups in the hydrogen phosphites to give secondary phosphine oxides.[317,323,579,639,735,736,891,1388,1389]

$$(BuO)_2PHO + 3\ C_8H_{17}MgX \longrightarrow (C_8H_{17})_2PHO$$

$$(BuO)_2PHO + 3\ PhCH_2MgCl \longrightarrow (PhCH_2)_2PHO + PhMe$$

Careful addition of Grignard reagent to the hydrogen phosphite in benzene gives first the magnesium salt,[441,1086]

$$EtMgBr + (EtO)_2PHO \longrightarrow (EtO)_2P(=O)MgBr + C_2H_6$$

and such salts have been used in place of sodium derivatives for reaction with allyl halides[1086,1102] or allenic ketones[1037] in the preparation of the corresponding phosphonates. Alternatively, the magnesium salt is made from the sodium derivative.[1044]

$$2\ R'MgX + (RO)_2PONa \longrightarrow R_2PONa + 2\ ROMgX$$

Phenyllithium was stated to react with triethyl phosphite to give an 80% yield of triphenylphosphine.[1387] Examples of the interaction of dialkyl hydrogen phosphites (R = Et, Bu) with aryllithium compounds have also been reported but no operational details were given.

$$(EtO)_3P + C_6H_5Li \longrightarrow Ph_3P$$

$$(EtO)_2PHO + o\text{-}MeOC_6H_4Li \longrightarrow (o\text{-}MeOC_6H_4)_2PHO$$

Again, the chlorine of phosphorochloridites is preferentially replaced in reactions with alkyl- or aryl-lithiums.[649]

$$R'Li + (RO)_2PCl \longrightarrow (RO)_2PR'$$

There are not enough data to warrant any general comments about the relative effectiveness of Grignard and lithium reagents in these systems, although at first sight advantages might appear in individual examples. (See also Ref. 166, 639.) The use of cadmium butyl, is mentioned.[1239]

C.9. Interaction with Carbon-Halogen Compounds: The Michaelis-Arbuzov Reaction

The interaction of trialkyl phosphites and alkyl halides, referred to as the Michaelis-Arbuzov reaction, was reported by Michaelis[859] and later exploited extensively by Arbuzov[70] and many other workers.

$$(RO)_3P + R'X \longrightarrow (RO)_2P(O)R' + RX$$

A stable 1:1 intermediate is obtainable in certain instances from triaryl phosphites[89,91,92,779,780] and in other special cases involving trialkyl phosphites and α,β-dihaloalkyl ethers,[4,16,17,23] although it is not yet clear whether this intermediate is a phosphonium compound or a phosphorane.[531] The overall reaction is one of the most important for creating a phosphorus-carbon bond and is discussed in detail in Chapter 12.

Contrary to previous observations,[36,69,134,205,316,495,746,850,851,858,1103,1106] α-halogenocarbonyl compounds can give a P-O bond rather than a P-C bond. This mode of function is called the Perkow reaction, or "anomalous reaction,"[1040,1103,1106-1108] and is illustrated as follows:

$$(EtO)_3P + MeCOCH_2Cl \xrightarrow{100^\circ} (EtO)_2P(=O){-}OC(Me){=}CH_2 + EtCl$$

Speculations on the mechanism have been analyzed,[196,289,791] and the following mechanisms have been suggested: nucleophilic attack of phosphorus on carbon to form the "Arbuzov intermediate" which then "decomposes" to give the phosphate and not the phosphonate product;[808,1038,1286] nucleophilic attack on the carbonyl oxygen.[35,669,691,747,889,1091,1329]

Other examples, including those of synthetic potential, have been described.[289,315,316,537,909,926,928] The peculiarity of fluorine compounds is shown thus.[1386]

$$C_2F_5COC_2H_5 + (MeO)_3P \longrightarrow (MeO)_2P(O)F + C_2F_5C(OMe){:}CFCF_3$$

C.10. Interaction with Polyhalogeno Compounds

It was shown some years ago that carbon tetrachloride is not an inert solvent in the trialkyl phosphite systems, for it can react in a way similar to that for the mono- and dihalogeno compounds.[180,238,658,664,729]

$$(RO)_3P + CCl_4 \longrightarrow Cl_3CP(O)(OR)_2 + RCl$$

Mixed alkyl phosphites form mixed or simple phosphonates when heated under reflux with carbon tetrachloride,[663] and there was a tendency for the smaller alkyl group to be eliminated as alkyl chloride. Usually, yields of less than 50% were obtained, and were attributed to the tendency of the phosphites to "rearrange," and to the formation of hexachloroethane.[655,663]

$$(RO)_3P + 2\ CCl_4 \longrightarrow (RO)_2P(O)Cl + C_2Cl_6 + RCl$$

Whereas the phosphite $(EtO)_2POPh$ gave ethyl chloride (100%) and the phosphonate, $Cl_3CP(O)(OEt)(OPh)$, the phosphite $(MeO)(PhO)_2P$ gave the chloridate, $(PhO)_2P(O)Cl$, and diphenyl benzyl phosphite gave the phosphonate, $(PhO)_2P(O)CH_2Ph$.[661]

It was postulated[662,663] that the lone pair of electrons on phosphorus initiated a chain reaction (see Chapter 12). The reaction system is light and peroxide catalyzed.[510]

Carbon tetrabromide reacted vigorously with triethyl phosphite, giving ethyl bromide and a considerable amount of undistillable material.[660] Triphenyl phosphite afforded the bromidate, $(PhO)_2P(O)Br$ at 100-20° (sealed tube). The mixed phosphite, $(MeO)_2POPh$, (cooled) gave methyl bromide, bromoform, and the bromidate $(MeO)(PhO)P(O)Br$. Other alkyl aryl mixed phosphites behaved similarly.

Chloroform could not be induced to react with trialkyl phosphites.[318,658] Radical-chain or ionic processes have been discussed.[227,254,259,261]

Against the contention that benzoyl peroxide initiated a radical reaction in the triethyl phosphite-chloroform system,[510] is the report[239,240] that benzoyl peroxide is rapidly consumed by triethyl phosphite "in an ionic fashion" in chloroform, and does not initiate the formation of diethyl trichloromethylphosphonate.[239] It has now been concluded[154] that the thermal reaction of triethyl phosphite with carbon tetrachloride in the dark, in vacuo, is heterolytic. Corresponding reactions of bromotrichloromethane, however, are less clearly defined; radical and ionic mechanisms apparently function.

The interactions of trialkyl phosphites with carbon tetrachloride in the presence of alcohols result in the formation of chloroform, a number of mixed trialkyl phosphates, and dialkyl trichloromethylphosphonate.

Mechanisms entailing the solvolysis of intermediate ions of the type $[(RO)_3PCCl_3]^+$ were discussed in detail.[163] In the presence of thiols[154,227,250] or thiophenols[937] trialkyl phosphites together with carbon tetrachloride or bromotrichloromethane give, amongst other products, the O,O-dialkyl-S-alkyl (or aryl) phosphorothioates. Reactions of trialkyl phosphites with diphenyldichloromethane[419] and with p,p'-bis(trichloromethyl) benzene[439] have also been studied.

Highly chlorinated ketones, such as decachloro-3-pentanone, undergo α,β-dechlorination with a trialkyl phosphite.[1070]

Triethyl phosphite gave a phosphonate on reaction with perfluorocyclobutene,[716,717] diphosphonates on reaction with 1,2-dichloroperfluorocycloalkenes,[413] and from chlorofluoroalkanes three main products were obtained.[1073]

Other triethyl phosphite-polyhalide systems have been reported.[1073]

Dialkyl hydrogen phosphites reacted rapidly with carbon tetrachloride in the presence of ammonia or "strong primary and secondary amines" at room temperature to yield phosphoramidates.[151,152,1409] Aniline[151] or imidazole,[975] however, required the presence of a tertiary base for reaction.

$$(RO)_2P(O)H + CCl_4 + \text{Base} \longrightarrow (RO)_2P(O)CCl_3 + \text{Amine HCl}$$

$$(RO)_2P(O)CCl_3 + R'R''NH \longrightarrow (RO)_2P(O)NR'R'' + CHCl_3$$

Pentachloroethane, or hexachloroethane may replace carbon tetrachloride, although chloroform or s-tetrachloroethane did not so react. A number of polychloro and mixed polychlorofluoro compounds were examined without revealing any advantage over carbon tetrachloride for the practical application of the process.[152] Carbon tetrabromide reacted vigorously with dibenzyl phosphite in the presence of "weak bases such as aniline," giving bromoform which was converted into methylene bromide. Results with bromoform, iodoform, and dichlorobromomethane were also given.[152]

The following dialkyl hydrogen phosphite systems have been described:

$$(RO)_2PHO + CF_3CF{:}CF_2 \xrightarrow{\;^{60}Co\;\;rays\;} phosphonate$$

$$(R = Me, Et, i\text{-}Pr)\,[616]$$

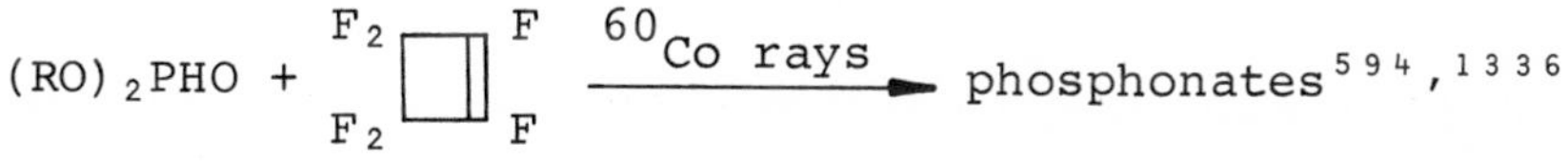

$$(RO)_2PHO + \underset{F_2}{\overset{F_2}{\square}}\overset{F}{\underset{F}{}} \xrightarrow{\;^{60}Co\;rays\;} phosphonates[594],[1336]$$

$$(EtO)_2PHO + CF_3NO \;(in\;CCl_4) \longrightarrow CF_3N(OH)P(O)(OEt)_2\,[715]$$

$$(MeO)_2PHO + (CF_3)_3CNO \longrightarrow (CF_3)_3CH(OH)P(O)(OMe)_2\,[715]$$

Other work has involved dialkyl phosphites with per-fluoronitroalkanes[361] and perfluoroketones.[618]   See also Ref. 685 for polyhalogeno systems.

C.11.   Interaction with Carbonyl Compounds

A considerable area of chemistry is being developed from the primary interaction of carbonyl compounds with alkyl and aryl phosphites.[112,179,193,194,231,234,235,490,605,759,821,1121,1158-1192,1241,1362] The essential mechanistic problem relates to whether the point of "attack" by the phosphorus atom is at oxygen or carbon. The first report indicates attack at carbon with the formation of a 1:1 adduct as follows:[3]

$$(RO)_3P + R'CHO \longrightarrow (RO)_3\overset{+}{P}CHR\overset{-}{-}O \longrightarrow (RO)_2P(O)CHR'OR$$

Attack at carbon has been confirmed for an unsubstituted aliphatic aldehyde such as propionaldehyde, although a 1:1 adduct was not detected and the 2:1 adduct (presumably) led to the formation of a 2,2,2-trialkoxy-1,4,2-dioxa-phospholane.[1181]

$$(RO)_3P + \underset{R'}{\overset{H}{C}}{=}O \longrightarrow (RO)_3\overset{+}{P}-\underset{R'}{\overset{H}{C}}-\overset{-}{O} \xrightarrow{R'CHO}$$

$$(RO)_3\overset{+}{P}-\underset{R'}{\overset{H}{C}}-O-\underset{R'}{\overset{H}{C}}-\overset{-}{O} \longrightarrow$$

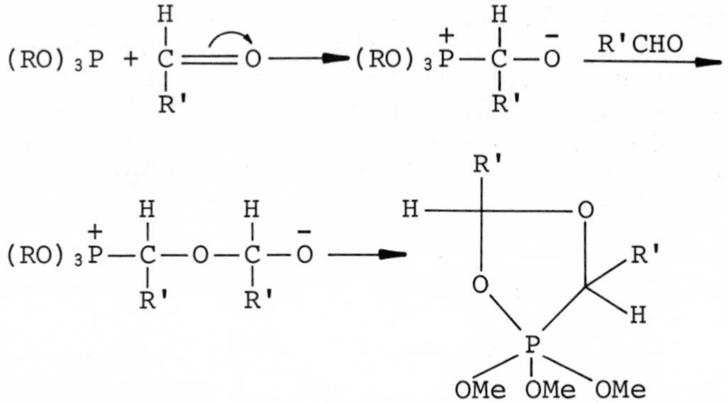

Translocation of alkyl was not observed, except for the 3:1 adduct.[1181]

$$(RO)_3\overset{+}{P}-\underset{R'}{\overset{H}{C}}-O-\underset{R'}{\overset{H}{C}}-\overset{-}{O} \xrightarrow{R'CHO} (RO)_2\overset{O}{\overset{\|}{P}}-\underset{R'}{\overset{H}{C}}-O-\underset{R'}{\overset{H}{C}}-O-\underset{R'}{\overset{H}{C}}-OR$$

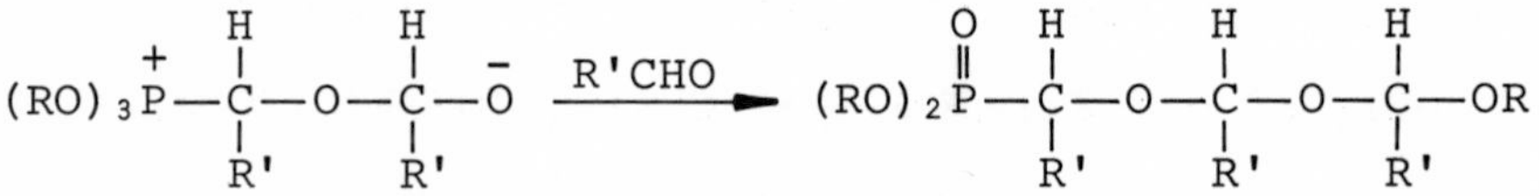

There is little reaction with unsubstituted aromatic aldehydes or ketones.

Acetophenone and tri-i-propyl phosphite, heated under reflux for 120 hr gave a number of nonphosphorus materials.[328] Furfural and triethyl phosphite at 160° (3 hr) merely gave triethyl phosphate and difurylethylene; and tri-i-propyl phosphite behaved similarly. Pyromucic and furylacrylic acids afforded the dialkyl hydrogen phosphites, and ethyl esters of the reactant acid. There was little interaction between triethyl phosphite and benzaldehyde at 160-70° during 7 hr; but at 220° a little stilbene and a small yield of the phosphonate

$$(EtO)_2P(O)\underset{OEt}{\overset{}{C}}HPh$$

were obtained.[141] However, aromatic aldehydes containing the strongly electron-attracting nitro group in the o- or p- (but not m-) positions react readily, a variety of products having been reported. Thus, trimethyl or triphenyl phosphite reacted with p-nitrobenzaldehyde to give 1:1 adducts of the type $(RO)_2P(O)CH(OR)C_6H_4NO_2$-p (R = Me or Ph), triethyl phosphite and p-nitrobenzaldehyde gave a 2:1 adduct formulated as $(EtO)_2P(O)CH(C_6H_4NO_2$-p$)OCH-(C_6H_4NO_2$-p$)OEt$ (together with p-nitrobenzyl p-nitrobenzoate), and from triethyl phosphite and o-nitrobenzaldehyde both the 2:1 adduct and a 1,4,2-dioxaphospholane were reported.[759] These products could result from the primary attack of phosphorus on carbon. Subsequent work has, however, provided strong evidence for attack at oxygen in these systems, with the formation of 1,3,2-dioxaphospholanes which on hydrolysis yield glycol phosphates.[1159,1163,1164,1188] Stabilization of the carbanion intermediate is thought to be due to the presence of o- or p-nitro substituents (see also Chapter 5B).

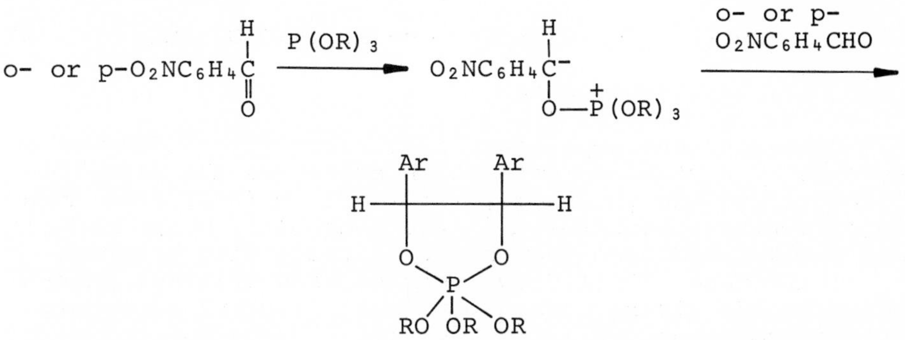

m-Nitrobenzaldehyde, the chlorobenzaldehydes, and those with electron-releasing groups were inert. The reaction of diacetyl with a trialkyl phosphite yields a phospholene, again by attack at oxygen.

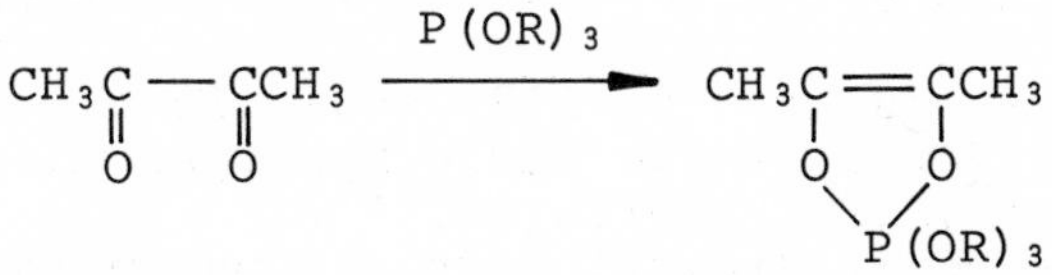

At higher temperatures diacetyl affords a product initially reported as $(RO)_2P(O)OC(CH_3){:}C(CH_3)OR$[756,757] but later amended to $(RO)_2P(O)C(OR)(CH_3)COCH_3$.[761,762] The interaction of p-quinones[758,760] with trialkyl phosphites affords a route to the preparation of the monoalkyl ethers of p-quinol via reductive O-alkylation;[801,1165,1167,1168] o-quinones form ring phosphorane adducts. A number of polycyclic o- and p-quinones have likewise been studied.[525,694,941-943,1162] Further studies include the reactions of trialkyl phosphites with indantrione,[943] α-brom-α-ketol phosphates,[1191] 2-acetoethyl acetate,[609,611] perfluorobenzaldehyde,[1187] and chloral.[292,687] The acyl phosphite-chloral system has also been described.[1120]

At 50° to 100° trialkyl phosphites (R = Me, Et, Pr, Bu) reacted with the ethyl ester of halogenated thiocarboxylic acid to afford dialkyl α-alkylthiovinyl phosphates, and as a byproduct, the phosphonate.[496]

$$(RO)_3P + XCH_2C(O)SEt \longrightarrow (RO)_2P(O)OC(SEt){:}CH_2$$

$$(RO)_3P + XCH_2C(O)SEt \longrightarrow (RO)_2P(O)CH_2C(O)SEt$$

$$(RO)_3P + CCl_3C(O)SEt \longrightarrow (RO)_2P(O)OC(SEt){:}CCl_2$$

(Energetic)

Cycloalkylene acyl phosphites react with aldehydes and ketones by a reaction path which preserves the ring;[955] Diacetyl and the phosphite, $(EtO)_2PCl$, or $(EtO)_2POAc$, react to give ring phosphates.[789] The phosphite, $(EtO)_2POAc$,[1118] and dialkyl hydrogen phosphites[839] react with ketenes.

The ketene, $(CF_3)_2C{:}CO$, reacted with triethyl phosphite at 0°, giving carbon monoxide, triethyl phosphate and the compound $(CF_3)_2C{:}C{:}C(CF_3)_2$.[145]

In the presence of formaldehyde compounds with an "active hydrogen atom" such as malonic, cyanoacetic, and acetoacetic esters react with trialkyl phosphites (R = Et, Pr, Ph) to give the alcohol and a phosphonate.

$$R'H + CH_2O + (RO)_3P \longrightarrow ROH + R'CH_2P(O)(OR)_2$$

Further references are 65, 112, 505. In the presence of triethylamine formaldehyde and a hydrogen phosphite give the phosphonate, $(RO)_2P(O)CH_2OH$.[1404]

The addition of dialkyl hydrogen phosphites, usually in the presence of sodium alkoxide, with various unsaturated compounds such as aldehydes, ketones, esters, etc., constitutes an area of considerable activity.[5-9,15,18,20,22,25,47,113,119,178,195,208,248,617,651,696,976,978,1085,1087-1090,1129,1131,1133,1134,1141]

C.12. Addition to Nonconjugated Olefins and Acetylenes

There are no reports of the addition of trialkyl phosphites to simple olefins, although dialkyl hydrogen phosphites add to, e.g., 1-hexene, 1-octene, in the presence of t-butyl peroxide; and photoinitiation is also effective.[1298]

$$(BuO)_2PHO + C_6H_{12} \xrightarrow[\text{16 hr at } 120°]{(t\text{-}BuO)_2} C_6H_{13}P(O)(OBu)_2$$

Unsaturated ethers also undergo addition to one or both of the double bonds.[1298]

$$(CH_2{:}CHCH_2)_2O + (BuO)_2PHO \xrightarrow[16\ hr,\ 130°]{(t\text{-}BuO)_2}$$

$$[(BuO)_2P(O)CH_2CH_2CH_2]\ O + CH_2{:}CHCH_2OCH_2CH_2CH_2P(O)(OBu)_2$$

Cyclization in the homolytic addition to 1,2-bis(vinyloxy)-ethane has been studied.[1330]

Addition to both of two remote double bonds in hesea-1,5-diene was also achieved by UV irradiation at 120° to 130° (90 to 100 hr) to give a diphosphonate, $(RO)_2P(O)$-$(CH_2)_6P(O)(OR)_2$.[1135]

In the acetylene series, trialkyl phosphites were first shown to add to t-acetylenic chlorides to yield an allenic phosphonate which underwent further reaction to give a diphosphonate. An Arbusov type intermediate was assumed.[1084]

$$(RO)_3P + HC{\equiv}CCClR'_2 \longrightarrow [(RO)_3\overset{+}{P}\text{-}CH{=}C{=}CR'_2] \xrightarrow{Cl^-}$$

$$(RO)_2P(O)CH{=}C{=}CR'_2 \xrightarrow{(RO)_3P} (RO)_2P(O)CHRC[P(O)(OR)_2]{=}CR'_2$$

Phenylacetylene undergoes uncatalyzed reaction with triethyl phosphite at 150° to give ethylene and diethyl β-styrylphosphonate by what is considered to be a cis elimination.[514]

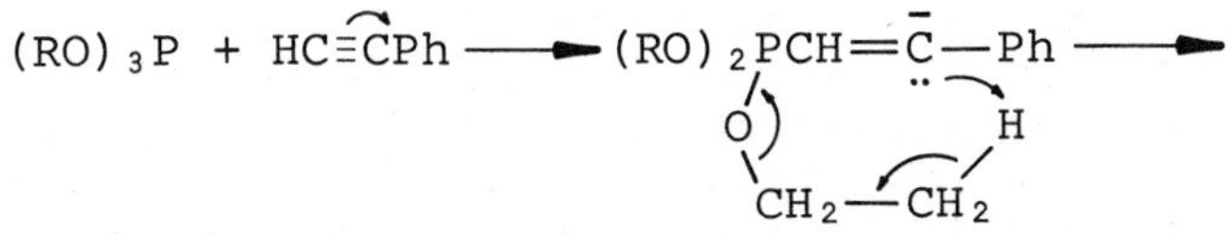

$$C_2H_4 + (RO)_2P(O)CH{=}CHPh$$

Nonterminal acetylenes and the higher alk-1-ynes are less reactive. Acetylene reacts with bis(2-chloroethyl) hydrogen phosphite in the presence of a radical initiator to give both mono- and diphosphonates.[1270]

$$CH{\equiv}CH + (ClCH_2CH_2O)_2PHO \longrightarrow CH_2{=}CHP(O)(OCH_2CH_2Cl)_2 \longrightarrow$$

$$[CH_2P(O)(OCH_2CH_2Cl)_2]_2$$

Diphosphonates have also been obtained by the reaction of 1-alkynes with dialkyl hydrogen phosphites, their thio analogs, or monoalkyl hydrogen alkylphosphonates in the presence of benzoyl or t-butyl peroxide.[1145]

C.13. Addition to Conjugated Systems

Conjugated dienes, or α,β-unsaturated carbonyl compounds,

undergo 1,4-addition with phosphites or halogenophosphites to yield phosphonates. An Arbuzov intermediate may be involved, e.g.,

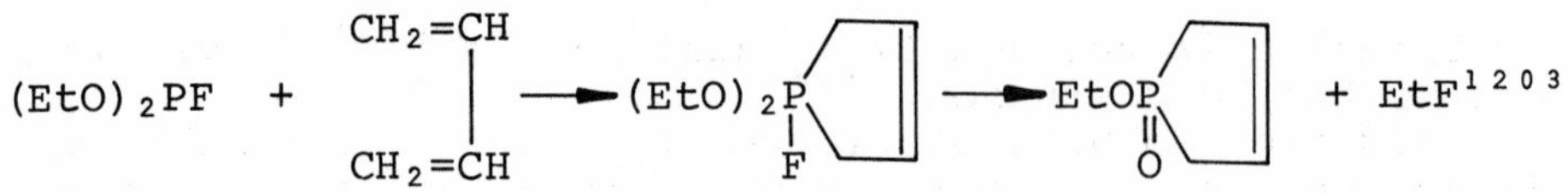

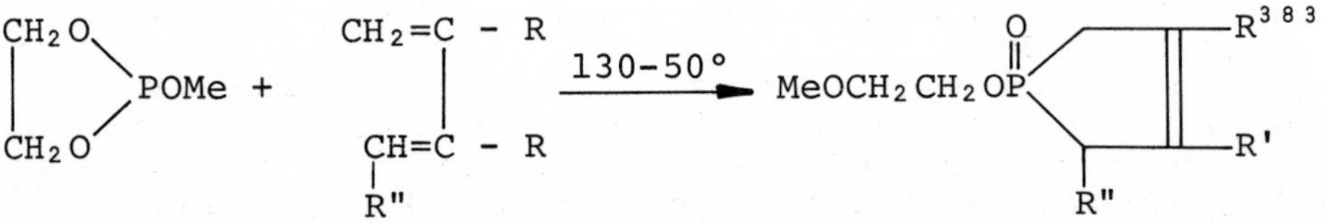

The additions of conjugated dienes to a variety of fluoro-, chloro-, bromo-, and isocyanatophosphites have been studied.[121,124,136,384,386,1064,1065,1112,1202,1206,1207,1363] α,β-Unsaturated carbonyl compounds undergo attack at the ethylenic double bond, again yielding phosphonates in many instances.[139,491,670,1364] Trimethyl phosphite adds to acrylic or methylacrylic acid at 120°[670] and cinnamic acid reacts similarly with triethyl phosphite.[668]

$$(EtO)_3P + PhCH{=}CHCO_2H \longrightarrow (EtO)_2P(O)CHPhCH_2CO_2Et$$

At 50° or less, maleic acid yields $(EtO)_2P(O)CH(CO_2H)$-CH_2CO_2Et.[668] In dioxan, or in the absence of solvent, acrolein or crotonaldehyde react with trialkyl phosphites to afford rather low yields of phosphonates containing a vinyl ether group.[669] The effect of phenol on a variety of such

$$(MeO)_3P + CH_2{=}CHCHO \longrightarrow (MeO)_2P(O)CH_2CH{=}CHOMe$$

additions has been studied; aldehydes such as acrolein give the corresponding diphenyl acetal $(RO)_2P(O)CH_2CH_2CH(OPh)_2$ and compounds having an electron-attracting substituent on the β-carbon atom undergo reductive dimerization.[531]

Phenylcyclobutendione[344] reacts with trialkyl phosphite to give the phosphonate but 2-benzylidene-1,3-indandione[125] gives a phosphorane (see also Chapter 5B).

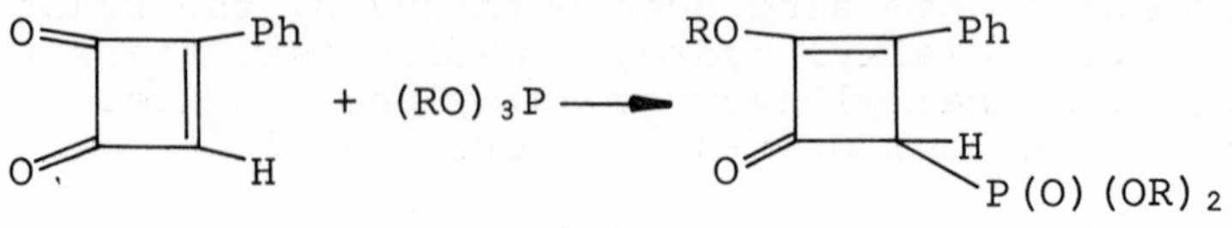

(R = Me, Et)

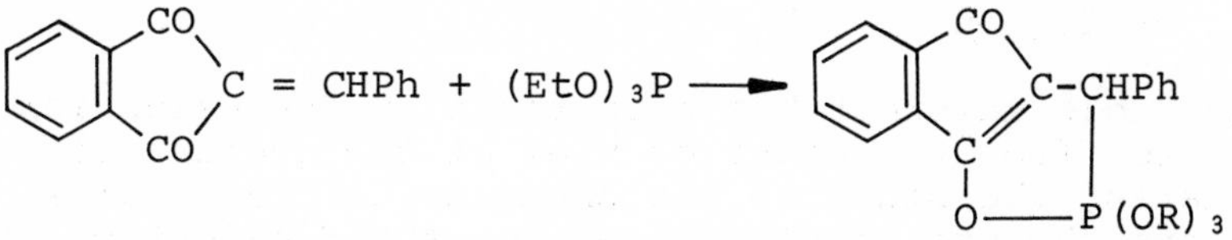

Trialkyl phosphites and ethylideneacetylacetone, ethylideneacetoacetic ester, and benzylideneacetoacetic ester provide other examples.[111,118,119]

Evidence for free-radical intermediates in the interaction of triethyl phosphite with substituted cyclohexadienones has been obtained by EPR studies.[652]

Compounds such as $CH_2{:}C(CN)CH_2P(O)(OEt)_2$, considered as polymerizable materials, were prepared from the trialkyl phosphite, e.g., $(EtO)_3P$, or from the sodium compound, e.g., $(BuO)_2PONa$, and an unsaturated nitrile, thus showing a preference for the Michaelis-Arbuzov reaction

$$CH_2{:}C(CH_2Br)CN + (EtO)_3P \longrightarrow CH_2{:}C(CN)CH_2P(O)(OEt)_2 + EtBr$$

path.[347] Reactions of chloridites with unsaturated aldehydes or ketones have also been studied.[989,1208]

Dialkyl hydrogen phosphites add to ethylenic compounds containing activating groups (CO, CO_2R, CN) in the α-position.[209,640] Sodium derivatives also react.[1132] In the presence of t-butyl peroxide dialkyl phosphites are stated to add to ethylenic silicon compounds.[831]

$$CH_2{:}CHSiMe(OEt)_2 \xrightarrow{(MeO)_2PHO} (MeO)_2POCH_2CH_2SiMe(OEt)_2$$

In an omnibus coverage the alkaline-catalyzed addition of dialkyl hydrogen phosphites to an olefinic double bond, or to an isocyanate group, is claimed.[778]

$$(EtO)_2PHO + CH_2{:}CHCO_2Me \xrightarrow[\substack{\text{amount of Na} \\ (60 - 70°)}]{\text{Small}} (EtO)_2P(O)CH_2CH_2CO_2Me$$

$$(EtO)_2PHO + \alpha\text{-}C_{10}H_7NCO \longrightarrow (EtO)_2P(O)CONHC_{10}H_7\text{-}\alpha$$

An alkylidene derivative of malonic ester reacts similarly in the presence of ethoxide,[1140] and 2-vinylpyridine gives the phosphonate $(RO)_2P(O)CH_2CH_2C_5H_4N$.[883] α,β-Acetylenic acids yield β-carboalkoxyvinylphosphonates by 1,4-addition of trialkyl phosphites.[697]

$$(RO)_3P + R'C{\equiv}CCO_2H \longrightarrow (RO)_3\overset{+}{P}CR'{=}C{=}C(OH)\overset{-}{O} \longrightarrow$$

$$(RO)_2P(O)CR'{=}C{=}C(OH)(OR) \longrightarrow (RO)_2P(O)CR'{=}CHCO_2R$$

C.14. Oxidation

Considerable interest has been taken in the oxidation of phosphites from the point of view of the preparation of phosphates, the study of the chemistry of free radical systems, and the use of phosphites for the deoxidation of organic compounds, an area covered profusely by the patent literature. The subject will be presented briefly under the following headings (see also Chapter 16).

C.14.1. Oxygen (Air) and Ozone

Oxidation by oxygen (air) must always be looked for in the handling of phosphites, and in certain examples the process may be one for the preparation of phosphates from the corresponding phosphites.[666,775]

$$(RO)_3P \xrightarrow{[O]} (RO)_3P{=}O$$

Trialkyl and triaryl phosphites are reported to be quantitatively oxidized to their respective phosphates by ozone, one, two, or all three of the oxygen atoms being functional in this way according to experimental factors.[714,1320] The complex, $(PhO)_3PO_3$, was obtained at -75° but trialkyl phosphites do not appear to form the adduct.[1412]

Speculations on mechanisms and comments by other workers have been presented in some detail.[1320] The primary product was deemed to have a ring structure

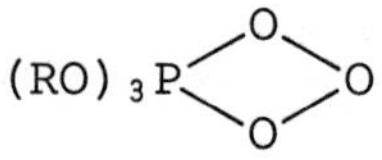

identifiable in the aryl, but not in the alkyl systems. It was observed that the oxygen formed during the decomposition of the adduct, $(PhO)_3P,O_3$, will undergo singlet oxygen reactions, and is a convenient source of singlet oxygen.[938,939] Electron paramagnetic resonance studies have been added.[1379] This topic may be linked with work on the interaction of ozone with the adducts of trimethyl phosphite or triphenyl phosphite with benzil,[1179] and of oxygen with adducts of phosphites with α-diketones. Air oxidation[567,1279] and the study of kinetic curves for the oxidation of polypropylene in the presence of diphenyl isooctyl phosphite are further examples.[1075] See also Ref. 966.

C.14.2. Peroxides

Hydrogen peroxide[1294] or organic peroxides[332,702,706,1310,1345] have been used. t-Butyl hydroperoxide oxidized phosphites to the corresponding phosphates in good yield.[1310] With a view to elucidating the effect of structural changes on the nucleophilicity of phosphites, oxidation by benzoyl peroxide was studied.[1345] A number of open chain and ring phosphites, including bicyclophosphites, were converted into the corresponding phosphates in good yields. Conclusions were tentatively reached; steric hindrance is less important for phosphites than for phosphines; the more strained the structure of the cyclic phosphites the slower the reaction; phenyl substituents decreased the reactivity. From a kinetic study of the trimethyl phosphite system in toluene, it was concluded that "the reaction was first order in each of the reactants." 1,1-Diphenylethane hydroperoxide has also been used.[706]

An indirect polarographic determination of phosphorus in phosphorous acid esters entails the interaction of the phosphite with hydroperoxides, e.g., i-propylbenzene hydroperoxide.[702]

Diacyl peroxides and alkyl hydroperoxides react with trialkyl phosphites by "ionic routes."[239,249] Diethyl peroxide and trialkyl phosphites were observed to react slowly at room temperature giving pentaalkoxyphosphoranes, e.g., $(EtO)_5P$,[333,335] (see Chapter 5B). Di-t-butyl and di-α-cumyl peroxides are stated to react slowly with triethyl phosphite by a homolytic mechanism.[1372] Rate measurements of these and other phosphite systems have been discussed.[250,1370] Triethyl, tributyl, tri(2-chloroethyl), and certain cyclic phosphites gave the corresponding phosphates in 70-90% yield (20 min at 130°) in the presence of di-t-butyl peroxide (0.1 mole/mole of phosphite) in chlorobenzene.[182] Low yields of phosphate were obtained if a phenoxy group was a constituent, e.g., $(EtO)_2POPh$.

There is relevance to the photo-induced oxidations of trialkyl phosphites,[252,1279] an attractive procedure for converting trialkyl phosphites into the phosphates; but there appears to be some confusion in the example of triphenyl phosphite.[250]

C.14.3. Oxides of Nitrogen

A number of open-chain phosphites, ring phosphites, and phosphorochloridites have been oxidized by addition of a solution of dinitrogen tetroxide in methylene chloride, chloroform, or carbon tetrachloride to the phosphite in the same solvent at -78° (or at ice-salt temperature) until a permanent faint green color persisted. Alternatively the gaseous tetroxide in nitrogen was passed into the phosphite at ice-salt temperature.[197,314,336,337,678] The yields of phosphates were about

$$(RO)_3P + N_2O_4 \longrightarrow (RO)_3PO + N_2O + N_2$$

70-75%. The nitroxides, such as CF_3NO or $(CF_3)_2CFNO$,[48,316,715] and the chlorides NOCl or NO_2Cl[126] have also been used.

Dialkyl hydrogen phosphites (1 mole) and pyridine (1 mole) in ligroin gave tetraalkyl pyrophosphates on addition of nitrosyl chloride in benzene at 25-30°.[518,883,884]

A kinetic study of the following system has been described.[754]

$$2 \text{ NO} + (EtO)_3P \longrightarrow N_2O + (EtO)_3PO$$

A study of the [18]O systems has been deemed to indicate a homolytic oxidation of dialkyl hydrogen phosphites by nitric oxide;[1236] a heterolytic process was suggested for the dialkyl phosphite-nitrosyl chloride system.[1237] Dialkyl hydrogen phosphites are reported to give high purity dialkyl phosphates when treated with an air stream containing dinitrogen tetroxide; the "by-product nitric oxide" can be oxidized for recyclization.[306,1058]

C.14.4. Mercuric Oxide

This reagent has been used in several investigations for oxidizing trialkyl phosphites to the phosphates,[156,337,678] but when yellow mercuric oxide was added to diethyl phosphite in acetone there was no apparent reaction.[678]

C.14.5. Epoxides

Ethylene oxide oxidized triethyl phosphite when heated in nitrogen under pressure (50 lb/sq. in. rising to 400 lb/ sq. in.) at 174° during 3 hr. Propylene oxide gave

$$\begin{array}{c} CH_2 \\ | \rangle O + (EtO)_3P \longrightarrow (EtO)_3PO + C_2H_4 \\ CH_2 \end{array}$$

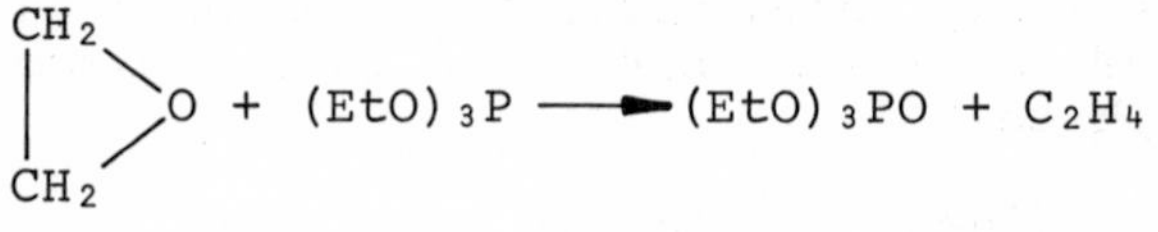

propylene.[1261]

C.14.6. Aqueous Systems

Operational details for oxidizing dialkyl phosphites, e.g., dibenzyl hydrogen phosphite, by alkaline permanganate, periodic acid, and hypochlorous acid have been described. Potassium iodate under basic conditions was ineffective. Monoalkyl phosphites were oxidized only by the permanganate reagent.[222,351] Other examples of permanganate oxidation have been reported.[1310a]

C.14.7. Lead Tetraacetate

This reagent oxidized trialkyl phosphites "instantly" in benzene at room temperature, giving 70-80% yields of the phosphates. Dialkyl phosphites had to be heated under reflux in benzene; and under these conditions with the reagent in excess, ammonium monoalkyl phosphites were oxidized.[350]

C.14.8. Sulfur Trioxide and Sulfur Dioxide

This reagent was added to tris(chloroethyl) phosphite in liquid sulfur dioxide cooled with Dry Ice-MeOH, to give a 94% yield of the phosphate, although tricresyl phosphite "gave no reaction."[626]

A detailed study of the oxidation of phosphorus(III) compounds, including trialkyl phosphites, ring chloridites, and ring bromidites has been described, and the relevant NMR data presented.[399,1291]

Sulfur dioxide was found to be difficult to remove from a reaction mixture containing phosphorus compounds,[1082] and this was attributed to the formation of "some kind of complex."[1081] At 100° trialkyl phosphites were stated to give the phosphorothionate and phosphate by a somewhat obscure sequence possibly as follows:[1081]

$$(EtO)_3P + SO_2 \longrightarrow (EtO)_3P\text{-}OSO \longrightarrow (EtO)_3PO + SO$$

$$(EtO)_3P + SO_2 \longrightarrow (EtO)_3PSO_2 \xrightarrow{(EtO)_3P} (EtO)_3PSO + (EtO)_3PO$$

After sulfur dioxide had been passed into diethyl hydrogen phosphite at 80° for several hours, the hydrogen phosphite was recovered in 87% yield; and triaryl phosphites were stated to be unreactive.[1081]

C.14.9. α-Halogenoacylamides

Whereas trialkyl phosphites were reported to give phosphonates (Michaelis-Arbuzov) with 2-monohaloacetamides,[1287] and with 2-monohaloacetates,[133,656,1287,1385] 2-chloro-N,N-dialkylacetoacetamides gave vinyl phosphates.[1382] Trihalo esters, aldehydes and ketones also afforded vinyl phosphates.[35,36,133,656,691,934,1038,1039,1391] α-Trihaloacetamides, however, afforded the simple phosphates.[1288]

$$Cl_3\overset{\overset{\displaystyle O}{\|}}{C}NR_2 + (R'O)_3P \xrightarrow{145\text{-}155°} Cl_2C{=}CClNR_2 + (R'O)_3PO$$

Isopropyl phosphite also gave a 37% yield of i-propyl

chloride, and ethyl phosphite a 25% yield of ethyl chloride, thus showing competition by other reaction sequences. When phenyl phosphite was heated even for 11 hr at 170-5° with $Cl_3CC(O)NMe_2$ or $Cl_3CC(O)NPh_2$, "only starting materials were recovered."[1288]

Monobromocyanoacetamide, or monobromomalonamide, in the presence of an alcohol at room temperature has been used as follows:[537,538,925,926]

$$(4\text{-}NO_2C_6H_4O)_2POEt \xrightarrow[+\ ROH]{NCCHBrCONH_2}$$

$$(4\text{-}NO_2C_6H_4O)_2P(O)OR + EtBr + NCCH_2C(O)NH_2$$

Dialkyl hydrogen phosphites were also oxidized by this process, the trivalent form being deemed the functional one.

$$(EtO)_2P(O)H \rightleftharpoons (EtO)_2POH \xrightarrow[+\ PhCH_2OH]{NCCHBrCONH_2}$$

$$(EtO)_2P(O)OH + NCCH_2C(O)NH_2 + PhCH_2Br$$

C.14.10. Phosphites as Reducing Agents

Phosphites are readily converted by oxidation processes to the phosphate or phosphonate structure,[249] and their use as antioxidants is common (see Section H). Some recent developments in their uses as reducing agents in organic synthesis include the deoxidation of nitro compounds, which provides a novel route for the preparation of heterocyclic compounds,[213,251,253,255-257,260,262,1303,1304] and their reactions with thionocarbonates or trithiocarbonates for the stereospecific synthesis of olefins.[308-310] Other typical reductions are of pyridine-N-oxide,[250,370] of quinones,[348,907,1165,1168] of N-sulfinylamines,[902] of mercuriated amides,[804] of mercuric carboxylates,[803] and of diethyl azodicarboxylate.[910] Debromination and desulfurizing by phosphites have also been reported.[352] Methyl phosphorodichloridite reduces trichloromethyltetrachlorophosphorane to the phosphonous dichloride, with the formation of methyl chloride and phosphoryl chloride.[1154]

C.15. Hot Atom Chemistry

In a study of the "hot atom" chemistry of phenylphosphorus compounds[712] triphenyl, tricresyl, and diphenyl hydrogen phosphites were irradiated in a nuclear reactor for 15 to 2100 min with a thermal neutron flux of 5×10^{12} neutrons/ cm^2/sec. The relative amounts of the ^{32}P-labeled gaseous products and the products soluble in an aqueous solution were estimated, and the chemical reactions of the recoil ^{32}P atom were discussed in relation to the results from other phenylphosphorus compounds.

In the $^{31}P\ (n,\gamma)\ ^{32}P$ reaction, the P atom has all its bonds broken, the primary retention of phosphorus compounds being negligibly small; and after a post-collision slowing down, ^{32}P-labeled products are formed.

C.16. Miscellaneous Reactions

Miscellaneous reactions of the trialkyl or dialkyl phosphites or their derivatives are given in Table 2. Other reactions which have been studied include those of trialkyl phosphites with mesoxalic acid,[1125] boric esters,[482] azo compounds,[489] allylamines,[1098] difluorodiazine,[320] Schiff's bases,[1142 1149] lactones,[745] carboxylic acids,[288] acid anhydrides,[305] and the formation of phosphite ketals and phosphatoalkyl phosphites.[268] Reaction also occurs with halogens plus alcohols,[414] $B_3N_3(Me)_6Cr(CO)_3$,[327] and with $Et_2NP(O)N_3$.[395] Photooxidation of trialkyl phosphites has been reported.[511] Complexes are formed with many metals, including platinum,[1222] nickel,[1012] manganese,[508] cobalt, rhodium, iridium,[790] uranyl,[288] etc. (see Chapter 3B). Dialkyl phosphites enter into reaction with vinylphosphonates,[1130] orthoformates,[922,1199] arylsulfonyl isocyanates,[1337] mercuric acetate,[393] sulfur and ammonia,[625] amines,[741] diisocyanates,[771] ketones and ammonia,[847] geranyl bromide,[773] phosphonates,[243] thiourea,[1315] thiamine,[679,1306,1307] Schiff's bases,[742] trichloroacetyl chloride,[1405] trialkyl chlorosilanes with quinoline and sulfur,[419] Mannich bases,[604,608,610,612,613] and chlorobutenes.[1327] Other applications involve the use of chloridites or dichloridites in peptide synthesis,[45,1402] and reactions with diglycidyl ether,[1114] mercaptans,[1369] thiophenols,[1369] α-oxosulphides,[927] etc. The reactions of the sodium derivatives of dialkyl phosphites with various halides,[676] diethyl bromomalonate (with or without ethanol or benzyl bromide),[1309] 2-bromo-1-(2'-pyridyl)ethane,[114] trialkylchlorosilanes,[285] lactones,[745] and phosphorochloridites (with ^{18}O labeling),[1235] have also been investigated.

Table 2. Miscellaneous Reactions of Trialkyl or Dialkyl Phosphites and Their Derivatives

Reactants	Products	Ref.
$(EtO)_3P$ + 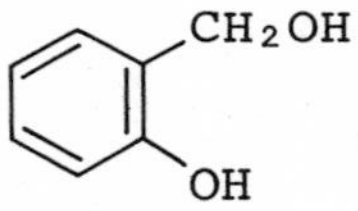	EtOH + 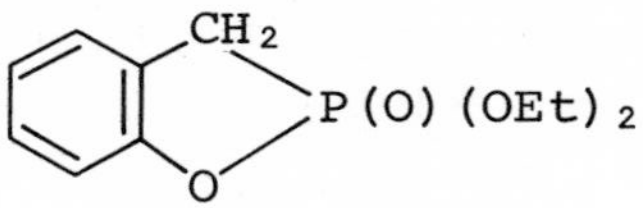	599,600
$(EtO)_3P + ClCH_2CH(OH)CH_2OR$	$(EtO)_2P(O)CH_2CH(OH)CH_2OR + EtCl$	549
	$(EtO)_2POCH(CH_2Cl)CH_2OR$	549
$(EtO)_3P + ClCH_2CH{:}CH_2$	$EtP(O)(OEt)_2 + CH_2{:}CHCH_2P(O)(OEt)_2$	95
$(EtO)_3P + ClCH_2CHO$	$CH_2{:}CHOP(O)(OEt)_2$	517
$(EtO)_3P + Cl_3CP(O)(OEt)_2$	the dichloromethylphosphonate	153
$(EtO)_3P + Et_3SiBr$	$EtBr + Et_3SiP(O)(OEt)_2 + (EtO)_2PHO$	120
$(RO)_2PCl + Cl_3CCH(OR)OH + Et_3N$	$(RO)_2P(O)CH(CCl_3)(OR')$	977
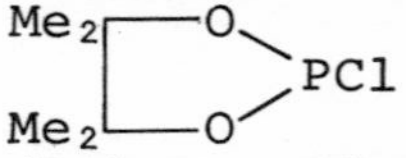 + $CH_2{:}CHCONH_2$	Cyclo-2-cyanoethylphosphonate	690
$(RO)_3P$ + R'-Se-Se-R'	$(RO)_2P(O)SeR'$	881
2,4-$(NO_2)_2C_6H_3SeBr + (i\text{-}PrO)_3P$	Ar-Se-Se-Ar + $(i\text{-}PrO)_2P(O)Br$ + i-PrBr	1041,1042
2 $C_{10}H_7TeI + (i\text{-}PrO)_3P$	Ar-Te-Te-Ar + $(i\text{-}PrO)_2P(O)I$ + i-PrBr	1041
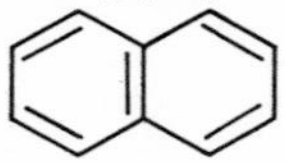 + $(EtO)_2PHO + (t\text{-}BuO)_2$	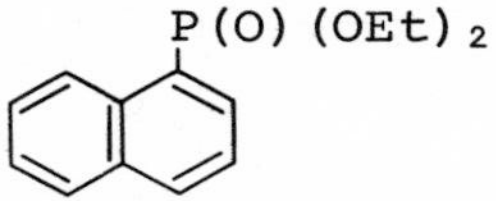	623
$(EtO)_2PHO + AcOCH_2CH_2CN$	$(EtO)_2P(O)CH_2CH_2CN$	633
$(RO)_2PHO$ + RSeCN	$(RO)_2P(O)SeR$	387

$(RO)_3P$ + $(RO)_2PSSH$	$(RO)_2PHO$ + $(RO)_2P(S)SR$	1137
$(RO)_3P$ + 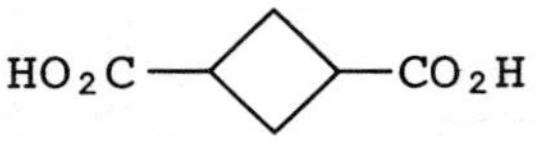	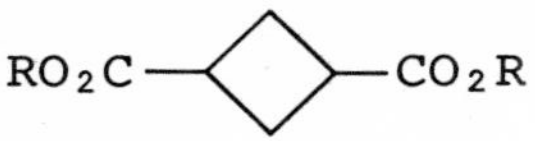	1030
$ROPCl_2$ + $(CH_3)_2C(OH)CH_2COCH_3$	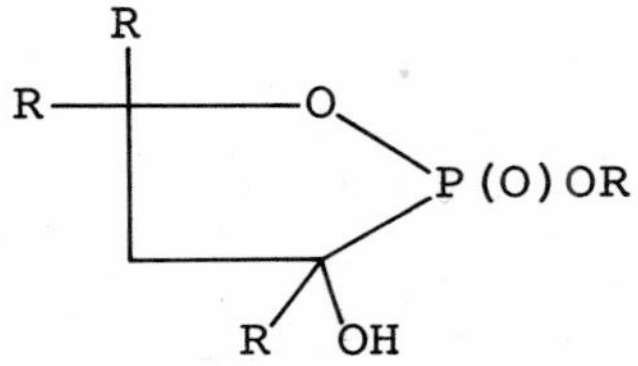	1220
$(EtO)_3P$ + Mg + 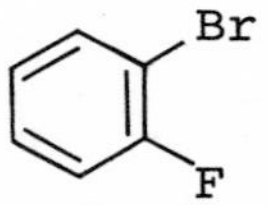	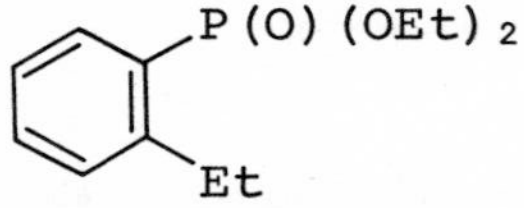	513
$(EtO)_2POAc$ or Et_2NPCl_2 + $HC(OEt)_3$	$(EtO)_2P(O)CH(OEt)_2$	750
$(EtO)_2PONa$ + $CBr_2(SO_2Et)_2$	$CH_2(SO_2Et)_2$	110
$(RO)_2PCl$ + $R'CO_2M$	$(RO)_2PO_2CR'$	1062
$(RO)_2PCl$ + $MeC(OEt)_3$	$(RO)_2P(O)CMe(OEt)_2$	921
$CCl_3CMe_2OPCl_2$ + $CH_2{:}CHCONH_2$	$ClCH_2CH_2CN$ and a tar	689

D. NITROGEN-PHOSPHORUS SYSTEMS

D.1. Preparative Procedures

D.1.1. Secondary Amine Systems

As for the phosphorus trichloride-hydroxy compound systems, so for the secondary amine systems a pattern of reactions having some measure of probability may be given for reference. The essential difference is that the amine itself inevitably constitutes a hydrogen chloride acceptor, and therefore there is no free hydrogen chloride to intervene.

$$2\ R_2NH + PCl_3 \longrightarrow R_2NPCl_2 + R_2NH_2Cl$$

$$2\ R_2NH + R_2NPCl_2 \longrightarrow (R_2N)_2PCl + R_2NH_2Cl$$

$$2\ R_2NH + (R_2N)_2PCl \longrightarrow (R_2N)_3P + R_2NH_2Cl$$

$$(R_2N)_3P + PCl_3 \longrightarrow R_2NPCl_2 + (R_2N)_2PCl$$

$$(R_2N)_2PCl + PCl_3 \longrightarrow 2\ R_2NPCl_2$$

In the absence of evidence to the contrary the 4-center, broadside mechanism is probably the simplest to postulate, e.g.,

$$
\begin{array}{ccc}
R_2N & \!\!\!\!\longrightarrow\!\!\!\! & P(NR_2)_2 \\
\Big\downarrow & & \Big\uparrow \\
Cl_2P & \!\!\!\!\longrightarrow\!\!\!\! & Cl
\end{array}
\qquad \longrightarrow \quad R_2NPCl_2 + (R_2N)_2PCl
$$

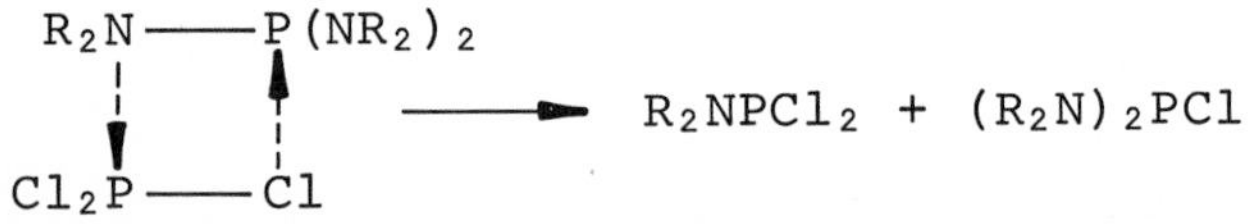

The relative rates of each step will depend upon the nature of R, the temperature, and the relative concentration of the reactants. The concentration of each species at a given stage of the operational procedure will further depend on the order in which the initial reagents are mixed.

IV. PROCEDURES FOR NITROGEN-PHOSPHORUS SYSTEMS

The following preparative procedures may be recognized.

IVa. The secondary amine is added to phosphorus trihalide alone or in a solvent such as ether, petroleum, or benzene; a tertiary base may be used.

IVb. Phosphorus trihalide is added to the secondary amine in a solvent;[753] a tertiary base, Et_3N or C_5H_5N, may be used to accept the hydrogen chloride.

In one recent investigation, a diamine in a considerable volume of ether was added dropwise to phosphorus trichloride in ether at -70°; and 2 mole equivalents of triethylamine were then added at once. After the mixture had been at room temperature for 1.5 hr, the amine hydrochloride was filtered off.[1183]

$$\begin{array}{c} \text{Me} \\ | \\ CH_2-NH \\ | \\ CH_2-NH \\ | \\ \text{Me} \end{array} + PCl_3 + 2\ Et_3N \longrightarrow \begin{array}{c} \text{Me} \\ | \\ CH_2-N \\ \qquad\qquad P-Cl \\ CH_2-N \\ | \\ \text{Me} \end{array}$$

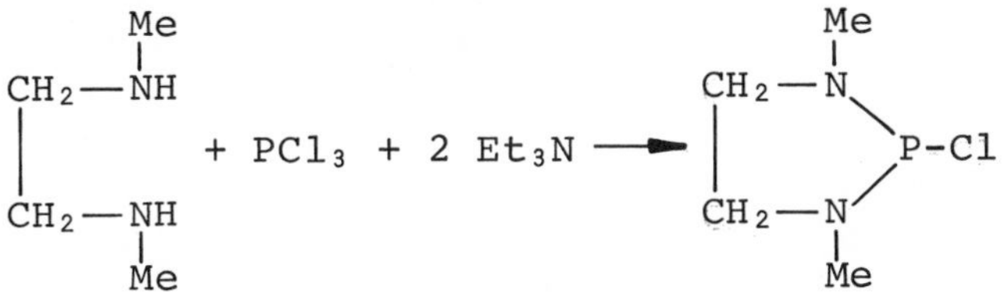

2-Chloro-1,3-dimethyl-1,3,2-diazaphospholane

IVc. The secondary amine hydrochloride is heated with phosphorus trichloride.[857]

$$R_2NH \cdot HCl + PCl_3 \longrightarrow R_2NPCl_2 + 2\ HCl$$

IVd. The aminophosphine is mixed with the appropriate amount of phosphorus trihalide.

$$(R_2N)_3P + PCl_3 \longrightarrow (R_2N)_2PCl + R_2NPCl_2$$

$$(R_2N)_2PCl + PCl_3 \longrightarrow 2\ R_2NPCl_2$$

A detailed study of the interactions of phosphorus trichloride with tris(dimethylamino)phosphine or tris(diethylamino)phosphine has been described.[1343] By a ^{31}P NMR procedure,[1342] and by observation of heat evolution on mixing, it was concluded that the primary exchange was rapid, but that each of the two chloro compounds did not "disproportionate," e.g., [2 $(Et_2N)_2PCl \rightarrow (Et_2N)_3P +$ Et_2NPCl_2] to a measurable extent. By this procedure, $(Me_2N)_2PCl$ was obtained in 91.5% yield.[983] The triphenyl phosphite-tris(diethylamino)phosphine system was also deemed to be nonrandom.[401]

IVe. The aminochlorophosphine is mixed with another secondary amine, probably in the presence of a tertiary amine, in a solvent.[990],[1183]

$$(Me_2N)_2PCl + \begin{array}{c} CH_2 \\ | \\ CH_2 \end{array}\!\!\!>\!\!NH \xrightarrow[\substack{\text{in benzene at} \\ \text{about } 0°}]{Et_3N} (Me_2N)_2PN(CH_2CH_2)$$

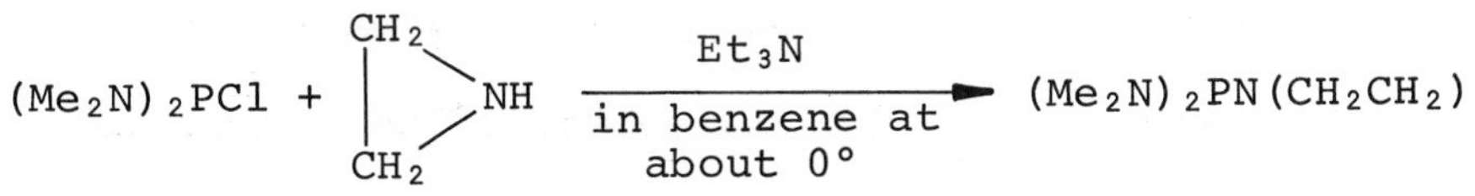

IVf. The secondary amine may be mixed with a phosphorochloridite, or dichloridite in a solvent, with or without a tertiary base, at about 0°.[11,106,226,857,1046,1194]

$$ROPCl_2 + 4\ R'_2NH \longrightarrow ROP(NR'_2)_2 + 2\ R'_2NH_2Cl$$

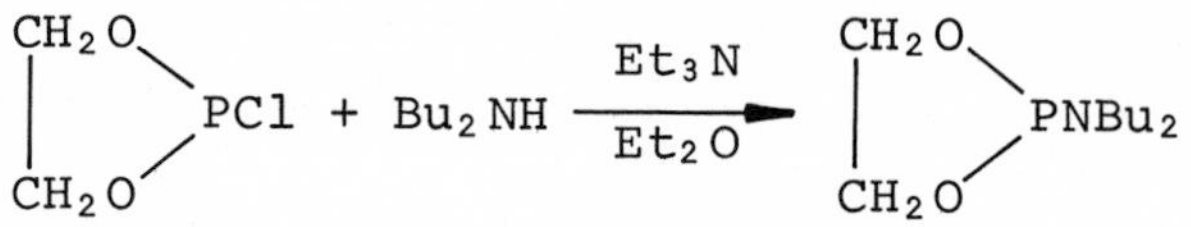

IVg. The aminochlorophosphine may be mixed with an alcohol or phenol or sodium derivative in the absence or presence of a tertiary amine in a solvent.[137,1183,1344]

$$R'_2NPCl_2 + 2\ ROH \xrightarrow[\text{solvent}]{Et_3N} R'_2NP(OR)_2$$

$$(Me_2CH)_2NPCl_2 + BuONa \longrightarrow (Me_2CH)_2NP(OBu)_2$$

IVh. The aminophosphine may be mixed with a phenol or alcohol, or amine.[964] In the examples of mixed aminophosphines, the particular amine removed will depend upon the relative electronic and steric requirements.[991]

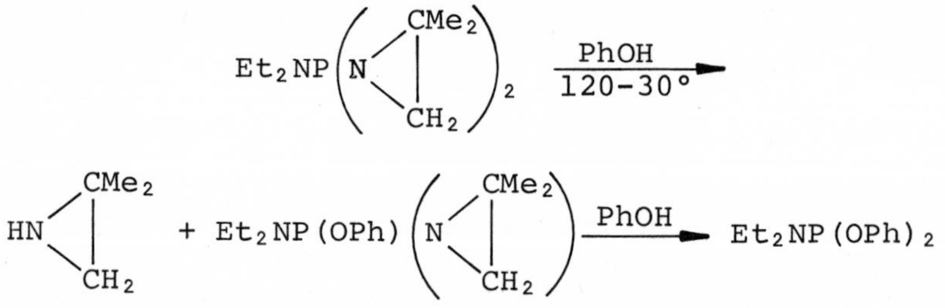

Carboxylic acid anhydrides may also be used.[381]

$$ROP(NEt_2)_2 \xrightarrow[<40°]{Ac_2O} ROP(NEt_2)(OAc)$$

IVi. Hydrogen halides, usually in a solvent such as ether, break the N-P bond and form the corresponding halophosphines.[984]

$$(Me_2N)_3P + 2\ HBr \longrightarrow (Me_2N)_2PBr + Me_2NH{\cdot}HBr$$

IVj. Ammonia and chloridites give "NH_2" compounds.[171]

$$(PhO)_2PCl + NH_3 \xrightarrow[\text{cool}]{Et_2O} (PhO)_2PNH_2 + NH_4Cl$$

$$MeOPCl_2 + NH_3 \xrightarrow{Et_2O} MeOP(NH_2)_2$$

IVk. Ammonia and amino dichloro compounds give "NH_2" compounds.[171]

$$Et_2NPCl_2 + NH_3 \xrightarrow{Et_2O} Et_2NP(NH_2)_2$$

$$MePhNPCl_2 + NH_3 \xrightarrow{Et_2O} MePhNP(NH_2)_2$$

D.1.2. Primary Amine-Phosphorus Trihalide Systems

The primary amine-phosphorus trichloride systems are much more complex, because each amine has now two normally "active" hydrogen atoms.

The first reported example of a properly characterized chlorine-free product from the interaction of phosphorus trichloride and a primary aliphatic amine is the crystalline solid, designated $P_4N_6Me_6$ and isolated in good yield from the interaction of phosphorus trichloride with methylamine in excess.[557,560,561] The stoichiometry was given as follows, the "reaction" being deemed to proceed "directly to the imide stage."

$$2\ PCl_3 + 9\ MeNH_2 = [P_2N_3Me_3]_2 + 6\ MeNH_2HCl$$

IR, NMR, and molecular weight measurements indicated a cage structure. The compound has remarkable properties,

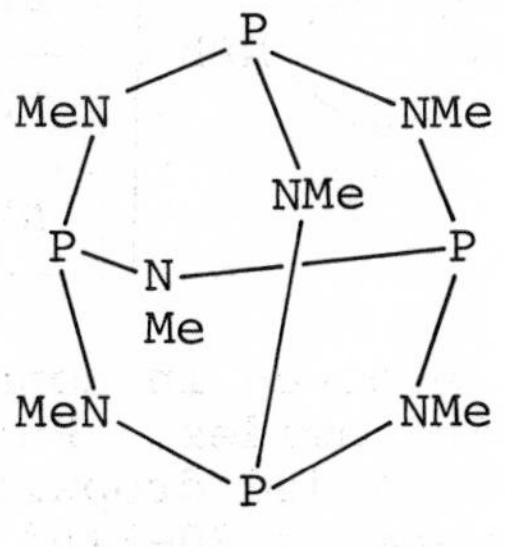

reacting with oxygen, sulfur, methyl iodide (see list). At -78° during 6 days hydrogen chloride converted 98.3% of the compound into the products shown.

$$P_4N_6Me_6 + 18\ HCl \longrightarrow 4\ PCl_3 + 6\ MeNH_2 \cdot HCl$$

Another procedure entailed the addition of the primary amine (2 mole) in petroleum to a cooled solution of phosphorus trichloride (1 mole), conditions tending to restrict the sequence to the first stage.[857]

$$2\ RNH_2 + PCl_3 \longrightarrow RNHPCl_2 + RNH_2{\cdot}HCl$$

$$(R = Et,\ Pr,\ i\text{-}Bu,\ Am)$$

It appears remarkable that these compounds could be distilled.

Phosphorus trichloride was added to allylamine and triethylamine in ether at -20° to afford the compound $(CH_2{:}CHCH_2NH)_3P$ in 78.2% yield. Reverse order of addition gave $CH_2{:}CHCH_2NHPCl_2$ as an undistillable oil in 55% yield.[1049]

By procedure (C), a modification of (A), phosphorus trichloride (1 mole) in petroleum was added dropwise to t-butylamine (8 mole) in petroleum at 0°.[562] The dimer $[P(Nt\text{-}Bu)(NHt\text{-}Bu)]_2$ was stated to be "formed directly," and the corresponding ethyl compound was mentioned. From i-butylamine, benzylamine,[857] and dodecylamine,[397] phosphorus trichloride in petroleum gave the tris-p-aminophosphines $(RNH)_3P$. Interaction of tris(dimethylamino)phosphine with aniline at 75-110° gave $Me_2NP(NHPh)_2$ and finally $(PhN{:}PNHPh)_2$.[1351]

In the earliest work[863] the compound PhN:PCl (probably a dimer) was obtained by heating aniline hydrochloride with phosphorus trichloride for several hours [procedure (D)]. The following scheme indicates the types of derivative immediately obtainable from the dimer.

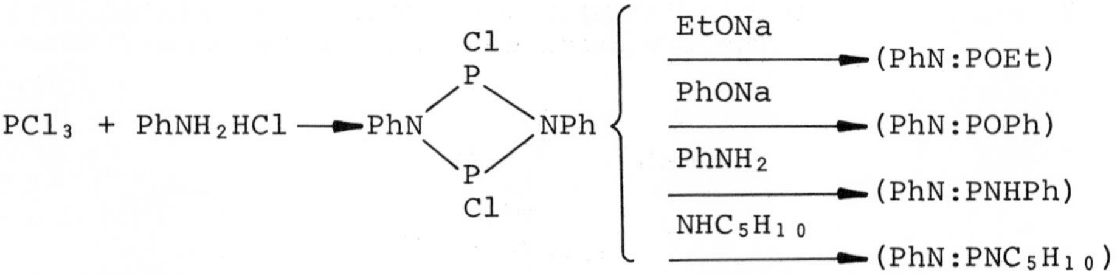

The next study[516] evolved in connection with the preparation of N-substituted amides. Phosphorus trichloride (1 mole) in toluene was added dropwise to the primary amine (4 mole) in toluene at 20-115° according to the degree of solubility of the amine [procedure (E)]. The base hydrochloride was separated as well as possible, and the final residue used with a carboxylic acid for the preparation of amides.

Heating aniline hydrochloride in phosphorus trichloride, in excess, afforded crystalline $PhN(PCl_2)_2$.[492] Addition of the amine (e.g., 4-$MeOC_6H_4NH_2$) to phosphorus trichloride in benzene afforded [(4-$MeOC_6H_4N$:P)$_2NC_6H_4OMe$-4], monomeric, deemed not to have a cage structure; and "some byproducts were formed."[492] 2-Aminobiphenyl and phosphorus trichloride gave a crystalline residue of the assumed $ArNHPCl_2$ compound which was cyclized in situ by aluminum chloride.[346]

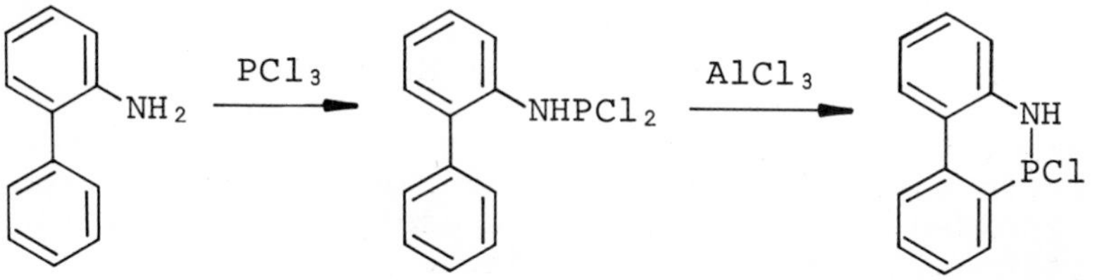

There appears to be an example of the subsequent interaction of phosphorus trichloride on the product amine hydrochloride. The presumed $ArNHPCl_2$ was obtained as a "dark sticky mass" by heating the amine (0.05 mole) in phosphorus trichloride (about 1 mole, i.e., considerable excess) whereupon "amine hydrochloride separated almost immediately but redissolved after 5 hr."[267]

$$RNH_2 \xrightarrow{PCl_3} RNHPCl_2 + [RNH_2HCl \xrightarrow{PCl_3} RNHPCl_2 + HCl]$$

Further work with monoalkylamino systems includes the formation of $[EtNPCl]_3$ from phosphorus trichloride and bis(trimethylsilyl)ethylamine,[2] the preparation of phosphorus-containing polymers from trialkyl or triaryl phosphites and diamines (e.g., hexamethylenediamine),[842] the reaction of aniline with the mixed anhydride, $(EtO)_2POP(O)(OEt)_2$, to give $(EtO)_2PNHPh$,[870] and the formation of alkylaminobis(dichlorophosphines) from phosphorus trichloride and the appropriate amine hydrochloride in sym-tetrachloroethane in the presence or absence of pyridine.[980] Antimony trifluoride reacts with the chlorophosphines to give the corresponding fluorides.[980]

$$RN(PCl_2)_2 \xrightarrow{SbF_3} RN(PF_2)_2$$

The reaction of tris(dimethylamino)phosphine with 1,2-dimethylhydrazine dihydrochloride in boiling benzene affords a bicyclic hydrazodiphosphine which is converted quantitatively to a monocyclic product on treatment with phosphorus trichloride.[1036]

$$(Me_2N)_3P + ClH \cdot HN\text{---}NH \cdot HCl \longrightarrow$$

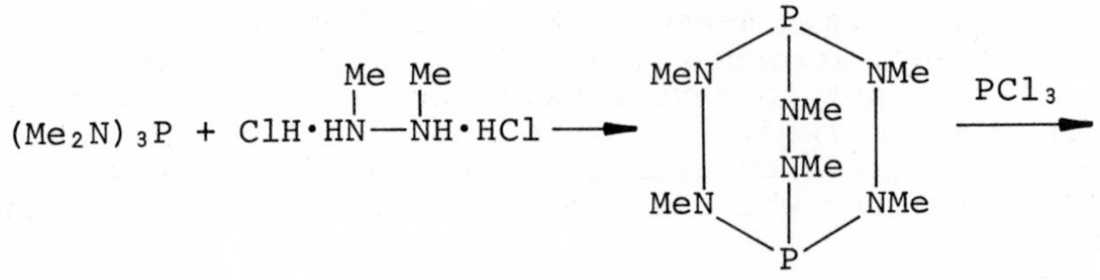

$$ClP[(NMe)_4]PCl + Me_2NHHCl$$

Phenylhydrazine reacts with a dialkyl phosphorochloridite to give the corresponding phenylhydrazophosphite $(RO)_2PNHNHPh$,[10] which gives the acetyl phosphite $(RO)_2POAc$ on treatment with acetic anhydride.[282]

D.1.3.   Tertiary Amine-Phosphorus Trihalide Systems

By contrast, the tertiary amine trimethylamine, forms only weakly bonded complexes with the phosphorus trihalides, X_3PNMe_3 (X = F, Cl, Br).[515,556,558,559,564,1031] Detailed vapor pressure studies have been described.

$$Cl_3PNMe_3 \rightleftharpoons PCl_3 + Me_3N$$

The apparent inability of triethylamine to complex with phosphorus trichloride was attributed to steric hindrance.[558]

D.1.4.   Amidoesters and Diamides of Phosphorous Acid.
Analogs of Dialkyl or Diaryl Hydrogen Phosphites

Previous to 1954 when the formation of $(PhNH)_2P(O)H$ was reported,[493] only one example, $(PhNHNH)_2P(O)H$,[862] appears to have been mentioned in the literature. Three further examples referred to as "phosphonic disecondary amides" were obtained by the hydrolysis of the triaminophosphine by phosphorous acid.[1377]

$$(R_2N)_3P \xrightarrow[\text{hydrolysis}]{H_3PO_3} (R_2N)_2P(O)H$$

Aqueous hydrolysis of the compound $(R_2N)_2POC(O)NR_2$ to $(R_2N)_2P(O)H$ has been mentioned.[1008] These systems have been given recent attention.[1413,1414,1416]

Hydrolysis of the corresponding monoalkoxy compound,[569] is not deemed to be satisfactory,[1416]

$$(R'_2N)_2POR \xrightarrow[\text{dioxan}]{H_2O} (R'_2N)(RO)P(O)H + R'_2NH$$

because attempts to prepare the initial monoalkoxy compound

gave mixtures.[1416]

Even when prepared from the phosphorodiamidous chloride, the diamidite product was still a mixture, and this observation led to the conclusion that "spontaneous disproportionation" was prevalent;[1416] however, see Refs. 1342 and 1343 and comments therein.

The most attractive procedure appears to be the following where yields of 50-80% are obtained:[1416]

$$RO-P\begin{array}{c}Cl\\[4pt]Cl\end{array} \ + \ 2\ R'_2NH \longrightarrow \begin{array}{c}R'_2N\\[4pt]RO\end{array}PCl \xrightarrow[Et_3N]{H_2O} \begin{array}{c}R'_2N\\[4pt]RO\end{array}P\begin{array}{c}O\\[4pt]H\end{array}$$

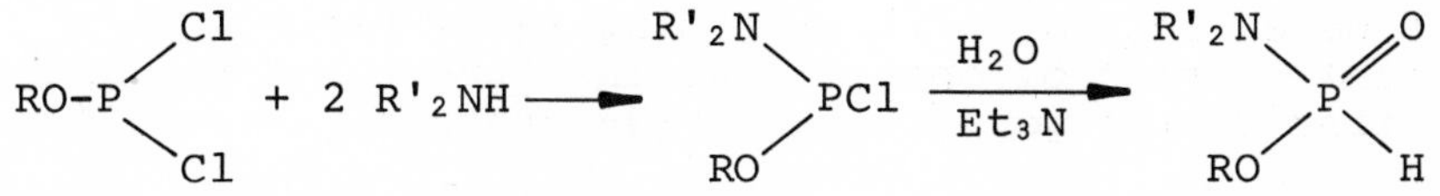

The reactions of secondary amines with O-alkylphosphorous-O,O-diethylphosphoric anhydrides at first appeared promising,[1413] but were later shown to give mixtures.[1416]

$$(EtO)_2\overset{O}{\underset{}{P}}-O-\overset{O}{\underset{H}{P}}(OR) \xrightarrow{\ 2\ R'_2NH\ }$$

$$R'_2N\overset{O}{\underset{}{P}}\begin{array}{c}H\\[4pt]OR\end{array} \ + \ R'_2N\overset{O}{\underset{}{P}}(OEt)_2 \ + \ \text{other products}$$

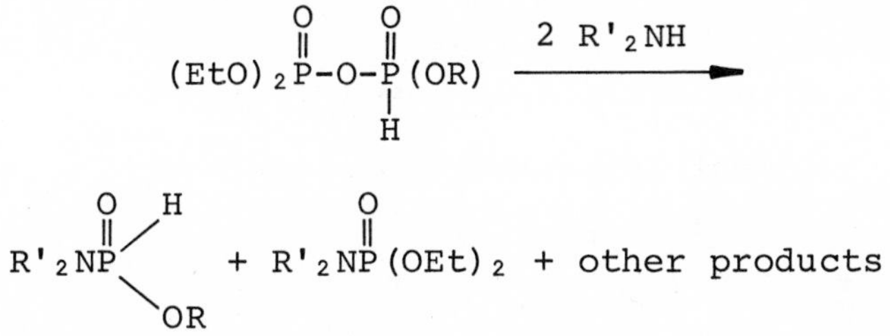

Symdialkyl pyrophosphites react similarly but they require purification by molecular distillation.[1416]

$$RO-\overset{O}{\underset{H}{P}}-O-\overset{O}{\underset{H}{P}}-OR \xrightarrow[\text{benzene } 10\text{-}15°]{2\ R'_2NH}$$

$$\begin{array}{c}R'_2N\\[4pt]RO\end{array}P\begin{array}{c}O\\[4pt]H\end{array} \ + \ [ROPH(O)O]^-\,[R'_2NH_2]^+$$

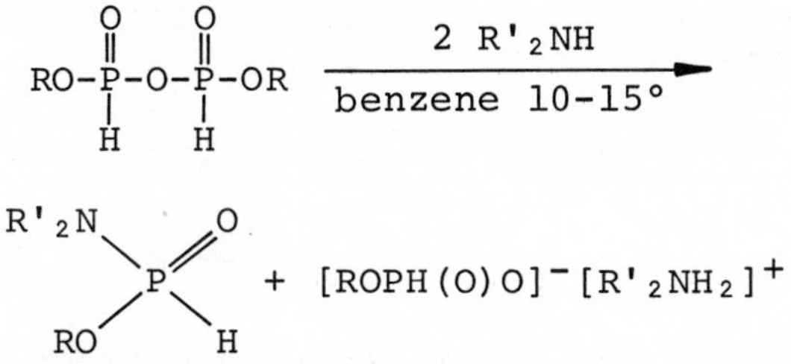

The monoalkyl phosphoroamidites, $(R'_2N)(RO)P(O)H$, may be purified by distillation at low pressure. They are thermally unstable (130-50°), but may be stored at room temperature. TLC data, and R_f values, of these and related compounds are available.[1416] The reaction of ethanolamine with a variety of phosphite esters yielded $\overset{+}{N}H_3CH_2CH_2OP(O)(H)O^-$, but from tris(dimethylamino)phosphine a cyclic product resulted.[1153]

$$2\ NH_2CH_2CH_2OH \ + \ (Me_2N)_3P \longrightarrow H_2NCH_2CH_2OP\begin{array}{c}O-CH_2\\[4pt]\diagdown\\ NH\end{array}\hspace{-4pt}\begin{array}{c}\\ CH_2\\[4pt]\end{array}$$

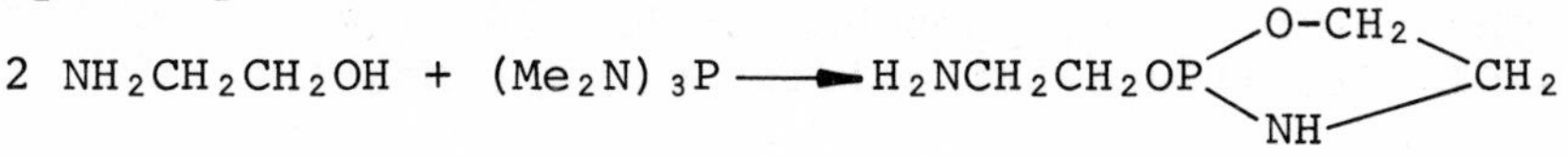

D.2. Reactions of Phosphorus-Nitrogen Systems

In the preparative section the mutual exchange of the groups NR_2, OR, OAr, and Cl, has been indicated.

The properties of the amino compounds of trivalent phosphorus can be considered from several aspects; but will be itemized in terms of the other ("attacking") reagent. As the details so far are in the main for the simpler compounds, e.g., $(Me_2N)_3P$, considerable caution must be exercised in interpreting these results as a general pattern of operational and structural factors relating to the general case R_2N- systems.

D.2.1. Interaction with Oxygen[11,1049]

Oxidation by elementary oxygen[1351] or by hydrogen peroxide[1302] yields the corresponding phosphoroamidates (see also Chapter 18).

$$(Me_2N)_3P \xrightarrow{O_2,\ 120°} (Me_2N)_3P{=}O$$

(About 20% conversion in 0.5 hr)

$$(R_2N)_3P \xrightarrow{H_2O_2\ aq.} (R_2N)_3P{=}O$$

(R = Et, Pr, Bu)

D.2.2. Interaction with Sulfur or Tellurium

Addition gives the corresponding thio-, or telluro-phosphoramidate,[506,1302,1351] e.g.,

$$(R_2N)_3P \xrightarrow{S} (R_2N)_3PS$$

$$(Me_2N)_3P \xrightarrow[150\text{-}60°\ (2\ hr)]{Te} (Me_2N)_3PTe$$

Similar reactions have been reported between sulfur and a variety of dialkylaminophosphines, including $(Bu_2N)_2PSCH_2Ph$,[1302] $(Me_2N)_2POC(O)NMe_2$,[1008] and $Et_2NPCl(SCH_2CHMeCl)$, the last compound adding sulfur in the presence of aluminum trichloride.[993] Diethylaminodichlorophosphine abstracted sulfur from ethylene sulfide at 80°.[993]

D.2.3. Interaction with Halogens or Halogenoalkanes

The formation of phosphonium halides $[(Me_2N)_3PX]^+X^-$ (X = Cl, Br, I) and trihalides, e.g., $[(Me_2N)_3PCl]^+Cl_3^-$, is reported.[987] Methyl iodide yields $[(R_2N)_3PMe]^+I^-$.[857,987]

D.2.4. Interaction with Hydrogen Halides

Hydrogen halide can cleave the N-P bond to give the phosphorohalidite.[984]

$$(Me_2N)_3P + HBr \xrightarrow[\text{(-78°, then warmed to room temperature)}]{Et_2O} (Me_2N)_2PBr + Me_2NHHBr$$

The alkylaminobis(difluorophosphines) are said to react quantitatively with hydrogen chloride.[980]

In the example of the mixed amino compounds, the question arises as to which amino group is removed; mixed compounds such as $(RO)_2PNR'_2$ and $ROP(NR'_2)_2$, where R is a group such as butyl, will most likely undergo dealkylation first.

D.2.5. Interaction with Azides

Reactions are exemplified as follows:

$$(Me_2N)_3P + PhN_3 \longrightarrow (Me_2N)_3P{:}NPh$$ [1351]

$$(RO)_2PNHPh + PhN_3 \xrightarrow{Et_2O} (RO)_2P(NHPh){:}NPh$$ [488]

$$(Me_2N)_3P + (Me_2N)_2P(X)N_3 \xrightarrow[MeCN]{Et_2O \text{ or}} (Me_2N)_2P(X)NP(NMe_2)_3$$ [1350]

D.2.6. Interaction with Carbon Disulfide and Isothiocyanates

Addition to one of the C=S bonds of carbon disulfide occurs, the nitrogen of the aminophosphine becoming attached to carbon and the phosphorus to sulfur.[1008,1010,1351]

$$(R_2N)_3P \xrightarrow{2\ CS_2} [R_2NC(S)S]_2PNR_2 \xrightarrow[\text{(reflux)}]{CS_2} [R_2NC(S)S]_3P$$

The first product shown was also obtained from dimethylaminodichlorophosphine as follows:[1351]

$$Me_2NPCl_2 + 2\ NaS\overset{\overset{S}{\|}}{C}NMe_2 \longrightarrow Me_2NP(S\overset{\overset{S}{\|}}{C}NMe_2)_2$$

Isothiocyanates react at the C=N bond.[1008,1010]

$$(Me_2N)_3P + PhN{:}C{:}S \longrightarrow (Me_2N)_2PN(Ph)C(S)NMe_2$$

With both reagents, the reaction of a P-N bond occurs more readily than that of a P-O bond, e.g.,

$$(Me_2N)_2POC_6H_4\text{-t-Bu-p} + 2\ CS_2 \longrightarrow [Me_2NC(S)S]_2POC_6H_4\text{-t-Bu-p}$$

$$Et_2NP(OEt)_2 + 3,4\text{-}Cl_2C_6H_3NCS \longrightarrow$$

$$(EtO)_2PN(3,4\text{-}Cl_2C_6H_3)C(S)NEt_2$$

D.2.7. Interaction with Carbon Dioxide

Addition to one of the C=O bonds occurs.[1008]

$$(R_2N)_3P \xrightarrow[20\text{-}40^\circ]{CO_2} (R_2N)_2POC(O)NMe_2$$

Again, with mixed amidites $(R_2N)_2POR'$ or $R_2NP(OR')_2$, the P-N bond reacts in preference to the P-O bond.

D.2.8. Interactions with Alcohols, Mercaptans, Phenols, Carboxylic Acids (or Anhydrides), and Amines

These processes correspond with transesterification, an amine being liberated.[965,973,1238,1284] Many examples are known, of which the following are representative examples:

$$(Me_2N)_3P + 3\ EtOH \longrightarrow (EtO)_3P + 3\ Me_2NH$$ [1351]

Trialkyl phosphites have been prepared by the alcohol-phosphorous tris(dialkylamide) system,[1247] and the kinetics of the interaction with butanol have been studied.[236] Mercaptans behave similarly.[1302]

$$(Bu_2N)_3P + PhCH_2SH \xrightarrow{140\text{-}50^\circ} (Bu_2N)_2PSCH_2Ph + Bu_2NH$$

Allyl alcohol and the triamide $(Me_2N)_3P$ gave a tar at 40-50°; but propargyl alcohol and the mixed compound $(RO)_2PNR_2'$ gave the phosphite $(RO)_2POCH_2C{\equiv}CH$ (in situ), which isomerized and eventually reacted with the liberated amine to give the phosphonate $(RO)_2P(O)CH{:}C(Me)NR'_2$.[68]

The following reaction sequence has been suggested for the phenol system, which eventually leads to the formation of triphenyl phosphite:[1047]

$$(R_2N)_3P \xrightarrow{HOPh} (R_2N)_3P(H)OPh \longrightarrow (R_2N)_2POPh + R_2NH$$

Even from mixed compounds such as $(MeO)_2PNEt_2$, the

amino group is replaced by a phenolic group,[1047] and such systems have been considered from the aspect of formation of mixed phosphites, e.g.,

$$(MeO)_2PNEt_2 \xrightarrow{HOC_6H_2Cl_3} (MeO)_2POC_6H_2Cl_3$$

The reaction of acetic acid with a mixed ester of this type gave N,N-dimethylacetamide and a viscous oil, but acid anhydrides yield acyl phosphites.[381,638]

$$(RO)_2PNEt_2 + (R'CO)_2O \longrightarrow (RO)_2POC(O)R' + R'C(O)NEt_2$$

Amines can take part in transamination, thus:

$$(Me_2N)_3P + PhNH_2 \xrightarrow{75\text{-}110°}$$

$$Me_2NH + Me_2NP(NHPh)_2 \xrightarrow[\text{+ petroleum}]{\text{boiling } C_6H_6} (PhN{:}PNHPh)_2$$

D.2.9. Interaction with Organometallic Compounds

The triamide $(Me_2N)_3P$ did not react with the methyl Grignard reagent, even in boiling butyl ether,[229] although chlorine is replaced in the corresponding dichloridite.[229,230,596]

$$Me_2NPCl_2 + MeMgBr \xrightarrow[-16°]{} Me_2NPMe_2$$

Organolithium compounds[984,1403] and sodium diphenylphosphine[1352] similarly attack the phosphorus-chlorine bond.

$$(Me_2N)_2PCl + Ph_2PNa \xrightarrow{C_6H_6 \text{ + petroleum}} Ph_2PP(NMe_2)_2$$

$$(Me_2N)_2PCl + BuLi \xrightarrow{Et_2O} (Me_2N)_2PBu$$

D.2.10. Interaction with Boron Compounds

Evidence points to coordination only from the phosphorus atom in adducts such as $(Me_2N)_3PBMe_3$ (obtained by direct interaction[563]), or $(Me_2N)_3PBH_3$, which is formed from the tris(dialkylamino)phosphine by reaction with diborane, acidified sodium borohydride, or the trimethylamine-borane complex.[1212]

Tris(dimethylamino)phosphine is deemed "a stronger Lewis base than trimethylamine with respect to borane," because it will replace the latter slowly at room temperature.[1212] The strengths of the electron donor function of the compounds $(Me_2N)_2PF$, $MeNHPF_2$, and $(MeNH)_2PF$ have been compared with reference to borane "BH_3" as the electron acceptor.[718]

Lithium aluminum hydride in ether at -40° reduces the chloridite and forms a borane adduct with the phosphine so produced.[985]

$$(Me_2N)_2PCl + LiBH_4 \xrightarrow[-40°]{Et_2O} (Me_2N)_2PH{\cdot}BH_3$$

A 1:1 adduct of dimethylaminodifluorophosphine and tetraborane is obtained by reaction of Me_2NPF_2 with B_4H_8CO.[1316] Halides of boron bring about mutual replacement reactions of the halogen and amino groups.[563]

$$(Me_2N)_3P + BCl_3 \xrightarrow[\text{rapid}]{0°} PCl_3 + Me_2NBCl_2$$

$$(Me_2N)_3P + BF_3 \longrightarrow PF_3 + Me_2NBF_2$$

D.2.11. Formation of Metal Complexes

It need only be mentioned here that complexes of the compounds $(Me_2N)_3P$, $(Me_2N)_2PCN$, and $(Me_2N)_2PCl$, with cadmium iodide, mercuric iodide, and nickel carbonyl have been described. The monoamino compound $Me_2NP(CN)_2$ with nickel carbonyl gave carbon monoxide and an oily mass[986] (see Chapter 3B).

In aluminum catalyzed reactions with ferrocene the following order of reactivity was shown: $PCl_3 << R_2NPCl_2 > (R_2N)_2PCl > (R_2N)_3P$, a reactive intermediate being deemed due to coordination of aluminum chloride with nitrogen and not phosphorus.[1281]

D.2.12. Miscellaneous Reactions and Uses of Phosphorus-Nitrogen Compounds

Special applications of organic phosphorus-nitrogen compounds include the uses of diethyl anilinophosphite, $(EtO)_2PNHPh$, chloridites, and tetraethyl pyrophosphites in relation to peptide syntheses,[43,46,1400] and the preparation of the compounds, $(R_4N)R'HPO_3$, R being a lipophilic radical of 8 to 20 carbon atoms, from dialkyl hydrogen phosphites and tertiary amines;[1411] they readily absorb atmospheric oxygen. Other miscellaneous reactions are listed in Table 3.

Reference to hexamethylphosphorous triamide and tetrahydrofuran as an effective medium for the interaction of halobenzenes with various nucleophiles (C.A. 68, 29346g) is incorrect. The original paper[272] relates to the use of hexamethylphosphoric triamide.

Table 3. Miscellaneous Reactions of Phosphorous Acid Amides and Their Derivatives

Reactants	Products	Ref.
2 $(Me_2N)_2PCl$ + 2 Na	$(Me_2N)_2P\text{-}P(NMe_2)_2$	983
$(Me_2N)_2PCl + Me_4Pb + AlCl_3$	$MeP(NMe_2)_2$	811
$Me_2NPCl_2 + AgCN$[a]	$Me_2NP(CN)_2$	984
$(R_2N)_3P$ + R'NCO	$(R_2N)_2PNR'CONR_2$ + polymer	1008
$(Me_2N)_3P + OCNSO_2F$	$(Me_2N)_3\overset{+}{P}\text{—}CO\overset{-}{N}SO_2F$	555
$(RO)_2C{=}NCl$ + 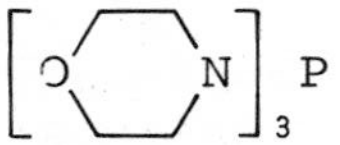	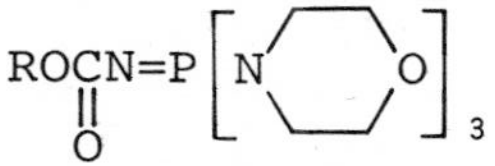	340,342
$(BuO)_2PN(i\text{-}Bu)SO_2Me$ + RI	$(RO)(RSO_2NR)P(O)R$	12
$EtOP(NMe_2)_2 + p\text{-}MeC_6H_4SO_2NCO$	$Me_2NP(OEt)N(p\text{-}MeC_6H_4SO_2)CONMe_2$	1010
$(RO)_2PNHC_6H_4X + H_2C{:}CR'CN$	$(RO)_2P(CH_2CHR'CN){:}NC_6H_4X$	1105
$AcOP(NEt_2)(OPr)$ + AcH	$AcOCHMeP(O)(OPr)(NEt_2)$ (75%) + $AcNEt_2$ (6.6%)	912
$(RO)_2PNEt_2 + [R'PO_2]$	$(RO)_2POP(=O)(R')NEt_2$	379
$(Et_2N)_3P$, $(Et_2N)_2POEt$, or $(Et_2N)P(OEt)_2$ + Mannich phenol bases	Derivatives of 1-oxa-2-phosphaindan	602,603
$(Me_2N)_3P + CCl_4$	$(Me_2N)_3\overset{+}{P}C\overset{-}{Cl_3}Cl + (Me_2N)_3PCCl_2$	578 and cited Refs.
$(Me_2N)_3P + CH_2(COOEt)_2$	$CH_2(CONMe_2)_2 + (EtO)_3P$	233
amidites + CH_3CO_2H	acetyl phosphites	963
$(R_2N)_3P$ + uracil	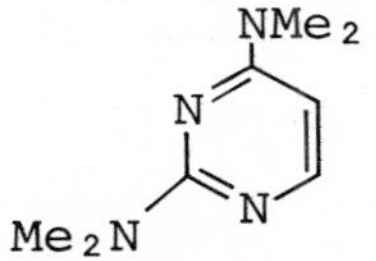	146

Table 3 (Continued)

Reactants	Products	Ref.
$(Me_2N)_3P$ + $(Me_2C{:}CO)_2$	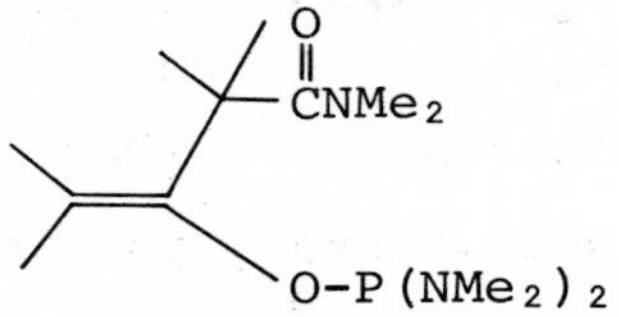 etc.	183
Et_2NPCl_2 + diethyl tartrate	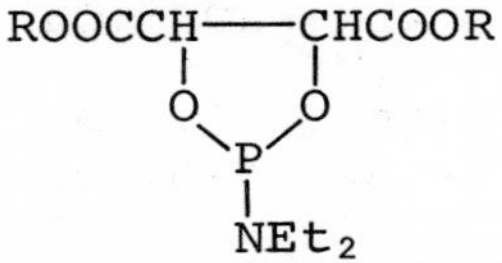	503
$(RO)_2PNHPh$ + Aromatic isocyanates	$ArN{:}C(NHPh)P(O)(OR)_2$	1104
$(RO)_2PNHPh$ + CCl_4 or $BrCCl_3$	$(RO)_2P(X){:}NPh$	578
$Et_2NP(OR)_2$ + $(R'O)_2PS_2H$	$Et_2NPH(O)(OR)$ + $(R'O)_2P(S)(SR)$	1138
$(RO)_2PNHPh$ + Aldehydes	$(RO)_2P(O)CHR'NHPh$	1148
Ph_2CCl_2 + $(Me_2N)_3P$	$Ph_2C{=}CPh_2$	419
$(Et_2N)_2POEt$ + trichloroethylideneacetone	$(Et_2N)_2P(O)OCMe{:}CHCH{:}CCl_2$	66
CH_2O/CH_2O (ring) $PN(CH_2CH_2Cl)_2$ + CCl_3CHO	$ClCH_2CH_2OP(O)(OCH{:}CCl_2)N(CH_2CH_2Cl)_2$	507
$(EtO)_2PN(CH_2CH_2Cl)_2$ + chloral	$CCl_2{:}CHOP(O)(OEt)N(CH_2CH_2Cl)_2$	65
$(EtO)_2PNBu_2$ + $ICH_2CH{-}CH_2$ (epoxide, O)	$BuP(O)(EtO)(NBu_2)$ + Et N,N-dibutylglycidylphosphonamidate	24
$(BuO)_2PNMe_2$ + EtI	$EtP(O)(OBu)(NMe_2)$	24
$(Me_2N)_3P$ + benzil	$PhC(O^-){=}CPhO\overset{+}{P}(NMe_2)_3$	1182

$(Et_2N)_3P$, $(Et_2N)_2POEt$, or $(Et_2N)P(OEt)_2$ + $NCCH_2CH_2OH$	$NCCH_2CH_2P(O)(NEt_2)_2$, etc.	607
$EtOP(NEt_2)_2$ + $CH_2{:}CHCO_2H$	$EtOP(NEt_2)HO$ + $EtOP(O)(NEt_2)CH_2CH_2CONEt_2$ + $(Et_2N)_2P(O)CH_2CH_2CO_2Et$	1143
$(RO)_2PNREt$ + $RCOMe$	$(RO)_2P(O)CMeRNEtR$	1124
$Me_2NP(OMe)_2$ + Cl_3CCHO	$Me_2NP(O)(OMe)OCH{:}CCl_2$	34
o-$C_6H_4O_2PNHPh$ + $R'CHO$	o-$C_6H_4OP(O)(NHPh)CHR'O$ (cyclic)	13
$(RO)_2PNEt_2$ + $R'CHO$	$Et_2NCHR'PO(OR)_2$	403

[a]Caution: only white silver cyanide should be used; discolored material can cause violent explosions.

D.2.13. Isocyanate Compounds

Isocyanates have been prepared by the interaction of sodium cyanate and a chloridite.[519]

$$(EtO)_2PCl + NaOCN \xrightarrow{\text{Heat}} (EtO)_2PNCO + NaCl$$

With an alkyl iodide they give the corresponding phosphonate.

$$(EtO)_2PNCO + RI \longrightarrow RP(O)(OEt)NCO + EtI$$

They undergo ready alcoholysis.[1296] Cyclization with $RC(:NCl)NH_2$ in benzene is accompanied by elimination of alkyl chloride.[722]

E. PREPARATION AND PROPERTIES OF SILYLPHOSPHITES

V. SILICON SYSTEMS

Alkoxytrimethylsilanes interact with phosphorus trichloride or phosphorus tribromide by a process of mutual exchange of alkoxyl and halogen which leads to the formation of phosphorodihalidites, halidites, or trialkyl phosphites according to reagent proportions and conditions.[109,286,322,394,485,843,1282]

$$Me_3SiOR + PX_3 \longrightarrow Me_3SiX + ROPX_2$$

$$ROPX_2 \xrightarrow{Me_3SiOR} (RO)_2PX \xrightarrow{Me_3SiOR} (RO)_3P$$

Alkyl groups having a highly reactive C-O bond such as in t-butyl, PhMeCH, or Ph_2CH, tend to give the alkyl chloride in accordance with the points already discussed (see Section Ia). Reaction is catalyzed by pyridine hydrochloride, a trace of which may be present in the alkoxysilane used, by virtue of its preparation with the use of pyridine.[322]

In other work[1358,1359] it was simply stated that phosphorus trichloride and ethoxytrimethylsilane did not lead to the formation of tris(trimethylsilyl) phosphite; although phosphorus tribromide in the presence of about 0.5 mole% of $ZnCl_2$, $FeCl_3$, or $SnCl_2$ gave alkyl bromide (75-100% yield) and the trialkylsilyl phosphite in yields about 30% or less.

$$3\ R_3SiOR' + PBr_3 \longrightarrow (R_3SiO)_3P + 3\ R'Br$$

Hexa-methyl- and hexa-ethyl-disiloxanes in the presence of catalytic $FeCl_3$ or $ZnCl_2$ gave the tris(trialkylsilyl) phosphites in yields up to 25% by PBr_3, but only 7% by PCl_3.[1358,1359]

In earlier work the interaction of phosphorus trihalides (Cl, Br) with alkoxysilicon and related compounds was examined from the aspect of the preparation of halosilanes or alkyl halides, and there was little or no mention of the phosphorus aspect. In the absence of oxygen, hexaethyldisiloxane and phosphorus trichloride gave, on heating, chlorotriethylsilane, but in the presence of oxygen a C-P bond was formed.[286] The reaction of a phosphorodichloridite with a trialkylsilyl acetate has also been described.[1014]

In an attempt to obtain tris(trialkylsilyl) phosphites $(R_3SiO)_3P$ from trialkylmethoxysilanes or trialkylchlorosilanes and phosphorous acid, the hydrogen phosphites, $(R_3SiO)_2PHO$, were obtained in yields of 50-70%. Raman spectra indicated the PH(:O) structure.[1357]

$$2\ R_2Si(OMe)_2 + P(OH)_3 \longrightarrow 2\ MeOH + (R_2SiO)_2PHO$$

$$2\ R_3SiCl + (HO)_2PHO \longrightarrow (R_3SiO)_2PHO + 2\ HCl$$

The hydrogen phosphites react with methylmagnesium iodide, methane being quantitatively evolved by $(Me_3SiO)_2PHO$ but less than quantitatively by $(MeEt_2SiO)_2PHO$, and others. The silyl esters were deemed to show "extremely high thermal and chemical stability (toward, e.g., HCl)" in comparison with their analogs $(R_3CO)_2PHO$.[1357] With trialkylchlorosilanes an exchange of groups occurs.[1355]

$$(Me_3SiO)_2PHO + Et_3SiCl \longrightarrow (Et_3SiO)_2PHO$$

Hexaalkyldisiloxanes and acetoxytrialkylsilanes also react with phosphorous acid to give hydrogen phosphites.[1022]

$$R_3SiOSiR_3 + H_3PO_3 \underset{(ZnCl_2\ \text{or}\ H_2SO_4)}{\overset{\text{catalyst}}{\rightleftharpoons}} (R_3SiO)_2PHO + H_2O$$

(in C_6H_6, MeC_6H_5, or $Me_2C_6H_4$, water azeotropically removed)

$$2\ R_3SiOAc + H_3PO_3 \rightleftharpoons (R_3SiO)_2PHO + 2\ AcOH$$

(AcOH distilled off as formed)

Tris(trialkylsilyl) phosphites have been obtained in rather poor yield, by the reaction of a trialkylchlorosilane with the sodium derivative of a bis(trialkylsilyl) hydrogen phosphite,[1021,1155,1355,1356,1359] or with the hydrogen phosphite itself in the presence of a tertiary base.[1018,1020]

Mixed silyl phosphites $(R_3SiO)_nP(OSiR'_3)_{3-n}$, which undergo disproportionation at 120-160°, are claimed by a procedure involving colloidal nickel and the bis(trialkylsilyl) hydrogen phosphite,[1018] while dialkyl trialkylsilyl phosphites have been prepared as follows:[285]

$$(EtO)_2PCl + Me_3SiONa \longrightarrow (EtO)_2POSiMe_3$$

$$Me_3SiCl + (EtO)_2PONa \longrightarrow (EtO)_2POSiMe_3$$

They add sulfur[285] and at 160° undergo both disproportionation and isomerization to the phosphonate, $RP(O)(OR)-(OSiR_3)$.[1018] Preparations of mixed esters from dialkyl phosphites and trialkylsilanes R_3SiH have been reported.[1016]

Mixed silyl alkyl hydrogen phosphites, $R_3SiO(R'O)PHO$, were obtained from the hydrogen phosphites at 160°.[1018]

$$(R'O)_2PHO + (R_3SiO)_2PHO \longrightarrow (R_3SiO)(R'O)PHO$$

A Michaelis-Arbusov type of reaction has been reported between tris(triethylsilyl) phosphite and acetyl chloride.[1017]

$$(Et_3SiO)_3P + AcCl \longrightarrow (Et_3SiO)_2P(O)Ac + Et_3SiCl$$

F.   THIOPHOSPHITES AND SELENOPHOSPHITES

Formally the thiophosphites may be considered alongside their oxygen analogs; but caution must be exercized in writing of this analogy and it is more helpful to take cognizance of the differences.

F.1.   Preparation and Properties of Thiophosphites

VIa.   INTERACTION OF THIOLS WITH PHOSPHORUS TRIHALIDE OR HALOGENOPHOSPHITES

As for the alcohol and phenol systems, it may be postulated that the thiophosphites can be built up by a sequence of 4-center "broadside" reactions. Addition of the thiol or thiophenol to the phosphorus trichloride held in excess

$$\longrightarrow RSPCl_2 \text{ etc. to } (RS)_3P$$

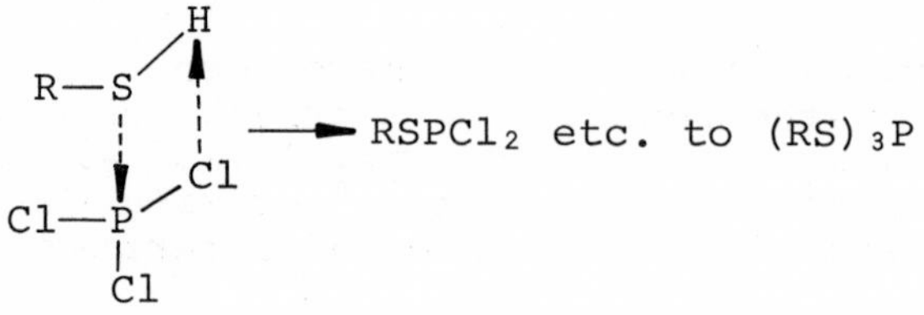

thus leads to the isolation of the thiodichloridite, $RSPCl_2$, and chloridite $(RS)_2PCl$.[159] In the reverse order of addition, or by continued addition of the thiol up to the 3:1 mole ratio, the trithiophosphite is formed but unlike the oxygen analogs does not undergo dealkylation by hydrogen halide.

Trithiophosphites were prepared, e.g., from phosphorus trichloride (1 mole) and butyl mercaptan (3.2 mole) at 100 to 300°F,[1378] and even the tri-t-alkyl trithiophosphites are claimed in good yields by the interaction of a tertiary alkyl mercaptan and phosphorus trichloride.[1258]

The mixed esters $ROP(SR^1)_2$ were obtained from the dichloridite and mercaptan.[904]

VIb. INTERACTION OF THIOLS AND PHOSPHORUS HALIDE OR HALOPHOSPHITE IN THE PRESENCE OF A TERTIARY BASE

The factors to be looked for are much the same as for the oxygen systems. In the proportions shown, the triester should result.[367]

$$3\ RSH + PCl_3 + 3\ R'_3N \longrightarrow (RS)_3P + 3\ R'_3NHCl$$

$$(RO)_2PCl + RSH + Et_3N \longrightarrow (RO)_2PSR + Et_3NHCl$$

A standard procedure for the preparation of mixed esters has been described.[905]

VIc. INTERACTION OF DIMETHYL DISULFIDE AND YELLOW PHOSPHORUS

This particular process has significance, because the product has been used to prepare other trialkyl thiophosphites by transesterification. A stirred mixture of finely divided yellow phosphorus, dimethyl disulfide, aqueous potassium hydroxide (trace), and acetone was heated to 40°. An exothermic reaction occurred. Trimethyl trithiophosphite was stated to be formed, and was used in the procedure described in Section VId.[1397] Other examples have been given.[1295]

VId. TRANSESTERIFICATION

This process has been described as a general one[1397] for preparing trialkyl trithiophosphites, $(RS)_3P$, mixed esters such as $(RS)(RO)_2P$, and even trialkyl phosphites, by the interaction of the trimethyl thiophosphite, $(MeS)_3P$, prepared as in Section VIc, and a mercaptan or alcohol.

It is claimed that by varying the proportions of reactants, mixed esters can be obtained. Although only the trimethyl thiophosphite was specifically cited as the common starting

$$3\ RSH + (MeS)_3P \longrightarrow (RS)_3P + 3\ MeSH$$

$$3\ ROH + (MeS)_3P \longrightarrow (RO)_3P + 3\ MeSH$$

material in the examples, an omnibus coverage for other trialkyl thiophosphites, $(RS)_3P$, as starting materials was claimed.

It is of interest to note that the interaction between alkyl mercaptans and triethyl phosphite does not appear to proceed by transesterification.

$$RSH + (EtO)_3P \longrightarrow RH + SP(OEt)_3$$

This reaction was stated to occur at elevated temperatures, or photochemically at room temperature[551] and was catalyzed by azobisisobutyronitrile; a radical process was suggested.[1371]

VIe. INTERACTION OF TRIALKYL TRITHIOPHOSPHITES WITH ALKYL OR ACYL CHLORIDES

In clear distinction from the behavior of trialkyl phosphites, which tend to react by the Michaelis-Arbuzov mechanism, the trithiophosphites are reported[353] to react by what may be provisionally described as a mutual exchange process. This process has been used for the preparation of thiophosphorochloridites and dichloridites, e.g.,

$$(EtS)_3P + RCl \xrightarrow{140\text{-}50°} (EtS)_2PCl + RSEt$$

$$(EtS)_2PCl + AcCl \xrightarrow{150°} EtSPCl_2 + EtSAc$$

Similar results were obtained with triphenyl trithiophosphite.

VI.f. INTERACTION OF PHOSPHORUS TRICHLORIDE WITH AN ALKYLENE SULFIDE

Thiophosphites of the general formula $(ClCHRCH_2S)_3P$ (R = H or alkyl, or aryl group) are stated to be prepared by the interaction of the alkylene sulfide and phosphorus trichloride in the presence of catalytic amounts of an organic base such as pyridine or its hydrochloride at 50-100°.[205] Likewise, the thiophosphites $(ArO)_nP(SCH_2CHRCl)_{3-n}$

are prepared from the aryl phosphorodichloridite or diaryl phosphorochloridite.

VIg. OTHER PROCEDURES

Trialkyl trithiophosphites have also been obtained by thermal decomposition of the corresponding thiophosphate,[845] e.g.,

$$(EtS)_3P{=}S \xrightarrow{120\text{-}30^\circ} (EtS)_3P + S$$

and by the interaction of phosphorus trichloride with certain sulfur containing organotin compounds.[1] Trialkyl dithiophosphites are said to be formed from mercaptides, tetraalkyldiamides of phosphorous acid, and acetic anhydride;[382] the interactions of alkali mercaptides with halophosphites have also been reported.[116]

VIh. MISCELLANEOUS REACTIONS OF THIOPHOSPHITES

As indicated above, the trithiophosphites are not dealkylated by hydrogen halides and do not undergo the Michaelis-Arbusov reaction with alkyl or acyl halides. Mixed cyclic esters do, however, react with 1,3-dienes much more readily than do their oxygen analogs, this being attributed to the lesser tendency of sulfur to be conjugated with the phosphorus atom.[385]

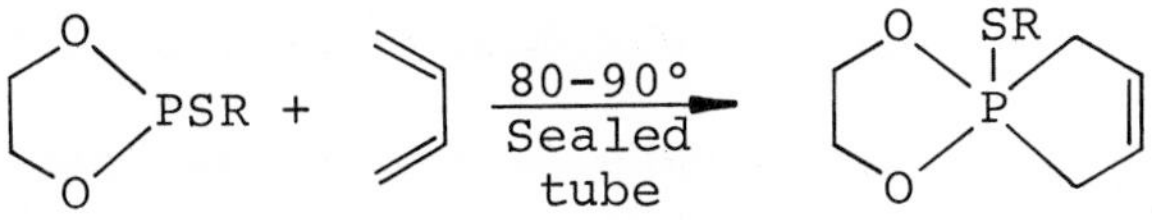

Alkyl-oxygen fission, rather than alkyl-sulfur fission occurs.[1204,1205]

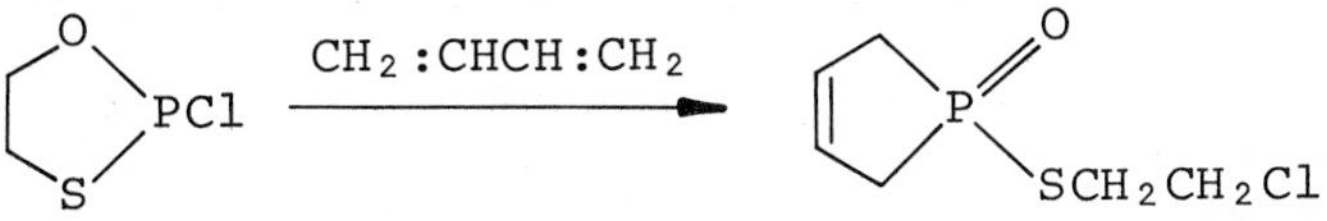

Other examples have been given.[363,738,739] The interaction of trimethyl thiophosphite with dimethyl disulfide, either alone or under the influence of γ-radiation, has been described.[355]

$$(MeS)_3P + Me_2S_2 \longrightarrow (MeS)_3PS$$

Dialkyl esters of N-chloroiminocarbonic acid react with

triaryl trithiophosphites as follows:[340]

$$(RO)_2C{=}NCl + (ArS)_3P \longrightarrow ROCON{=}P(SAr)_3$$

The chloridites $(RS)_nPCl_{3-n}$ and sodium cyanate afford the mixed compounds, $(RS)_2PNCO$, and $RSP(NCO)_2$.[1273]

VIi. FORMATION AND PROPERTIES OF HYDROGEN THIOPHOSPHITES, $(RO)_2PHS$

1. Dialkyl thiophosphites may be obtained by the interaction of hydrogen sulfide and chloridites in the presence of a tertiary base such as pyridine.[730,864,865,877]

$$(RO)_2PCl + H_2S + C_5H_5N \longrightarrow (RO)_2PHS$$

$$(BuO)(EtO)PCl + H_2S + C_5H_5N \longrightarrow (BuO)(EtO)PHS$$

The "stronger base" triethylamine also gave tetraethyl thiopyrophosphite from the corresponding chloridite.

$$(EtO)_2PCl + H_2S + Et_3N \longrightarrow [(EtO)_2P]_2S + (EtO)_2PHS$$

$$(EtO)_2PCl + (EtO)_2PHS + Et_3N \longrightarrow [(EtO)_2P]_2S$$

When the hydrogen sulfide was used in excess the diphosphorus compound could not be isolated, for it gives the hydrogen thiophosphite under such conditions.[865]

$$[(EtO)_2P]_2S + H_2S \longrightarrow 2\ (EtO)_2PHS$$

Tetraethyl thiopyrophosphite reacts vigorously with water, and is oxidized in air.

$$[(EtO)_2P]_2S + H_2O \longrightarrow (EtO)_2PHO + (EtO)_2PHS$$

$$[(EtO)_2P]_2S + O_2 \longrightarrow [(EtO)_2P(O)]_2S$$

It might have potential as a reagent in peptide synthesis, since it forms amides in the presence of a carboxylic acid and a primary amine.

2. Passage of hydrogen sulfide into a mixture of triethyl phosphite and triethylamine at 10° caused a change of refractive index and distillation after 2 days gave the hydrogen thiophosphite in 92.5% yield; but attempted distillation at a higher temperature (130-40°) gave mainly the hydrogen phosphite, results attributed to a two-course decomposition of the assumed intermediate $[(EtO)_3\overset{+}{P}H]SH^-$.[29,30]

Other examples have been reported.[591]

$$Et_3NH^+SH^- + (EtO)_3P \longrightarrow [(EtO)_3PH^+]SH^- \begin{cases} \xrightarrow{130\text{-}40°} (EtO)_2PHO + EtSH \\ \xrightarrow{90°} (EtO)_2PHS + EtOH \end{cases}$$

Several dialkyl thiophosphites, $(RO)_2PHS$, were prepared by the interaction of phosphorus heptasulfide, P_4S_7, and an alcohol (in excess)[636] and the extent to which the concurrent formation of the dialkyl phosphorothioate can be suppressed has been the subject of several investigations.[39,936,1069,1335] Methanol afforded the hydrogen phosphite $(MeO)_2PHO$ (50% yield) as the main product, whereas ethanol, propan-1-ol, and butan-1-ol gave good yields of the dialkyl hydrogen thiophosphites.[936]

The structures of the dialkyl hydrogen thiophosphites were discussed in detail;[637] and it was concluded that all properties of these compounds indicate the structure $(RO)_2P(S)H$. They do not react with cuprous chloride, sulfur, or alkyl halides as the trialkyl phosphites do. The sodium derivative, which can be prepared by adding the calculated amount of the thiophosphite to sodium ethoxide, might be deemed to be $(RO)_2PSNa$, but with water is instantly hydrolyzed. Aqueous-ethanolic sodium hydroxide slowly hydrolyzes the dialkyl thiophosphites to the monoalkyl ester, $ROP(S)(ONa)H$, which give the corresponding acid-ester on acidification.[637] The following reactions of the sodium salts[42,635,637,845,1122,1146] are thought to be due to the function of tervalent phosphorus.

$$(EtO)_2PSNa + S \xrightarrow[\text{(Then + } H_2O)]{\text{Exothermic}} (EtO)_2P(S)SH$$

$$(EtO)_2PSNa + RHal \longrightarrow RP(S)(OEt)_2 + NaHal$$

In this respect an analogy has been drawn between the chemical behavior of 0,0,0-trialkyl phosphites and that of the sodium salt of 0,0-diethyl thiophosphite.[40]

With p-chlorobenzenesulfenyl chloride the following reaction occurs:

$$(EtO)_2PSNa + p\text{-}ClC_6H_4SCl \longrightarrow (EtO)_2P(S)SC_6H_4Cl\text{-}p$$

Reaction with the sulfinyl chloride is also described.[38]

Addition of an equivalent amount of aqueous silver

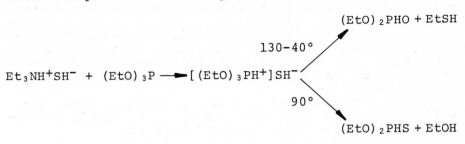

nitrate and a little aqueous ammonia solution to diethyl hydrogen thiophosphite in ethanol gives a colorless precipitate of the silver "salt."[637]

In contrast to the "salts" of diethyl hydrogen phosphite, those of the corresponding hydrogen thiophosphite have relatively low solubility in dioxane, benzene, or diethyl ether, and this is believed to be the cause of their slow reactions.[1147]

Analogously with the hydrogen phosphite system, diethyl hydrogen thiophosphite reacted readily with chlorine at -10°.[637] Thiocyanogen reacted similarly.[879]

$$(EtO)_2PHS + Cl_2 \longrightarrow (EtO)_2P(S)Cl + HCl$$

It has been pointed out[868] that dialkyl hydrogen phosphites and dialkyl hydrogen thiophosphites react in the same way as trialkyl phosphites with sulfenyl chlorides such as EtP(EtO)(O)S-Cl, by a mechanism which appears to involve the tervalent form of the hydrogen phosphite or thiophosphite. In the presence of a hydrogen chloride acceptor at 100°, trichloromethyl sulfenyl chloride gave O,O-di-Et S-trichloromethyl phosphorodithioate.[391]

The reaction of diethyl hydrogen thiophosphite with sulfur monochloride and pyridine has been reported to give a product containing the P(S)-O-P(O) link.[528,845]

Nitrosyl chloride reacts with simple dialkyl thiophosphites, either in the presence of pyridine,[885] or alone,[41] as follows:

$$(EtO)_2PHS + C_5H_5N + NOCl \longrightarrow [(EtO)_2P(S)]_2$$

(In ligroin)

$$(EtO)_2PHS + NOCl \longrightarrow (EtO)_2P(S)OP(O)(OEt)_2$$

The addition of dialkyl thiophosphites to 2-vinylpyridine,[822] and to unconjugated olefinic double bonds in ultraviolet light,[1135] yields phosphonates.

In the presence of sodium alkoxide, diethyl hydrogen thiophosphite added to ylidene derivatives of malonic ester, and acetylacetone, thus showing analogy with the hydrogen phosphite systems.[1140]

α-Mercurated carbonyl compounds react thus:[988]

$$(PrO)_2PHS + Hg(CH_2CHO)_2 \longrightarrow HgS + (PrO)_2POCH{:}CH_2$$

Dialkyl hydrogen thiophosphites [$(RO)_2PHS$] react with acyl type disulfides,[878] and disulfides, R_2S_2, afford the phosphates, $(RO)_2P(O)(SR)$.[1045] Interactions with unsaturated carboxylic acids[1128] and with propargyl aldehyde[1109] have been reported.

F.2. Selenium Systems

The preparation of a number of dialkyl hydrogen seleno-phosphites and their outstanding properties have been described.[744] Hydrogen selenide reacts readily with dialkyl phosphorochloridites when it is slowly passed into the benzene solution containing also a tertiary base such as pyridine or triethylamine.[866] Hydrogen selenide, and

$$(RO)_2PCl + H_2Se + R'_3N \longrightarrow (RO)_2P\underset{H}{\overset{Se}{\big<}} + R'_3NHCl$$

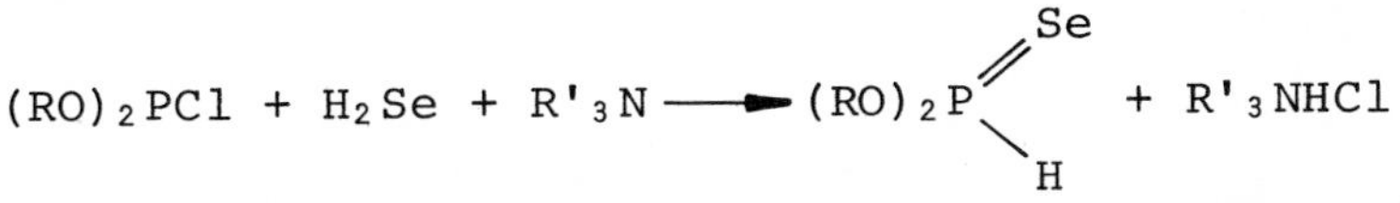

indeed these organoselenium compounds, have a repulsive odor, and special facilities and disciplines are needed for comfortable work. The toxicity needs consideration. The following schemes illustrate their reactivities in comparison with those of the corresponding dialkyl hydrogen phosphites.

$$(RO)_2P(Se)H + NaOH \longrightarrow \underset{NaO}{\overset{RO}{\big>}}P\underset{H}{\overset{Se}{\big<}}$$

(in aqueous ethanol)

$$(EtO)_2P\underset{H}{\overset{Se}{\big<}} \xrightarrow{Na} (EtO)_2PSeNa + 1/2\ H_2$$

$$(RO)_2P\underset{H}{\overset{Se}{\big<}} \begin{cases} \xrightarrow{Cl_2} (RO)_2P\underset{Cl}{\overset{Se}{\big<}} \\ \\ \xrightarrow{SO_2Cl_2} (RO)_2P\underset{Cl}{\overset{Se}{\big<}} \\ \\ \xrightarrow{(SCN)_2} (RO)_2P\underset{N=C=S}{\overset{Se}{\big<}} \end{cases}$$

$$(RO)_2PSeNa + R'X \longrightarrow (RO)_2\underset{\overset{\|}{Se}}{P}R' + NaX$$

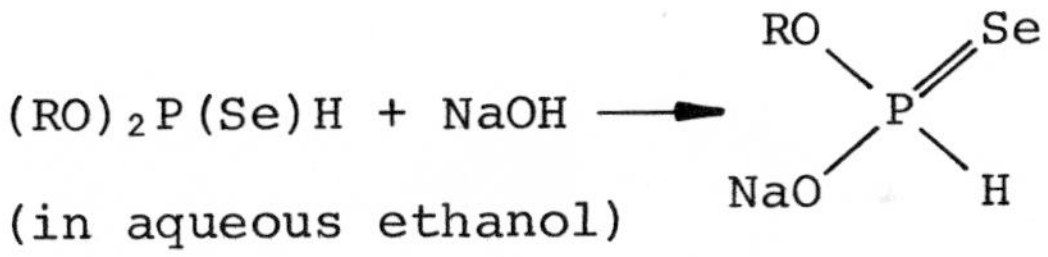

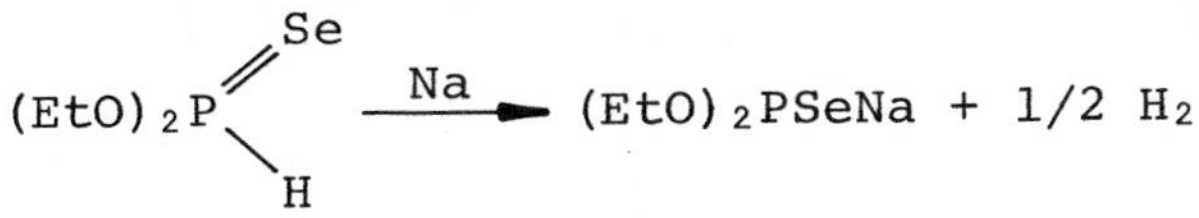

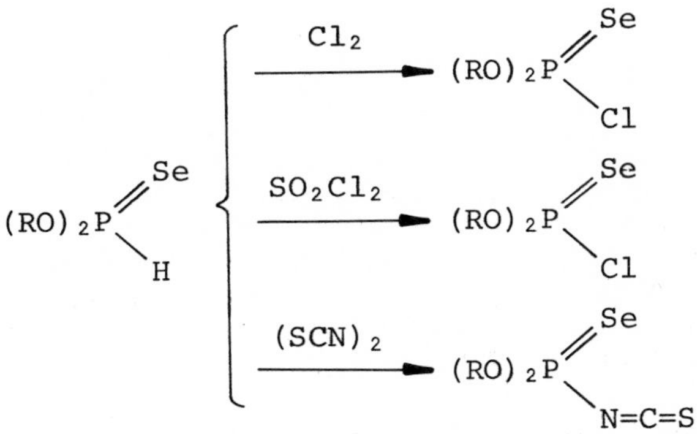

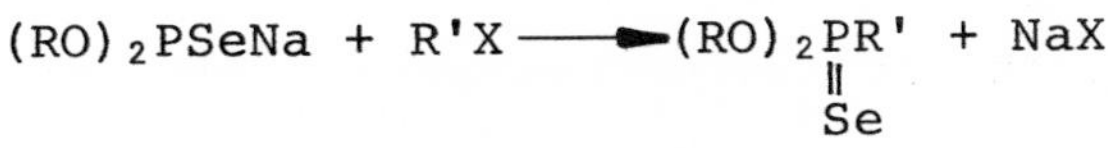

(X = halogen)

This last reaction is an example of the Michaelis-Becker reaction as shown by the sodium salts of the corresponding phosphites, and thiophosphites. In the hydrocarbon solvents, petroleum or benzene, no reaction was observed "under standard conditions"; but in dimethoxyethane, the phosphonate product was obtained in over 60% yield. Addition to carbonyl groups and certain olefins is also analogous, and is catalyzed by sodium ethoxide, which is deemed to form the "effective nucleophile" $(RO)_2P\text{-SeNa}$.

$$(EtO)_2P\overset{Se}{\underset{H}{\diagup}} \quad + \quad \overset{H}{\underset{O}{}}C\text{-}\underset{}{\bigcirc}\text{-}NO_2 \quad \xrightarrow{EtONa}$$

$$(EtO)_2P\overset{Se}{\diagup}\underset{HO}{\diagdown}CHC_6H_4NO_2\text{-}1,2$$

$$(EtO)_2P\overset{Se}{\underset{H}{\diagup}} \quad + \quad \overset{CH_2}{\underset{CH\text{-}CN}{||}} \quad \xrightarrow{EtONa} \quad (EtO)_2\underset{Se}{\overset{||}{P}}\text{-}CH_2CH_2CN$$

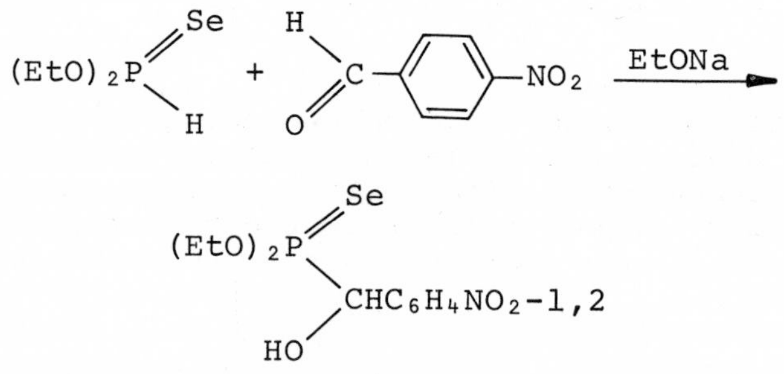

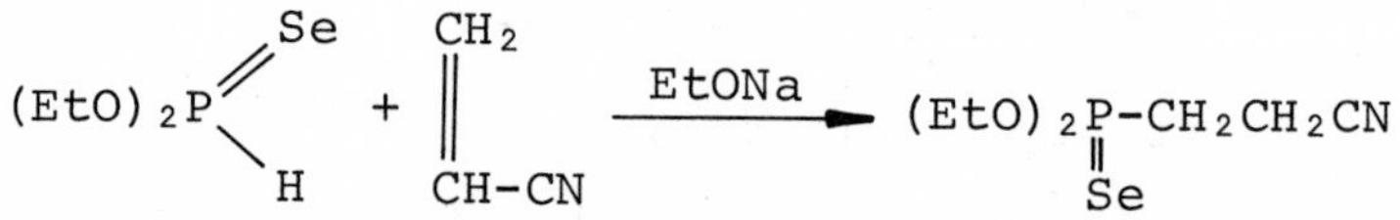

From the similarity of the chemistry of the selenophosphites with that of their oxygen and sulfur analogs, the potentiality of the selenophosphites in the synthesis of organophosphorus compounds containing selenium clearly emerges.

A detailed infrared study[744] revealed the P-H stretch at 2440 to 2390 cm^{-1}, in accordance with the P(Se)H structure. The ^{31}P NMR spectra show that the P(Se)H form accounts for practically all of the particular selenophosphite,[744] in accordance with similar conclusions for the dialkyl hydrogen phosphites,[935] and thiophosphites.[836]

Although the selenium phosphite, $(PhSe)_3P$, was deemed to be formed from phosphorus trichloride and the hydrogen selenide, it was not isolated, because of its spontaneous oxidation to the P=O compound.[1042]

$$3\ PhSeH + PCl_3 \longrightarrow (PhSe)_3P \xrightarrow{[O]} (PhSe)_3PO$$

In addition, the following system has been described:[1224]

$$i\text{-}PrOC(Se)SeK \xrightarrow{POCl_3} [i\text{-}PrOC(Se)Se]_2 + PCl_3 \longrightarrow$$

$$[i\text{-}PrOC(Se)Se]_3P$$

G. SPECTROSCOPIC, PHYSICAL, AND THERMOCHEMICAL DATA

G.1. Infrared and Raman Spectra

The IR spectra of trialkyl phosphites exhibit a moderately broad envelope near 1030 cm^{-1} which has been associated with the P-O-C linkage.[324] Although there is general agreement that a band near 1000 cm^{-1} is characteristic of the P-O-C structure, there is divergence of opinion on the assignment to P-O or O-C linkages. The following general assignments for tervalent phosphorus have been suggested:[1318] P-O-CH_3 (1015 to 1034 cm^{-1}); P-O-C_2H_5 (1008 to 1042 cm^{-1}); P-O-$CH(CH_3)_2$ (950 to 978 cm^{-1}). The peak for propyl esters is usually near 1000 cm^{-1} and for butyl is near 1025 cm^{-1}. A series of ethyl compounds $(EtO)_2PX$ all showed maxima between 1015 and 1032 cm^{-1}, for X = OEt, OC_2H_4OEt, $OC_2H_4NEt_2$, OC_2H_4SEt, F, Cl, NHR, -O-P^{III}, and -O-P^{V}. Absorption maxima between 1000 and 1190 cm^{-1} were also recorded for a range of 1,3,2-dioxaphosphorinanes. In aromatic esters the absorption shifts to somewhat lower frequency, being given as 914 to 994 cm^{-1} (100 compounds), or as 919 to 935 cm^{-1} for aryl phosphorodichloridites.[1001]

Dialkyl hydrogen phosphites have been studied by infrared and by Raman spectroscopy and exist virtually exclusively in the phosphonate form $(RO)_2PHO$. Reports of bands due to hydroxyl were due to the use of impure materials.[301,408,570,642,1242,1392] A P-H stretching vibration occurs in the region 2400 to 2450 cm^{-1} for aliphatic open chain esters[108,174,176,855,1003,1263,1264] and for cyclic esters (phospholanes and phosphorinanes, etc.).[1415] The sharpness of the band has led to the view that intermolecular hydrogen bonding is absent,[174,176,354,1263,1264] contrary to earlier observations.[108,855] This conclusion is supported by a number of studies.[174,175,237,323,408,571,807,844,890,1317]

Against a background of infrared data on a large number of organophosphorus compounds,[1002,1004] detailed analyses of the $(RO)_2P(O)H$, $(RO)_2P(O)D$ (R = Me and Et) system have been presented.[1003] Exchanging H by D has little effect on the various modes connected with the Me-O, and Et-O groups, but (P-H) (R = Me), 2438 cm^{-1}, became 1773 cm^{-1} for (P-D), the corresponding change for R = Et being 2439 to 1771 cm^{-1}, $\nu(P\text{-}H)/\nu(P\text{-}D) = 1.37$.

P-H bending vibrations (displacement of H in the O=P-H plane) have been assigned to the region 970 to 980 cm^{-1}.[1003] These are the most intense bands for dialkyl hydrogen phosphonates and have been accepted as a strong diagnostic feature,[1003] in contrast to earlier opinion[287] which considered P-H deformations to be difficult to detect in the presence of the stronger P-O vibrations. P-H deformations perpendicular to the O=P-H plane are thought to give rise to absorption at 1031 to 1150 cm^{-1}.[287]

P=O stretching vibrations are a strong feature of the hydrogen phosphonates and occur at 1250 to 1266 cm^{-1} for dialkyl esters,[324,1003,1318] at 1275 to 1280 cm^{-1} for mixed alkyl aryl esters,[1318] and at 1282 cm^{-1} for diaryl esters.[1318] Solutions of dialkyl phosphonates in phenol or methanol show P=O shifts to lower frequency of about 34 and 10 cm^{-1}, respectively, because of hydrogen bonding. A series of cyclic esters (phospholanes, phosphorinanes, etc.) showed P=O stretching absorptions in the range 1270 to 1295 cm^{-1} for liquid films or Nujol mulls; the increase of frequency observed in carbon tetrachloride or carbon disulfide and the slight decrease (usually) in chloroform were interpreted in terms of intermolecular association, not involving the P-H bond, in the neat esters.[1415] Monoalkyl phosphites RO(HO)P(O)H absorb at 1200 to 1212 cm^{-1}.[1318] Compounds containing P-NH are susceptible to a reduction in the P=O frequency because of hydrogen bonding.[1318]

The P-O-C skeleton of the hydrogen phosphonates gives rise to complex vibrations; a broad envelope near 1050 cm^{-1} was attributed to this system.[324] There has been some disagreement over assignment to the P-O and O-C bonds.[1318,1319] A band near 795 cm^{-1} has been attributed to a P-O stretching mode and bands in the 900 to 1200 cm^{-1} region to skeletal vibrations associated with the C-O bonds.[855] In connection with studies on metal derivatives of the dialkyl hydrogen phosphites, frequencies in the parent esters $(RO)_2PHO$ were given as 976 ± 2 cm^{-1} (C-O symmetric stretch), 1109 to 1033 cm^{-1} (C-O asymmetric stretch), 725 to 745 cm^{-1} (P-O symmetric stretch), and 773 to 805 cm^{-1} (P-O asymmetric stretch).[241] Assignments in cellulose phosphites have been made.[435-437] Other studies have been reported.[420,568,950,1271,1393]

G.2. Ultraviolet Spectroscopy

Absorption maxima at about 260 nm have been reported for di- and trialkyl phosphites.[328a,893a,1262,1263] It has been clearly shown, however, that pure triethyl phosphite is transparent in this region and that absorption could be due to traces of diethylaniline used in the preparation (Method Ib).[524a] Absorption due to traces of pyridine has also been demonstrated.[893b,896a] Vapor phase UV studies on trimethyl phosphite reveal an intense absorption at 189 nm, which is attributed to an $n \rightarrow \delta^*$ transition of phosphorus.[524b]

G.3. NMR Spectra

^{31}P NMR spectroscopy was recently discussed in detail and a list of NMR data then available was given for all types of organophosphorus compounds, including phosphites and their derivatives.[321] The NMR spectra of nuclei other than ^{31}P in phosphorus-containing compounds have also been discussed and references to ^{31}P NMR spectra appended.[836]

The ^{31}P chemical shift for simple trialkyl phosphites is approximately -140 ppm relative to 85% phosphoric acid and frequently gives rise to a poorly resolved multiplet resulting from long-range coupling between phosphorus and the alkyl group (J_{PH} ca. 5 to 10 Hz). For triaryl phosphites the chemical shift is at slightly higher field, in the region -125 to -130 ppm. Esters having phosphorus in a monocyclic system (derivative of glycols, etc.) usually show chemical shifts between -125 and -135 ppm while the "cage-structure" of bicyclic esters (derivative of, e.g., trihydroxymethylmethane) gives rise to values around -90 ppm. Attachment of one or more atoms of chlorine or bromine to phosphorus results in a marked downfield shift. The dialkyl phosphorochloridites thus have chemical shifts in the range -160 to -170 ppm and the alkyl or aryl phosphorodichloridites in the range -170 to -180 ppm. Chemical shifts of ca. -190 to -200 ppm are reported for PhOPBrCl and $PhOPBr_2$. With fluorine there is a change to higher field, the difluoridites having chemical shifts at ca. -110 ppm and coupling constants J_{PF} of about 1300 Hz.[321]

Trialkyl trithiophosphites $(RS)_3P$ and tris(dimethylamino)phosphines $(R_2N)_3P$ have chemical shifts which are not substantially different from those of the trialkyl phosphites and lie in the region -120 to -130 ppm. Again, the signal for the corresponding chloridites and bromidites is shifted to lower field (-180 to -210 ppm for the thio compounds and -160 to -170 ppm for the amino analogs). The effect of fluorine in the amino compounds appears to be quite different from that in the oxygen compounds, the chemical shift for bis(dimethylamino)fluorophosphine (-150 ppm) being at lower field than for tris(dimethylamino)phosphine.[321] A study has been made of the ^{31}P, ^{19}F, and ^{1}H NMR spectra of a number of fluorides of the types $(RO)_nPF_{3-n}$ and $(R_2N)_nPF_{3-n}$, the chemical shifts and coupling constants J_{PH}, J_{PF}, J_{HF} being determined. Some long range J_{HF} values for separation up to five bonds are given.[1209]

The hydrogen phosphites show clear evidence of phosphonate structure, with a ^{31}P chemical shift which is usually in the region 0 to 12 ppm, and a large coupling constant J_{PH} of ca. 700 Hz.[321] ^{1}H NMR spectra of dialkyl phosphonates are complicated by coupling from phosphorus and a doublet is observed for the phosphorus-bonded hydrogen.[396,806] Metal derivatives (Li, Na, K) of the hydrogen phosphites in aqueous solution have ^{31}P chemical shifts

(-140 to -155 ppm) consistent with their being regarded as derivatives of tervalent phosphorus, i.e., $(RO)_2P\text{-}O^-M^+$, rather than having the phosphonate structure.[913] For further work see Refs. 313, 338, 432, 628, 828, 832, 835, 916, and 1277.

G.4. Mass Spectroscopy

Ionization-dissociation schemes have been reported for dialkyl hydrogen phosphites, $(RO)_2PHO$ (R = Me, Et, Pr, i-Pr, Bu, allyl),[527] and for the triesters, $(RO)_3P$ (R = Me, Et, Pr, i-Pr, Bu, phenyl).[1006a] The trialkyl esters present some practical difficulties because of their tendency to undergo hydrolysis or oxidation and the possibility of peaks arising from impurities must be remembered. Parent-ion peaks appear in relatively high abundance for the di- and trimethyl esters, for triethyl phosphite and for triphenyl phosphite, but are weak or insignificant for other members of both series. In addition to the expected ions resulting from simple bond scissions, the fragmentation patterns of both di- and trialkyl esters are characterized by a high degree of hydrogen migration from the alkyl moiety to phosphorus. A particularly stable ion at m/e 83, attributed to $HP(OH)_3^+$, is observed in the mass spectra of all di- and trialkyl phosphites, except the methyl esters.

G.5. Miscellaneous Physical Properties

A review has been made of the parachor, viscosity, molecular weight, boiling point, dipole moment, molecular refraction, and magnetic rotation data for di- and trialkyl phosphites, with particular reference to the structure of dialkyl hydrogen phosphites.[354] A review of the atomic refraction of phosphorus in esters and chloridites reflects the earlier interests of the Russian workers,[629,630] and further data have been given on the dipole moments of trialkyl phosphites.[223] Magnetic susceptibilities have been measured for about twenty compounds P(XYZ) in which X = SR, R, Cl, OR, and the magnetizations of the P-C, P-O, P-S, and P-Cl bonds determined for symmetrical molecules PX_3.[1032] Reference has also been made to the electron-donor capacity of trialkyl phosphites (by infrared),[371] the molecular refraction of dialkyl phosphites,[1152] and the magnetooptics of trialkyl thiophosphites.[1354]

G.6. Thermochemistry

Owing to the difficulty of ensuring complete combustion in bomb calorimetry, reaction calorimetry had to provide most of the data on the thermochemistry of organic phosphorus compounds.[284,947]

The heats of formation $\Delta H_f°$ were computed to be $\Delta H_f°$ [P-(OMe)$_3$, liquid] = -175.0, and $\Delta H_f°$ [P(OEt)$_3$, liquid] = -204.1 kcal/mole. The mean bond dissociation energies were D(P-OR) given as 79 to 82 kcal/mole. From the heats of oxidation of triisopropyl phosphite by aqueous hydrogen peroxide, and of diethyl hydrogen phosphite by aqueous iodine plus bicarbonate, together with certain hydrolytic data for chloridates, etc., the following estimates were made:[947] D(P-H), 78 kcal/mole and D(P=O), 134.1 kcal/mole. A value exceeding 140 kcal/mole for D(P=O) was obtained, from the oxidation of triethyl phosphite by hydrogen peroxide in ethanol (see Ref. 947 for citation). The following estimates of bond dissociation energies quoted in Ref. 947 may be given here for reference. Great caution should always be exercised in considering thermochemical data on such systems.

D(P - F), 117; D(P - Cl), 76.3; D(P - Br), 61.7;

D(P - H), 76.3 kcal/mole

Bomb calorimetry has been used for trialkyl phosphites.[974] The following standard enthalpies of formation were determined: (BuO)$_3$P(l), - 261.9 kcal/mole; (PrO)$_3$P(l), -232.9; (AmO)$_3$P(l), -290.8; ($C_6H_{13}O$)$_3$P(l), -319.8; (PhO)$_3$P(s), -75 kcal/mole.

For other data on bond-dissociation energies see Ref. 919.

Heats of solution for C_2H_2 and CO in hexamethylphosphorus triamide were found to be 6570 and 3830 ca., respectively.[1243]

ACKNOWLEDGEMENT

The authors are grateful to D. R. Hepburn, V. K. Maladkar, and J. E. Weekes for their invaluable assistance in checking the references and the list of compounds.

H. TECHNOLOGICAL ASPECTS*

(MeO)$_3$P. Interacts with maelic anhydride in the presence of acetic acid.[129] For production of pesticidal compositions,[91,118] and for dehalogenating such compounds as trans-BzCHBrCHBrBz.[136,141] Promotes the photochemical reaction of mono- and diolefins with hydrogen sulfide.[362]

(EtO)$_3$P, (ClCH$_2$CH$_2$O)$_3$P. Used to produce stabilizers

*References are given in a separate list. See Technical References.

for poly(vinyl chloride).[265]

$(EtO)_3P$. Reacts with iodomethylated polystyrene to give a solid linear phosphorus-containing polymer.[502] Used in the polymerization of olefinic compounds by peroxide catalysts,[128] and in the preparation of phosphorus-containing polyurethanes and polyols.[599]

$(RO)_3P$. To esterify organophosphinic acids.[133]

$(RO)_3P$, e.g., $(EtO)_3P$. Used to prepare parasiticidal, insecticidal, and surface active agents.[151,153,465,483,523]

$(RO)_3P$-CuCN complexes. As bactericides.[435]

$(RO)_3P$, Me, Et, Et_2 tetrahydrofurfuryl, Et ethylene. Give pesticides with $X_2HCCOCH_2CONR_2$, X = halogen and R = H or alkyl or aryl.[48]

$(RO)_3P$. Used to prepare insecticides,[359,408] and drug intermediates.[424,594]

$(RO)_3P$, R = Et, Bu, $(RO)_2PONa$. Give pesticides with perchloromethyl mercaptan.[51,102]

$(RO)_3P$. With alkylxanthates give biological toxicants.[53]

Oleyl methyl phosphite. Reacts with sulfur to give an insecticide.[522]

$(Br\text{-}PrO)_2(ClBr\text{-}PrO)P$. For production of flame-resistant polystyrene sheet.[8]

$(RO)_3P$, Et, Bu. For flame-resistant vinyl polymers having good low temperature flexibility and plasticizer retention.[25]

$(ClCH_2CH_2O)_3P$, $(ClCH_2CHClCH_2O)_3P$. Form high molecular weight products at 180-300°, usable for reducing flammability of polymers.[26,27]

$(RO)_3P$. For polyurethane foams.[123]

Triethylene phosphite. Used to prepare intermediates for synthetic resins, especially polyurethanes.[531]

Phosphites produced from a trialkyl phosphite and a glycol, $HO(CH_2)_nOH$ (n=2-10). Are claimed to have general application as antistatic and fireproofing agents, stabilizers, and plasticizers for polymerization and other systems.[564]

Several phosphites. Were prepared to study their synergistic action with the antioxidant.[352]

$(ClCH_2CH_2O)_3P$. Used to prepare such as 1-[bis(2-chloroethoxy)phosphinyl]ethyl bis(2-chloroethyl)phosphite for oxidation with ozone to give phosphates usable as additives for lubricants and gasoline.[52]

$(RO)_3P$. Used to prepare oil additives, hydraulic fluids, and plasticizers.[615] Recommended for removal of sulfur from petroleum.[22]

$(RO)_3P$, R = Me, Et, Pr, i-Pr, Bu, ethylhexyl, nonyl, dodecyl, mixed esters, and $(RO)_2PONa$. Used for interaction with a polyhalogenated linear hydrocarbon.[407]

$(RO)_3P$, R = Me, Et, Ph. Stabilize mono- and dichlorobenzene used as working fluids.[60]

$(RO)_3P$. Used for preparation of insecticides, and plasticizers for polymers and lubricants.[130]

$(RO)_3P$, P = Me, Et, Bu. With an acid anhydride, e.g., Ac_2O, in the presence of a trace of boron trifluoride (ether complex) afford insecticides for flypaper and sprays.[101]

Tris(1-ethynylcyclohexyl) phosphite. Named as a parasiticide, plant-growth control material, and corrosion inhibitor.[364]

$(C_{18}H_{37}O)_3P$. Used to prepare polymer stabilizers.[563]

$(RO)_3P$ (alkyl, aryl or arylalkyl). Implement pigmentation of caprolactam by titanium dioxide.[207]

A phosphite from propylene oxide and PBr_3. Used in flame proofing of synthetic resins.[419]

$(RO)_3P$, R = Me to octyl. Produces polybutadiene having a high proportion of cis-1,4 structure.[30]

$(RO)_3P$. Used in the stabilization of aliphatic, cycloaliphatic, and ether alcohols required as humectants, solvents, plasticizers, elastomers, and components of polyurethanes and epoxy polymers.[160]

May be used in conjunction with arylsulfenyl chlorides for the preparation of arylthio esters of acylamino acids for peptide formation.[401]

Nonvolatile phosphites. Used in connection with furfuryl alcohol binders for foundry cores.[155]

$(RO)_3P$. Reduce polyhalocyclopentadienes.[67]

Phosphorous acid esters of the anilide and the 2-naphthylamide of 2-hydroxy-3-naphthoic acid. Are dyes for wool and cotton fibers.[506]

Phosphites from ether-alcohols. Used to prepare dyes containing phosphorus.[576,577]

$(RO)_3P$ R = Me, Et, i-Pr, $ClCH_2CH_2$, $CH_2{:}CHCH_2$. Used to prepare chlorocarbon, $C_{10}Cl_8$, usable as a deep blue dye or pigment, or, after treatment, as a pesticide.[373]

$[(BrCH_2)_2C(CH_2Cl)CH_2O]_3P$. Used in production of flameproof plastics and cellulose materials.[79]

$(CH_2XCHXCH_2O)_3P$, X = halogen. Are named as plasticizers and flammability reducers for plastics.[524]

Phosphites obtained from trimethylolpropane or pentaerythritol or similar compounds. Omnibus usefulness as stabilizers, antioxidants, fire retardants for polyolefins, poly(vinyl chloride), polyurethanes, is claimed.[34,35,517]

$(RO)_3P$, $(ArO)_3P$. Used as sensitizers for photographic emulsions.[589]

Tris(2,4-dichlorophenoxyethyl) phosphite. Is named for weed control for strawberry plantings.[229]

$C_{24}H_{15}Cl_{18}PO_3$. Used as an additive to a transaxle fluid (ATF).[525]

Polyphosphites. Used to produce extractants and ion-exchange resins.[468]

Poly(amidoethyl) phosphites. Used for flame proofing phenolic resins of the resol type.[181]

Cyclic phosphites. Used in production of high molecular weight foams.[146]

Complex phosphite-phosphonate esters. "Negative" substituents in β-position appear necessary for cleavage desired for production of preignition inhibiting agents.[54]

$[(EtO)_2POCH_2CH_2]_2S$. Used as precursor to insecticides.[483]

$(PhO)_3P$. Has influence on the catalytic addition of HCN to nonactivated olefins.[76] Mentioned in a short review on industrial applications of phosphorus compounds.[203]

$(PhO)_3P$. When complexed with nickel in the presence of a suitable organic halide is a very active initiator of free-radical polymerization.[31] Named as an extremely good inhibitor of chicken liver fluoroacetanilidase.[417]

$(PhO)_3P$, $(CH_3C_6H_4O)_3P$, polyphosphites. Inhibit color formation during certain esterification processes.[567]

$(PhO)_3P$. Is a constituent of a heat-stabilizing mixture for polymers.[150,578]

$(ArO)_3P$. Used to prepare nickel complexes for the dimerization of dienes.[23] Used for esterifying natural amino acids such as hippuric acid.[539]

1,2-$C_6H_4O_2POAr$ (Ar = α-Naphthyl, β-Naphthyl, or 2,6-di-t-Bu-4-Me-C_6H_2). Prepared for use as "acid stabilizers" of polymer materials.[198]

Tri(mixed mono- and dinonylphenyl) phosphite, 2,2'-di-t-Bu-4, 4'-isopropylidenediphenol bis(p-nonylphenyl) phosphite. Are permissible as antioxidants and (or) stabilizers for food-contact materials, e.g., rubber articles for repeated use, and for poly(vinylchloride) resins used for manufacture of rigid bottles.[13-15]

$(MeO)_2PHO$. Reacts with imidazolines oxazolines to afford compositions useful as additives for antiicing and antiknock gasoline fuels and for pesticides, plastics, and asphalt.[124,125] Used in the oxidation of hydrocarbons.[466] Gives lignin phosphite with lignin.[634] For preparation of ester stabilizers and fuel additives (to reduce autoignition in leaded fuels).[80,134] Referred to in catalytic (BF_3) isomerization.[41] Gives fire resistance to polymers containing nitrile groups.[510a] Improves thermal stability of nylon fibers.[112]

$(EtO)_2PHO$. Reacts with $(MeO)_2CHNMe_2$ to give a phosphonate active in the synthesis of carboxylic acids.[191]

$(EtO)_2PHO$. Used in preparation of fire-resisting polysiloxanes, and foams.[441,557] Used in conjunction with compounds containing C:C and C:N bonds.[327] Used to prepare polymer intermediates and textile-treating agents.[478] Reacts with a thioether, R (or Ar)-SCH_2Cl to give an insecticide.[535] Used in the purification of dialkyl terephthalates.[272]

$(BuO)_2PHO$. Used in production of wetting agents, softeners, adhesives, corrosion inhibitors, flame-proofing agents, etc.[38] Used to prepare selective herbicides.[114]

"Partial" phosphites from pentaerythritol. Are functional in the formation of flame-resistant spirobi(m-dioxane) resins.[192-195]

$(RO)_2PHO$. Used to prepare phosphonates containing fluorine,[175] and to prepare diluents for epoxy resins.[499]

$(RO)_2PHO$, R = Am, octyl, dodecyl, hexadecyl. For production of corrosion inhibitors.[210]

$(RO)_2PHO$ [and $(RO)_3P$]. Used in the synthesis of metal-complexing polymers containing phosphorus.[293,294,330]

$(RO)_2PHO$, R = Et, Bu. For preparation of surface active agents.[230]

$(RO)_2PHO$. Give chloridates for phosphate production.[273] Interaction with borazole has been examined in connection with formation of applicable inorganic polymers.[178]

$(RO)_2PHO$. Used to produce pesticides claimed to have negligible toxicity to warm-blooded animals. Even therapeutic properties are claimed.[585] Used to prepare insecticides[103] and hydrocarbon oil additives.[237] To prepare diphosphonates as plasticizers.[326]

$(RO)_2PHO$. Are converted into textile aids, pesticides, plant protectives, and material for flame-proof plastics by condensation with aldehydes and ketones in systems involving also acetic acid and diketene.[437] Condensed with thiourea to give lubricants, pesticides, and pharmaceuticals.[532] To prepare oil additives, etc.[237] Used to condense with a 1-oxophosphonic acid ester.[533] Give phosphonylated formals when heated with orthoformates.[505]

$(ClCH_2CH_2O)_2PHO$. Produced a considerable number of mutations for eggs and larvae of insects.[503] Used to impregnate wood for after-treatment with ammonia to give fire-retarding properties.[185]

$(MeO)_3P$, $(RO)_2PHO$. Used to impregnate wood for similar purpose.[607]

$(ClCH_2CH_2O)(MeO)PHO$, $(C_6H_{11}O)(MeO)PHO$, $(Cl_3CCH_2O)(MeO)PHO$. For use in preparing pesticides.[360]

$(ArO)_2PHO$. Used in production of curing agents for melamine-formaldehyde and urea formaldehyde resins.[78]

Trialkyl phosphite-nickel compounds. Relate to the decomposition of nickel coatings.[276]

$(EtO)_3P$ and 2-hydroxyethyl ethylene phosphite. Are named as constituents of metal cleaning compositions.[304]

A mixture of dimethyl hydrogen, or bis(2-ethylhexyl)

hydrogen phosphite, and $(EtO)_2PHO$-EtOPHO(OH). Used in conjunction with zinc oxide for coating metals with what is believed to be an amorphous polymer $[-ZnOP(HO)O-]_n$.[501]

Dialkyl hydrogen, and trialkyl phosphites. Are constituents of surface coating removers.[184]

"Ester amide" of phosphorous acid. Mentioned with reference to reduction of adhesion of paraffin and other deposits in pipelines, valves, and tanks.[436]

$(PhO)_3P$, $(EtO)_3P$, $(EtO)_2PHO$. Are constituents of a production mixture for organosulfate surfactants.[439]

Reaction products of phosphorus trichloride and ethylene oxide. Are named as low-foaming surfactants.[418]

Polycondensates from phosphorus trichloride and such as hexamethylene diamine. May be applicable as ion exchangers.[248]

Tris(1-butoxyethoxy-2-propyl), tris(2-hexoxyethyl), tris(2-butoxyethyl), and tris(ethoxyethyl) phosphites. Are stated to produce high-gloss and stable emulsion polishes based on natural or synthetic waxes. Leveling was good, and a uniform coating was obtained.[36]

A bicyclic phosphite $EtC(CH_2O)_3P$, in conjunction with a cobalt compound. Is a catalyst for the production of aldehydes from olefins and carbon monoxide at high temperatures and pressures.[457]

$(RO)_3P$. Are used in the preparation of alkylaminofluorenes.[458]

$(RO)_3P$, R = Me, i-Pr. Are used to produce desulfurized α-6-deoxytetracycline from the mercapto compound.[477]

Phosphite links. The insolubilization of cellulose ethers and esters by phosphorus trichloride probably involves phosphite links.[406]

PCl_3 gives "esters." Mentioned in connection with the treatment of a mineral oil fraction (b. 200-350°).[74]

Phosphorus compounds obtained by the interaction of white phosphorus and phenols or polyhydric alcohols. An omnibus claim of applicability is made.[292]

Triallyl phosphite. The copolymerization with sulfur dioxide, gives phosphorus containing polysulfones.[323]

Phosphites. Are involved in the phosphorylation of cellulose.[474,476]

$(RO)_3P$. Stabilize aromatic carbocyclic amines.[249]

Phosphites. Those volatile enough to be removable at the end are used as esterification catalysts for the higher fatty or rosin acids by mono- or polyhydric alcohols.[209]

Aryl, haloaryl, and nitroaryl phosphites. Form aqueous emulsions with hydrocarbon or vegetable oil solvent which are stable if pH is adjusted to about 8.[220]

$(RO)_3P$, R = Me, Et, Pr. Are used to modify organopolysiloxanes.[271]

$(RO)_3P$. Are used to prepare dialkoxyphosphorylarsines from the thiocyanates of secondary arsines.[305] Are probably involved in the use of PCl_3 as a latent curing agent for polymeric compositions.[308]

"Esters of phosphorous acid." Are solvents for nonaqueous electrolytes used in electrolytic capacitors.[516]

$(RO)_3P$, $(ArO)_3P$, e.g., diphenyl isodecyl phosphite. Are used as sensitizers for photographic emulsions.[588]

$(EtO)_3P$, $(BuOCH_2CH_2O)_3P$, and $(CH_2{:}CHCH_2O)_3P$. Are oxidized by air or oxygen in the presence of a metal oxide catalyst (e.g., Al_2O_3) to give phosphates for use as plasticizers, and as additives for gasoline, lubricant, and functional liquids.[242]

2,2-Di-Me-1,3-propanediol cyclic hydrogen phosphite and the phosphite. Are named as flame-proofing plasticizers for cellulose acetate, as stabilizers, lubricants, or pesticides.[379]

Hydroxylactone phosphites. Were prepared for use as inhibitors in gastric secretion.[281] Nondialyzable alkali-soluble derivatives of hydroxysteroids, which have prolonged hormonal function, have high molecular weights and can contain phosphite, O-P-O, groups in place of phosphate or thiophosphate groups.[4]

"A reaction product of acetyl chloride and phosphorous acid or a derivative." Present in permanent waving and wool creasing compositions.[49]

$(EtO)_2PCl$. Used for preparation of insecticides.[135,534] Used for forming a P-N link with L-arginine for work on hypertensive decapeptide.[538]

$(RO)_2PCl$, obtained from the pentaerythritol-PCl_3 system. Mentioned as intumescent agent, a gasoline additive, heat and light stabilizer for vinyl and vinylidene resins, a flame-retardant, antioxidant and chemical intermediate.[173]

$ROPCl_2$. Used to prepare fungicides.[58]

H.1. Stabilizers for Poly(vinyl Chloride)

Bridged-ring phosphites such as trimethylolethane, -propane, -butane, or pentaerythritol phosphites. Stabilize resins against heat degradation.[1]

Such phosphites as trioxetyl phosphite, prepared from triphenyl phosphite and hydroxyoxetane, $\overline{CH_2CH(OH)CH_2O}$. Are stated to be especially useful.[16]

Tridodecyl phosphite and a phenol. Are likewise used.[17,20]

Several phosphites. Were prepared for use in stabilization, and also for chlorinated paraffins or rubbers without affecting their mechanical properties.[40]

Phosphites. Are used to stabilize rigid or semirigid chlorine containing vinyl compositions against light.[77]

FadeOmeter data for 192 hr, and heat stability data are given for the many phosphites.[111]

Phosphites. Are used as chelating agents to improve thermal stability.[169,224]

Phosphites prepared from a polyhydric alcohol, such as pentaerythritol, sorbitol, mannitol, trimethylolethane, and trimethylolpropane, or a mixture of these; diphenyl decyl phosphite, and hydrogen phosphites. Are used as stabilizers.[83]

Phosphites. Give with vinyl chloride copolymers having stabilizing function.[206]

$(PhO)_2POCH_2CHEtBu$ and $PhOP(OCH_2CHEtBu)_2$. Are more effective than trialkyl or triaryl phosphites as stabilizers at temperatures up to 340°F.[208]

Phosphites and cyclic phosphites, e.g., $Ph\overline{OPOCH_2CH_2O}$ and $(Et\text{-}3\text{-oxetanylidene-}CH_2O)_3$. Are useful as antioxidants and plasticizers for vinyl chloride polymers.[216,221]

Polyphosphites of the formula $C[CH_2O(C_3H_6O)_nP(OR)_2]_4$, where R = isodecyl, nonylphenyl, Ph, or Cl_5C_6, which can vary within the molecule. Have good thermal and hydrolytic stability and are usable as stabilizers for poly(vinyl chloride) resins and polyolefins.[343,344,346]

Phosphite from 4,4'-isopropylidenedicyclohexanol and triphenyl phosphite in the presence of catalytic diphenyl hydrogen phosphite. Poly(vinyl chloride) bottles, suitable as food containers, are made by using a phosphite.[356,357]

Phosphites such as triisoctyl phosphite. Are used for color-stable vinyl resin solutions.[253,254,278,347,348,515]

Mixed alkylaryl phosphites. Are superior in activity to either aliphatic or aromatic phosphites.[400]

Esters of phosphorous acid. Are mentioned for improvement of mechanical properties.[414]

Polyphosphites based on erythritol, $Y(OCH_2)_2C(CH_2O)_2Y'$, Y and Y' are $[P(OR)_2]_2$ or POR groups. As stabilizers, were compared with $(PhO)_3P$ and $(OctO)P(OPh)_2$.[461]

Many other phosphites. Are considered useful as stabilizers.[189,232,250,358,367-369,399,453,480,494,497,540,555,556,583]

$(RO)_2PHO$. The examination of the polymerization of vinyl chloride in various solvents showed that of 30 compounds listed only dialkyl phosphites had the effect of producing marked crystallinity and high syndiotacticity. The molecular weights of the polymers were low, presumably due to the formation of a dialkyl phosphonate with the polymer chain.[572,573]

$(EtO)_3P$, $(PhO)_3P$, $(PhO)_2POOct$, $(C_9H_{19}O)_3P$. Phosphorylated poly(vinyl chloride) resulted from the formation of phosphonate units, $-CH_2C[P(OR)_2O]H$.[635]

$(EtO)_3P$. Interaction with vinyl 9-chlorostearate, appeared to be "quite complex"; one reaction appeared to involve addition, giving $Me(CH_2)_5CH(Cl)(CH_2)_7-CO_2CH_2CH_2P(O)(OEt)_2$ and ethylene.[355]

$PhOP(OCH_2)_2C(CH_2O)_2POPh$ (A). Used for preparation of stabilizers,[219] as is also the following.

$PhOP(OCH_2)_2C[CH_2OP(OPh)_2]_2$.[219] Compound (A) and 1,4-dichlorobutane in the presence of aluminum chloride gives a soft rubbery gel, which burns in a flame, but is self-extinguishing.[42] Compound (A) and a tetraphosphite, $[(PhO)_2POCH_2]_4C$, are named as stabilizers for vinyl chloride resins, as well as antipreignition gasoline additives, and antioxidants for lubricating oils and for natural and synthetic rubber.[218]

$RCO_2MOP(OR^1)(OR^2)$ in which M is Ba, Sr, or Ca. Are named as stabilizers. Five examples were given.[277]

H.2. Other Vinyl Systems

The phosphites of poly(vinyl alcohol). Are made from phosphorous acid and the alcohol.[426]

$(RO)_2PHO$. Are used for the phosphorylation of poly(vinyl alcohol) to give flame resistance.[449]

Phosphites. Are involved as catalysts and plasticizers for vinylcarbazole polymerization.[566]

$(ClCH_2CH_2O)_3P$. Reacts with diacetoxysilane and poly(vinyl alcohol) to give products which do not support combustion and smoldering, are water repellent, and rot resistant in boiling water.[603]

$(RO)_3P$ + Group IIIA metal halide. Are catalysts for crystalline poly(vinyl ethers).[626]

$(OctO)_2PHO$. Named as a constituent of a copolymer of ethylene and vinyl acetate for interlayer safety glass.[319]

Bis(2-ethylhexyl) hydrogen phosphite. Used as a suspending agent in the suspension polymerization.[288]

Tris(nonylphenyl) phosphite. Used to plasticize poly(vinyl carbazoles).[107]

$(RO)_3P$. Named for production of cation-exchange resins involving chloromethylated vinyl aromatic copolymers.[385] Stabilizer and antioxidant toxicity has been reviewed.[586] For vinyl resins.[252]

Tetraphosphites. Are useful as flame retardants and stabilizers for vinyl halide resins, monoolefin polymers, natural and synthetic rubbers, polyurethanes, and polyesters.[497]

H.3. Polyester, Polyether, and Related Systems

$(ArO)_3P$. Polymerization is effected by process A.5

involving a polyphenol. The polymers can be cured by heating with paraformaldehyde. A rubbery gel, stable at 250°, was obtained from triphenyl phosphite, hydroquinone, and sodium phenate at 185°. A laminate from impregnated asbestos could be fabricated.[6]

$(PhO)_3P$. Used with pentaerythritol to effect transesterification polymerization. The physical properties of the polymers vary from brittle solids to viscous liquids. Cross linking can occur by the cleavage of the spiro pentaerythritol phosphite. Hydrolytic instability limits applicability.[157]

An alkyl phosphite. Fiber-forming poly(ethylene terephthalate) is produced by the condensation of di-Me terephthalate with ethylene glycol in the presence of an alkyl phosphite.[395]

$(EtO)_2PHO$. An ether-ester type rubber-like polymer was obtained by the transesterification process, i.e., involving bis(β-hydroxyethyl) polysulfide.[61] Treating the resulting oligomer with a diisocyanate produced phosphorus and sulfur-containing polyurethanes.[62]

$(MeO)_2PHO$, $(EtO)_2PHO$. Polyesters are prepared by the transesterification, i.e., with glycol such as $(CH_2OH)_2$.[94]

$(C_3H_5O)_3P$. Improves cross-linked acrylic acid copolymers.[605,606] Are used as constituents in a polyether resin mixture for curing at 150° for laminations, surface coatings, adhesives, etc. $R^1OP(OR^2)(OR^3)$ in which R^1, R^2, R^3 are like or unlike, alkyl aralkyl, aryl, cycloalkyl or alkenyl groups, having 1 to 18 carbon atoms. Preference is for aryl groups such as Ph or C_6H_4Me. The operation mixture has low viscosity and can be poured into moulds for conversion.[84]

$(RO)_3P$, $(RO)_2PHO$. Undergo polycondensation with poly(alkylene oxides) or with polyepoxy compounds to give products of significance in photographic technology.[179]

$(RO)_3P$. Are used to form a stable 1-package glycidyl polyether curing system.[545]

$(RO)_3P$, $(ArO)_3P$. Are used in the preparation of unsaturated polyester resins.[604]

$(PhO)_3P$. Is, in the presence of toluenesulfonic acid, a catalyst for the production of polyesters from adipic acid and trimethylolethane.[460]

Mixed esters such as didodecyl phenyl phosphite. Are used as plasticizers in the incorporation of titanium dioxide in the spinning of fiber forming polyesters.[504]

$(C_8H_{17}O)_3P$, $(C_6H_{11}O)_3P$, $(PhO)_3P$. Used in the catalytic polymerization of unsaturated aldehydes, especially acrolein.[553,554]

$(EtO)_2PHO$, Et_2 ethylene pyrophosphite. Used for synthesis of oligoamides.[537]

$(CH_2{:}CHCH_2OCH_2)_2CHOPH(O)(OEt)$. Prepared in connection with the synthesis of phosphorus-containing oligomers having allyl groups.[303]

$(ClCH_2CH_2O)_2PHO$, $(FCH_2CH_2O)_2PHO$, $(PhO)_2PHO$. Are used to form the phosphites (RO)(R'O)PHO in which R' has a cellulose structure.[475]

$(EtO)_2PHO$. Used to produce polyphosphites for reaction with disulfides, sulfenyl chlorides, sulfur, and nitrogen oxides.[470]

Polyphosphites. Were obtained from the following combinations: α-Me-glucoside--$PrOP(NEt_2)_2$; α-Pr-glucoside--$PrOP(NEt_2)_2$; α-Pr-glucoside--$C_8H_{17}OP(NEt_2)_2$; α-Me-glucoside--$(PhO)_2PHO$; Ph-glucoside--$PrOP(NEt_2)_2$; Ph-glucoside--$BuOP(NEt_2)_2$; Ph-glucoside--$(PhO)_2PHO$. Oxidation with nitrogen dioxide gives polyphosphates, e.g. poly(α-Me-glucopropyl phosphate).[472] The system α-Me-glucose--$EtOP(NEt_2)_2$ and addition of sulfur to give polyphosphorothiolates are also described.[473]

$(EtO)_2PHO$, $(PhO)_3P$. Used for the phosphorylation of cellulose to give a flame-resistant material which can be used as an ion-exchange resin.[467]

$(MeO)_2PHO$, MeOPHO(OH). The effect of solvents on the phosphorylation of viscose cellulose fabric has been investigated.[486]

$(MeO)_2PHO$. Transesterification and other aspects have been described, and the mixed ring phosphite $(CH_2)_nO_2PNMe_2$ has been used for phosphorylation of cellulose.[484,485,487,488]

$(EtO)_2PHO$. Linear poly(alkylene phosphites) are prepared by transesterification. The molecular weight of the polymers based on the named diol were: 1,5-pentanediol, 80 000; 1,6-hexanediol, 50 000; diethylene glycol, 67 000; triethylene glycol, 50 000; diethanolamine, 150 000; p-bis(hydroxymethyl)benzene, -; dianhydrosorbitol, 70 000. Pentafluoropentamethylene 1,5-glycol and bis(2-chloro-ethyl) hydrogen phosphite gave a polymer having a molecular weight of 143 000. Oxidation with nitrogen dioxide gave an elastic polymer and addition of sulfur gave a spongy, rubbery mass.[469] Poly(α,ω-alkylene phosphites) appear to be analogously produced.[471]

$(MeO)_3P$, $(MeO)_2PHO$, $(ClCH_2CH_2O)_2PHO$, $(PhO)_3P$. Are transesterification reagents to phosphorylize novolak resin. The aminophosphines, $(Me_2N)_3P$, $(Et_2N)_3P$, and $Me_2N\underline{POCHMeCH_2CH_2O}$ are stated to be more efficient than the phosphites.[148]

$(MeO)_2PHO$. Used with ethanolamine, $HOCHR'CHRNH_2$, where R' and R can be H or Me, form phosphites which are treated with a partly condensed resin-forming aldehyde such as urea-formaldehyde to give flame-resistant resinous complexes.[95]

$C_6H_{13}OP(NEt_2)_2$. Reacted with $(H_2NCH_2CH_2S)_2$ to give a rubber-like product, giving in turn a white powder with chloroform, and then a cross-linked polymer by reaction with carbon tetrachloride. The linear and cross-linked polymers are considered as protective agents against radiation.[430]

$(RO)_3P$. Hydroxylated polyesters containing phosphorus are produced by the interaction with a polyol such as diethylene glycol, a butanediol or an amino-alcohol, $MeN(CH_2CH_2OH)_2$. They are usable as plasticizers for halogenated polymers, and as lubricants.[442]

$(ClCH_2CH_2O)_3P$. Used in connection with carbohydrates to produce the polyesters.[442]

H.4. Stabilizers for and Stabilization of Polymers

Phosphites. Color-stabilize polycarbonates.[63,170,295,309,320] Help to improve surfaces of articles made from a polyester.[106]

Polyphosphites. Are stabilizers for polyesters.[104]

$(i\text{-}OctO)_3P$. In conjunction with benzotriazole gives light resistance to a chlorinated polyester.[174]

$(EtO)_2PHO$. Interaction with (hydroxyphenyl)propane produces a linear polyester, m. 125-30°, which does not burn in an open flame. At 200° paraformaldehyde gave a brown nonflammable resin, nonfusible up to 200°, and insoluble in organic solvents.[176]

$(PhO)_2POC_{10}H_{21}$, $PhOP(OC_{10}H_{21})_2$, $(PhO)_3P$. Counteract the deletereous effects of metal oxide pigments in polycarbonates.[177]

di-Ph, ditolyl, bis(p-t-Bu-phenyl), bis(p-hexylphenyl), and bis(p-nonylphenyl). Stabilize polycarbonates.[410]

Bicyclic phosphites such as trimethylolethane and trimethylolpropane bicyclic phosphites. Form with polycarboxylic acids, polymers useful as parasiticides, antistatic and flame-proofing agents.[90]

4-Hydroxy-3,5-di-t-Bu-benzyl dimethyl phosphite. Is named for polyester stabilization.[182]

$p\text{-}RC_6H_4OPCl_2$, $(p\text{-}RC_6H_4O)_2PCl$. React with the phenolic compounds to give products useful for stabilizing polyesters, natural rubber, butyl rubber, etc.[187]

$(ClCH_2CH_2O)_3$. Increases storage life of peroxide-catalyzed unsaturated polyester compounded mixtures.[196]

$(RO)_3P$, e.g., $R = C_3H_5$. React with monoalkyl maleates, e.g., monoallyl maleate, to give polyester resins containing phosphorus.[234]

$(MeO)_3P$. Is a constituent of a mixture leading to the formation of fire-retardant polyesters containing phosphorus and halogen.[66]

$(MeO)_2PHO$. Is used in conjunction with a poly(alkylene)

maleate to produce a flame-resistant polyester resin.[109]

Esters of phosphorus acid, and preferably P-N compounds. Are used in the production of fiber and film-forming polyesters.[236]

$(RO)_2PHO$, $(PhO)_3P$. Improve weather stability of unsaturated polyester resins.[246]

$(PhO)_3P$ and a thiophosphite. Improve the color of fiber-forming polyester.[262]

$(RO)_2PHO$, e.g., $(MeO)_2PHO$. Give, with polyols and organic acids, polyesters for reaction with polyisocyanates to obtain plastic foams of low flammability.[274]

$(PhO)_3P$. Has a marked effect on the irradiation degradation of poly(methyl methacrylate); and it is named as a heat stabilizer for a poly(ester-ether).[306,307]

$(RO)_3P$, (R = Ph, Et, i-Pr, i-Bu, i-Am). Polytransesterification, i.e., involving a phosphite and 2,5-di-t-Bu-hydroquinone, or bisphenol A, avoids the residual nonremovable deleterious chlorine compounds which result when chloridites are used to prepare polyphosphites for stabilization.[280]

$(PhO)_3P$. Used in the production of poly(ethylene terephthalate) (A),[282] and in the study of the mechanism of thermostabilization of (A),[636] and for stabilizing (A).[463]

Other phosphites. Are named as stabilizers for (A).[301,321,587]

Polyphosphites, e.g., poly(diphenylolpropane) phosphites. Are named as stabilizers.[519]

H_3PO_3. Is a constituent of a polymerization mixture of ethylene glycol and terephthalic acid.[438]

$CH_2{:}C(R)C(O)OAOP(OR')(OR'')$ (A is a straight-chain, branched, or cyclic alkylene group). Used to produce hard, clear, colorless, flame-resistant polymers useful as coatings and textile finishes.[329]

Tris(nonylphenyl), tris(t-Bu-phenyl), triphenyl phosphites. Are condensed with a phenol and an aldehyde to produce stabilizers for plastics.[420]

$(PhO)_3P$. The effects on the molecular weight distribution of poly(methyl methacrylate) have been assessed.[312]

Acyl phosphites. Effect the phosphorylation of macromolecular polyols.[428]

$(MeO)_2PHO$. Interaction with hexitols produces nonflammable polymers.[429]

$(ClCH_2CH_2O)_3P$. Used with α,β-unsaturated acids or polyhydroxy compounds to produce polyesters for further function leading to flame-resisting polymers having good mechanical properties.[431-434] Used in production of flame-resistant foams.[443]

Several phosphites. Are used for the color stabilization of poly(methyl methacrylate) and copolymers.[445]
Tris(nonylphenyl) phosphite. Is used for a wide variety of polymers based on butylated hydroxytoluene and lauryl thiodipropionate.[507]
$Me_2C(CH_2O)_2POR$, in which R is alkyl, phenyl, or alkylphenyl. Are used for cellulose esters.[508,509]
$(C_{10}H_{21}O)_3P$. Is named as a stabilizer for polypropylene glycol.[490]
$(MeO)_3P$. Interaction with ethanolamine produces compounds of the type $RP(OH)[NHCH_2CH_2OP(OH)]_nR'$, in which R and R' may be $NHCH_2CH_2OH$ or $OCH_2CH_2NH_2$, to incorporate with phenolformaldehyde resins to produce fireproof melamine laminates and related products.[498]
O-(Norbornen-2-ylmethyl) phosphites. Are named for flame-retardant polyesters and polyurethanes.[500]

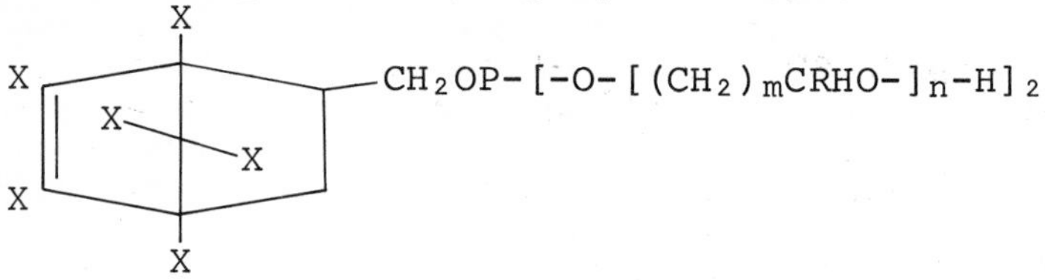

X = Cl or Br; m = 1 or 2, n = 0-20

R = H or a lower alkyl group

Polyphenyl phosphites, $(RC_6H_4O)_yP(OA)_z$. Are named as antioxidant stabilizers for polymers.[560]
$(EtO)(C_3H_5O)PHO$. Is named in the report on the kinetic study of copolymerization involving poly(1,3-butanediol fumarate).[561]
$(EtO)_3P$. Is used in the oligomerization of methyl methacrylate and acrylic acids; the presence of $(EtO)_2P(O)$- and Et groups in the oligomer was revealed by IR and molecular weight.[551]
Other phosphites. Are stabilizers for acrylic elastomers.[569]
$(PhO)_3P$, $(p\text{-}CH_3C_6H_4O)_3P$. Are named as stabilizers for acrylonitrile and acrylate esters against polymerization.[82]
$(PhO)_3P$. Is named for the vulcanization products of alkyl vinyl polyethers.[328]
$(EtO)_2PHO$, $(Et_2N)_3P$. Are used for polyacrylonitrile.[579]
$(ClCH_2CH_2O)_3P$. Used to improve the polymerization of acrylic esters especially for dental uses.[75]
$(m\text{-}NH_2C_6H_4O)_3P$. Hardens polyepoxy compounds.[9]
A considerable number of phosphites. Are described in a detailed and extensive specification.[43,285] These are for the stabilization of epoxy compounds.[20,386,387,536,614]

Phosphites. When mixed in an epoxide resin composition, are stabilized by alkaline earth metal oxides.[85]

$(MeO)_2PHO$. Used to prepare flame-retardant epoxy resins.[92]

$(RO)_3P$. Are used to increase flexural strength, decrease brittleness, lower viscosity and ensure shorter hardening times of epoxide resins.[110]

$(PhO)_3P$. Is named for the curing of polyepoxide compositions.[226,228] This and other phosphites are stabilizers.[235]

$R^1OP(OR^2)(OR^3)$, $R^1 = R^2 = Me$, or when $R^1 = Ph$, $R^2 + R^3 = CH_2CH_2$. Are used in resin compositions to decrease the viscosity of the solutions.[245]

$(PhO)_3P$, $(RO)_2PHO$. Are named for fire-proofing compositions.[247,256]

Several phosphites. Are used in the stabilization of ethylcellulose compositions against thermal degradation,[115] and to react with chlorocellulose to effect the phosphorylation of cotton cellulose.[629] They are stabilizing ingredients for ethylcellulose hot melt compositions.[311]

Phosphorus esters. Are condensation catalysts for polyether thioethers.[137]

Cyclic phosphites, $ROP(OCR'_2)_2$, in which R = alkyl or aryl, and R' is H or halogen. Used to prepare thermally stabilized copolymers of trioxane.[512]

$(RO)_3P$, $(RO')(RO)_2P$, where R is an aminoalkyl group, and R' is an alkyl or aryl group, e.g., $(Et_2NCH_2CH_2O)_3P$. Are used for formaldehyde polymers.[389]

Trioctadecyl phosphite. Is named for polyaldehydes.[464]

Tris(nonylphenyl) phosphite. Is named for the stabilization of lactone polymers.[549]

$(ArO)_3P$, e.g., $(PhO)_3P$. Used to produce noncombustible heat-hardened resins.[117]

$(PhO)_3P$. Is used to inhibit ester exchange and to control molecular-weight distribution.[544]

Novalak resin phosphites, AOPHO(OMe), and $AOP(NEt_2)_2$ (A is a novalak-resin residue), and the cyclic ethylene phosphite of novalak. Are used in the synthesis of novalak resins.[427]

$(ClCH_2CH_2O)_2PHO$ and bis[2-(chloromethyl)ethyl]hydrogen phosphite. Were produced on a large scale for use as flame-proofing hardeners for epoxy resins.[332]

H.5. Stabilization and Modification of Polyolefins and Related Polymers

$[(RO)_2PO]_2R'$. The correlation between the structure of these and the efficiency as stabilizers against thermal oxidation and photodegradation of bulk polypropylene has been described.[2]

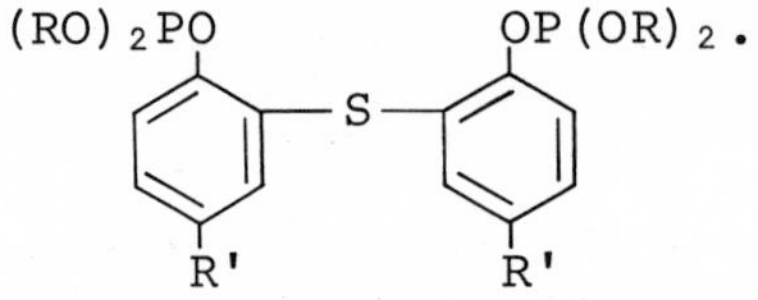

The tensile strength and elongation at break were measured before and after aging. Certain compounds allowed complete degradation, even during molding, but the compound p-$AcOC_6H_4OP(OC_6H_4CH_2CHMe_2)_2$ appeared to give maximum protective action.[2]

An organic phosphite. Is a constituent of another stabilizer system.[18,19]

ROP(OR')(OR"), where R is a C_{1-30} hydrocarbon group, R' and R" are hydrogen, a metal atom, or a C_{1-30} hydrocarbon group. Are used for improving the thermal stability of polyolefins.[24]

$(BuO)_2PHO$, (2-Et-hexylO)$_2$PHO. Used as a decolorizer in connection with the stabilization of polypropylene.[39]

Certain phosphites. Inhibit the thermoxidative degradation of polypropylene with an efficiency at concentration > 4 × 10^{-2} mole/kg better than that of 2,6-di-t-Bu-4-Me-phenol (Topanol A).[45,46,64]

Trilauryl phosphite. Is a constituent of systems leading to the production of polyolefin fibers, films, etc., stabilized to heat, light, and aging.[69]

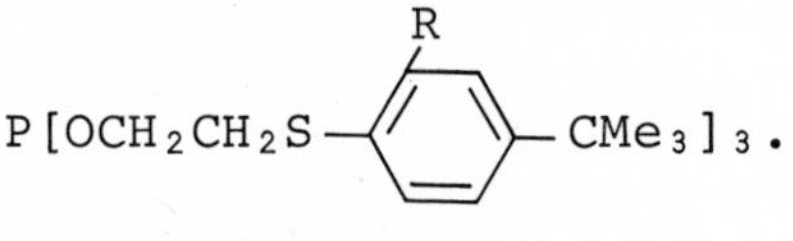

Active in stabilizing polyethylene and polypropylene at 200°.[70]

Tris(β-laurylthioethyl) phosphite. Is a constituent of a mixture leading to the production of stabilized polyethylene.[71]

Phenolic phosphites. Are used in combination with thiodipropionates, to stabilize polypropylene against light and heat.[73]

Diphenyl isoctyl phosphite, and tris(nonylphenyl) phosphite. The phosphites named are stated to have synergistic function. Free-radical oxidation processes are retarded.[89]

Tris(octylthioethyl) phosphite, and tris[bis(octylthio) isopropyl] phosphite. Are named as being used in the production of stabilized polyolefins.[143,145] Other phosphites are also claimed.[147]

[BuLi - $(BuO)_3P$] catalyst. Diene-styrene copolymers are produced at a higher rate.[152]

Phosphites. In combination with phenolic antioxidants show enhanced stabilizing properties.[156]

$(RO)_3P$, R contains C_8 to C_{22}. Are similarly used.[172]

Tris(nonylphenoxyethyl) phosphite. Used for the light stabilization of polypropylene.[211]

$(RO)_3P(RO)_2PHO$. Stabilize low-pressure polyolefins.[212-215,223,225,227,231]

$(RO)_2PHO$. Are recommended as stabilizers for antistatic polystyrene molding compositions.[97]

$(PhO)_2P(OC_{10}H_{21})$. Is a constituent in a polyolefin stabilizer.[99]

$PhOP(OC_{10}H_{21})_2$. Is likewise named.[100]

Tris(alkylaryl) phosphites. Are used in the stabilization of solid polyolefins against heat and ultraviolet light.[116]

Phosphites. Used with α-olefin polymers or copolymers to prepare fibers, films, or pressed products having resistance to degradation due to heat, light and oxygen.[238]

$(RO)_3P$, R are C_{1-8} alkyl groups. Are used in connection with metal-titanium halide catalysts for producing polypropylene of high crystallinity and d. 0.91 to 0.92.[105]

$(EtO)_3P$. Named in connection with the production of polypropylene.[628]

$(EtO)_2PHO$. In conjunction with lauryl peroxide can thermally degrade polyethylene to a hard wax which was 100% emulsifiable. Triethyl phosphite without catalyst at 380° functioned similarly. The products are useful as emulsifying agents, wax substitutes, and self-polishing coatings.[204]

$(C_{18}H_{37}O)_3P$. Is named as a stabilizer.[257]

$C_{12}H_{25}OP(OR)O(CH_2)_nO$— 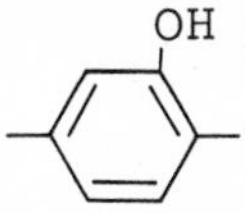 —COPh. Are prepared from trilauryl phosphite and are ultraviolet absorbers.[264]

Ring phosphites. Are named as stabilizers.[267]

An alkaryl phosphite. Is used in the modification of styrene polymers.[275]

2,4,6-tri-t-Bu-phenyl pyrocatechol phosphite. The mechanism of its antioxidative effect has been investigated with reference to isotactic polypropylene and high pressure polyethylene at 200° and 300 mm pressure of oxygen.[279]

Tris(nonylphenyl) phosphite, and a phosphite from pyrocatechol. A heat and light stabilizer for isotactic polypropylene involves the use of these.[296,297,300]

Phosphites. The use of organophosphorus compounds as stabilizers for polymers has been reviewed under 130 references.[297] The effectiveness of such stabilizers has been assessed by the interaction of α,α-diphenyl-β-picrylhydrazyl and a phosphite, the effectiveness decreasing in the order trialkylphenyl phosphites > mixed phosphites > diphosphites.[298] The rate of reaction was studied by EPR and photocolorimetry; interaction with free radicals was deemed to be the stabilizing function.[299]

$(i\text{-}OctO)_3P$. Is named as a stabilizer.[403]
$[(RO)(R'O)PO]_2Q$, where R and R' are alkyl groups, and Q is a divalent phenolic or hydrogenated phenolic residue. Are stabilizers.[339,341,342,349]
A mixture of 2,2'-methylenebis(6-t-Bu-4-Me-phenol) and tris(nonylphenyl) phosphite, or diphenyl isooctyl phosphite. Is named as a stabilizing additive.[354]
Other phosphites. Are named for antistatic compositions.[513,514]
Phosphites. Are stabilizers for polyolefins and poly(vinyl chlorides).[284,337]
Mixed allyl phosphites, prepared from $CH_2{:}CHCH_2OPCl_2$. Used in the modification of polyolefins.[324]
$(PhO)_2POCH_2C[CH_2OP(OPh)_2]_3$. Are named as color stabilizers for acrylonitrile-styrene polymers.[366]
$(BuO)_3P$. Is named in connection with the production of dyeable polypropylene.[374]
Several phosphites. Were assessed as heat stabilizers for low and high pressure polyethylene and polyethylene-polypropylene copolymer.[377]
$(C_8H_{17}O)_2PHO$. Assists in reducing blistering in an extruded film, and is a constituent in a synergistic polymer stabilizer system.[396,397]
A number of phosphites. Are stated to enhance light fastness and brilliancy of colored polypropylene fibers.[402] Some were tested as stabilizers for polypropylene and cellulose acetate butyrate plastic (etrol).[450]
Phosphites such as 2-ethylhexyl bis(octylphenyl), $(PhO)_2PHO$, bis(2-ethylphenyl)octylphenyl, bis(p-dodecylphenyl). Are named.[454,456]
ROP(OR')(OR"). Are used in conjunction with hydrocarbyl borates as stabilizers for polyolefins and other types of polymer.[421,422]
2,2'Thiobis(4-t-amylphenol)bis(dicyclohexyl) phosphite. Deemed the most effective stabilizer for polypropylene; the effectiveness of alkyl-aromatic phosphites declined with increasing alkyl chain length.[479]
Phosphites. Are used in conjunction with phenolic antioxidants.[491]
Polyphosphites. Are deemed suitable as flame-proofing agents, stabilizers, antistatic agents,[493] plasticizers, and antioxidants for polyolefins, and for isotatic polypropylene.[459]
Some phosphites. Are named for poly-α-olefins for electrical insulation.[587]
$[(Ar^2O)_2POAr^1]_2S$. Was mixed with polypropylene to give a pressed sheet for stability tests which included those on triphenyl phosphite-dilauryl thiodipropionate.[530]
$(PhO)_3P$, $(PhO)_2PHO$, and others. Are heat stabilizers

for halogenated polyolefins.[542] The effect of diphenyl decyl phosphite is enhanced when used in conjunction with certain barium and cadmium compounds, and a polyhydric alcohol.[543]

Tris(nonylphenyl) phosphite. Low pressure polyolefins are stabilized against UV light by a synergistic system.[546]

Mixed phosphites. Are used in conjunction with dibutyltin maleate for heat stabilization.[552]

$(PhO)_3P$ and nickelbis(2-Et-hexyl) phosphite. Are named for light stabilization.[559]

$(ArO)_3P$. Give with formaldehyde phosphorus containing polymers which are stated to be stabilizers for polypropylene and styrene-butadiene rubber.[571]

$(PhO)_3P$, $PhOP(OC_{10}H_{21})_2$. Enhance the stabilizing action of certain phosphites.[574]

$(ArO)_3P$. Are costabilizers of isotactic polypropylene.[575,581,582,598]

Phosphites, diphosphites, cyclic phosphites. The stabilizing efficiency has been compared with that of other antioxidants by a new rheological method.[602,609-611]

$(PhO)_3P$, $(CH_3C_6H_4O)_3P$. In conjunction with bisphenol A, are involved in the catalytic cyclooligomerization of conjugated aliphatic diolefins.[93,378]

$(PhO)_2PHO$. Is named for the production of polyolefins containing phosphorus.[270]

Tris(2-ethylhexyl) phosphite. In conjunction with nickel stearate, is named with reference to dyeable polyolefins.[313] In conjunction with a sulfurized phenol prevents discoloration.[171]

$(PhO)_3P$, $(EtO)_3P$, $(m\text{-}MeC_6H_4O)_3P$, $(o\text{-}MeC_6H_4O)_3P$. The catalytic composition for the polymerization of α-olefins may contain a phosphite promotor.[591,592]

$(RO)_2PHO$. Are reagents used with polyalkene wax.[597]

$4\text{-}t\text{-}BuC_6H_4OPO_2C_6H_4\text{-}o > o\text{-}C_6H_4O_2POPh > o\text{-}C_6H_4O_2PHO > (C_9H_{19}C_6H_4O)_3P$. The period of inhibition of oxidation of propylene increased linearly with the concentration of each of the phosphites; the slope of the lines increased in the order shown.[350]

The phosphites, $1,2\text{-}C_6H_4O_2POR$, where R is such as Bu or $2,6\text{-di-}t\text{-Bu-}4\text{-}MeC_6H_2$; $(ArO)_3P$, where Ar = Ph, p-octyl-C_6H_4, p-dodecylC_6H_4, and p-t-BuC_6H_4. Are named for increasing the thermal and hydrolytic stability of terephthalate melts.[393,394]

Three phosphites. Only the aromatic one protected polypropylene against sunlight and heat.[3]

Nickel-aryl phosphites. Are usable for inhibiting the polymerization of butadiene during the dimerization to 1,5-cyclooctadiene.[322]

Phosphites. Synergism was observed for mixtures of

phosphites with certain monosulfides, but there was antagonism with disulfides and mercaptans.[351]

Cyclic esters such as $[(CH_2O)_2POCH_2]_2C(CH_2OH)_2$. Form polyesters and used as stabilizers, plasticizers, flame retardants for polyolefins and oils.[120]

$(ArO)_3P$. The function of phenolic phosphites in the prevention of degradation of polypropylene by heat, oxidation, processing, and radiation appears to be as free radical scavengers and peroxide decomposers. They are active and synergistic with dithiopropionate esters.[460]

H.6. Natural and Synthetic Rubber and Related Compositions

$(RC_6H_4O)_mP(OR')_n$. Synthetic rubber (GRS) is stabilized against thermal degradation.[65] The Mooney plasticity is stabilized during heat (100°) aging.

Aryl phosphites. Polybutadiene or styrene-butadiene copolymers are stabilized against thermal cyclization by addition of a triaryl phosphite, which contains p-OH groups and certain tertiary alkyl groups.[139]

2,4,6-tri-t-Bu-phenyl pyrocatechyl phosphite. A manometric study of the antioxidant effect on butyl rubber has been reported; the 2,6-di-t-butyl analog, and α-naphthyl pyrocatechyl phosphite are less effective.[353]

$(PhO)_3P$, alkylated triphenyl phosphites. Are named for stabilizing chloroprene polymers.[481]

Neoalkyl phosphites. Are useful for stabilizing natural rubber and other polymers.[495]

Aryl esters of "pyrocatechin phosphorous acid." Are used as nondiscoloring preservatives for natural and synthetic rubber.[10]

Certain phosphites. Are named for the stabilization of 1,4-cis-polyisoprene against light and air oxidation.[371]

Alkaryl phosphites. Are named for diene rubbers.[405]

Tris[2-(α-methylbenzyl)-4-methylphenyl] phosphite. Is named for natural and synthetic diene rubbers.[302]

$(PhO)_3P$, nonylphenyl dioctylphenyl phosphite. Are named for stabilizing diene rubbers.[440,511]

Tris(nonylphenyl) phosphite. For stabilizing synthetic rubbers.[398]

Phosphites from an alkylphenol and bisphenol A derivative. Stabilize natural and synthetic rubber against oxidation.[409,511,520,601]

Certain phosphites. Further to an earlier disclosure that triaryl phosphites may be used for stabilizing synthetic rubber, several phosphites were prepared for the same purpose by the interaction of phosphorus trichloride and a substituted hydroquinone.[254,255]

$(RO)_3P$. As antifatigue agents for butadiene-styrene rubber vulcanizates, the following phosphites, $(RO)_3P$, were tested: R = i-Pr, Bu, i-Bu, i-Am, octyl, dodecyl, p-t-Bu-C_6H_4, and p-dodecylC_6H_4, the last showing special promise. The antifatigue action is enhanced as the length of the p-R group is increased.[580]

$[2,4-(Bu)(C_{12}H_{25})C_6H_3O]_2PHO$. Was compounded with styrene-butadiene rubber.[596]

$(PhO)_3P$. Is named for the cyclooligopolymerization of 1,3-butadiene.[621]

$(ArO)_3P$, $(ArO)_2PHO$. Oligomers of conjugated dienes are prepared by the use of catalytic systems which include as constituents a phosphite such as tri-Ph, tri-p-tolyl, tri-o-tolyl, tri-α-naphthyl, tris(2,4,6-tri-Me-phenyl), and a hydrogen phosphite such as di-Ph, di-α-naphthyl, di-p-tolyl, bis(3-Et-phenyl), and Ph p-tolyl.[637]

$(MeO)_2PHO$, $(EtO)_2PHO$. The phosphorylation of natural rubber was effected by the hydrogen phosphites in the presence of benzoyl peroxide. The products were nonflammable and could be vulcanized.[632,633]

Tris(nonylphenyl) phosphite. A direct UV spectrophotometric method has been described for determining this in styrene-butadiene rubber.[68]

Phosphites. The compounds: p-$C_6H_4(NHPO_2C_6H_4$-o$)_2$, the N,N'-di-Me, N,N'-di-i-Pr, N,N'-di-MeEtCH, N,N'-di-MePhCH analogs, o-$C_6H_4O_2PNHPh$, the p-anisyl analog of the latter, and the p-phenethyl analog of it were tested as stabilizers for rubber; the last two compounds increased the stability by 100%.[287]

Phosphites. Are named for natural and synthetic rubber.[492]

H.7. Polyurethanes and Related Polymers

$XP(OH)[NHCH_2CH_2OP(OH)]_nX'$, $XPH(:O)[NHCH_2CH_2OPH(:O)]_nX'$ (X and X' are terminal groups such as $NHCH_2CH_2OH$, $OCH_2CH_2NH_2$, or OR). Polycondensation products, e.g., a polyamidoethyl phosphite, of monoethanolamine and phosphorous acid, or a dialkyl hydrogen phosphite, $(RO)_2PHO$, R = such as Me, Et, were described as having the tautomeric general formulas shown.[5] From the product and a resol-type phenolic resin, a phenoplast complex for flame proofing is prepared.

$(ClCH_2CH_2O)_3P$. Associated with a urethane group self-extinguishing synthetic resin foams are produced.[29]

$(EtO)_2PHO$. Flame-resistant polyesters containing phosphorus are prepared.[28]

Certain phosphites. Are named as stabilizers.[32,33]

Pentaerythritol phosphites. Are named as retardants in foams and elastomers.[37]

$(ClCH_2CH_2O)_3P$. Can be used to stabilize the color of polyacrylonitrile solutions.[44]

A product from the ethylene glycol-phosphorus trichloride-ethylene oxide system. Is a precursor for the production of flame-resistant polyurethane foams.[56] Plasticizers and flame-retardant additives for polymer systems have a similar origin.[57]

Polyhydroxy compounds containing dialkyl phosphite groups. Are used in preparing flame-resistant polyurethane foams.[59,81]

Tris(2,3-dibromopropyl) and tris(2,3-dichloropropyl) phosphites. Are named as flame-retarding agents for polyurethane foams.[96]

$(C_8H_{17}O)_3P$. Used in the stabilization against oxidative degradation.[149] Phosphites containing PhO groups were stated to be ineffective.

$(PhO)_3P$. With N-pentachlorophenyl-ethanolamine gives phenylbis(N-pentachlorophenylaminoethyl) phosphite. Named for the production of fire-proof foams.[489]

$(HORO)_2POROP(OROH)_2$. React with polyisocyanates to produce flame-resistant polyurethanes containing phosphorus.[138] From the same source is the disclosure that bis(2-chloroethyl) hydrogen phosphite is used for transesterification with 1,6-hexanediol to give the compound $[(ClCH_2CH_2O)HPO_2(CH_2)_3]_2$ used in the production of poly addition products of high molecular weight containing phosphorus and nitrogen.[140] A product from the interaction of tris(2-chloroethyl) phosphite and diethylene glycol is used in the improvement of the adhesion of polyurethane varnish compositions.[142]

A poly phosphite. With a polyisocyanate gives stable, flame-resistant polyurethane foams, coatings, or fabricated articles.[158,159,161-164]

$(PhO)_3P$. With 1,2,6-hexanetriol gives a polymer which is used in the same connection. From the same source come details on the production of polyphosphites.[165-167] One purpose behind this work is the search for flame-retardant additives for polyurethanes and other types of polymers. Site functionality is deemed of importance.[168] Transesterification by means of a phosphite (RO)(R'O)(R"O)P and a vicinal glycol and a polyalkylene glycol, in the presence of a diaryl hydrogen phosphite (as catalyst) or of a catalytic alkaline reagent, affords many named phosphites.[168]

Bis(nonylphenyl)2-chloropropyl phosphite. Is named for cellular polyurethane stabilization.[183]

Polyol phosphites. Flame-resistant tobacco smoke filters are prepared from phosphorus containing foamed polyurethanes by means of a polyol phosphite in turn obtained by means of triphenyl phosphite and catalytic diphenyl hydrogen phosphite.[199] There is a similar approach to phosphorus containing phenol-aldehyde resins suitable for nonburning paints.[200]

Oxypropylenated sucrose phosphites. Are prepared by the $(PhO)_3P$ + $(PhO)_2PHO$ (catalyst) system, and are used to stabilize polymers.[202]

$(ClCH_2CH_2O)_2PHO$. Used in the production of flame-proofed polyurethanes.[205]

Polyhydroxy compounds containing trivalent phosphorus. Are used.[391]

$(RO)_2PHO$, $(RO)(OH)PHO$. Used for producing phosphorus containing polyurethane foams.[233]

$(ClCH_2CH_2O)_3P$ and analogs. Are likewise involved.[240]

2-Hydroxyethyl phosphite. Is used.[283]

Tris(9-phenylnonyl) phosphite, and other ω-phenyl-substituted alkanyl analogs, e.g., tris(6-Ph-5-Et-1-hexyl) phosphite. Are named for plasticized polyamides and polyurethanes.[258-261]

A hydroxyalkyl phosphite is prepared, by transesterification of a diol, $HO(C_3H_6O)_aCH_2RCH_2O(C_3H_6O)_bH$ (R is alkylene), with triphenyl phosphite. Used for incorporation to produce heat-sealable urethane foams.[333]

Complex phosphites. Improve combustion resistance.[335,336]

A number of diphosphites. Are used as stabilizers.[338] A number of diphosphites related to the following example are described.[345]

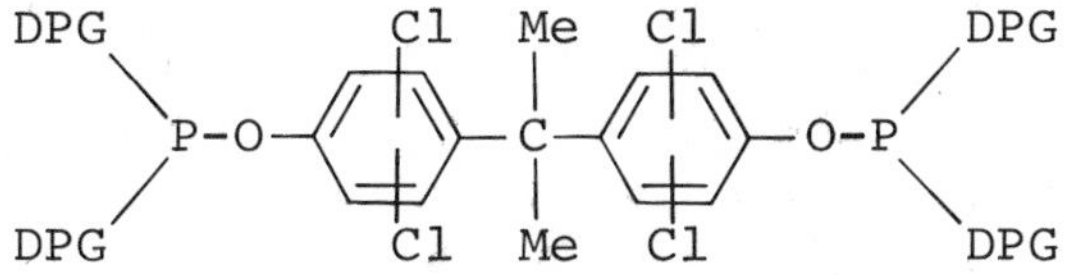

(DPG = dipropyleneglycol)

2,4,6-Tribromophenyl, 2,4,6-trichlorophenyl, and 2,4,5-trichlorophenyl bis(dipropylene glycol) phosphites. React with toluene diisocyanates to give flame retardant polyurethane.[340,345]

$(XRO)_3P$ (X = H or Cl or alkoxyl, and R = alkylene), tris-(2-ethylhexyl) and tridecyl phosphites. Are used to prevent the yellowing of polyurethane.[376]

A polyphosphite. Was selected for stabilization against discoloration by UV light and atmospheric action.[388]

Tris(p-nonylphenyl) phosphite. Was stated to give stability to UV light, but not to atmospheric action.[388]

Neocarboxylate tetrol diphosphites and polymers. Are used for rigid foams.[496]

$(RO)_3P$. React with aromatic diisocyanates to give phosphorus containing polyisocyanates.[325] The interactions of phosphites with diamines or aminoalcohols produce polymers containing P-O-C and P-N-C linkages, and further interaction with such as 2,4-tolylene diisocyanate produce polyureas.[381] 1,6-Hexanediamine and triethyl phosphite at 125-250° afford ethanol and an extremely viscous oil, solid at 25°. The IR spectrum showed the following structure,

$$\left(-\underset{\displaystyle OEt}{\overset{}{P}}NH(CH_2)_6NH-\right)_n$$

$(RO)_2PHO$. Used to produce polymers containing phosphorus and hydroxyl groups.[382]

Phosphites based on polyhydroxy compounds. Are incorporated into foamed or unfoamed urethanes.[593]

Phosphorylated polyols. Are stated to be material for many uses, including the manufacture of fire-proof polyurethane foams, by the interaction of isocyanates with reactive terminal hydrogen atoms.[595]

$(PhO)_3P$ and diisodecylpentaerythrityl diphosphite. In conjunction with a phenolic compound improve color stability.[263]

$(C_8H_{17}O)_2PHO$, $(PhO)_2PHO$. In conjunction with a somewhat complex phenolic compound are used for the stabilization of nylon 66 filaments.[268]

$(ArO)_3P$, $(MeO)_3P$. Stabilize organic isocyanates[562] and isocyanate compositions.[619]

Certain aromatic phosphites. Stabilize polyamides.[310,416,423]

$(EtO)_3P$, $(PhO)_3P$. Used to prepare lactams for the production of polyamides.[446]

Several phosphites. The stabilizing effects on nylon 7 have been assessed; the effect is a synergistic one.[510]

$[HO(RO)_m][HO(R'O)_n]PHO$ in which R and R' are ethylene groups optionally substituted with alkyl and (or) haloalkyl groups, and m and n are 1 to 10. Are used in the production of flame-proofed polyurethane foams.[570]

$(ClCH_2CH_2O)_3P$, $(Cl_3CCH_2O)_3P$. Used for interaction with cyanuric chloride.[590]

$(HOCH_2CH_2OCH_2CH_2O)_3P$. Used to prepare phosphorus-containing polyethers suitable as additives to polyurethane foams.[627]

$[HO(ClC_3H_5O)_x]_2PHO$. Used in the preparation of flame-retardant foams.[331]

$(EtO)_2PHO$. Reacts with $H_2NC_5H_{10}OH$ at 200° to give a product (white crystals at 25°) that reacts readily with diisocyanates to give polymers of high molecular weight suitable for production of transparent, tough, flame-resistant films.[380]

$(PhO)_3P$. A phenoxy resin of molecular weight 30 000, is heated with this phosphite in the presence of diphenyl hydrogen phosphite to give a flame-resistant additive for polyurethanes.[201]

Complex phosphites. Produce polyurethane foams by the interaction with organic polyisocyanates.[222]

A dihydroxyalkylene phosphite. Used to produce foams.[126]

H.8. Oil Additives (Lubricants)

A phosphite from cyclohexanol or methyl lactate. Stabilize liquid hydrocarbon lubricants and fuels against oxidation and polymerization.[11] The accompanying patent relates to the interaction of phosphorus trichloride and "OH-substituted" carboxylic ester to produce mixtures of mono-, di-, and triesters.[12]

$(ClCH_2CH_2O)_3P$, $(ClCH_2CH_2O)_2PHO$. Used as additives for extreme pressure lubricants.[47,50] Preferred compounds are 1-[bis(haloalkoxy)phosphinyl]alkyl bis(haloalkyl) phosphites and 2-[1-(dialkyloxyphosphinyl)hydrocarbyloxy]-1,3,2-dioxaphospholanes.[55]

$(PhO)_2POC_{12}H_{25}$ and others. Improve the thermal stabilities and autogenous ignition temperatures of high temperature lubricating oils.[72]

$(RO)_2PHO$, in which R = Bu, pentyl, cyclopentyl, 2-ethylhexyl. Are used in connection with molybdenum pentachloride to inhibit corrosion by mineral lubricating oils.[121]

$(PhO)_3P$, $(CH_3C_6H_4O)_3P$, bis(3-carbomethoxy-4-hydroxyphenol) thioether ditolyl phosphite. Are used in connection with lubricating fluids having extreme-pressure properties and resistance to oxidation at high temperatures.[131,132]

$(BuO)_2PHO$. Help with load-carrying properties of lubricating oil.[154]

$(i\text{-}C_8H_{17}O)_3P$. Has been tested in connection with antiwear and antiseizure properties of oils.[108]

Trilauryl phosphite and others. Are claimed as useful oil additives, effective, e.g., in the cold rolling of aluminum.[7]

Phosphites and mixed phosphites. Are named as corrosion inhibitors for polyalkene glycol lubricants.[290 291]

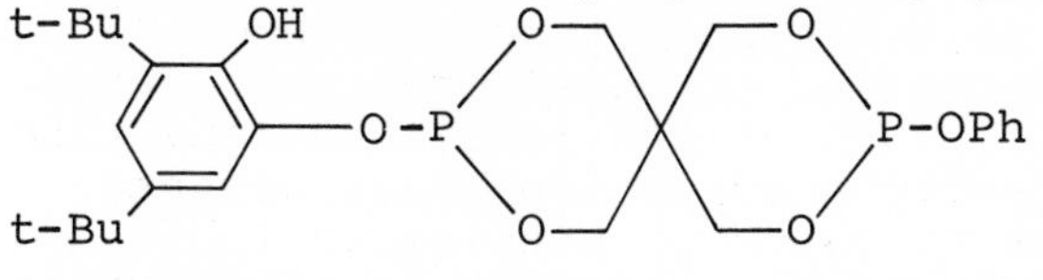

and analogs. Used as antioxidants for lubricants, fats, and plastics.[122]

$(BuO)_3P$. Is named as an antioxidant for wool-lubricating oil in spinning.[188]

Tris(alkylphenyl) phosphites. Improve quality of certain lubricants.[314]

Certain phosphites. Increase the thermal stability, anticorrosive, and flame-retardant properties of mineral and synthetic oils.[315-317] Antiseizing and anticorrosive properties were evaluated.[318]

Low-molecular weight polymers obtained by the interaction of phosphorus trichloride and trimethylolpropane. Are interacted with chloral to form lubricating oil additives. High chlorine content is correlated with maximum lubricity.[334]

Dibutyl p-(5-phenyl-2-thienyl)phenyl phosphite. Increased the failure load for extreme-pressure lubricating oil.[363]

$(BuO)_3P$. Is named for compositions used to prevent leakage of hydrocarbon oils through cellulose gaskets.[365]

$Me_2C($ 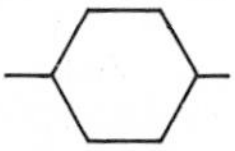 $-OPX_2)_2$ (cyclohexyl). Used as additives to extreme-pressure lubricants to improve wear and oxidation properties.[375]

$(i\text{-}PrO)_2PHO$ and phosphites. Are named.[383,618]

$(BuO)_2PHO$, $(PhO)_3P$. Stabilize hydraulic fluids.[392]

$(ArO)_2PHO$ (Ar can be o-cresyl, 4-nonylphenyl, 4-MeO-phenyl, 4-Cl-phenyl). Are named as effective lubricating oil additives. Bis(nonylphenyl) hydrogen phosphite was stated to be an outstanding load-carrying and antiwear agent for low and high temperature oils.[411,412]

Tristearyl phosphite. Is named.[413]

$(RC_6H_4O)_2POC_6H_4OP(OC_6H_4R)_2$, where R is an alkyl group of C_{1-18}. Are named.[425]

$(BuO)_2PHO$. Is a reagent for the preparation of lubricant additives and pesticides.[444]

$(s\text{-}BuO)_2PHO$ and $(MeO)_2PHO$. In conjunction with alkylphenols, these give synergistic mixtures for stabilizing lubricants.[447,448]

Phosphites. Were prepared for testing as lubricating (motor) oil antioxidants and corrosion inhibitors. Alkylphenyl dicyclohexyl phosphite improved thermal stability; alkylphenyl 2-i-Pr-4-HO-phenyl and alkylphenyl 2,5-di-t-Bu-4-HO-phenyl phosphites were effective as anticorrosives, but less effective for thermal stability. Additives containing both N and P improved stability in both aspects.[451,452]

Phosphites. Are named as wear-inhibiting compounds.[541] Products of the interaction of phosphorus trichloride and certain epoxides, such as ethylene, butylene, and higher homologs, are reported to impart excellent oiliness and extreme-pressure properties to lubricating oils.[558] Triaryl phosphites, phosphite esters of aromatic hydroxy-substituted thioethers, or disulfides, or dimorpholinylphenylmethane are named as additives.[588] Used for sludge inhibition and antiwear.[251] The properties of polyglycerol monoester as an extreme-pressure agent are improved by heating with phosphorus trichloride.[613]

High-molecular weight polyphosphites. Are useful as stabilizers for lubricants, plastics and elastomers.[518]

$(PhO)_3P$, etc. The life of an ash-free detergent in mineral lubricating oils is increased by using a combination of a bisphenol and a phosphite or a hydrogen phosphite.[547]

$C[CH_2OP(OPh)_2]_4$. Is used in lubricating oils for refrigerating systems.[550]

The products of the interaction of phosphorus trichloride with an alcohol, C_{6-24}, or an alkanethiol, and with a thiophenethiol. Greatly improved the oxidation resistance of mineral oil lubricants.[616,617]

An "acidic" alkyl or cycloalkyl phosphite. Is used for foam inhibition of gear oil.[21]

$(RO)_3P$. Are constituents of anticorrosive oil or grease for metal surfaces.[521] In detailed study of the function of phosphites in the load-carrying capacity and wear reduction of lubricants, the activity of organic phosphites is stated to depend upon the structure of the hydrocarbon group. The lower trialkyl trithiophosphites, e.g., tripropyl, are more effective than such as tri-n-octyl trithiophosphite. Tri(trichloroethyl) and tri(trichloro-t-Bu) phosphites reduced wear to lower than 0.8 mm at 300 kg loads. The significance of the mechanism of the decomposition of phosphites, $4(C_nH_{2n+1}O)_3P \rightarrow 3H_3PO_4 + PH_3 + 12C_nH_{2n}$, and trithiophosphites (H_2S is also produced) has been discussed. Films of metal phosphides or sulfides appear to form.[526-529]

$(EtO)_3P$. Used in the production of lubricant additives comprising ethylene-propylene copolymers containing phosphonamide groups.[548]

$(PhO)_3P$. Was used in a study of the mechanism of corrosion of rubbing surfaces by lubricating oil additives.[630]

$(PhO)_3P$. Was used in a study of the chemical activity of antiseizing additives; a radio indicator method was used.[631]

H.9. Oil Additives (Fuel Oils)

$(C_6H_{13}O)_3P$ and others. In conjunction with a ketone and phosphorus trichloride, are used to produce preignition additives for leaded gasoline.[404]

Mono-, di-, and trialkyl, or aryl, phosphites. Decrease the corrosion of steel by leaded aircraft-engine fuels.[612]

$(C_8H_{17}O)_3P$. Is a constituent of a composition which causes a significant lowering of carbureter deposit formation from motor fuel.[113] Organophosphorus compounds are used in leaded gasoline to control surface ignition and spark-plug misfiring; and the rating of trialkyl phosphites has been assessed.[197]

Phosphites of aminoalkyl amides. Are additives to prevent emulsion formation in fuel-injection systems.[286]

Aromatic cyclic, and the branched alkyl cyclic phosphites. Improve leaded gasoline.[482]

Mixed phosphites. Are additives for lead-free hydrocarbon fuels.[622]

H.10. Sulfur Compounds

$(EtS)_3P$. Interaction with compounds of the type $X_2HCCOCH_2CONR_2$ (X = halogen, R = H or alkyl or aryl) gives pesticides.[48]

$(BuS)_3P$, $(PhS)_3P$. Usable as additives to lubricating and fuel oils, as flame-proofing agents, antioxidants, biocides, and defoliants.[625]

$(RS)_3P$ R = Me, Et, Pr, Bu, tolyl, cycloaliphatic groups. Are rocket fuels.[86]

$(RS)_3P$. Induce potentiation and neurotoxicity.[88]

$(t\text{-}BuS)_3P$. Named as a defoliant.[186]

$(HSRS)_3P$, R = a hydrocarbyl group of C_{1-20}. For the preparation of polysulfides.[568]

$(XRS)_3P$, X = H or Cl or alkoxyl, and R = alkylene. Used with stearic acid to prevent yellowing of polyurethane.[376]

$(C_{12}H_{25}S)_3P$. Enhances light fastness and brilliancy of colored polypropylene.[402]

$(EtS)_3P$, $(PhS)_3P$. Have been tested in connection with antiwear and antiseisure properties of lubricating oils.[108]

$(PrC_6H_4S)_3P$ and others. Have been tested as stabilizers for lubricating oils. The sulfur addition products, e.g., $(t\text{-}AmC_6H_4S)_3PS$ have also been tested.[372]

$(RS)_3P$, $(RS)_2PHS$. Used in conjunction with a bis phenol increase the life of an ash-free detergent in mineral lubricating oils.[547]

Thiophosphites. Used to prepare phosphonothioates containing fluorine.[175] Used in the stabilization of vinyl chloride, vinylidene chloride polymers, and for chlorinated paraffins or rubbers.[40] In combination with phenolic antioxidants, show enhanced stabilizing properties.[156] Can be used for stabilizing ethers and ether alcohols against oxidation, giving materials useful as humectants, solvents, plasticizers, elastomers, and as components of polyurethanes, and epoxy polymers.[160] Reagents used in connection with polyalkene wax.[597]

$(i\text{-}PrS)_3P$, $(i\text{-}PrS)_2POPh$, $(PhCH_2S)_3P$, $(PhCH_2S)_2POPh$, $(PhCH_2S)(PhO)PHO$, $(C_8H_{17}S)_3P$, $(C_8H_{17}S)_2PHO$,

$(C_{12}H_{25}S)_3P$, $(C_{12}H_{25}S)_2PHO$. Used as chemical intermediates, stabilizers,[217] for polyolefins, additives for fuels and lubricants, and cotton defoliants.[243, 244]

Trilauryl trithiophosphite. Used to stabilize organic isocyanates and isocyanate compositions (synergistic protective effect in conjunction with a phenolic antioxidant).[619] Useful lubricating oil additive.[7, 383] Named as a stabilizer for polypropylene.[87,266, 491] Decreases molecular weight of crystalline polypropylene.[600]

Trilauryl trithiophosphite, and other phosphites. Named for stabilizing low pressure polyethylenes, and polymers of α-olefins.[623,624]

$(RS)_3P$ R = Me, Et, Pr, Bu, vinyl, allyl, butenyl. Used in conjunction with hydrogen peroxide as hypergolic propellants.[98]

XSP(Y)OR, R is a hydrocarbyl group, X is an alkyl group (containing a thioether group as a substituent, and Y is OR or NR_2, prepared from a sodium mercaptan and $(RO)_2PCl$, or $(R_2N)_2PCl$. Used as pesticides: $EtSCH_2CH_2SP(OEt)_2$ (for red spiders and their eggs); $EtSCH_2CH_2SP(OEt)(NEt_2)$ [from $(Et_2N)(EtO)PCl$] (for rats, plant lice, and mites); $EtSCH_2SP(OEt)_2$ (for red spiders, their eggs, and plant lice).[361]

$(R'S)_2POR$. Are named as effective nematocides.[190]

H.11. Phosphorus-Nitrogen Compounds

Me_2NPCl_2. For the preparation of dialkylhalophosphines and aminoalkylphosphonates.[370]

$(Me_2N)_3P$. As catalyst for the polycondensation of a polyhydric alcohol and thiodiglycol to give intermediates for synthetic resins, especially polyurethanes.[531] A constituent of metal cleaning compositions.[304] Useful in the stabilization of polyformaldehyde.[269] In conjunction with a tetrafluoroborate stabilizes poly(2,6-dimethyl-p-phenylene ethers).[239]

$(Et_2N)_3P$. Used in the interfacial polymerization of formaldehyde.[565] Used in conjunction with such as $(MeO)_2P(S)SCH_2CONHMe$ for production of insectofungicides.[390] Named in connection with the production of polypropylene.[628] A constituent of a polymerization composition.[415]

$(RR'N)_3P$. Named for the production of anion-exchange resins involving chloromethylated vinyl aromatic copolymers.[384]

$(RR'N)_xP(OR'')_y$, e.g., $(Me_2N)_3P$. Used in curing epoxy resins.[608]

$(R_2N)_3P$. Used in connection with titanium catalysts for producing polypropylene of high crystallinity.[105]

$(RR'N)_3P$ (NHR' could be in place of RR'N). Named in connection with the antistatic treatment of plastics.[144]

s-Hexamethylene-diaminophosphite. Named in connection with the production of profiled plastic strips which are weather and light stable.[119]

O,O-diethyl N,N-diethylphosphoroamidite. Used for the preparation of pesticides.[48]

Amides and imides of phosphorous acid. Catalyze the polymerization of formaldehyde to high molecular weight poly(oxymethylenes).[127]

Polycondensates from PCl_3 and such as hexamethylene diamine. Applicable as ion exchangers.[248]

$(R_2N)_3P$, R = Me, Et, Pr, Bu, especially Me. Improve leaded gasoline with regard to preignition and octane number.[462]

Di-Et N-(2,4-diMe-phenyl) phosphoramidite. Is an additive for motor gasoline, improving the octane number, and minimizing combustion zone deposits without lowering the octane number.[241]

"N,N',N"-(trioctylphenyl)phosphorous triamide" "Di-Bu-N-phenyl amidophosphite." Named as antiknock additives for gasoline.[289]

$PhOP(NCO)_2$. Named in connection with pigmented polyurethane coating compositions having improved viscosity stability.[620]

I. LIST OF COMPOUNDS

I.1. Phosphites

I.1.1. Difluorophosphites

TYPE: $ROPF_2$

CH_3OPF_2. ^{31}P -111 ppm, J_{PF} 1275 Hz.[628]

$F_2POCH_2CH_2OPF_2$. $(CH_2OP)_2Cl_4 + SbF_3$. b_{180} 50°, n_D^{26} 1.3523, ^{19}F NMR,[1249] ^{31}P -112.0 ppm, J_{PF} 1295 Hz.[1209,1249]

$PrOPF_2$. $ROPCl_2 + SbF_3$. b. 44.5°, n_D^{20} 1.3400, ^{19}F NMR,[1249] ^{31}P -111.5 ppm, J_{PF} 1287 Hz.[1209,1249]

$CH_2{:}CHCH_2OPF_2$. $ROPCl_2 + SbF_3$. b. 42°, ^{19}F NMR,[1249] ^{31}P -111.9 ppm, J_{PF} 1290 Hz.[1209,1249]

$BuOPF_2$. $ROPCl_2 + SbF_3$. b. 75°, n_D^{20} 1.3580,[1249] ^{31}P -111.9 ppm, J_{PF} 1288 Hz.[1209]

$PhOPF_2$. $PhOPCl_2 + SbF_3$. b_{60} 58°, n_D^{27} 1.4575, ^{19}F NMR,[1249] ^{31}P -110.1 ppm, J_{PF} 1326 Hz.[1209,1249]

1,4-$C_6H_4(OPF_2)_2$. $-PCl_2 + SbF_3$. b_{12} 59°, n_D^{23} 1.4488, ^{19}F NMR,[1249] ^{31}P -109.8 ppm, J_{PF} 1328 Hz.[1209,1249]

I.1.2. Monofluorophosphites with P in Ring System

$\overline{OCH_2CH_2OP}F$. $(RO)_2PCl + SbF_3$.[1064,1201,1249] b_{170} 48°,[1249] b_{18} 26°,[1201] d_4^{20} 1.3552,[1201] $n_D^{23.5}$ 1.4003,[1249] n_D^{20} 1.4039, MR_D 19.90 (20.56),[1201] ^{19}F NMR,[1249] ^{31}P -124.4 ppm, J_{PF} 1223 Hz,[1209,1249] ^1H NMR.[432]

$\overline{OCH(Me)CH_2OP}F$. $(RO)_2PCl + SbF_3$. b_{100} 44°, d_4^{20} 1.2226, n_D^{20} 1.4035, MR_D 24.78 (25.18),[1064,1201] IR.[1201]

$\overline{OCH(Me)CHMeOP}F$. $(RO)_2PCl + SbF_3$. b_{16} 28°, d_4^{20} 1.1568, n_D^{20} 1.4020, MR_D 29.08 (29.79).[1201]

$\overline{OCH(Me)CH_2CH_2OP}F$. $(RO)_2PCl + SbF_3$. b_{16} 37°, d_4^{20} 1.1857, n_D^{20} 1.4160, MR_D 29.22 (29.79).[1201]

$\overline{OCH_2.C(Et)(Bu)CH_2OP}F$. $(RO)_2PCl + SbF_3$. b_1 61°, d_4^{20} 1.1241, n_D^{20} 1.4765, MR_D 52.28 (52.88),[1201] ^1H NMR.[1201]

$\overline{OCH_2(CH_2)_2CH_2OP}F$. $(RO)_2PCl + SbF_3$. b_{16} 38°, d_4^{20} 1.2180, n_D^{20} 1.4450, MR_D 30.16 (29.80),[1201] IR,[1201] ^1H NMR.[1201]

$\overline{OCH_2(CH_2)_4CH_2OP}F$. $(RO)_2PCl + SbF_3$. b_1 66°, d_4^{20} 1.0840, n_D^{20} 1.4270, MR_D 39.94 (39.03).[1201]

$\overline{OCH_2(CH_2)_8CH_2OP}F$. $(RO)_2PCl + SbF_3$. b_2 80°, d_4^{20} 1.1041, n_D^{20} 1.4798, MR_D 57.09 (57.50).[1201]

1,2-$C_6H_{10}O_2PF$. $(RO)_2PCl + SbF_3$. b_1 34°, d_4^{20} 1.2140, n_D^{20} 1.4586, MR_D 36.93 (36.83),[1201] ^1H NMR.[1201]

1,2-$C_6H_4O_2PF$. $(RO)_2PCl + SbF_3$,[1201,1249] or + NaF.[1249] b_6 36.5°,[1249] b_6 38°,[1201] d_4 1.3592,[1201] n_D^{25} 1.5092, n_D^{27} 1.5080,[1249] n_D^{20} 1.5160, MR_D 35.13 (35.43),[1201] ^{19}F NMR,[1249] ^{31}P -123.1 ppm, J_{PF} 1305 Hz, ^1H NMR,[1201] J_{POCCH} ca. 1 Hz, J_{POCCCH} < 0.5 Hz.[1209,1249]

3-Me-1,2-$C_6H_3O_2PF$. $(RO)_2PCl + SbF_3$. b_2 58°, d_4^{20} 1.3045, n_D^{20} 1.5170, MR_D 39.94 (calc. 40.04).[1202]

4-Me-1,2-$C_6H_3O_2PF$. $(RO)_2PCl + SbF_3$. b_7 84°, d_4^{20} 1.3150, n_D^{20} 1.5220, MR_D 39.92 (calc. 40.04).[1202]

1,2-$C_6H_4C(:O)OPF$ (ring: C_6H_4–O–P). $(RO)_2PCl + KSO_2F$. $b_{0.15-0.2}$ 44-7°, n_D^{25} 1.5390.[1249]

I.1.3. Dichlorophosphites

TYPE: $ROPCl_2$

CD_3OPCl_2. Ib. b_{60} 31-2°, d^{24} 1.3892, n_D^{24} 1.4682.[357]

$MeOPCl_2$. Ia. b_{758} 95-6°, d_4^0 1.4275, d_4^{20} 1.3980, n_D^{20} 1.47725,[299,740] ^{31}P -180.5, -181.0 ppm.[396,935]

$C_6H_5CH(CO_2Et)OPCl_2$. Ia. b_2 105-8°, d_4^{10} 1.2827, d_4^{21} 1.2720, n_D^{21} 1.5259, α_D^{18} - 117.5° (l = 10 cm) from the (-)-mandelate α_D^{16} - 131.0°.[450]

EtOPCl$_2$. Ia.[301,854,1259] Id.[277] b. 117.5-9°,[78] b. 117-8°,[301,854] d_0^0 1.316,[854] d_4^0 1.30526,[1322] d_4^0 1.3083, d_4^{20} 1.2857, d_4^{117} 1.1831,[740] n_D^{20} 1.47176,[740] $n_D^{24.5}$ 1.46409,[1410] b_{30} 30°,[1259] b. 117.5°,[1259] ^{31}P -177, -177.0 ppm.[401,826]

ClCH$_2$CH$_2$OPCl$_2$. Ic. b. 172.5°, b_{11} 59.5-60.5°, d_0^{20} 1.4688, d_4^{20} 1.4675, n_D^{20} 1.5051,[643] ^{31}P -177.6 ppm.[826]

Cl$_3$CCH$_2$OPCl$_2$. Ia. $b_{0.1}$ 42°, n_D^{22} 1.5140.[459]

MeOCH$_2$CH$_2$OPCl$_2$. Ia. b_{14} 63-5°, d^{20} 1.2997, n_D^{20} 1.4756.[951]

EtOCH$_2$CH$_2$OPCl$_2$. Ia. b_{10} 67°, d^{20} 1.2407, n_D^{20} 1.4685.[951]

NCCH$_2$CH$_2$OPCl$_2$. Ia. b_9 110-2°, d^{20} 1.3919, n_D^{20} 1.4971.[774]

NCCHMeOPCl$_2$. Ia. b_{11} 67-8°, d^{20} 1.3359, n_D^{20} 1.4805.[774]

EtO$_2$C·CHMe·OPCl$_2$. Ia. b_{15} 88°, d_4^{17} 1.2745, d_4^{35} 1.2522, n_D^{19} 1.4654, α_D^{16} + 201.3° (l = 10 cm) (from ethyl (+)-lactate).[477]

MeCH(Cl)(C$_6$H$_5$)OPCl$_2$. Ia. $b_{0.05}$ 88-90°, d^{20} 1.3738, n_D^{20} 1.5640.[1063]

Cl$_2$POCH$_2$CH$_2$OPCl$_2$. Ia. b_2 84-5°, d_{20}^{20} 1.5689, d_4^{20} 1.5655, n_D^{20} 1.5280.[1226]

PrOPCl$_2$. Ia. b_{755} 143-5°,[740] b_{13} 40°,[94] d_4^0 1.2495, d_4^{20} 1.2278, d_4^{144} 1.1121, n_D^{20} 1.46604.[740]

CH$_2$:CH·CH$_2$OPCl$_2$. Ia. $b_{742.5}$ 140.5°,[1077] b_{753} 137°,[1074] d_0^0 1.29003, d_0^{18} 1.2685.[1077]

CH⋮CCH$_2$OPCl$_2$. Ib. b_{20} 43°, d_4^{20} 1.3465, n_D^{20} 1.5010.[585]

NCCHEtOPCl$_2$. Ia. b_8 78-9°, d^{20} 1.2868, n_D^{20} 1.4800.[774]

i-PrOPCl$_2$. Ia. b_{20} 40°,[1255] ^{31}P -174.4 ppm.[826]

MeCH(CH$_2$Cl)OPCl$_2$. Ic. b_{13} 65°, d^{20} 1.3801, n_D^{20} 1.5060.[1117]

(ClCH$_2$)$_2$CHOPCl$_2$. Ia. b_{12} 95-6°, d_4^{20} 1.5118, n_D^{20} 1.5210, MR$_D$ 46.47 (calc. 46.04),[776] b_1 49-51°, d^{20} 1.5296, n_D^{20} 1.5222.[1221]

MeCH(CCl$_3$)OPCl$_2$. Ia. b_{25} 130°, b_{70} 140°, b_{758} 223-4°, d_4^{20} 1.5870.[550]

NCCMe$_2$OPCl$_2$. Ia. b_{11} 78-80°, d^{20} 1.2760, n_D^{20} 1.4773.[774]

BuOPCl$_2$. Ia.[76,449,740,1259] Id.[449] b. 157°,[740] b_{11} 49-50°,[1259] b_{750} 161°,[1259] b_{14} 56°,[449] b_{10} 47°,[467] b_{18} 66-7°,[76] d_4^0 1.1923,[740] d_4^{16} 1.1801,[449] d_4^{20} 1.166,[467] d_4^{20} 1.1657,[740] n_D^{20} 1.46086,[740] ^{31}P -178.8 ppm.[826]

CH$_2$=CH·CHClCH$_2$OPCl$_2$. Ia. b_1 49-51°, d^{20} 1.3593, n_D^{20} 1.5068.[1063]

NCCHPrOPCl$_2$. Ia. b_{10} 92-4°, d^{20} 1.2295, n_D^{20} 1.4765.[774]

Me(Et)CHOPCl$_2$. Ia.[467,1255] b_{14} 50°.[1255] b_{22} 61-4°, d_4^{20} 1.173, n_D^{23} 1.4655.[467]

i-BuOPCl$_2$. Ia. b. 154-6°,[854] b_{12} 42.5°,[467] d_0^0 1.191,[854] d_4^{22} 1.169, n_D^{20} 1.4642.[467]

i-PrCH(CN)OPCl$_2$. Ia. b_8 83-4°, d^{20} 1.2410, n_D^{20} 1.4780.[774]

Me$_2$(CCl$_3$)COPCl$_2$. Ia.[481] Ib.[19] $b_{0.2}$ 66°,[481] b_{16} 118°,[19,688] d_4^{20} 1.5280, n_D^{20} 1.5236, MR$_D$ 55.53 (calc. 55.58),[19] d 1.5269, n_D^{20} 1.5238.[688]

Cl$_3$CMeEtOPCl$_2$. Ib. b_8 121-2°, d^{20} 1.4913, n_D^{20} 1.5262.[368]

AmOPCl$_2$. Ia.[278] b$_2$ 53°,[278] b$_{14}$ 71-3°,[951] b$_{16}$ 83°,[278] d^{20} 1.1345, n_D^{20} 1.4650,[951] n_D^{20} 1.4675, ^{31}P -176.5 ppm.[278]
i-AmOPCl$_2$. Ia.[854,875] b. 173°,[854] b. 178°,[740] d^0 1.109,[854] d_4^0 1.1563, d_4^{20} 1.1364, n_D^{20} 1.45566,[740] b$_{750}$ 175-6°, b$_{10}$ 56°.[875]
i-BuCH(CN)OPCl$_2$. Ia. b$_{10}$ 95-6°, d^{20} 1.2020, n_D^{20} 1.4770.[774]
PrMeCHOPCl$_2$. Ia. b$_{16}$ 74-6°, b$_{8.5}$ 57°, b$_{0.2}$ 26-7°, n_D^{20} 1.4655, ^{31}P -174.5 ppm.[278]
i-PrMeCHOPCl$_2$. Ia. ^{31}P -174.0 ppm.[278]
Me$_3$CCH$_2$OPCl$_2$. Ia. b$_{10}$ 47.5-48°.[475]
(ClCH$_2$)$_3$CCH$_2$OPCl$_2$. Ia. b$_{1.5}$ 120°, d^{20} 1.5298, n_D^{20} 1.5320.[777]
n-C$_6$H$_{13}$OPCl$_2$. Ia. b$_{24}$ 104°, d_0^0 1.371, $n_D^{16.5}$ 1.4669,[1200] $n_D^{26.5}$ 1.4699.[1193]
Me$_3$CCH(Me)OPCl$_2$. b$_3$ 52°.[1253]
(CH$_2$CH$_2$)$_2$C(CCl$_3$)OPCl$_2$. Ia. b$_5$ 127°, d^{20} 1.5215, n_D^{20} 1.5425.[19]
Cyclo-C$_6$H$_{11}$OPCl$_2$. Ia.[1024,1255] b$_{0.3}$ 54-5°, d_4^{20} 1.2122, n_D^{20} 1.5012, MR$_D$ 47.225 (47.94),[1024] b$_3$ 70°.[1255]
(CH$_2$)$_5$C(CN)OPCl$_2$. Ia. b$_{10}$ 124-6°, d^{20} 1.2818, n_D^{20} 1.5118.[1035]
2-MeC$_6$H$_{10}$OPCl$_2$. Ia. b$_{0.5}$ 82-3°, d_4^{20} 1.2041, n_D^{20} 1.4910, MR$_D$ 51.84 (52.19).[1024]
CH$_2$(CH$_2$CH$_2$)$_2$C(CCl$_3$)OPCl$_2$. Ia. b$_1$ 129-30°, m. 55-6°, d^{20} 1.5091, n_D^{20} 1.5528.[19]
CH$_2$=CH·CHCH$_2$CHClCHOPCl$_2$. Ic. b$_1$ 96-8°, d^{20} 1.3069,
CH$_2$CH$_2$ (ring bond from CH to CHOPCl$_2$)
n_D^{20} 1.5256.[444]
n-C$_7$H$_{15}$OPCl$_2$. Ia. b$_{12.5}$ 107°, d_0^0 1.1138, n_D^{14} 1.4720.[1193]
n-C$_8$H$_{17}$OPCl$_2$. Ia. b$_{12}$ 123-4°, d_0^0 1.0919, n_D^{21} 1.4682.[1193]
2-C$_8$H$_{17}$OPCl$_2$. Ia. b$_1$ 75-6°, b$_2$ 83-4°, b$_{17}$ 118-9°, d_4^{18} 1.0749, d_4^{25} 1.0683, n_D^{16} 1.4666, n_D^{9} 1.4669, α_D^{18} -34.5°, from (+)-ROH α_D^{18} +8.0° (l = 10 cm).[450]
C$_6$H$_5$OPCl$_2$. Ia.[63,274,982,1259] Ia (plus MgCl$_2$).[1326] Id.[299] IIId.[416] b$_{10}$ 90°,[1326] b$_{10}$ 91°,[1259] b$_{11}$ 90°,[63] b$_{11}$ 90-2°,[274] b$_{20-30}$ 100-30°,[787] b$_{190}$ 177°,[1326] b. 216 (decomp.),[982] n_D^{20} 1.5588,[1326] d_{18}^{18} 1.348,[982] d_4^{20} 1.3543,[63,299,982] d_4^{20} 1.3539,[1326] ^{31}P -173.0, ca. -179 ppm.[826]
1,3-C$_6$H$_4$(OPCl$_2$)$_2$. Ia. b$_{56}$ 240°, d_0^{18} 1.5696.[713]
1,4-C$_6$H$_4$(OPCl$_2$)$_2$. Ia. m. 65°, b$_{65}$ 200°.[713]
2-ClC$_6$H$_4$OPCl$_2$. Ia.[273,274] Ia (plus MgCl$_2$).[1326] b$_{10}$ 111-2°,[1326] b$_{13}$ 116-20°,[273,274] b$_{190}$ 197°, n_D^{20} 1.5736, d_4^{20} 1.4686.[1326]
4-ClC$_6$H$_4$OPCl$_2$. Ia.[274,1299] Ia (plus MgCl$_2$).[1326] b$_{10}$ 113-3.5°,[1326] b$_{12}$ 118-20°,[274,1326] b$_{12}$ 128-30°,[274,1299] b$_{190}$ 199°, n_D^{20} 1.5749, d_4^{20} 1.4714.[1326]
2,4-Cl$_2$-C$_6$H$_3$OPCl$_2$. Ia (plus MgCl$_2$). b$_{10}$ 134°, b$_{190}$ 220°, d_4^{20} 1.5651, n_D^{20} 1.5860,[1326] ^{31}P -184.8, -182.8 ppm.[826]

2,5-Cl_2-$C_6H_3OPCl_2$. Ia (plus $MgCl_2$). b_{10} 138°, d_4^{20} 1.5672, n_D^{20} 1.5869.[1326]
3,4-Cl_2-$C_6H_3OPCl_2$. Ia (plus $MgCl_2$). b_{10} 136°, d_4^{20} 1.5736, n_D^{20} 1.5894.[1326]
2,4,5-Cl_3-$C_6H_2OPCl_2$. Ia (plus $MgCl_2$). b_{10} 154-5°, b_{190} 240°, d_4^{20} 1.6556, n_D^{20} 1.6007.[1326]
2,4,6-Cl_3-$C_6H_2OPCl_2$. Ia (plus $MgCl_2$). b_{10} 156°, d_4^{20} 1.6563, n_D^{20} 1.6012.[1326]
2-$MeC_6H_4OPCl_2$. Ia. b_{11} 116°,[1299] b_{11} 106°.[219]
3-$MeC_6H_4OPCl_2$. Ia. b_{12} 114°.[219]
4-$MeC_6H_4OPCl_2$. Ia. b_{11} 118°,[1299] ^{31}P -175.9 ppm.[826]
4-$MeOC_6H_4OPCl_2$. Ia. b_{13} 130-1°,[57] b_{13} 135°, n_D^{21} 1.568.[360]
1,3-$EtC_6H_4OPCl_2$. Ia + $MgCl_2$. b_{10} 115°, d_4^{20} 1.2590, n_D^{20} 1.5474.[1326]
2-CH_2:$CHCH_2C_6H_4OPCl_2$. Ib. b_{10} 126-7°, d^{20} 1.2503, n_D^{20} 1.5535.[421]
2Cl-4-t-C_4H_9-$C_6H_3OPCl_2$. Ia (plus $MgCl_2$). b_{10} 154°, d_4^{20} 1.2870, n_D^{20} 1.5510.[1326]
2Br-4-t-C_4H_9-$C_6H_3OPCl_2$. Ia (plus $MgCl_2$). b_{10} 165°, d_4^{20} 1.4669, n_D^{20} 1.5676.[1326]
2-t-Bu-4-$MeC_6H_3OPCl_2$. Ia. b_{25} 156-64°.[945]
2,6-di-t-$BuC_6H_3OPCl_2$. Ib. b_1 143-6°.[1266]
2,6-di-t-Bu-4-$ClC_6H_2OPCl_2$. Ib. $b_{2.6}$ 156-66°.[1266]
2,6-di-t-Bu-4-$MeC_6H_2OPCl_2$. Ib. $b_{0.5}$ 117-20° (solidifies on standing).[1266]
2,4,6-tri-t-$BuC_6H_2OPCl_2$. Ib. $b_{0.5}$ 119-24°.[1266]
2,6-di-t-Bu-4-$C_9H_{19}C_6H_2OPCl_2$. Ib. $b_{2.5}$ 148-54°.[1266]
2,4,6-tris(phenethyl)$C_6H_2OPCl_2$. Ib.[1266]
1-$C_{10}H_7OPCl_2$. Ia.[769] $b_{0.05}$ 86-8°, $b_{0.65}$ 108-10°,[204] b_{15} 174-6°, d_0^{15} 1.0776,[769] ^{31}P -184 ppm.[826]
2-$C_{10}H_7OPCl_2$. Ia.[769] $b_{0.1}$ 101-3°, $b_{0.6}$ 119°,[204] b_{15} 179-81°, d_0^{15} 1.0781.[769]

I.1.4. Monochlorophosphites with Two Discrete Radicals

TYPE: $(RO)_2PCl$

$(MeO)_2PCl$. Ia + SO_2Cl_2.[845] n_D^{25} 1.4107 (crude),[845] ^{31}P -169 ppm.[826]
$(EtO)_2PCl$. Ia + SO_2Cl_2.[845] Ib.[873,1195] Ih.[80,211] Ij.[80] b. 153-5°,[80] b_{10} 36.5-7°,[1195] b_{25} 53.5-4.5°,[873] b_{30} 62-5°, b_{30} 63-5°,[80,1076] d_0^0 1.0962,[80] d^{16} 1.0632, d_0^{20} 1.0747,[80] d^{20} 1.0876,[1195] n_D^{20} 1.4350,[80] n_D^{20} 1.4360,[1076] n_D^{20} 1.4370,[873,1195] ^{31}P -165, -164 ppm,[401,628] UV.[328a]
$(ClCH_2CH_2O)_2PCl$. Ic. $b_{4.5}$ 101-3°, d_0^{20} 1.4019, d_4^{20} 1.4007, n_D^{20} 1.4950.
$(CF_3CH_2O)_2PCl$. Id. b_{15} 60°.[786]
$(Cl_3CCH_2O)_2PCl$. Ia. $b_{0.1}$ 95-6°, n_D^{22} 1.5167.[459]

$(NCCHMeO)_2PCl$. Ia. b_{10} 140-2°, d^{20} 1.1844, n_D^{20} 1.4575.[774]
$(EtO_2C \cdot CHMeO)_2PCl$. Ia. b_{15} 155-60°, α_D^{16} + 160.4°, (l = 1 dm) [from ethyl (+)-lactate].[477]
$(PrO)_2PCl$. Ia + SO_2Cl_2.[845] Ib.[873,1195] Ih.[94] b_8 65-5°,[94] b_{12} 69-70°,[1195] b_{15} 71°,[271] b_{17} 80°,[873] d_0^0 1.0626,[94] d^{20} 1.0420,[1195] d_4^{20} 1.093,[271] n_D^{20} 1.4420,[94] n_D^{20} 1.4386,[1195] n_D^{20} 1.4418,[271] n_D^{20} 1.4426,[873] n_D^{25} 1.4159 (crude),[845] $[P]_M$ 974 μ radian.[271]
$(CF_2HCF_2CH_2O)_2PCl$. Ia. b_1 65°.[786]
$(NCCHEtO)_2PCl$. Ia. b_{11} 152-5°, d^{20} 1.1470, n_D^{20} 1.4612.[774]
$(i\text{-}PrO)_2PCl$. Ib. b_{12} 62-4°, n_D^{20} 1.4242,[873] ^{31}P -165.4 ppm,[826] UV.[1263]
$[MeCH(CH_2Cl)O]_2PCl$. Ic. b_2 88-9°, d^{20} 1.2763, n_D^{20} 1.4820.[1117]
$[(ClCH_2)_2CHO]_2PCl$. Ia. $b_{0.2}$ 126-7°, d_4^{20} 1.4891, n_D^{20} 1.5193, MR_D 65.77 (calc. 66.04),[776] b_1 138°, d^{20} 1.4901, n_D^{20} 1.5180,[701] b_1 122-4°, d^{20} 1.4905, n_D^{20} 1.5183.[1221]
$(Cl_3CCHMeO)_2PCl$. Ia. b_{25} 210°.[550]
$(NCCMe_2O)_2PCl$. Ia. b_{11} 139-40°, d^{20} 1.1417, n_D^{20} 1.4557.[774]
$(BuO)_2PCl$. Ia.[449] Ia + SO_2Cl_2.[845] Ib.[873,1195] Id.[449] Ih.[76] b_6 91.5-2.5°,[76] b_{10} 96-8°,[1195] b_{12} 99-102°,[449] b_{25} 115-7°,[873] d_0^{15} 1.014,[76] d^{20} 1.0091,[1195] n_D^{20} 1.445,[76] n_D^{20} 1.4472,[1195] n_D^{20} 1.4454,[873] n_D^{25} 1.4291 (crude).[845]
$(CH_2{=}CH \cdot CHClCH_2O)_2PCl$. Ia. $b_{1.5}$ 111-2°, d^{20} 1.2570, n_D^{20} 1.5025.[1063]
$(NCCHPrO)_2PCl$. Ia. b_3 138-40°, d^{20} 1.1176, n_D^{20} 1.4630.[774]
$[i\text{-}PrCH(CN)O]_2PCl$. Ia. b_3 127-8°, d^{20} 1.1089, n_D^{20} 1.4620.[774]
$(t\text{-}BuO)_2PCl$. Ib. ^{31}P -170.3 ppm.[827]
$[Me_2C(CCl_3)O]_2PCl$. Ia.[481] Ib.[19] $b_{0.2}$ 128-9°,[481] b_5 171°, d_4^{20} 1.5192, n_D^{20} 1.5265, MR_D 84.50 (calc. 84.64).[19]
$(C_5H_{11}O)_2PCl$. Ib. b_1 91.5-2°, d^{20} 0.9868, n_D^{20} 1.4415.[1195]
$[i\text{-}BuCH(CN)O]_2PCl$. Ia. b_1 138-40°, d^{20} 1.0808, n_D^{20} 1.4623.[774]
$[(ClCH_2)_3CCH_2O]_2PCl$. Ia. b_1 192°, d^{20} 1.4970, n_D^{20} 1.5340.[777]
$(Cl_3CCMeEtO)_2PCl$. Ib. m. 69-70°, $b_{0.14}$ 162°, d^{20} 1.4683, n_D^{20} 1.5310.[368]
$[(CH_2CH_2)_2C(CCl_3)O]_2PCl$. Ib. b_3 193-4°, m. 57-8°, d_4^{20} 1.4966, n_D^{20} 1.5450.[19]
$[(CH_2)_5C(CN)O]_2PCl$. Ia. b_{10} 203-7°, d^{20} 1.1810, n_D^{20} 1.5050.[774]
$(C_6H_{13}O)_2PCl$. Ib.[1063,1198] b_8 143-5°,[1198] b_8 145-7°,[1063]

d^{20} 0.9678,[1198] d^{20} 0.9688,[1063] n_D^{20} 1.4510,[1198] n_D^{20} 1.4435.[1063]

$(CH_2{=}CH{\cdot}CHCH_2CHClCHO)_2PCl$ (ring closed by CH_2CH_2 between the CH and CHO carbons). Ic. b_1 154-6°, d^{20} 1.2131, n_D^{20} 1.5231.[444]

$(n\text{-}C_8H_{17}O)_2PCl$. Ih. Undistillable oil.[1193]

$(2\text{-}C_8H_{17}O)_2PCl$. Ia. b_2 135-40°, n_D^{20} 1.4430.[450]

$(PhO)_2PCl$. Ia.[63,982,1076] $(PhO)_3P + BCl_3$.[416] Id.[299] $b_{0.1-0.2}$ 110-30°,[787] b_1 165-74°,[498] b_{10} 179-81°,[407] b_{11} 172°,[63] b_4 146.5-48°,[407] b_{15} 110°,[1076] b_{221} 265-70°, b_{731} 295°, d_{18}^{18} 1.221,[982] d_0^{20} 1.2471,[407] n_D^{20} 1.5721,[659] n_D^{20} 1.5760,[1076] n_D^{25} 1.5789,[407] ^{31}P ca. -159, -157.3 ppm.[628,826]

$(2\text{-}ClC_6H_4O)_2PCl$. Ia. b_{10} 205-10°.[273,274]

$(4\text{-}ClC_6H_4O)_2PCl$. Ia. b_{11} 225-7°, b_{12} 205-15°.[274]

$(2\text{-}MeC_6H_4O)_2PCl$. Ia. b_{11} 190°, b_{11} 195-6°.[1299]

$(3\text{-}MeC_6H_4O)_2PCl$. Ia. b_{11} 198°.[219]

$(4\text{-}MeC_6H_4O)_2PCl$. Ia.[1063,1299] b_{11} 206-8°,[1299] b_1 161-4°, d^{20} 1.1994, n_D^{20} 1.5684.[1063]

$(2\text{-}MeOC_6H_4O)_2PCl$. Ia. b_{13} 235°, n_D^{21} 1.586.[360]

$(2\text{-}CH_2{:}CHCH_2C_6H_4O)_2PCl$. Ib. b_7 192°, d^{20} 1.1572, n_D^{20} 1.5665.[421]

$(2,6\text{-di-t-BuC}_6H_3O)_2PCl$. Ia.[1267] Ib.[1266]

$(2,6\text{-di-t-Bu-4-ClC}_6H_2O)_2PCl$. Ia.[1267] Ib.[1266] $b_{2.6}$ 156-66°,[1266,1267] m. 147-8.5°,[1267]

$(2,6\text{-di-t-Bu-4-MeC}_6H_2O)_2PCl$. Ia.[1267] Ib.[1266] m. 113-5°.[1266,1267]

$(2,4,6\text{-tri-t-BuC}_6H_2O)_2PCl$. Ia. m. 173-4°.[1267]

$(1\text{-}C_{10}H_7O)_2PCl$. Ia. $b_{0.03}$ 174°, $b_{0.35}$ 210°, n_D^{20} 1.6706.[204]

$(2\text{-}C_{10}H_7O)_2PCl$. Ia. $b_{0.03}$ 181-3°, $b_{0.08}$ 206°.[204]

TYPE: (RO)(R'O)PCl

(MeO)(iso-BuO)PCl. Ih. b_{17} 62-72° (crude).[1193]

(EtO)(BuO)PCl. Ib.[877] Ih.[1193] b_9 72-5°,[1193] b_{16} 80-2°.[877]

$(EtO)(n\text{-}C_7H_{15}O)PCl$. Ih. b_{11} 122.5-4°, d_0^0 1.0249, $n_D^{12.5}$ 1.4468, n_D^{20} 1.4411.[1193]

(EtO)(MeCClPhO)PCl. Ia. $b_{0.03}$ 101-2°, d^{20} 1.2213, n_D^{20} 1.5278.[1063]

$(ClCH_2CH_2O)(NCMe_2CO)PCl$. Ib. b_{13} 135-7°, d^{20} 1.3215, n_D^{20} 1.4835.[774]

(PrO)(BuO)PCl. Ih. b_9 85-6°, d_0^0 1.1212, n_D^{20} 1.4421.[1193]

$(Cl_3CCMe_2O)(Cl_3CCMeEtO)PCl$. Ib. $b_{0.1}$ 132°, $b_{0.22}$ 141-2°, d^{20} 1.4963, n_D^{20} 1.5293.[368]

$(Cl_3CCMe_2O)[(CH_2CH_2)_2C(CCl_3)O]PCl$. Ib. b_4 178-9°, m. 39-40°, d_4^{20} 1.5152, n_D^{20} 1.5365.[19]

$(Cl_3CCMe_2O)(C_6H_5O)PCl$. Ib. $b_{0.1}$ 114°, d^{20} 1.3820, n_D^{20} 1.5445.[688]

$(Cl_3CCMe_2O)[CH_2(CH_2CH_2)_2C(CCl_3)O]PCl$. Ib. b_2 181-2°, d_4^{20} 1.5047, n_D^{20} 1.5421.[19]

$(Cl_3CMeEtO)[(CH_2CH_2)_2C(CCl_3)O]PCl$. Ib. $b_{0.22}$ 169-70°, d^{20} 1.5130, n_D^{20} 1.5427.[686]

I.1.5. Monochlorophosphites with P in a Ring System.

OCH_2CH_2OPCl. Ia.[270,979,1226] Ib.[105] b_{10} 41.5°,[105] b_{15} 45-6°,[979] b_{47} 66-8°,[105] b_{42} 65-6°,[270] d_0^{20} 1.4172,[105] d_4^{20} 1.4199, d_{20}^{20} 1.4229,[1226] n_D^{20} 1.4894,[1226] n_D^{20} 1.4915,[105] n_D^{25} 1.4897,[270] b_{45} 66°,[1076] n_D^{20} 1.4831,[1076] ^{31}P -166.6, -167, -168.4 ppm,[628,826] ^1H NMR.[432]

$OCHMeCH_2OPCl$. Ia.[799,979,1415] b_{25} 58°, b_{50} 75-5.8°,[799] b_{12} 43-4°,[1415] n_D^{25} 1.4707,[799] n_D^{20} 1.4725,[1415] b_{15} 74-7°,[979] ^1H NMR.[493a]

$OCHMeCHMeOPCl$. Ia.[979,1415] Ib.[1027] b_{14} 54-6°,[1027] 57-8°,[1415] b_{15} 66-7°,[979] n_D^{25} 1.4700,[1027] n_D^{20} 1.4658,[1415] ^1H NMR.[432]

$OCH(CH_2OMe)CH_2OPCl$. Ib. b_9 78.5-9.2°, d_0^{20} 1.2984, n_D^{20} 1.4722.[105]

$OCMe_2CMe_2OPCl$. Ib.[82,1415] b_{13} 81.5-2°,[82] b_{12} 77-8°,[1415] d_0^{20} 1.1562,[82] n_D^{25} 1.4720,[108] n_D^{20} 1.4714,[1415] ^1H NMR.[493a]

$OCMeEtCMeEtOPCl$. Ib. $b_{0.2}$ 57°, n_D^{20} 1.4816.[1415]

$OCEt_2CEt_2OPCl$. Ib. $b_{0.02}$ 62-3°, n_D^{20} 1.4910.[1415]

$OCH_2CH_2CH_2OPCl$. Ia.[799,979,1415] b_{15} 66.5-7.5°,[799] b_9 54-5°,[1415] n_D^{25} 1.4884,[799] n_D^{20} 1.4884,[1415] b_{15} 66-7°,[979] ^{31}P -153, -153.9 ppm.[628,826]

$OCHMeCH_2CH_2OPCl$. Ib.[105,1027] Ia.[1415] b_{12} 65°,[105] b_{10} 55-7°,[1027] b_{20} 82-4°,[1415] d_0^{20} 1.2496,[105] n_D^{20} 1.4765,[105] 1.4670,[1027] 1.4776.[1415]

$OCH_2CMe_2CH_2OPCl$. Ia.[1415] Ib (PCl_3 in 50% excess). b_{13} 66°,[365] b_{25} 83-4°, b_{20} 78-9°,[1415] n_D^{22} 1.4746,[365] n_D^{20} 1.4757,[1415] ^{31}P -146.7 ppm.[826]

$OCH_2CMe[CHMe(OMe)]CH_2OPCl$. Ia.[210] Ib.[225] $b_{0.02}$ 63°, d_4^{20} 1.1940, n_D^{20} 1.4700.[210,225]

$OCH_2C(i\text{-}Pr)[CHMe(OMe)]CH_2OPCl$. Ia.[210] Ib.[225] $b_{0.01}$ 62.5°, d_4^{20} 1.1714, n_D^{20} 1.4810.[210,225]

$OCH_2CEtBuCH_2OPCl$. b_1 81°, d_4^{20} 1.1050, n_D^{20} 1.4830, MR_D 58.09 (57.83),[1201] ^1H NMR.[1201]

$OCH_2C(i\text{-}Pr)[CHMe(i\text{-}Pr)]CH_2OPCl$. Ia.[210] Ib.[225] $b_{0.02}$ 80-2°, d_4^{20} 1.1256, n_D^{20} 1.4742.[210,225]

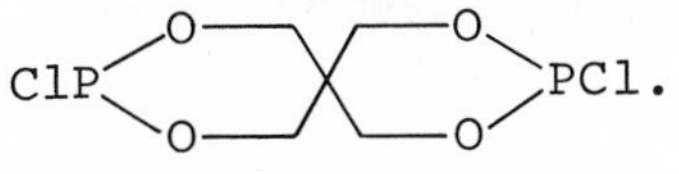

Ia. m. 123-5°, b_8 147-8°,[674] ^{31}P -148.1 ppm.[826]

$\overline{OCH_2CH_2CH_2CH_2OP}Cl$. Ib.[105] Ia.[979,1415] b_8 74-5.5°,[105] $b_{0.2}$ 51-3°,[1415] d_0^{20} 1.2858,[105] n_D^{20} 1.5010,[105] 1.4994.[1415]

$\overline{OCH_2CH_2OCH_2CH_2OP}Cl$. Ib. b_{15} 104-5°, d_0^{20} 1.2693, n_D^{20} 1.5165.[105]

$\overline{OCH_2(CH_2)_4CH_2OP}Cl$. $b_{0.16}$ 78°, d_4^{20} 1.2070, n_D^{20} 1.4880, MR_D 43.54 (44.00),[1201] ^1H NMR.[1201]

$\overline{OCH_2(CH_2)_8CH_2OP}Cl$. b_2 130°, d_4^{20} 1.1039, n_D^{20} 1.4860, MR_D 62.13 (62.47),[1201] ^1H NMR.[1201]

1,2-$C_6H_{10}O_2PCl$. b_{20} 80°, d_4^{20} 1.2914, n_D^{20} 1.5140, MR_D 42.09 (41.80),[1201] ^1H NMR.[1201]

1,2-$C_6H_4O_2PCl$. Ia.[55,96,97,629] Id.[55,56] Needles, m. 30°,[54,96] b_{10} 80°,[54,96] b_{10} 81-2°, b_6 71-2°,[629] b_{13} 86°,[97] b_{16} 91°,[55] b_{65} 140°,[713] b_{32} 110°,[1076] n_D^{20} 1.5711.[1076] See also Ref. 60. ^{31}P -167.0 ppm.[826]

4-Me-1,2-$C_6H_3O_2PCl$. Ia. b_{11} 102°, m. 22-4°,[62] b_{13} 109°,[60] b_9 87.5-8°, d^{20} 1.3308, n_D^{20} 1.5635.[708]

4-t-Bu-1,2-$C_6H_3O_2PCl$. Ia. b_{13} 134-5°, d^{20} 1.2085, n_D^{20} 1.5415.[708]

4-t-Am-1,2-$C_6H_3O_2PCl$. Ia. b_{13} 144-5°, d^{20} 1.1890, n_D^{20} 1.5364.[708]

3,5-di-t-Bu-1,2-$C_6H_2O_2PCl$. Ia. b_{16} 157°, d^{20} 1.1183, n_D^{20} 1.5258.[708]

3,5-di-t-Am-1,2-$C_6H_2O_2PCl$. Ia. b_{13} 157.5-8°, d^{20} 1.0950, n_D^{20} 1.5240.[708]

1,2-$\overline{OC_6H_4CO_2P}Cl$. II.[63,245,247,1400] m. 36-7°, b_{11} 127°,[63] b_{11} 127-8°,[1400] b_9 123-4°,[245,247] ^{31}P -151.2 ppm.[826]

2,2'-$\overline{OC_6H_4-C_6H_4OP}Cl$. Ia.[60,1347] b_{12} 195°, m. 63°,[60] m. 62-3°, b_8 178°, d^{20} 1.3563, n_D^{20} (supercooled) 1.6347.[1347]

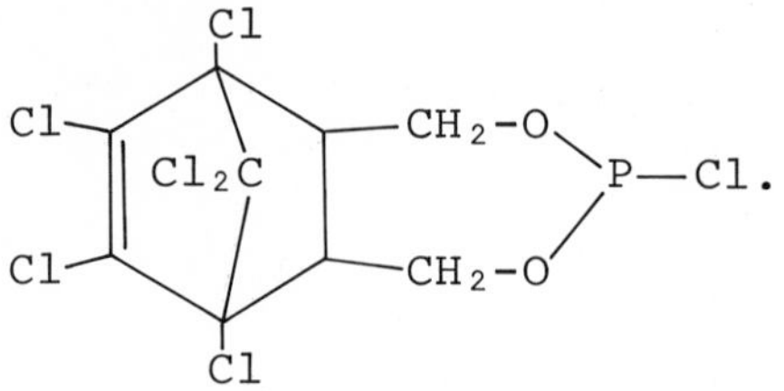

Ia. m. 65-8°.[509]

I.1.6. Dibromophosphites

TYPE: $ROPBr_2$

$MeOPBr_2$. Ib. b_{20} 68-70°, d^{24} 2.4324, n_D^{24} 1.6233.[357]
CD_3OPBr_2. Ib. b_{15} 62-4°, d^{24} 2.4076, n_D^{24} 1.6137.[357]
$BrCH_2CH_2OPBr_2$. Ic. b_2 79-80°, d_0^{20} 2.3786, n_D^{20} 1.6106.[1227]

$MeCH(CF_3)OPBr_2$. Ia. b. 156-7°, f. 48°.[1305]
n-$BuOPBr_2$. Ia. $b_{0.01}$ 38°, n_D^{20} 1.5441.[463]
i-$BuOPBr_2$. Ia.[463,896] b_{20} 85-7°,[896] b_5 64°,[463] d_4^{17} 1.673,[896] d_4^{20} 1.768, n_D^{20} 1.5391.[463]
s-$BuOPBr_2$. Ia. $b_{0.1}$ 38°, n_D^{20} 1.5409.[463]
$Me_3CCH_2OPBr_2$. Ia. $b_{0.3}$ 30°, n_D^{20} 1.5201, ^1H NMR.[575]
$PhOPBr_2$. Ia.[1299] IIId.[417] b_{11} 130-2°,[1299] b_{11} 120-130°, n_D^{20} 1.6202,[417] ^{31}P ca. -200 ppm.[826]

I.1.7. Monobromophosphites with Two Discrete Radicals

TYPE: $(RO)_2PBr$

$(BrCH_2CH_2O)_2PBr$. Ic. $b_{1.7}$ 114-5.5°, d_0^{20} 2.1133, d_4^{20} 2.1109, n_D^{20} 1.5671 (not completely pure).[1227]
$(BuO)_2PBr$. Ib. $b_{0.05}$ 50°, n_D^{20} 1.4690.[463]
(i-BuO)$_2$PBr. Ib. $b_{0.4}$ 64°, n_D^{20} 1.4535.[463]
(s-BuO)$_2$PBr. Ib. $b_{0.05}$ 32-4°, n_D^{20} 1.4598.[463]
$(PhO)_2PBr$. Ia. b_{11} 189-92°,[1299] ^{31}P -176, -177 ppm.[826]

I.1.8. Monobromophosphites with P in a Ring System

$\overline{OCHMeCH_2OP}Br$. Ia. $b_{0.08}$ 48°, d_4^{20} 1.6790, n_D^{20} 1.5269, IR, ^1H NMR.[1363]

$\overline{OCHMeCHMeOP}Br$. Ia. $b_{0.08}$ 45°, d_4^{20} 1.5711, n_D^{20} 1.5095, IR, ^1H NMR.[1363]

$\overline{OCH_2 \cdot CEtBuCH_2OP}Br$. Ia. $b_{0.08}$ 92°, d_4^{20} 1.3100, n_D^{20} 1.5040, IR, ^1H NMR.[1363]
1,2-$C_6H_4O_2PBr$. Ia (30% yield). Ib (66% yield). b_9 99.5-100°, $b_{10.5}$ 102°, d_4^{20} 1.719, n_D^{20} 1.6148,[21] ^{31}P -195.6 ± 2.0 ppm.[826]

I.1.9. Iodophosphites

TYPE: $(RO)_2PI$

$(EtO)_2PI$. $(EtO)_2PCl$ + LiI (in Et_2O). $b_{0.01}$ 45°, d^{20} 1.599.[1349]

I.1.10. Trialkyl or Triaryl Phosphites

TYPE: $(RO)_3P$

$(MeO)_3P$. Ib.[898,1408] Ih.[69,70,129] b_{23} 22°,[129] b. 111-2°, b_{745} 110-1.5°,[898] b_{760} 111-2°,[70] d_0^0 1.0790, d_0^{20} 1.0540,[69,70] d_0^{20} 1.0520,[129] n_D^{20} 1.4095,[129] 1.4090.[1408] Salts with cuprous halides: A_2 CuI, m. 69-70°; A CuCl, m. 190-2°; A CuBr, m. 180-2°; A CuI, m. 175-7°;[69,70] Adduct with AuCl, m. 100-1°;[792] A AuCl $2NH_3$, m. 75-6°.[788] See also Ref. 276. ^{31}P -139.6 to

-141 ppm,[826] ^1H NMR,[496a] UV,[328a,524b,893a,1264] IR,[855] Raman,[855] mass spect.[1006a]

$(NCCH_2O)_3P$. Ib. Could not be distilled.[774]

$(EtO_2CCH_2O)_3P$. Ib. b_1 176-7°, d_0^{20} 1.2104, n_D^{20} 1.4491.[351,1196]

$(i\text{-}BuO_2CCH_2O)_3P$. Ib. b_2 185-215°.[349]

$(Bu_2NCOCH_2O)_3P$. Ib. Undistillable oil.[349]

$(PhCH_2O)_3P$. Ib.[478] Ie.[281] $(Me_2N)_3P$ + ROH.[1192] $b_{0.02}$ 185°,[478] n_D^{14} 1.5749,[478] IR,[1192] n_D^{20} 1.5550,[281] $b_{0.02}$ 142-8°,[1192] ^1H NMR,[1192] ^{31}P -138.8 ppm,[826] -138.6 ppm.[1192]

$(Ph_2CHO)_3P$. Ib. Residue (20-5°/0.1 mm). Waxy solid, m. 65-6°. HCl at -10° quickly gave Ph_2CHCl (91.2% yield based on 3 groups).[478]

$(EtO)_3P$. Ib.[129,404,819,837] Ih.[69,70,80,620] Ij.[849,1071] b_{11} 48-9°,[80] b_{12} 48.2°,[621] b_{12} 49°,[129] b_{14} 52°,[404] b_{19} 55°,[819] b_{19} 57.5°,[404] b_{740} 155-6°,[898] b_{755} 154.5-5.5°,[819] b_{757} 157.9°,[87] b. 154-5°,[849] d_0^0 0.9777,[69,70] d_0^0 1.0028,[1071] d_0^{17} 0.9605,[69,70] d_0^{20} 0.9687,[129] d_4^{20} 0.96867,[87] d_4^{20} 0.9665,[849] $n_D^{17.5}$ 1.4140,[404] n_D^{20} 1.41309,[87] n_D^{20} 1.4134,[80] n_D^{20} 1.4135,[129] n_D^{20} 1.4136, MR_D 42.75 (calc. 42.97).[849] Salts with CuCl, liquid: CuBr, m. 27-8°; CuI, m. 109-10°,[69,70] m. 111-2°;[80] AgCl, m. 4.5-5.5°; AgBr, m. 40-0.5°; AgI, m. 81-3°.[72] See also Refs. 737, 908, and 1408. ^{31}P -136.9 to -140 ppm,[826] ^1H NMR,[1277] UV,[328a,524b,893a] IR,[176,324,855] Raman,[855] mass spect.[1006a]

$(ClCH_2CH_2O)_3P$. Ib (poor).[646] Ic.[643,795,1374] 97% yield, highly pure without distillation,[795] $b_{2.5}$ 112-5°,[643] b. 98°,[1374] d_0^{26} 1.3453, d_4^{26} 1.3443, n_D^{26} 1.4818,[643] n_D^{20} 1.4876,[795] n_D^{25} 1.4825,[1374] ^{31}P -138.7, -139 ppm.[935,1342]

$(BrCH_2CH_2O)_3P$. Ic. Undistillable without isomerization.[1227]

$(Cl_3CCH_2O)_3P$. Ia. Ib. $b_{0.1}$ 125-7°, $b_{0.05}$ 122°,[459] b. 263°,[329] n_D^{22} 1.5178.[459]

$(CF_3CH_2O)_3P$. Ia. b_{743} 130-1°, d^{25} 1.4866, n_D^{20} 1.3224.[749] Ib. b. 32-4°, n_D^{20} 1.3235.[786]

$(NH_2CH_2CH_2O)_3P$. Ib. Ie. m. 103-5°.[390,565]

$(NCCHMeO)_3P$. Ib. b_2 152-4°, b_5 166-8°, d^{20} 1.1188, n_D^{20} 1.4470.[774]

$(NCCH_2CH_2O)_3P$. Ib. Could not be distilled.[774]

$(MeOCH_2CH_2O)_3P$. Ib. b_5 138.5-40°, d_0^{20} 1.0960,[129] n_D^{25} 1.4401,[1275] n_D^{20} 1.4402.[129]

$(EtSCH_2CH_2O)_3P$. Ib. b_2 181-3° (decomp.), $b_{0.0007}$ 87-90°, d^{20} 1.1184, n_D^{20} 1.5213, MR_D 94.42 (calc. 94.68).[848]

$(EtO_2CCHMeO)_3P$. Ib. b_2 150-5°, n_D^{10} 1.4382, α_D^{16} + 79.5° (l = 1 dm) [from ethyl(+)-lactate].[477]

$(PhCH_2CH_2O)_3P$. Ib. $b_{0.05}$ 162-71°, n_D^{25} 1.5550,[478] b_1 256-8°,[100] d_0^{20} 1.1089,[100] n_D^{20} 1.5596,[100] γ^{20} 41.27.[100]

$[MeCH(Ph)O]_3P$. Ib. Residue (60°/0.1 mm), d_4^0 1.117, n_D^{21} 1.5440, Decomp. at 120-30°/0.05 mm to styrene.[478] 3HCl at -10° was rapidly absorbed to give MeCHClPh (94%).[478]

$(PhOCH_2CH_2O)_3P$. Ia + $(CH_2)_2O$. d^{20} 1.1950, n_D^{20} 1.5565.[201]

$(2\text{-}ClC_6H_4OCH_2CH_2O)_3P$. Ib. Pot residue (165°/0.65 mm), n_D^{20} 1.5717.[1338]

$(2,4\text{-}Cl_2C_6H_3OCH_2CH_2O)_3P$. Ib. Pot residue (135°/0.53 mm), n_D^{20} 1.5875,[530,1338] n_D^{25} 1.5795.[201]

$(2,4,5\text{-}Cl_3C_6H_2OCH_2CH_2O)_3P$. Ib. Pot residue, $n_D^{26.5}$ 1.5920.[1338]

$(ClCH_2CHPhO)_3P$. Ic. n_D^{25} 1.5760 (not distilled).[893]

$(PhCH_2OCH_2CH_2O)_3P$. Ib. $b_{0.05}$ 195°, n_D^{20} 1.5360.[351]

$[MeCH(C_{10}H_7\text{-}2)O]_3P$. Ib. Residue (20°/0.4 mm). HCl gave $MeCHClC_{10}H_7\text{-}2$ (76.3% based on 3 groups).[478]

$[PhCH_2CH(Ph)O]_3P$. Ib. Viscous residue (80°/0.1 mm), n_D^{26} 1.5870. Decomp. at 165°/0.0005 mm.[478]

$(PrO)_3P$. Ib.[129,898,1292] Ie.[899] Ih.[69,70] Ij.[849] b_{8-10} 83°,[69,70] b_{10} 82-4°,[129] b_{10} 83°,[129,849] b_{12} 86-7°,[1292] b_{13} 89-9.5°, b_{24} 103°,[898] b_{10} 83°, b. 206-7°,[78] d_0^0 0.9705,[69,70] d_0^{20} 0.9522,[129] d_4^{20} 0.9525,[849] $d_0^{21.5}$ 0.9503,[69,70] n^{20} 1.4265,[69,70] n^{20} 1.4290,[849] MR 56.41 (calc. 56.82).[849] Salts: CuCl, liquid; CuBr, liquid; CuI, m. 64-5°.[69,70] ^{31}P -137.9 ppm,[826] UV,[893a,896a] mass spect.[1006a]

$(CF_2HCF_2CH_2O)_3P$. Ib. b. 80°.[786]

$(MeCHClCH_2O)_3P$. Ic. b. 107-8°, n_D^{25} 1.4720, d_{25}^{25} 1.2157,[1374] ^{31}P -142.8, -144.0 ppm.[826]

$(ClCH_2CHClCH_2O)_3P$. Ic. $b_{0.1}$ about 190°, d^{22} 1.5172,[1234] ^{31}P -141.7 ppm.[826]

$(ClCH_2CHBrCH_2O)_3P$. Ic. (decomp. on distillation),[1234] ^{31}P -140.2, -141.5 ppm.[826]

$(CH_2{:}CClCH_2O)_3P$. Ib. $b_{0.0001}$ 92.2-3°, d^{20} 1.2922, n_D^{20} 1.4992.[142]

$(HC{\equiv}CCH_2O)_3P$. Ib. d^{20} 1.1445, n_D^{20} 1.4935,[918] ^{31}P -135 ppm.[826]

$(CH_2{:}CH{\cdot}CH_2O)_3P$. Ib. b_3 71-2.5°, b_9 85-6°, d_0^0 1.0534, d_0^{18} 0.9983,[654] ^{31}P -138.5 ppm.[826] Gave the methanephosphonate with MeI, under reflux for 8 hr.[654]

$(\overline{OCH_2CH}{\cdot}CH_2O)_3P$. Ib.[1380]

$(NCCHEtO)_3P$. Ib. b_2 162-4°, d^{20} 1.0810, n_D^{20} 1.4515.[774]

$(PhCH_2CH_2CH_2O)_3$. Ib. $b_{0.05}$ 195-200°, n_D^{18} 1.5480.[478]

$(iso\text{-}PrO)_3P$. Ib.[404] Ih.[69,70] b_{8-10} 60-1°,[69,70] $b_{12.5}$ 65-7°, b_{11} 63-4°,[78] d_0^0 0.9361, $d_0^{18.5}$ 0.9187.[69,70] Salts: CuCl, m. 112-4°; CuBr, m. 149-50°; CuI, m. 184-5°.[69,70] See also Refs. 276 and 1408. ^{31}P -136.9 to -138 ppm,[826,935,1342] ^1H NMR,[496a] UV,[893a,1263] mass spect.[1006a]

$[(Cl_3C)_2CHO]_3P$. Ib. White rhombic, m. 161°.[464]

$[MeCH(CH_2Cl)O]_3P$. Ic. $b_{0.03}$ 98-9°, d^{20} 1.2285, n_D^{20} 1.4710.[1117]

$(NCCMe_2O)_3P$. Ia.[774] Ib.[774] $b_{0.3}$ 130-1°,[283] b_4 153-4°,[774] d_4^{12} 1.082,[290] d^{20} 1.0749,[774] n_D^{12} 1.4467,[290] n_D^{20} 1.4462,[774] n_D^{20} 1.4468.[774] Reacted with CuCl only with difficulty on heating: adduct, m. 181°.[774] ^{31}P -143.3, -143.5 ppm.[826]

$[MeOCH_2CH(CH_2Cl)O]_3P$. Ib. Viscous oil. d_4^{20} 1.2786, n_D^{20} 1.4732.[764]

$[EtOCH_2CH(CH_2Cl)O]_3P$. Ib. Viscous oil. d_4^{20} 1.2018, n_D^{20} 1.4698.[764]

$[BuOCH_2CH(CH_2Cl)O]_3P$. Ib. Viscous oil. d_4^{20} 1.1781, n_D^{20} 1.4694.[764]

$[BuSCH_2CH(CH_2Cl)O]_3P$. Ia. d_4^{20} 1.1615, n_D^{20} 1.5132.[765]

$[AmSCH_2CH(CH_2Cl)O]_3P$. Ia. d_4^{20} 1.1195, n_D^{20} 1.5049.[765]

$[C_6H_{13}OCH_2CH(CH_2Cl)O]_3P$. Ib. Viscous oil. d_4^{20} 1.1034, n_D^{20} 1.4662.[764]

$[n\text{-}C_6H_{13}SCH_2CH(CH_2Cl)O]_3P$. Ia. d_4^{20} 1.0940, n_D^{20} 1.5034.[765]

$[C_6H_5OCH_2CH(CH_2Cl)O]_3P$. Ib. Viscous oil. n_D^{20} 1.5517.[764]

$[PhCH_2CH(Me)O]_3P$. Ib. $b_{0.05}$ 181°, n^{19} 1.5364, α_D^{21} + 21.12° (1 = 1) (from ROH α_D^{16} + 26.72°). Gave ROH α_D^{16} + 26.32° after hydrolysis with aqueous potassium hydroxide.[478]

$[4\text{-}MeC_6H_4OCH_2CH(CH_2Cl)O]_3P$. Ib. Viscous oil. n_D^{20} 1.5456.[764]

$[n\text{-}C_7H_{15}SCH_2CH(CH_2Cl)O]_3P$. Ia. d_4^{20} 1.0825, n_D^{20} 1.5026.[765]

$[n\text{-}C_8H_{17}SCH_2CH(CH_2Cl)O]_3P$. Ia. d_4^{20} 1.0705, n_D^{20} 1.5012.[765]

$[n\text{-}C_9H_{19}SCH_2CH(CH_2Cl)O]_3P$. Ia. d_4^{20} 1.0601, n_D^{20} 1.5004.[765]

$[n\text{-}C_{10}H_{21}SCH_2CH(CH_2Cl)O]_3P$. Ia. d_4^{20} 1.0465, n_D^{20} 1.4994.[765]

$[n\text{-}C_{12}H_{25}SCH_2CH(CH_2Cl)O]_3P$. Ia. d_4^{20} 1.0250, n_D^{20} 1.4932.[765]

$[(PhCH_2)_2CHO]_3P$. Ib. Residual liquid (15°/0.1 mm), decomp. at 120-80°/0.2 mm.[478]

$(BuO)_3P$. Ib.[100,449,898] Ie.[899] Ih.[79] Ij.[849] b_6 114-6°, b_8 119.5-20°, b_{10} 120°,[79,892] b_{12} 122°,[449] b_{12} 122-3°,[898] b_{13} 125-6°, b_{16} 125°,[449] b_{18} 127-8°,[449] b_{26} 137°,[849] b_8 121.5°,[100] d_0^0 0.9309,[79] d_4^{14} 0.9324,[449] d_0^{17} 0.9201,[79] d_4^{20} 0.92530,[898] d_4^{20} 0.9259,[449] d_4^{20} 0.9267,[849] d_4^{23} 0.9247,[449] d_0^2 0.9133,[100] n_D^{16} 1.4339,[449] n_D^{19} 1.4321,[899] n_D^{20} 1.4320,[849] n_D^{20} 1.4327,[100] γ^{20} 27.67 [P] 619.6,[100] MR_D 70.28 (calc. 70.67).[849] See also Refs. 201, 276, 768, 908, and 1408. IR,[324] ^{31}P -137.7 to -142.6 ppm,[826,913,1342] ^1H NMR,[496a, 1277] mass spect.[1006a]

$(iso\text{-}BuO)_3P$. Ib.[129] Ih (poor).[69,86] Ij.[849] $b_{4.5}$ 100.5°,[69,129] b_{10} 135-6°,[78] b_{12} 107°, b. 234-5°,[86] b_{14} 112°,[849] d_0^0 0.919,[69] d_0^0 0.9184-0.9196,[86] d_4^0 0.9193,[86] d_0^{20} 0.9040,[129] d_4^{20} 0.904,[69] d_4^{20} 0.9036,[86]

d_4^{20} 0.9060,[849] d_{20}^{20} 0.9052,[86] n_D^{20} 1.4330, MR_D 70.71 (calc. 70.67).[849] CuI salt, m. 48°.[86] UV.[893a]

$(s\text{-}BuO)_3P$. Ib. b_{10} 101°, n_D^{20} 1.4286,[497] ^{31}P -138.8 to -139.8,[826] -140 ppm,[497] J_{POCH} 9.5 Hz,[826] α_D^{23} + 1.68 from ROH α_D^{20} -5.21 (l = 1), $\alpha_D^{23.5}$ -1.81 from ROH α_D^{20} + 10.73 (l = 1).[497]

$(t\text{-}BuO)_3P$. Ib.[236,733,827] b_4 65-6°,[236,733] n_D^{25} 1.4229,[236,733] m. 4-5°,[827] ^1H NMR,[827] ^{31}P -138.3, -138.2 ppm.[826,827] IR.[827]

$(MeCH{:}CHCH_2O)_3P$. Ib. b_1 98-9°, d^{20} 0.9757, n_D^{20} 1.4680.[1092]

$(C_3F_7CH_2O)_3P$. Ia. b_{16} 97.5-8°, d^{20} 1.6618, n_D^{20} 1.3143.[749]

$(NH_2C_4H_8O)_3P$. Ib. Ie. Also by $P(NEt_2)_3$ + ROH. n_D^{20} 1.4873.[390,565]

$[PrCH(CN)O]_3P$. Ib.[774] Ig (as byproduct).[673] $b_{0.15}$ 140-5°,[673] b_2 168.9°,[774] d^{20} 1.0421,[673] d^{20} 1.0433,[774] n_D^{20} 1.4530.[673,774]

$[i\text{-}PrCH(CN)O]_3P$. Ib.[774] Ig (as byproduct).[673] $b_{0.15}$ 135-40°,[673] b_2 163-4°,[774] d^{20} 1.048,[673] d^{20} 1.0475,[774] n_D^{20} 1.4540,[673] n_D^{20} 1.4545.[774]

$(AmO)_3P$. b_8 126-7°, d^{20} 0.9170, n_D^{20} 1.4303.[1059]

$(i\text{-}AmO)_3P$. Ih. b. 270-5°, d_0^{15} 0.9005,[620,1156] UV.[1264]

$(PrMeCHO)_3P$. Ib. $b_{0.5}$ 118-9°, n_D^{20} 1.4319, ^{31}P -140 ppm. α_D^{20} + 1.315 from ROH α_D^{20} -5.66 (l = 1).[279]

$(i\text{-}PrMeCHO)_3P$. Ib. $b_{0.1}$ 83-84°, n_D^{20} 1.4361, ^{31}P -141 ppm.[278]

$(EtMeCHCH_2O)_3P$. Ib. $b_{0.5}$ 146-8°, n_D^{20} 1.4371, ^{31}P -138 ppm.[278]

$(Me_3CCH_2O)_3P$. Ib. $b_{0.15}$ 80°, m. 55-7°,[278] IR,[176] ^{31}P NMR -137,[278] -137.7, -137.9,[826] -139 ppm,[575] J_{POCH} 6.2 Hz.[826]

$(Et_2CHO)_3P$. Ib. $b_{0.25}$ 94°, n_D^{20} 1.4389, ^{31}P -142 ppm.[577]

$[NCCH(CH_2CHMe_2)O]_3P$. Ib. b_2 182-5°, d^{20} 1.0138, n_D^{20} 1.455.[774]

$(C_6H_{13}O)_3P$. Ib. b_3 167-8°, d_0^{20} 0.9002, n_D^{20} 1.4405, γ^{20} 27.86, [P] 853.5. Purer by Ie [from $(EtO)_3P$]. b_2 157-7.5°, d_0^{20} 0.8981, n_D^{20} 1.4428, γ^{20} 27.82, [P] 856.1 (close to structure requirement for 2 parallel chains).[100] b. 138-9°, n_D^{20} 1.4419,[1059] ^{31}P -138.2 ppm.[826]

$[NH_2(CH_2)_6O]_3P$. ROH + $P(NEt_2)_3$. Oil.[390]

$[CH_3(CH_2)_2CH(NH_2)(CH_2)_2O]_3P$. Ib.[565]

$[(CH_2)_5C(CN)O]_3P$. Ib. m. 75°.[774]

$(C_7H_{15}O)_3P$. Ie. $b_{0.7}$ 164-5°, d^{20} 0.8940, n_D^{20} 1.4465.[1059a]

$(C_8H_{17}O)_3P$. Ib. b_1 210°, d_0^{20} 0.8936, n_D^{20} 1.4489, [P] 1084.1. By Ie [from $(EtO)_3P$]. b_2 212-4°, d_0^{20} 0.8907, n_D^{20} 1.4475, γ^{20} 28.90, [P] 1089.8 (close to 2-chain parallelism).[100] See also Ref. 201. ^{31}P -140 ppm.[1342]

$(2\text{-}C_8H_{17}O)_3P$. Ib. b_2 162-4°, d_4^{22} 0.8843, n_D^{22} 1.4449, α_D^{16} + 0.8° from ROH + 8.0° (l = 1 dm).[450]

Triisooctyl phosphite. Ie. b_{9-10} 75-90°, d^{29} 0.900, n_D^{25} 1.4520.[548]

$(BuEtCHCH_2O)_3P$. Ie.[281] Ib.[276] $(Me_2N)_3P$ + ROH.[1247] b_1 155-6°,[1059] b_1 153°,[1247] n_D^{20} 1.4520,[281] n_D^{20} 1.4476,[1059] d^{20} 0.9114,[1059] ^{31}P -137 ppm,[826] IR.[176]

$(CH_2{=}CH\cdot CHCH_2CHClCHO)_3P$ (with CH_2CH_2 bridging the marked carbons). Ic. Purified at 200°/0.15 mm for < 20 min (isomerizaton occurs on longer heating). d^{20} 1.1638, n_D^{20} 1.5220.[444]

$(C_9H_{19}O)_3P$. Ie [from $(EtO)_3P$]. b_1 226.5-7.5°, b_3 223-5°, d_0^{20} 0.8879, d^{20} 0.8873. n_D^{20} 1.4505, n_D^{20} 1.4495, γ^{20} 29.39, [P] 1208 (very close to 2-chain parallelism).[100]

$(C_{10}H_{21}O)_3P$. Ib. b_2 263°, b_3 254-5°, d^{20} 0.8857, d_0^{20} 0.8856, n_D^{20} 1.4500, n_D^{20} 1.4557, γ^{20} 29.76, [P] 1325.7. By Ie [from $(EtO)_3P$]. $b_{0.5}$ 236.5-7.5°, d_0^{20} 0.8842, n_D^{20} 1.4518,[100] n_D^{25} 1.4556,[428] γ^{20} 29.99, [P] 1330.2 (very close to 2-chain parallelism).[100] ^{31}P -138.7 ppm.[826]

Triisodecyl phosphite. Ie. n_D^{25} 1.4556.[428]

$(C_{16}H_{33}O)_3P$. Ie.[130,131,1059a] $b_{0.4}$ 288-90°, m. 51° (from hexane),[130,131] m. 53-4°.[1059a]

Tris(2-hexyldecyl) phosphite. Ib. $b_{0.006}$ 230°, d^{20} 0.878, n_D^{20} 1.409.[224]

$(C_{18}H_{37}O)_3P$. Ie. m. 40°.[428]

Tris(2-octyldecyl) phosphite. Ib. $b_{0.011}$ 300°, d^{20} 0.8745, n_D^{20} 1.4644.[224]

$(2\text{-}MeC_6H_{10}O)_3P$. Ib. n_D^{20} 1.4842.[1023]

$(3\text{-}MeC_6H_{10}O)_3P$. Ib. n_D^{20} 1.4890.[1023]

$(4\text{-}MeC_6H_{10}O)_3P$. Ib. n_D^{20} 1.4874.[1023]

$(4\text{-}i\text{-}PrC_6H_{10}O)_3P$. Ib. n_D^{20} 1.4893.[1023]

$(4\text{-}t\text{-}BuC_6H_{10}O)_3P$. Ib. n_D^{20} 1.4882.[1023]

$(4\text{-}t\text{-}AmC_6H_{10}O)_3P$. Ib. n_D^{20} 1.4861.[1023]

Tri-2-decahydronaphthyl phosphite. Ih. m. 75°.[575]

Tri-l-menthyl phosphite. Ib. m. 44-5°.[894,897]

(1,2:5,6-di-O-isopropylidene-α-D-gluco·O)$_3$P. $(Et_2N)_3P$ + ROH. m. 56-62°.[971]

(1,2:3,4-di-O-isopropylidene-α-D-galacto·O)$_3$P. $(Et_2N)_3P$ + ROH. m. 55-7°.[971]

Tris(dipropylene glycol)phosphite. Ie. d_4^{25} 1.097, n_D^{25} 1.4610.[1151]

Tris(polypropylene glycol 425) phosphite. Ie. d_4^{25} 1.028, n_D^{25} 1.4535.[1151]

Tris(polypropylene glycol 1025) phosphite. Ie. d_{15}^{25} 1.022, n_D^{25} 1.4515.[1151]

Tris(polypropylene glycol 2025) phosphite. Ie. d_{15}^{25} 1.006, n_D^{25} 1.4501.[1151]

$(PhO)_3P$. Ia.[63,129,570,982,1374] Ib.[595,899,1292] $b_{0.01}$ 129-30°, $b_{0.05}$ 117-36°,[908] $b_{0.8}$ 161°,[570] b_1 183-

4°,[1374] b_1 209-10°,[498] b_5 200-1°,[1292] b_{11} 220°,[63] b_{12} 228°,[129] b_{18} 235°,[64] b > 360°,[982] d_{18}^{18} 1.184,[982] d_0^{20} 1.1844,[129] d^{20} 1.188,[570] d_{25}^{25} 1.183,[1374] n_D^{20} 1.591,[570] n_D^{25} 1.5890,[1374] m. 25°,[570,1292] m. 17-22°,[498] ^{31}P -125 to -129 ppm,[826] IR,[176,324] mass spect.[1006a] Salts with cuprous halides: 2A CuCl, m. 70°; 2A CuBr, m. 73-4°; 2A CuI, m. 73-5°; A CuCl, m. 95-6°; A CuBr, m. 90.5-1.5°; all crystallize from Et_2O.[70] 2A $PtCl_2$, m. 155°.[1225]

$(2\text{-}ClC_6H_4O)_3P$. Ia. b_3 230°, n_D^{25} 1.6041.[1374]

$(4\text{-}ClC_6H_4O)_3P$. Ia.[666,859,1374] m. 49°,[666,859] $b_{0.08}$ 119-46°,[908] $b_{1.5}$ 207,[1374] b_{15} 290-7°,[859] b_3 230-1°, b_{15} 290°,[666] m. 48-50.[1374] CuCl complex, m. 73°.[666]

$(2,4\text{-}Cl_2C_6H_3O)_3P$. Ia. b_3 254-5°, m. 56°. Adds CuCl, MeI, S.[657]

$(2,4,6\text{-}Cl_3C_6H_2O)_3P$. Ia. m. 183-4°. CuCl complex, m. 176.[666]

$(2\text{-}NH_2C_6H_4O)_3P$. Ib. Ie. m. 145°.[390,565]

$(3\text{-}NH_2C_6H_4O)_3P$. Ib. Ie. m. 115-9°. As oil, n_D^{20} 1.6448.[390]

$(3\text{-}NH_2\text{-}2\text{-}ClC_6H_3O)_3P$. Ib. Ie. m. 160°.[565]

$(3\text{-}NH_2\text{-}6\text{-}ClC_6H_3O)_3P$. Ib. m. 160°.[390]

$(2\text{-}O_2NC_6H_4O)_3P$. Ia. m. 126°.[666]

$(4\text{-}O_2NC_6H_4O)_3P$. Ia.[666,1299] Ib.[595] Needles, m. 170-1°,[595,1299] m. 170°[666] (decomposes above m.).[1299]

$(2\text{-}MeC_6H_4O)_3P$. Ia.[219,570,1299,1374] b_1 193-4°,[1374] b_{11} 238°,[219] b_{11} 248°,[1299] $b_{0.8}$ 182-8°, d^{20} 1.138,[570] d_{25}^{25} 1.1195,[1374] n_D^{20} 1.578,[570] n_D^{25} 1.5760.[1374]

$(3\text{-}MeC_6H_4O)_3P$. Ia.[219,570,1374] b_1 188°,[1374] b_7 235-8°, b_{10} 240-3°, b_{12} 248-50°,[219] b_5 217°, d^{20} 1.127,[570] d_{25}^{25} 1.1195,[1374] n_D^{20} 1.577,[570] n_D^{25} 1.5734.[1374]

$(4\text{-}MeC_6H_4O)_3P$. Ia.[570,859,1299,1374] b_1 194°,[1374] b_{10} 250-5°,[859] b_{11} 285°,[1299] $b_{0.8}$ 198-200°, d^{20} 1.128,[570] d_{25}^{25} 1.1107,[1374] n_D^{20} 1.575,[570] n_D^{25} 1.5734,[1374] ^{31}P -127.6 ppm.[935]

$(3\text{-}NH_2\text{-}4\text{-}MeC_6H_3O)_3P$. Ib. Crystals.[390]

$(2\text{-}MeOC_6H_4O)_3P$. Ia.[360] Older preparation by Ih[160] is faulty. m. 59°, b_{13} 275-80°.[360]

$(p\text{-}CNC_6H_4O)_3P$. Ib. m. 129-30°.[595]

$(p\text{-}CH_3SO_2C_6H_4O)_3P$. Ib. m. 185-8°.[595]

$(p\text{-}Me_2NC_6H_4O)_3P$. Ib. m. 71°.[595]

$(p\text{-}CH_3\cdot O\cdot COC_6H_4O)_3P$. Ib.[595]

$(i\text{-}PrC_6H_4O)_3P$. Ia. n_D^{20} 1.5464.[815]

Tripseudocumyl phosphite. Ia. b_{16} 270-4°, d_0^{17} 1.097.[859]

$(2\text{-}CH_2{:}CHCH_2C_6H_4O)_3P$. Ib. b_7 245°, d^{20} 1.1030, n_D^{20} 1.5712,[421] MR_D 128.10 (calc. 127.75).[421]

$(4\text{-}tert\text{-}BuC_6H_4O)_3P$. Ia.[768,815,923,1374] b_8 288-94°,[923] b_1 253-4°,[1374] b_{10} 250-60°, n_D^{20} 1.5469,[768] n_D^{20} 1.5456,[815] d^{20} 1.0570,[768] m. 75°,[1374] m. 75-6°.[923]

$(4\text{-}t\text{-}AmC_6H_4O)_3P$. Ia.[768,815] b_{10} 280-5°,[768] b. 284°,[592] n_D^{20} 1.5420,[592] n_D^{20} 1.5390,[815] n_D^{20} 1.5415, d^{20}

1.0388,[768] d^{20} 1.0457.[592]
$(4\text{-s-}C_6H_{13}C_6H_4O)_3P$. Ia. n_D^{20} 1.5263.[815]
$(2\text{-}PhC_6H_4O)_3P$. Ia. b_5 336-40°, m. 95°.[923]
Tri-o-cyclohexylphenyl phosphite. Ia. b_8 324-9°, n_D^{60} 1.5580.[923]
$(4\text{-s-}C_7H_{15}C_6H_4O)_3P$. Ia.[768,815] b_2 280-90°, n_D^{20} 1.5266,[768] n_D^{20} 1.5232,[815] d^{20} 0.9950.[768]
$(4\text{-s-}C_8H_{17}C_6H_4O)_3P$. Ia. n_D^{20} 1.5159.[815]
$(4\text{-}Me_3CCH_2CMe_2C_6H_4O)_3P$. Ia. b_{10} 354-7°, n_D^{60} 1.5205.[923]
$(4\text{-s-}C_9H_{19}C_6H_4O)_3P$. Ia. n_D^{20} 1.5133.[815]
$(4\text{-}PhCMe_2C_6H_4O)_3P$. Ib. m. 64°.[1334]
$(4\text{-}PhCMe_2\text{-}2\text{-}BrC_6H_3O)_3P$. Ib. m. 56°.[1334]
$(4\text{-}PhCMe_2\text{-}2\text{-}NO_2C_6H_3O)_3P$. Ib. m. 95°.[1334]
$(4\text{-}PhCMe_2\text{-}2,6\text{-}Cl_2C_6H_2O)_3P$. Ib. m. 181°.[1334]
$(4\text{-s-}C_{10}H_{21}C_6H_4O)_3P$. Ia. n_D^{20} 1.5047.[815]
$[2,4\text{-}(Me_2EtC)_2C_6H_3O]_3P$. Ia. d^{20} 0.9861, n_D^{20} 1.5238.[592]
Tris(distyrylphenyl) phosphite. Ia. d_{25}^{25} 1.122, n_D^{25} 1.6000.[1339]
$(1\text{-}C_{10}H_7O)_3P$. Ia.[58,204] Ib.[595] m. 91° (from xylene),[58,595] m. 90-1°,[204] $b_{0.04}$ 250°.[204]
$(2\text{-}C_{10}H_7O)_3P$. Ia.[58,204] Ib.[595] m. 94° (from xylene),[58] m. 93-5°,[204] m. 93-4°,[697] $b_{0.04}$ 257°,[204] m. 97°.[595]
Tri-1-(2,4-dibromo)naphthyl phosphite. Ia. m. 289°; does not add bromine or chlorine.[58]
Tri-2-(1,6-dibromo)naphthyl phosphite, Ia. m. 245°; does not add bromine or chlorine.[58]
Tri-1-anthryl phosphite. Ia. Decomp. 182-90°; does not add bromine or chlorine.[58]

I.1.11. Triacyl Phosphites

TYPE: $(RCO_2)_3P$

$(C_6H_5CO_2)_3P$. II ($RCO_2Na + PCl_3$). m. 93-4°.[1061]

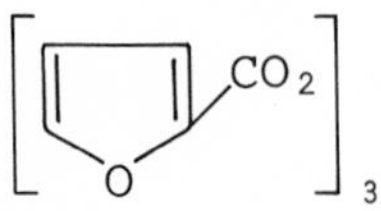

P. II ($RCO_2Na + PCl_3$). m. 123.5-5°.[1061]

I.1.12. Mixed Phosphites

TYPE: $(RO)_2(R'O)P$

$(MeO)_2POEt$. Ib ($EtOPCl_2$). b_{14} 28-9°, b_{759} 122-4°, d^{20} 1.0259, n_D^{20} 1.4110.[663]
$(MeO)_2POPr$. Ie. b_{14} 47-8°, d_4^{20} 0.9939, n_D^{20} 1.4165, MR_D 38.45 (calc. 38.34).[84]
$(MeO)_2POCMe_2CN$. Ib. b_{11} 78-80°, d^{20} 1.0489, n_D^{20} 1.4240.[774]
$(MeO)_2POBu$. Ib ($BuOPCl_2$). b_{15} 65-6°, d_0^0 1.0010, n_D^{20}

1.4215.[663]

$(MeO)_2POCH(CH_2Cl)CH_2OCH_2CH_2Cl$. Ib. $b_{0.5}$ 90-2°, d_4^{20} 1.2550, n_D^{20} 1.4688.[687]

$(MeO)_2POCMeEtCCl_3$. Ib. $b_{0.1}$ 78-9°, d^{20} 1.3130, n_D^{20} 1.4857.[368]

$(MeO)_2POPh$. Ib. b_{12} 86°, d^{20} 1.1218, n_D^{20} 1.4940,[661] ^{31}P -135.2 ppm, J_{POCH} 9.7 Hz.[826]

$(MeO)_2PO$-2,4,5-$Cl_3C_6H_2$. IVh. b_2 126-8°, d^{20} 1.4072, n_D^{20} 1.5400.[1047]

$(MeO)_2POC_6H_4CH_2CH{:}CH_2$-2. Ib. b_{12} 107-8°, d^{20} 1.0768, n_D^{20} 1.4986.[421]

$(MeO)_2POC_{10}H_7$-1. Ib. b_2 138°, d^{20} 1.1861, n_D^{20} 1.5808.[49]

$(MeO)_2POC_{10}H_7$-2. Ib. b_2 145-7°, d^{20} 1.1766, n_D^{20} 1.5867.[665]

$(EtO_2CCH_2O)_2POEt$. Ib. b_2 146°, d_0^{19} 1.1513, n_D^{19} 1.4412.[1197]

Dibenzyl 1,2:3,4-di-O-isopropylidene-α-D-galactopyranosyl phosphite.

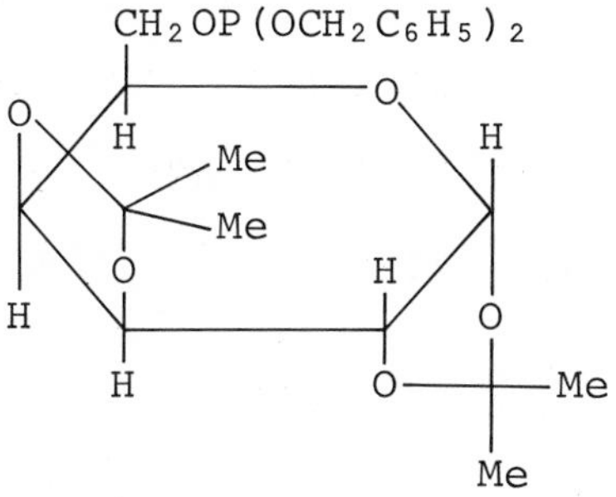

IVh. $b_{0.015}$ 190-200°, n_D^{20} 1.5372.[1060]

Dibenzyl 1,2:5,6-di-O-isopropylidene-α-D-glucofuranosyl phosphite.

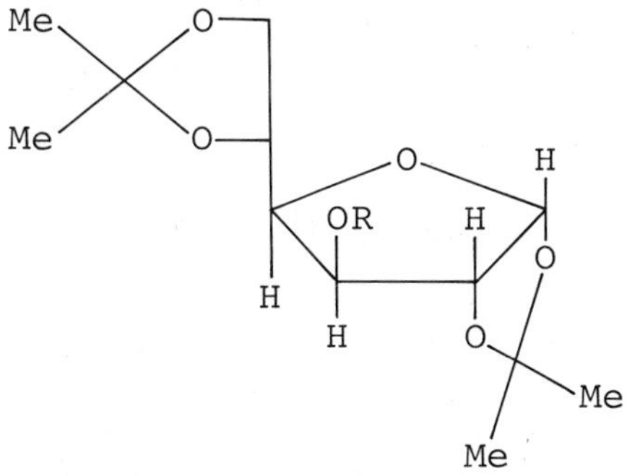

IVh. n_D^{20} 1.6245.[1060]

R = $PO(CH_2C_6H_5)_2$

$(EtO)_2POCH_2CH_2F$. Ib. $b_{1.5}$ 41-2°, n_D^{20} 1.4208, d^{20} 1.0980.[634]

(EtO)$_2$POCH$_2$CH$_2$NMe$_2$. b$_{14}$ 91-2°, d^{20} 0.9665, n_D^{20} 1.4330.[633]

(EtO)$_2$POCH$_2$CH$_2$SEt$_2$. IVh. b$_3$ 88-90°, d^{20} 1.0227, n_D^{20} 1.4635.[1047]

(EtO)$_2$POCH$_2$CH$_2$CN. Ib. b$_{10}$ 119-20°, d^{20} 1.0579, n_D^{20} 1.4352.[774]

(EtO)$_2$POCH$_2$CH$_2$Ac. Ib. b$_{0.04}$ 57-8°, d^{20} 1.0398, n_D^{20} 1.4367.[1219]

(EtO)$_2$POCH$_2$CH$_2$O$_2$CC(CH$_3$):CH$_2$. Ib.[1076,1126] b$_1$ 83-4°, d_4^{20} 1.0625,[1126] n_D^{20} 1.4430, d 1.1097,[1076] n_D^{20} 1.4483.[1126]

(EtO)$_2$POC(Cl)MePh. Ib. Or from (EtO)$_2$PCl + styrene oxide. b$_3$ 123-5°, d^{20} 1.1325, n_D^{20} 1.5037.[1112]

(EtO)$_2$POCH(CCl$_3$)C$_6$H$_4$Cl-p. Ib. b$_{0.1}$ 131-2°, d^{20} 1.3367, n_D^{20} 1.5275.[33]

(EtO)$_2$POPr. Ie. b$_{23}$ 72-4°, d_4^{20} 0.9547, n_D^{20} 1.4150, MR$_D$ 47.27 (calc. 47.59).[663]

(EtO)$_2$POCH$_2$C≡CH. Ib. b$_{0.03}$ 34-5°, d^{20} 1.0245, n_D^{20} 1.4360, MR$_D$ 45.02 (calc. 45.59),[1101] ^{31}P -137 ppm.[826]

(EtO)$_2$POCH$_2\overline{\text{CHCH}_2\text{O}}$. Ib.[1218,1380] b$_{0.5}$ 51-4°,[1218] b$_{8-9}$ 88-90°,[1380] d^{20} 1.0757, n_D^{20} 1.4380,[1218] n_D^{20} 1.4451.[1380]

(EtO)$_2$POi-Pr. Ib (i-PrOPCl$_2$). b$_{20}$ 61-3°.[581] ^{31}P -137.3, -137.1 ppm, J_{PH} 8.6 Hz.[826]

(EtO)$_2$POCH(CH$_2$Cl)Me. b$_{10}$ 84-5°, d^{20} 1.0761, n_D^{20} 1.4355.[1117]

(EtO)$_2$POCH(CH$_2$Cl)$_2$. Ib. Ic. b$_{1.5}$ 78-5°, d^{20} 1.1856, n_D^{20} 1.4565.[1111]

(EtO)$_2$POCH(CH$_2$NH$_2$)$_2$. Ib. Ie. Oil.[390,565]

(EtO)$_2$POCMe$_2$CN. Ib. b$_{10}$ 89-91°, d^{20} 1.0108, n_D^{20} 1.4190.[774]

(EtO)$_2$POCH$_2$CHMeAc. Ib. b$_{0.05}$ 60-2°, d^{20} 1.0206, n_D^{20} 1.4365.[1219]

(EtO)$_2$POCHMeCH$_2$Ac. Ib. b$_{0.05}$ 58-9°, d^{20} 1.0112, n_D^{20} 1.4358.[1219]

(EtO)$_2$P·OCMe=CH·CO$_2$Me. (EtO)$_2$PCl + AcCH$_2$CO$_2$Me + Et$_3$N. b$_3$ 93-6°, d^{20} 1.0778, n_D^{20} 1.4580.[647]

(EtO)$_2$P·OCMe=CHAC. (EtO)$_2$PCl + AcCH$_2$Ac + Et$_3$N. b$_2$ 85-7°, d^{20} 1.0450, n_D^{20} 1.4620.[647]

(EtO)$_2$POCH(CH$_2$Cl)CH$_2$OCH$_2$CH$_2$Cl. Ib. b$_{0.15}$ 96-8°, d_4^{20} 1.1830, n_D^{20} 1.4622.[687]

(EtO)$_2$OBu. Ib. b$_{13}$ 72-4°, d_0^0 0.9749, d^{20} 0.9561, n_D^{20} 1.4262.[663]

(EtO)$_2$PO trans-CH$_2$CH:CHMe. Ib (EtO)$_2$PCl. b$_{12}$ 60-1°.[581]

(EtO)$_2$POCH(Me)CH:CH$_2$. Ib (EtO)$_2$PCl. b$_{0.5-1}$ 44-6°.[581,785]

(EtO)$_2$POCH$_2$CHClCH:CH$_2$. Ib. Or from (EtO)$_2$PCl + butadiene oxide. b$_{15}$ 107-8°, d^{20} 1.0737, n_D^{20} 1.4515.[1112]

(EtO)$_2$POCH(CH$_2$Cl)CH:CH$_2$. Ib. b$_{15}$ 104-5°, d^{20} 1.0707, n_D^{20} 1.4492.[1112]

(EtO)$_2$POCMe$_2$CH$_2$Ac. Ib.[1217,1219] b$_{0.05}$ 73-5°,[1219] b$_{0.1}$ 76-7°,[1217] n_D^{20} 1.4380,[1217,1219] d^{20} 1.0085.[1217,1219]

(EtO)$_2$POCMe$_2$CH$_2$COMe. Ib. b$_{0.06}$ 56-8°, d^{20} 0.9376, n_D^{20} 1.4630.[1219] (sic)(cf. above compound).

$(EtO)_2POAm$-i. Ie. b_{15} 83-7°, d_4^{20} 0.9357, n_D^{20} 1.4230, MR_D 56.71 (calc. 56.83).[663]

$(EtO)_2POCMeEtCCl_3$. Ib. $b_{0.18}$ 89-90°, d^{20} 1.2430, n_D^{20} 1.4768.[368]

$(EtO)_2POCH_2\overline{CH\cdot CH_2\cdot O\cdot CHMe\cdot O}$. Ib. $b_{0.5}$ 79°, d_4^{20} 1.0764, n_D^{20} 1.4390, MR_D 58.15 (calc. 57.94).[138]

$(EtO)_2PO\overline{CH\cdot CH_2\cdot O\cdot CHMe\cdot O\cdot CH_2}$. Ib. b_2 97°, d_4^{20} 1.1017, n_D^{20} 1.4490, MR_D 57.91 (calc. 57.94).[138]

$(EtO)_2POC_6H_{11}$. Ib ($C_6H_{11}OPCl_2$). b_{20} 122-5°.[581]

$(EtO)_2PO2$-C_8H_{17}. Ib [$(EtO)_2PCl$]. $b_{0.5-1}$ 74-80°, α_D^{26} + 4.8 (c 10.2, EtOH) from ROH α_D^{27} + 9.2 (c 10.0 EtOH).[581]

$(EtO)_2(EtBuCHCH_2O)P$. IR.[176]

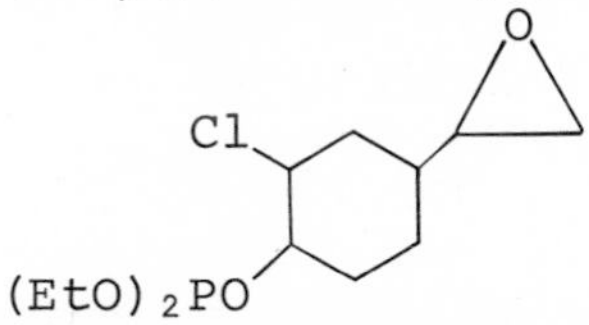

Ic. $b_{0.012}$ 103-4°, n 1.4790, d 1.1491.[933]

$(EtO)_2PO$-exo-norbornyl. Ib [$(EtO)_2PCl$]. $b_{0.3}$ 64°.[581]

$(EtO)_2PO$-endo-norbornyl. Ib [$(EtO)_2PCl$]. $b_{0.3}$ 62-4°.[581]

D(-)-pantolactone 2-(diethyl phosphite). Ie. b_{15} 151-5°, $b_{0.1}$ 88°, n_D^{25} 1.4536.[809]

$(EtO)_2POPh$. Ib.[661] IVh.[1197] b_{11} 111-2°,[661] b_{12} 113-5°,[1149] d^0 1.0848, d^{20} 1.0630,[661] d^{20} 1.0550, n_D^{20} 1.4931,[1149] n_D^{20} 1.4885.[661]

4-$ClC_6H_4OP(OEt)_2$. IVh. b_7 128-9°, d^{20} 1.1576, n_D^{20} 1.5050.[1149]

4-$MeC_6H_4OP(OEt)_2$. IVh. b_7 116-9°, d^{20} 1.0444, n_D^{20} 1.4932.[1149]

$(EtO)_2POC_6H_4CH_2CH{:}CH_2$-2. Ib. b_{12} 112°, d^{20} 1.0350, n_D^{20} 1.4903.[421]

$(EtO)_2POC_{10}H_7$-1. Ib. b_2 150-1°, d^{20} 1.1354, n_D^{20} 1.5721.[49]

$(EtO)_2POC_{10}H_7$-2. Ib. b_2 151-3°, d^{20} 1.1134, n_D^{20} 1.5640.[665]

$(FCH_2CH_2O)_2POEt$. Ib. $b_{0.5}$ 47-8°, n_D^{20} 1.4200, d^{20} 1.1772.[634]

$(ClCH_2CH_2O)_2POCH(CN)Me$. Ic. b_{10} 166-8°, d^{20} 1.2773, n_D^{20} 1.4735.[774]

$(ClCH_2CH_2O)_2POCMe_2CN$. Ib. b_{11} 160-3°, d^{20} 1.2350, n_D^{20} 1.4685.[774]

$(NCCH_2CH_2O)_2POCH_2CH{=}CH_2$. Ib. $b_{8.5}$ 165-70°, d^{20} 1.0989, n_D^{20} 1.4580.[774]

$(NCCH_2CH_2O)_2POBu$. Ib. b_8 190-2°, d^{20} 1.0837, n_D^{20} 1.4475.[774]

$(MeOCH_2CH_2O)_2POAm$. Ib. b_{14} 153-60°, d^{20} 0.9569, n_D^{20} 1.4395.[951]

$(EtOCH_2CH_2O)_2POAm$. Ib. $b_{13.5}$ 156-7°, d^{20} 0.9484, n_D^{20} 1.4366.[951]

$(EtNHCH_2CH_2O)_2POEt$. $b_{0.1}$ 71-3°.[924]

$(CH_2{:}CHCH_2NHCH_2CH_2O)_2POCH_2CH{:}CH_2$. By $ROP(NMe_2)_2$ + ROH. Oil.[390]

$(PhNHCH_2CH_2O)_2POEt$. EtOP:NPh + $PhNHCH_2CH_2OH$.[908] $b_{0.005}$ 100-4°,[924] $b_{0.009}$ 114-6°.[908]

$(CH_2{:}CHOCH_2CH_2O)_2POC_6H_5$. Ib. b_1 147-50°, d^{20} 1.1251, n_D^{20} 1.5066,[692] IR.[1271]

$(CH_2{:}CHOCH_2CH_2O)_2PO\text{-}1\text{-}C_{10}H_7$. Ib. $b_{0.5}$ 173-6°, d^{20} 1.1595, n_D^{20} 1.5620.[692]

$(CH_2{:}CHOCH_2CH_2O)_2PO\text{-}2\text{-}C_{10}H_7$. Ib. $b_{0.5}$ 177-9°, d^{20} 1.1579, n_D^{20} 1.5613.[692]

$(PrO)_2POCH_2CO_2Pr$. Ib. b_8 139-40°, $b_{0.25}$ 116-7°, d_0^{20} 1.001, n_D^{22} 1.4267, compound with CuCl, CuBr.[1197]

$(PrO)_2POEt$. Ib $(EtOPCl_2)$. b_{17} 85.7°, d_0^0 0.9720, n_D^{20} 1.4215.[663]

$(PrO)_2POCH_2CH_2Ac$. Ib. $b_{0.02}$ 77-8°, d^{20} 1.0069, n_D^{20} 1.4391.[1219]

$(PrO)_2POCClPhMe$. Ic, from styrene oxide. b_3 140-1°, d^{20} 1.0960, n_D^{20} 1.4973.[1112]

$(PrO)_2POCH_2CH_2O_2CC(CH_3){:}CH_2$. Ib. b_2 118-9°, d^{20} 1.0543, n_D^{20} 1.456.[1126]

$(PrO)_2POCH(CCl_3)C_6H_4Cl\text{-}p$. Ib. $b_{0.1}$ 144-6°, d^{20} 1.2863, n_D^{20} 1.5190.[33]

$(PrO)_2POCH_2\overline{CHCH_2O}$. Ib. $b_{0.5}$ 77-9°, d^{20} 1.0373, n_D^{20} 1.4408.[1218]

$(PrO)_2POCH(CH_2Cl)Me$. Ic. b_{10} 108-10°, d^{20} 1.0353, n_D^{20} 1.4380.[1117]

$(PrO)_2POCH(CH_2Cl)_2$. Ia. b_5 117-9°, d^{20} 1.1392, n_D^{20} 1.4566.[1111]

$(PrO)_2POCMe_2CN$. Ib. b_{10} 103-4°, d^{20} 0.9845, n_D^{20} 1.4290.[774]

$(PrO)_2POCH(CH_2Cl)CH_2OMe$. Ic. b_2 102-4°, d^{20} 1.0668, n_D^{20} 1.4456.[1111]

$(PrO)_2POCH(CH_2Cl)CH_2OEt$. Ib. Ic. b_7 131-2°, d^{20} 1.0468, n_D^{20} 1.4430.[1111]

$(PrO)_2POCH_2CHMeAc$. Ib. $b_{0.01}$ 82-3°, d^{20} 0.9933, n_D^{20} 1.4400.[1219]

$(PrO)_2POCH(CH_2Cl)CH_2OCH_2CH_2Cl$. Ib. $b_{0.15}$ 119-20°, d_4^{20} 1.1380, n_D^{20} 1.4614.[687]

$(PrO)_2POBu$. Ie. b_{24} 110-5°, d^{20} 0.9276, n_D^{20} 1.4265.[377]

$(PrO)_2POCMe_2CH_2Ac$. Ib.[1217,1219] $b_{0.01}$ 76-8°,[1219] $b_{0.03}$ 76-8°,[1217] n_D^{20} 1.4400,[1217,1219] d^{20} 0.9865.[1217,1219]

$(PrO)_2POAm\text{-}i$. Ie. b_5 64-6°, d_4^{20} 0.9223, n_D^{20} 1.4290, MR_D 66.07 (calc. 66.07).[84]

$(PrO)_2POCMeEtCCl_3$. Ib. $b_{0.24}$ 112-3°, d^{20} 1.1880, n_D^{20} 1.4748.[368]

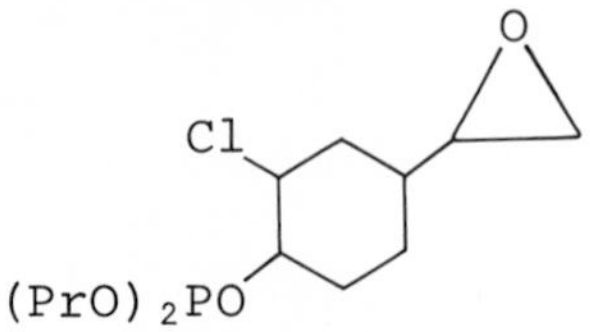

Ic. $b_{0.007}$ 110-2°, n 4764, d 1.1112.[933]

$(PrO)_2POPh$. Ib. b_7 122-3°, d^0 1.0550, d^{20} 1.0370, n_D^{20} 1.4865.[661]

$(PrO)_2POC_6H_4CH_2CH{:}CH_2$-2. Ib. b_{12} 140°, d^{20} 0.9977, n_D^{20} 1.4815.[421]

$(PrO)_2POC_{10}H_7$-1. Ib. b_2 162-3°, d^{20} 1.0913, n_D^{20} 1.5497.[49]

$(PrO)_2POC_{10}H_7$-2. Ib. b_2 165-6°, d^{20} 1.0781, n_D^{20} 1.5501.[665]

$(CH_2{:}CHCH_2O)_2POCH_2CH_2CN$. Ib. $b_{8.5}$ 140-5°, d^{20} 1.0691, n_D^{20} 1.4636.[774]

$(CH_2{:}CHCH_2O)_2POCMe_2CN$. Ib. b_9 114-16°, d^{20} 1.0230, n_D^{20} 1.4425.[774]

$[CH_2{:}C(NH_2)CH_2O]_2POCH_2CH{:}CH_2$. Ib. Ie.[565]

$(CH_2{:}CHCH_2O)_2POC_{10}H_7$-2. Ib. b_3 170-2°, d^{20} 1.1173, n_D^{20} 1.5732.[665]

$(OCH_2CHCH_2O)_2POMe$. Ib. $b_{0.5}$ 94-6°, d^{20} 1.2235, n_D^{20} 1.4648.[1218]

$(OCH_2CHCH_2O)_2POEt$. Ib.[1218,1380] $b_{0.2}$ 104-6°,[1380] $b_{0.5}$ 108-10°,[1218] n_D^{20} 1.4628,[1380] n_D^{20} 1.4608,[1218] d^{20} 1.1763.[1218]

$(OCH_2CHCH_2O)_2POPr$. Ib. $b_{0.5}$ 115-7°, d^{20} 1.1456, n_D^{20} 1.4602, $b_{0.1}$ 109-10°.[1218]

$(OCH_2CHCH_2O)_2POBu$. Ib.[1218,1380] $b_{0.3}$ 120-5°, n_D^{20} 1.4584,[1380] n_D^{20} 1.4600,[1218] d^{20} 1.1234.[1218]

$(OCH_2CHCH_2O)_2PO$i-Bu. Ib. $b_{0.4}$ 122-5°, n_D^{20} 1.4609.[1380]

$(OCH_2CHCH_2O)_2PO$s-Bu. Ib. $b_{0.2-0.3}$ 115-8°, n_D^{20} 1.4599.[1380]

$(OCH_2CHCH_2O)_2POCH_2CH{:}CH_2$. Ib. $b_{0.5}$ 107-8°, d^{20} 1.1758, n_D^{20} 1.4734.[1218]

$(OCH_2CHCH_2O)_2POPh$. Ib. $b_{0.005}$ 129-30°, d^{20} 1.2327, n_D^{20} 1.5219.[1218]

$(OCHMeOCH_2CHCH_2O)_2POEt$. Ib. $b_{0.5}$ 125°, d_4^{20} 1.1533, n_D^{20} 1.4550, MR_D 72.99 (calc. 72.84).[138]

$(CH_2OCHMeOCH_2CHO)_2POEt$. Ib. b_1 137-9°, d_4^{20} 1.1916, n_D^{20} 1.4688, MR_D 72.99 (calc. 72.84).[138]

(i-PrO)$_2$POEt. Ib ($EtOPCl_2$). b_{18} 65.5-6.5°, b_{25} 75-7°, d_0^0 0.9471, n_D^{20} 1.4150.[581,663]

(i-PrO)$_2POCH_2CH_2Ac$. Ib. $b_{0.09}$ 55-6°, d^{20} 0.9936, n_D^{20} 1.4310.[1219]

(i-PrO)$_2POCH_2CH_2O_2CC(CH_3){:}CH_2$. Ib. $b_{1.5}$ 92-3°, d^{20} 1.0250, n_D^{20} 1.4465.[1126]

$(Me_2CHO)_2POCH(CCl_3)C_6H_4Cl$-p. Ib. $b_{0.5}$ 143-5°, d^{20} 1.2843, n_D^{20} 1.5173.[33]

(i-PrO)$_2$POPr. Ie. $(RO)_2POAc$ + ROH. b_7 70-1°, n_D^{20} 1.4430.[430]

(i-PrO)$_2POCH_2CHMeAc$. Ib. $b_{0.05}$ 64-6°, d^{20} 0.9780, n_D^{20}

1.4330.[1219]
MeCH(CH_2Cl)OP(OCHMe)$_2$. Ic. b_3 78-80°, d^{20} 1.1527, n_D^{20} 1.4546.[1117]
(i-PrO)$_2$POCH(CH_2Cl)Me. Ic. b_{10} 97-8°, d^{20} 1.0310, n_D^{20} 1.4310.[1117]
(i-PrO)$_2$POCMe$_2$CN. Ib. b_{11} 99-101°, d^{20} 0.9731, n_D^{20} 1.4221.[774]
(i-PrO)$_2$POCH(CH_2Cl)$CH_2OCH_2CH_2$Cl. Ib. $b_{0.18}$ 111-2°, d_4^{20} 1.1270, n_D^{20} 1.4558.[687]
(i-PrO)$_2$POBu. Ib ($BuOPCl_2$). b_{10} 76-8°, d_0^0 0.9405, n_D^{20} 1.4170.[663]
(i-PrO)$_2$POCMe$_2CH_2$Ac. Ib.[1217,1219] $b_{0.05}$ 79-80°,[1217] $b_{0.05}$ 77-9°,[1219] n_D^{20} 1.4353, d^{20} 0.9742.[1217,1219]
(i-PrO)$_2$POCMeEtCCl$_3$. Ib. $b_{0.08}$ 82°, d^{20} 1.1824, n_D^{20} 1.4703.[368]
(i-PrO)$_2$POPh. Ib. b_{10} 117-8°, d^0 1.0365, d^{20} 1.0168, n_D^{20} 1.4825.[661]
(i-PrO)$_2$POC(C_6H_4Cl-p)$_2$(C_6H_4Cl-o). (RO)$_2$OAg + Cl-CAr$_3$. m. 128-9°.[90]
(i-PrO)$_2$POC$_{10}H_7$-2. Ib. b_2 158-60°, d^{20} 1.0774, n_D^{20} 1.5484.[665]
(MeCH(CH_2Cl)O)$_2$P(OEt). Ic. b_{10} 118-9°, d^{20} 1.1547, n_D^{20} 1.4520.[1117]
[($ClCH_2$)$_2$CHO]$_2$POMe. Ib. b_1 138-9°, d^{20} 1.4010, n_D^{20} 1.4970.[701]
[($ClCH_2$)$_2$CH]$_2$POEt. Ib. b_1 145°, d^{20} 1.3568, n_D^{20} 1.4928.[701]
[($ClCH_2$)$_2$CHO]$_2$POPr. Ib. $b_{0.5}$ 141-2°, d^{20} 1.3128, n_D^{20} 1.4882.[701]
[($ClCH_2$)$_2$CHO]$_2$POBu. Ib. $b_{0.5}$ 145-6°, d^{20} 1.2803, n_D^{20} 1.4845.[701]
[($ClCH_2$)$_2$CHO]$_2$POAm. Ib. b_2 156°, d^{20} 1.2365, n_D^{20} 1.4711.[701]
[($ClCH_2$)$_2$CHO]$_2$POC$_6H_{13}$. Ib. b_2 179-80°, d^{20} 1.2447, n_D^{20} 1.4807.[701]
[($ClCH_2$)$_2$CHO]$_2$POC$_8H_{17}$. Ib. b_1 181-2°, d^{20} 1.2017, n_D^{20} 1.4745.[701]
[($ClCH_2$)$_2$CHO]$_2$POPh. Ib. b_2 206-7°, d^{20} 1.3808, n_D^{20} 1.5362.[701]
[($ClCH_2$)$_2$CHO]$_2$POBuC$_6H_4$. Ib. b_1 193-5°, d^{20} 1.2899, n_D^{20} 1.5251.[701]
[($ClCH_2$)$_2$CHO]$_2$PO-2,6,4-t-Bu$_2$MeC$_6H_2$. Ib. m. 64.5-6°.[701]
(NCCMe$_2$O)$_2$POMe. Ib. b_9 139-41°, d^{20} 1.0653, n_D^{20} 1.4382.[774]
(NCCMe$_2$O)$_2$POEt. Ib. b_8 136-9°, d^{20} 1.0471, n_D^{20} 1.4330.[774]
(NCCMe$_2$O)$_2$POCH$_2CH_2$Cl. Ib. b_{10} 165-7°, d^{20} 1.1386, n_D^{20} 1.4525.[774]
(NCCMe$_2$O)$_2$POPr. Ib. b_{10} 147-9°, d^{20} 1.0342, n_D^{20} 1.4371.[774]
(NCCMe$_2$O)$_2$POCH$_2$CH=CH$_2$. Ib. b_{10} 153-6°, d^{20} 1.0546, n_D^{20}

1.4450.[774]
$(NCCMe_2O)_2PO$-i-Pr. Ib. b_{12} 142-3°, d^{20} 1.0261, n_D^{20} 1.4360.[774]
$(NCCMe_2O)_2POBu$. Ib. b_{11} 155-6°, d^{20} 1.0150, n_D^{20} 1.4371.[774]
$(NCCMe_2O)_2PO$-i-Bu. Ib. b_7 143-4°, d^{20} 1.0190, n_D^{20} 1.4380.[774]
$(NCCMe_2O)_2PO$-i-Am. Ib. b_9 154-5°, d^{20} 1.0021, n_D^{20} 1.4405.[774]
$(NCCMe_2O)_2POC_6H_{11}$. Ib. b_7 165-6°, d^{20} 1.0556, n_D^{20} 1.4588.[774]
$(NCCMe_2O)_2POPh$. Ib. b_{11} 186-7°, d^{20} 1.1035, n_D^{20} 1.4852.[774]
$(BuO)_2POEt$. Ib $(EtOPCl_2)$. b_{10} 120.5-2°, d_0^0 0.9510, n_D^{20} 1.4285,[663] b_{18} 82-4°.[908]
$(BuO)_2POCH_2CH_2CN$. Ib. b_8 150-3°, d^{20} 1.0002, n_D^{20} 1.4425.[774]
$(BuO)_2POCH_2CH_2Ac$. Ib. $b_{0.05}$ 90-1°, d^{20} 0.9830, n_D^{20} 1.4420.[1219]
$(BuO)_2POCH_2CH_2O_2CC(CH_3):CH_2$. Ib. $b_{0.08}$ 112°, d^{20} 1.0135, n_D^{20} 1.4534.[1126]
$(BuO)_2POPr$. Ie. b_2 94-6°, d_4^{20} 0.9242, n_D^{20} 1.4285, MR_D 65.85 (calc. 66.07).[663]
$(BuO)_2POCH_2\overline{CHCH_2O}$. Ib.[1218,1380] $b_{0.1}$ 83-5°,[1218] b_1 103-5°,[1380] d^{20} 1.0112, n_D^{20} 1.4400,[1218] n_D^{20} 1.4401.[1380]
$(BuO)_2POCH_2CHMeAc$. Ib. $b_{0.02}$ 93-5°, d^{20} 0.9882, n_D^{20} 1.4460.[1219]
$(BuO)_2PO$iso-Pr. Ib. b_6 92-3°, n_D^{20} 1.4256.[189]
$(BuO)_2POCH(CH_2Cl)Me$. Ic. $b_{1.5}$ 86-8°, $b_{1.5}$ 87-8°, d^{20} 1.0198, d^{20} 1.201, n_D^{20} 1.4433, n_D^{20} 1.4430.[1117]
$(BuO)_2POCMe_2CN$. Ib. b_{14} 134-5°, d^{20} 0.9629, n_D^{20} 1.4328.[774]
$(BuO)_2POCH(CH_2Cl)CH_2OCH_2CH_2Cl$. Ib. $b_{0.18}$ 139-40°, d_4^{20} 1.1090, n_D^{20} 1.4612.[687]
$(BuO)_2PO$-tert-Bu. Ib. $b_{2.5}$ 73-3.5°, n_D^{20} 1.4298.[189]
$(BuO)_2POCMe_2CH_2Ac$. Ib.[1217,1219] $b_{0.05}$ 100-2°,[1219] $b_{0.1}$ 101-2°,[1217] n_D^{20} 1.4427, d^{20} 0.9675.[1217,1219]
$(BuO)_2POCH_2\overline{CMeCH_2OCMe_2OCH_2}$. From ketal and $(RO)_3P$. $b_{0.2}$ 125-35°.[426]
$(BuO)_2POAm$. Ie.[377,378] Ib.[189] b_1 97-101°,[377] b_6 116-7°,[189 378] n_D^{20} 1.4300,[377] n_D^{20} 1.4310.[189,378]
$(BuO)_2POCMeEtCCl_3$. Ib. $b_{0.07}$ 110°, d^{20} 1.1516, n_D^{20} 1.4738.[368]
$(BuO)_2POC_6H_{11}$. From -POAc + $C_6H_{11}OH$ + Et_3N.[378] Ib.[189] $b_{2.5}$ 107-9°, n_D^{20} 1.4503.[189,378]
$(BuO)_2POC_7H_{15}$. Ie. b_2 142-3°, d_4^{20} 0.9030, n_D^{20} 1.4355, MR_D 84.58 (calc. 84.54).[663]
$(BuO)_2POPh$. Ib.[189,1047] b_2 128-9°,[189] b_5 148-50°, d^{20} 0.9972, n_D^{20} 1.4858,[1149] n_D^{20} 1.4840.[189]

$(BuO)_2PO\text{-}p\text{-}MeC_6H_4$. Ib. $b_{1.5}$ 130-1°, n_D^{20} 1.4840.[189]

$(BuO)_2POC_6H_4CH_2CH{:}CH_2\text{-}2$. Ib. b_{10} 143°, d^{20} 0.9806, n_D^{20} 1.4750.[421]

$(BuO)_2POC_{10}H_7\text{-}1$. Ib. b_3 189°, d^{20} 1.0559, n_D^{20} 1.5350.[49]

$(BuO)_2POC_{10}H_7\text{-}2$. Ib. $b_{2.5}$ 188-9°, d^{20} 1.0545, n_D^{20} 1.5420. Rearranged internally at 200-10°.[665]

$(BuO)_2POCH_2CH_2OP(BuO)_2$. Ib.[189] Ie.[378] $b_{0.01}$ 140°, n_D^{20} 1.4445.[189,378]

$(BuO)_2POCHCHMeCH_2N(Me)CH_2CHMe$. Ib. $b_{0.02}$ 110-20°, n_D^{20} 1.4545.[189]

$(BuO)_2POC(C{:}CH)CHMeCH_2N(Me)CH_2CHMe$. Ib.[189] Ie.[378] $b_{0.01}$ 140°, n_D^{20} 1.4840.[189,378]

$(CH_2{:}CHCH_2CH_2O)_2POC_6H_5$. Ib. $b_{0.7}$ 122-5°, d^{20} 1.039, n_D^{20} 1.5055, MR_D 76.07.[703]

$(CH_2{=}CH\text{-}CHCl\cdot CH_2O)_2POEt$. $(EtO)_2PCl$ + butadiene oxide. b_1 107-8°, d^{20} 1.1571, n_D^{20} 1.4765.[1063]

$(i\text{-}BuO)_2POEt$. Ib ($EtOPCl_2$). b_5 86-9°, d_0^0 0.9462, n_D^{20} 1.4258.[663]

$(i\text{-}BuO)_2POCH(CCl_3)C_6H_4Cl\text{-}p$. Ib. $b_{0.1}$ 157-9°, d^{20} 1.2362, n_D^{20} 1.5100.[33]

$(i\text{-}BuO)_2POCH_2CHMeAc$. Ib. $b_{0.05}$ 87-9°, d^{20} 0.9723, n_D^{20} 1.4390.[1219]

$(i\text{-}BuO)_2POCH_2CHCH_2O$. Ib. $b_{0.05}$ 84-6°, n_D^{20} 1.4376.[1380]

$(i\text{-}BuO)_2POCMe_2CN$. Ib. b_{11} 115-7°, d^{20} 0.9412, n_D^{20} 1.4241.[774]

$(i\text{-}BuO)_2POCH(CH_2Cl)CH_2OCH_2CH_2Cl$. Ib. $b_{0.19}$ 134-5°, d_4^{20} 1.0980, n_D^{20} 1.4580.[687]

$(i\text{-}BuO)_2POCMe_2CH_2Ac$. Ib.[1217,1219] $b_{0.05}$ 87-90°,[1219] $b_{0.5}$ 89-90°,[1217] n_D^{20} 1.4383, d^{20} 0.9605.[1217,1219]

$(i\text{-}BuO)_2POCMeEtCCl_3$. Ib. $b_{0.16}$ 116-7°, d^{20} 1.1451, n_D^{20} 1.4711.[368]

$(i\text{-}BuO)_2POC(CN)(CH_2)_5$. Ib. b_9 160-2°, d^{20} 1.0034, n_D^{20} 1.4580.[774]

$(i\text{-}BuO)_2POC_{10}H_7\text{-}2$. Ib. $b_{1.5}$ 176-8°, d^{20} 1.0501, n_D^{20} 1.5400.[665]

$[i\text{-}PrCH(CN)O]_2POBu\text{-}i$. Ib. b_{10} 167-70°, d^{20} 1.0039, n_D^{20} 1.4452.[774]

$(s\text{-}BuO)_2POCH_2CHCH_2O$. Ib. $b_{0.5}$ 86-8°, n_D^{20} 1.4401.[1380]

$(s\text{-}BuO)_2POCMeEtCCl_3$. Ib. $b_{0.2}$ 108°, d^{20} 1.1618, n_D^{20} 1.4737.[368]

$[Me_2(CCl_3)CO]_2POMe$. Ib. b_4 152-4°, d_4^{20} 1.4317, n_D^{20} 1.5040, MR_D 85.82 (calc. 86.03), $(R'O)_2POR\text{-}CuCl$ complex m. 182-3°.[19]

$[Me_2(CCl_3)CO]_2POEt$. Ib. b_4 160-0.5°, d_4^{20} 1.4190, n_D^{20} 1.5028, MR_D 89.79 (90.64); -CuCl m. 178° (decomp.).[19]

$[Me_2(CCl_3)CO]_2POPr$. Ib. b_3 168-9; d_4^{20} 1.3853, n_D^{20} 1.4998, MR_D 94.13 (95.26); -CuCl m. 138° (decomp.).[19]

$[Me_2(CCl_3)CO]_2PO\text{-}i\text{-}Pr$. Ib. b_4 166°, d_4^{20} 1.3808, n_D^{20} 1.4986, MR_D 94.13 (95.26); -CuCl m. 184° (decomp.).[19]

$[Me_2(CCl_3)CO]_2POBu$. Ib. b_3 171.5-2.5°, d_4^{20} 1.3440, n_D^{20} 1.4997, MR_D 100.20 (99.88); -CuCl m. 162° (decomp.).[19]

$[Me_2(CCl_3)CO]_2PO$-i-Bu. Ib. $b_{2.5}$ 166-7°, d_4^{20} 1.3397, n_D^{20} 1.4960, MR_D 99.88 (99.88); -CuCl m. 191° (decomp.).[19]

$[Me_2(CCl_3)CO]_2PO$-s-Bu. Ib. b_4 171-3°, d_4^{20} 1.3596, n_D^{20} 1.5010, MR_D 99.04 (99.88); -CuCl m. 172° (decomp.).[19]

$[Me_2(CCl_3)CO]_2PO$-n-C_8H_{17}. Ib. b_4 202-3°, d_4^{20} 1.2536, n_D^{20} 1.4890, MR_D 118.01 (118.35); -CuCl, a syrup.[19]

$[Me_2(CCl_3)CO]_2PO$-2-furfuryl. Ib. m. 60-1°, decomp. on distillation; -CuCl m. 136° (decomp.).[19]

$[Me_2(CCl_3)CO]_2POC_6H_{11}$. Ib. b_4 198°, crystallized after distillation, m. 46-7°, d_4^{20} 1.3721, n_D^{20} 1.5082, MR_D 105.00 (105.19); -CuCl m. 182° (decomp.).[19]

$[Me_2(CCl_3)CO]_2POC_6H_5$. Ib. b_4 208°, d_4^{20} 1.4101, n_D^{20} 1.5405, MR_D 106.12 (105.52); -CuCl m. 156°.[19]

$[Me_2(CCl_3)CO]_2POCH_2CCl_3$. Ib. $b_{0.002}$ 133-4°, m. 71-2°.[367]

$(AmO)_2POCH_2CH_2OMe$. Ib. b_{13} 156-61°, d^{20} 1.0180, n_D^{20} 1.4393.[951]

$(AmO)_2POCH_2CH_2OEt$. Ib. $b_{14.5}$ 158-63°, d^{20} 0.9857, n_D^{20} 1.4371.[951]

$(AmO)_2POC_{10}H_7$-2. Ib. $b_{2.5}$ 191-2°, d^{20} 1.0357, n_D^{20} 1.5342.[665]

(i-AmO)$_2$POMe. Ie. b_4 87-8°, d_4^{20} 0.9232, n_D^{20} 1.4305, MR_D 66.19 (calc. 66.07).[663]

(i-AmO)$_2$POCMe$_2$CN. Ib. b_{11} 145-7°, d^{20} 0.9378, n_D^{20} 1.4320.[774]

(i-AmO)$_2$POBu. Ib. b_4 114-4.5°, d_0^0 0.9303, n_D^{20} 1.4350.[663]

(i-AmO)$_2$POCMeEtCCl$_3$. Ib. $b_{0.1}$ 120°, d^{20} 1.1195, n_D^{20} 1.4718.[368]

(i-AmO)$_2$POC$_{10}$H$_7$-1. Ib. b_1 172-3°, d^{20} 1.0451, n_D^{20} 1.5343.[49]

(i-AmO)$_2$POC$_{10}$H$_7$-2. Ib. b_3 195-6°, d^{20} 1.0300, n_D^{20} 1.5330.[665]

$(CCl_3CMeEtO)_2POPr$. Ib. $b_{0.22}$ 154-5°, d^{20} 1.3740, n_D^{20} 1.5136.[686]

$(CCl_3CMeEtO)_2POBu$. Ib. $b_{0.13}$ 153-4°, d^{20} 1.3280, n_D^{20} 1.5062.[686]

$(CCl_3CMeEtO)_2PO$-i-Bu. Ib. $b_{0.12}$ 143°, d^{20} 1.3302, n_D^{20} 1.5030.[686]

$(C_6H_{13}O)_2POCH(CH_2Cl)CH_2OCH_2CH_2Cl$. Ib. $b_{0.012}$ 163-5°, d_4^{20} 1.0450, n_D^{20} 1.4583.[687]

$(C_6H_{13}O)_2POC_{10}H_7$-1. Ib. b_3 228-30°, d^{20} 1.0220, n_D^{20} 1.5272.[49]

$(C_6H_{13}O)_2POC_{10}H_7$-2. Ib. b_2 216-7°, d^{20} 1.0184, n_D^{20} 1.5300.[665]

(cyclo-$C_6H_{11}O)_2POPr$. Ib. n_D^{20} 1.5175.[1024]

(cyclo-$C_6H_{11}O)_2POCMe_2CN$. Ib. b_3 160-2°, d^{20} 1.0488, n_D^{20} 1.4735.[774]

(cyclo-$C_6H_{11}O)_2POBu$. Ib. n_D^{20} 1.5169.[1024]

(cyclo-$C_6H_{11}O)_2POAm$. Ib. n_D^{20} 1.5162.[1024]

(cyclo-$C_6H_{11}O)_2POC_6H_{13}$. Ib. n_D^{20} 1.5005.[1024]

$(cyclo-C_6H_{11}O)_2POC_7H_{15}$. Ib. n_D^{20} 1.4961.[1024]
$(cyclo-C_6H_{11}O)_2POC_8H_{17}$. Ib. n_D^{20} 1.4934.[1024]
$(cyclo-C_6H_{11}O)_2POC_9H_{19}$. Ib. n_D^{20} 1.4895.[1024]
$(cyclo-C_6H_{11}O)_2POC_{10}H_{21}$. Ib. n_D^{20} 1.4856.[1024]
$[(CH_2)_5C(CN)O]_2POBu$. Ib. $b_{0.4}$ 160-2°, d^{20} 1.0735, n_D^{20} 1.4815.[774]
$[(CH_2)_5C(CN)O]_2POBu$-i. Ib. b_1 183-6°, d^{20} 1.0701, n_D^{20} 1.4805.[774]
$(C_7H_{15}O)_2POBu$. Ie. b_3 167-8°, d_4^{20} 0.8997, n_D^{20} 1.4410, MR_D 98.15 (calc. 98.39).[663]
$(2-MeC_6H_{10}O)_2POC_6H_4$-4-Pr. Ib. n_D^{20} 1.5156.[1024]
$(2-MeC_6H_{10}O)_2POC_6H_4$-4-Bu. Ib. n_D^{20} 1.5142.[1024]
$(2-MeC_6H_{10}O)_2POC_6H_4$-4-Am. Ib. n_D^{20} 1.5110.[1024]
$(2-MeC_6H_{10}O)_2POC_6H_4-4-C_6H_{13}$. Ib. n_D^{20} 1.4984.[1024]
$(2-MeC_6H_{10}O)_2POC_6H_4-4-C_7H_{15}$. Ib. n_D^{20} 1.4920.[1024]
$(2-MeC_6H_{10}O)_2POC_6H_4-4-C_8H_{17}$. Ib. n_D^{20} 1.4864.[1024]
$(2-MeC_6H_{10}O)_2POC_6H_4-4-C_9H_{19}$. Ib. n_D^{20} 1.4845.[1024]
$(2-MeC_6H_{10}O)_2POC_6H_4-4-C_{10}H_{21}$. Ib. n_D^{20} 1.4810.[1024]
$(C_8H_{17}O)_2PO-4-i-PrC_6H_4$. Ib. d^{20} 0.936, n_D^{29} 1.4742.[217]
$(C_8H_{17}O)_2PO-4-t-BuC_6H_4$. Ib. d^{20} 0.936, n_D^{29} 1.4750.[217]
Diisooctyl cresyl phosphite. Ia.[547] Ie.[548] d^{29} 0.954, n_D^{25} 1.4814.[547,548]
$(EtBuCHCH_2O)_2POEt$. IR.[176]
$(EtBuCHCH_2O)_2POC_6H_5$. Ib. $b_{0.06}$ 148-56°, n_D^{26} 1.4791, d^{20} 0.964.[168]
$(EtBuCHCH_2O)_2POC_6H_4Me$-p. Ib.[168,217] $b_{0.05}$ 143-8°,[168] $n_D^{30.7}$ 1.4738,[217] $n_D^{30.9}$ 1.4749,[168] d^{20} 0.946,[217] d^{20} 0.954.[168]
$(EtBuCHCH_2O)_2POC_6H_4CMe_2CH_2CMe_3$-p. Ib. n_D^{25} 1.4824, d^{20} 0.935 (undistilled).[168]
$(C_9H_{19}O)_2POPh$. Ib. d^{20} 0.933, $n_D^{21.9}$ 1.4714.[217]
$(C_9H_{19}O)_2PO-4-MeC_6H_4$. Ib. d^{20} 0.936, $n_D^{31.9}$ 1.4730.[217]
$(C_9H_{19}O)_2PO-4-C_8H_{17}C_6H_4$. Ib. d^{20} 0.920, n_D^{32} 1.4756.[217]
$(C_{10}H_{21}O)_2POPh$. Ib.[217] Ie.[428] d^{20} 0.936, n_D^{32} 1.4732,[217] n_D^{25} 1.4772.[428,1150]
$(C_{10}H_{21}O)_2PO-4-MeC_6H_4$. Ib. d^{20} 0.940, $n_D^{32.2}$ 1.4764.[217]
$(C_{10}H_{21}O)_2PO-4-C_8H_{17}C_6H_4$. Ib. d^{20} 0.922, $n_D^{32.5}$ 1.4771.[217]
$(C_{12}H_{25}O)_2PO-4-MeC_6H_4$. Ib. d^{20} 0.913, n_D^{33} 1.4759.[217]
$(C_{18}H_{37}O)_2PO-4-i-PrC_6H_4$. Ib.[217]
$(C_{18}H_{37}O)_2PO-4-C_8H_{17}C_6H_4$. Ib.[217]
$(C_6H_5O)_2POMe$. Ib. b_{11} 169.5-70.5°, d^0 1.1825, d^{20} 1.1643, n_D^{20} 1.5568.[661]
$(C_6H_5O)_2POCH_2C_6H_5$. Ib. b_{14} 197.5-8°, d^0 1.2086, d^{20} 1.1674, n_D^{20} 1.5836.[661]
$(C_6H_5O)_2POEt$. Ie.[1265] b_{15} 79-81°,[908] b_4 150-5°, n_D^{20} 1.5540, d^{20} 1.1383.[1265]
$(C_6H_5O)_2POCH_2CH_2OCH{=}CH_2$. Ib. $b_{0.5}$ 143-5°, d^{20} 1.1562, n_D^{20} 1.5505.[692]
$(C_6H_5O)_2POCH_2CH_2O_2CC(CH_3){:}CH_2$. Ib. n_D^{20} 1.5295, d 1.1420.[1076]

$(C_6H_5O)_2POPr$. Ie. b_5 165-71°, n_D^{20} 1.5411, d^{20} 1.1096.[1265]

$(C_6H_5O)_2POCH_2CH{:}CH_2$. Ib.[661,1099] b_6 161°,[1099] b_{10} 186-7°,[661] d^{20} 1.1380,[1099] d^0 1.1497, d^{20} 1.1321, n_D^{20} 1.5515,[661] n_D^{20} 1.5550.[1099]

$(C_6H_5O)_2PO\text{-}i\text{-}Pr$. Ie. b_1 155-7°, n_D^{20} 1.5432, d^{20} 1.1088.[1265]

$(C_6H_5O)_2POCMe_2CN$. Ib. b_3 172-4°, d^{20} 1.1448, n_D^{20} 1.5341.[774]

$(C_6H_5O)_2POCH(CH_2Cl)CH_2OCH_2CH_2Cl$. Ib. $b_{0.013}$ 173-4°, d_4^{20} 1.2590, n_D^{20} 1.5485.[687]

$(C_6H_5O)_2POCH_2\overline{CHOCMe_2OC}H_2$. Ie. $b_{0.0005}$ 125-30°, d^{20} 1.1890, n_D^{20} 1.5445.[1057]

$(C_6H_5O)_2POBu$. IVh. b_7 176-8°, n_D^{20} 1.5369.[1047]

$(C_6H_5O)_2POCMe_2CCl_3$. Ib. $b_{0.12}$ 160-5°, $b_{0.15}$ 158-9°, d^{20} 1.3126, n_D^{20} 1.5628.[688]

$(C_6H_5O)_2POCH_2CH_2CH{:}CH_2$. Ib. $b_{0.6}$ 149-51°, d^{20} 1.1110, n_D^{20} 1.5453, MR_D 81.90.[703]

$(C_6H_5O)_2POCH_2CH{:}CHMe$. Ib. $b_{0.001}$ 117-8°, d^{20} 1.1190, n_D^{20} 1.5495.[1099]

$(C_6H_5O)_2POCHMeCH{:}CH_2$. Ib. $b_{0.001}$ 115-7°, d^{20} 1.1146, n_D^{20} 1.5435.[1099]

$(C_6H_5O)_2POCMeEtCCl_3$. Ib. $b_{0.09}$ 158°, d^{20} 1.3040, n_D^{20} 1.5645.[368]

$(C_6H_5O)_2POCH_2CHMeBu$. Ia. $b_{0.15}$ 148-56°, $n_D^{27.4}$ 1.5207, d^{20} 1.054.[168]

$(C_6H_5O)_2PO\text{-}i\text{-}C_8H_{17}$. Ia.[547] Ie.[548] d^{29} 1.035, n_D^{25} 1.5190.[547,548]

$(C_6H_5O)_2POC_9H_{19}$. Ib. d^{20} 1.033, $n_D^{31.5}$ 1.5158.[217]

$(C_6H_5O)_2POC_{10}H_{21}$. Ib.[217] Ie.[428] d^{20} 1.025,[217] n_D^{25} 1.5180,[428,1150] $n_D^{32.1}$ 1.5138.[217]

$(C_6H_5O)_2PO\text{-}2,6\text{-di-}t\text{-}BuC_6H_3$. Ig. b. 200-2°.[1266]

$(C_6H_5O)_2POC_{10}H_7\text{-}1$. Ib. b_4 245-8°, d^{20} 1.209, n_D^{20} 1.6317.[49]

$(p\text{-}ClC_6H_4O)_2POC_{10}H_7\text{-}1$. Ib. b_1 228-30°, d^{20} 1.320, n_D^{20} 1.6322.[49]

$(3\text{-}NH_2C_6H_4O)_2POEt$. ROH + $ROP(NEt_2)_2$. Oil, n_D^{20} 1.5983.[390]

$(4\text{-}NO_2C_6H_4O)_2POEt$. Ih.[926] Ib.[595] m. 65°,[926] m. 64-5°.[595] Hygroscopic and unstable to storage.

$(4\text{-}NO_2C_6H_4O)_2POPh$. Ib. m. 85-6°.[595]

$(o\text{-}MeC_6H_4O)_2POCH_2CH{=}CH_2$. Ib. $b_{0.05}$ 120-7°, n_D^{32} 1.5453.[168,217]

$(o\text{-}MeC_6H_4O)_2PO$(3-cyclohexenyl). Ib. $n_D^{25.5}$ 1.5516 (undistilled).[168]

$(o\text{-}MeC_6H_4O)_2PO$(2-ethylhexyl). Ib. $b_{0.05}$ 146-50°, $n_D^{27.2}$ 1.5164, d^{20} 1.027.[168]

$(p\text{-}MeC_6H_4O)_2PO$(2-ethylhexyl). Ib. $n_D^{30.8}$ 1.5191, d^{20} 1.032.[168]

$(4\text{-}MeC_6H_4O)_2POC_9H_{19}$. Ib. d^{20} 1.009, n_D^{32} 1.5095.[217]

$(4\text{-}MeC_6H_4O)_2POC_{10}H_{21}$. Ib. d^{20} 1.005, $n_D^{32.4}$ 1.5095.[217]

$(o\text{-}MeC_6H_4O)_2POC_{10}H_7\text{-}1$. Ib. $b_{0.5}$ 213-6°, d^{20} 1.159, n_D^{20}

1.6025.[49]
$(m\text{-}MeC_6H_4O)_2POC_{10}H_7\text{-}1$. Ib. b_2 236-8°, d^{20} 1.163, n_D^{20} 1.6022.[49]
$(p\text{-}MeC_6H_4O)_2POC_{10}H_7\text{-}1$. Ib. b_2 238-40°, d^{20} 1.162, n_D^{20} 1.6022.[49]
$(2\text{-}CH_2{:}CHCH_2C_6H_4O)_2POEt$. Ib. $b_{2.5}$ 178°, d^{20} 1.0876, n_D^{20} 1.5450.[421]
$(2\text{-}CH_2{:}CHCH_2C_6H_4O)_2POBu$. Ib. b_7 196-8°, d^{20} 1.0521, n_D^{20} 1.5300.[421]
$(4\text{-}i\text{-}PrC_6H_4O)_2POC_8H_{17}$. Ib. d^{20} 0.994, n_D^{28} 1.5100.[217]
$(4\text{-}t\text{-}BuC_6H_4O)_2POC_8H_{17}$. Ib. d^{20} 0.991, $n_D^{29.5}$ 1.5091.[217]
$(p\text{-}Me_3CCH_2CMe_2C_6H_4O)_2POCH_2CHEtBu$. Ib. n_D^{20} 1.5037, d^{20} 0.960 (undistilled).[168]
$(4\text{-}C_8H_{17}C_6H_4O)_2POC_9H_{19}$. Ib. d^{20} 0.947, n_D^{26} 1.5010.[217]
$(4\text{-}C_8H_{17}C_6H_4O)_2POC_{10}H_{21}$. Ib. d^{20} 0.953, n_D^{26} 1.5019.[217]
(p-nonyl $C_6H_4O)_2POCH_2CH{=}CH_2$. $n_D^{25.5}$ 1.5100 (undistilled).[168]
Bis(1,2:5,6-di-O-isopropylidene-α-D-gluco) 1,2:3,4-di-O-isopropylidene-α-D-galacto phosphite. $ROP(NEt_2)_2$ + R'OH. m. 57-60°.[971]

TYPE: $(RO)_2POR'P(O)(OR'')_2$

$(CH_3O)_2P_\alpha OCH(CH_3)P_\beta(O)(OCH_3)_2$. $^{31}P_\alpha$ -141.8, $^{31}P_\beta$ -24.8 ppm.[826]
$(ClCH_2CH_2O)_2P_\alpha OC(CH_3)(CN)P_\beta(O)(OCH_2CH_2Cl)_2$. $^{31}P_\alpha$ -140.5, $^{31}P_\beta$ -13.1 ppm.[826]
$CH_3CH(Cl)CH_2OP_\alpha\{OCH(CH_3)P_\beta(O)[OCH_2CH(Cl)CH_3]_2\}_2$. $^{31}P_\alpha$ -144.0, $^{31}P_\beta$ -21.8 ppm.[826]
$[CH_3CH(CH_3)O]_2P_\alpha OCH(p\text{-}C_6H_4OCH_3)P_\beta(O)[OCH(CH_3)CH_3]_2$. $^{31}P_\alpha$ -138.5, $^{31}P_\beta$ -16.5 ppm.[826]
$[ClCH_2CH(Br)CH_2O]_2P_\alpha OCH(CH_3)P_\beta(O)[OCH_2CH(Br)CH_2Cl]_2$. $^{31}P_\alpha$ -143.1, $^{31}P_\beta$ -22.8 ppm.[826]
$[CH_3CH(Cl)CH_2O]_2P_\alpha OCH(CH_3)P_\beta(O)[OCH_2CH(Cl)CH_3]_2$. $^{31}P_\alpha$ -140.3, $^{31}P_\beta$ -21.4 ppm.[826]
$[CH_3CH(Cl)CH_2O]_2P_\alpha OCH[CH(CH_3)_2]P_\beta(O)[OCH_2CH(Cl)CH_3]_2$. $^{31}P_\alpha$ -144.0, $^{31}P_\beta$ -21.6 ppm.[826]
$[CH_3CH(Cl)CH_2O]_2P_\alpha OCH(C_3H_7)P_\beta(O)[OCH_2CH(Cl)CH_3]_2$. $^{31}P_\alpha$ -136.0, $^{31}P_\beta$ -20.4 ppm.[826]

TYPE: $(RO)_2(R'CO_2)P$

$(EtO)_2POC(O)Me$. $(EtO)_2PCl$ + $R'CO_2M$.[1063] $(EtO)_2PNR_2$ + Ac_2O.[189,638] b_2 52-2.5°,[638] b_6 59-60°,[189] b_9 67-9°,[1063] d^{20} 1.0588,[638] d^{20} 1.0710,[1063] n_D^{20} 1.4208,[638] n_D^{20} 1.4207,[1063] n_D^{20} 1.4160.[189]
$(EtO)_2POC(O)Et$. $(RO)_2PNEt_2$ + Ac_2O. b_{10-12} 80-4°, d^{20} 1.0450, n_D^{20} 1.4250.
$(EtO)_2POC(O)CMe{:}CH_2$. $(EtO)_2PCl$ + KO_2CR. b_4 63°, d_4^{20} 1.0465, n_D^{20} 1.4390.[1126]
$(EtO)_2POC(O)Ph$. $(RO)_2PNEt$ + Ac_2O. $b_{0.5}$ 113-5°, d^{20}

1.1220, n_D^{20} 1.4962.
$(EtO)_2POC(O)NMe_2$. $(EtO)_2PNMe_2$ + CO_2; anhydride + $P(NR_2)_3$. $b_{0.15}$ 71°, n_D^{20} 1.4471.[1008]
$(EtO)_2POC(O)NEt_2$. Anhydride + $P(NR_2)_3$. $b_{0.45}$ 91°, n_D^{20} 1.4440.[1009]
$(PrO)_2POC(O)Me$. $(RO)_2PNEt_2$ + Ac_2O. b_6 88-9°, d^{20} 0.9945, n_D^{20} 1.4234.
$(PrO)_2PO_2CCMe{:}CH_2$. $(PrO)_2PCl$ + KO_2CR. b_2 75-6°, d_4^{20} 1.0172, n_D^{20} 1.4411.[1126]
$(BuO)_2POC(O)Me$. $(RO)_2PCl$ + $R'CO_2M$.[1063] $(BuO)_2PNR_2$ + Ac_2O.[189] b_1 88-9°,[189] b_8 110-1°, d^{20} 0.9796,[1063] n_D^{20} 1.4314,[189] n_D^{20} 1.4277.[1063]
$(BuO)_2POC(O)Et$. IV. b_1 76-8°, d_4^{20} 0.9703, n_D^{20} 1.4329.[381]
$(BuO)_2POC(O)Pr$. IV. b_1 100-3°, d_4^{20} 0.9657, n_D^{20} 1.4343.[381]
$(BuO)_2PO_2CCMe{:}CH_2$. $(BuO)_2PCl$ + KO_2CR. $b_{0.85}$ 89°, d_4^{20} 0.9906, n_D^{20} 1.4432.[1126]
$(i\text{-}BuO)_2POAc$. $(RO)_2PNEt_2$ + Ac_2O. b_7 100°, n_D^{20} 1.4300.[189]
$(C_6H_{13}O)_2POC(O)Me$. $(RO)_2PCl$ + $R'CO_2M$.[1063] b_6 154-5°, d^{20} 0.9422, n_D^{20} 1.4341.
$(C_6H_{13}O)_2POC(O)Et$. IV. b_1 106-7°, d_4^{20} 0.9531, n_D^{20} 1.4398.[381]
$(C_6H_{13}O)_2POC(O)Pr$. IV. b_1 120-1°, d_4^{20} 0.9432, n_D^{20} 1.4411.
$(C_6H_5O)_2POC(O)Me$. $(RO)_2PCl$ + $R'CO_2M$.[1063] $b_{1.5}$ 184-5°, d^{20} 1.1967, n_D^{20} 1.5692.
$(p\text{-}MeC_6H_4O)_2POC(O)Me$. $(RO)_2PCl$ + $R'CO_2M$.[1063] $b_{0.0001}$ 90-3° 1.1444, n_D^{20} 1.5491.

TYPE: $(RCO_2)_2(R'O)P$

$(MeCOO)_2POBu$. II. $b_{0.07}$ 70-3°, n_D^{20} 1.4261.[246]
$(EtCOO)_2POBu$. II. $b_{0.01}$ 84-8°, n_D^{20} 1.4377.[246]
$(i\text{-}PrCOO)_2POBu$. II. $b_{0.07}$ 80-4°, n_D^{20} 1.4279.[246]

TYPE: (RO)(R'O)(R"O)P

(MeO)(EtO)POPr. Ie. b_{45} 60-1°, d_4^{20} 0.9732, n_D^{20} 1.4160, MR_D 42.84 (calc. 42.98).[663]
(MeO)(PrO)POAm-i. Ie. b_5 75-6°, d_4^{20} 0.9289, n_D^{20} 1.4275, MR_D 57.62 (calc. 56.83).[663]
$(MeO)(PhO)(CCl_3CMe_2O)P$. Ib. $b_{0.15}$ 129-30°, d^{20} 1.3195, n_D^{20} 1.5280.[688]
$(EtO)(CCl_3CMe_2O)(CCl_3CMeEtO)P$. Ib. $b_{0.11}$ 126-7°, d^{20} 1.3934, n_D^{20} 1.5067.[367,686]
$(EtO)(CCl_3CMeEtO)(1\text{-}CCl_3\text{-cyclopentyl-}O)P$. Ib. $b_{0.18}$ 165-6°, d^{20} 1.4020, n_D^{20} 1.5191.[686]
$(EtO)(PhO)(CCl_3CMe_2O)P$. Ib. $b_{0.12}$ 132-4°, d^{20} 1.2867, n_D^{20} 1.5219.[688]
$(ClCH_2CH_2O)(PhO)(CCl_3CMe_2O)P$. Ib. d^{20} 1.3561, n_D^{20} 1.5308.[688]

$(PrO)(CCl_3CMe_2O)(CCl_3CMeEtO)P$. Ib. $b_{0.13}$ 133-4°, d^{20} 1.3704, n_D^{20} 1.5050.[686]
$(PrO)(PhO)(CCl_3CMe_2O)P$. Ib. $b_{0.15}$ 141-3°, d^{20} 1.2583, n_D^{20} 1.5191.[688]
$(i\text{-}PrO)(CCl_3CMe_2O)(CCl_3CMeEtO)P$. Ib. $b_{0.15}$ 133°, d^{20} 1.3850, n_D^{20} 1.5088.[686]
$(BuO)(PhO)(CCl_3CMe_2O)P$. Ib. $b_{0.1}$ 134-5°, d^{20} 1.2365, n_D^{20} 1.5165.[688]
$(s\text{-}BuO)(CCl_3CMe_2O)(CCl_3CMeEtO)P$. Ib. $b_{0.2}$ 154°, d^{20} 1.3603, n_D^{20} 1.5035.[686]
$(s\text{-}BuO)(PhO)(CCl_3CMe_2O)P$. Ib. $b_{0.2}$ 138-40°, d^{20} 1.2394, n_D^{20} 1.5165.[688]
$(i\text{-}BuO)(CCl_3CMeEtO)(1\text{-}CCl_3\text{-cyclopentyl-}O)P$. Ib. $b_{0.2}$ 172-3°, d^{20} 1.3598, n_D^{20} 1.5170.[686]
$(CCl_3CMe_2O)(AmO)(PhO)P$. Ib. $b_{0.12}$ 145-7°, d^{20} 1.2176, n_D^{20} 1.5134.[688]
$(CCl_3CMe_2O)(t\text{-}AmO)(PhO)P$. Ib. d^{20} 1.2270, n_D^{20} 1.5153.[688]
$(CCl_3CMe_2O)(PhO)(1\text{-}C_{10}H_7O)P$. Ib. m. 51-3°.[688]
Benzyl 1,2:5,6-di-O-isopropylidene-α-D-gluco 1,2:3,4-di-O-isopropylidene-α-D-galacto phosphite. $(RO)(R'O)PNEt_2 + C_6H_5CH_2OH$. $b_{0.001}$ 180-90° (bath).[971]

I.1.13. Mixed Diphosphites and Triphosphites

TYPE: $[(RO)_2PO]_2R'$

1,4-$[(MeO)_2PO]_2C_6H_{10}$-cyclo. Ie. m. 20-5° (cis), m. 103-4° (trans).[326]
1,2-$[(EtO)_2PO]_2C_6H_4$. Ib. b_2 146-7°, d^{20} 1.1179, n_D^{20} 1.4917.[1094]
1,3-$[(EtO)_2PO]_2C_6H_4$. Ib. b_4 155-6°, d^{20} 1.1118, n_D^{20} 1.4922.[1094]
1,4-$[(PrO)_2PO]_2C_6H_4$. Ib. b_2 181-2°, d^{20} 1.0632, n_D^{20} 1.4863.[1094]
1,4-$[(BuO)_2PO]_2C_6H_4$. Ib. $b_{0.04}$ 163-5°, d^{20} 1.0297, n_D^{20} 1.4818.[1094]
4,4'-$[(C_{12}H_{25}O)_2POC_6H_4]_2CMe_2$. Ie. As liquid residue d^{25} 0.932, n_D^{25} 1.4780.[1228]
[1,4-$(EtO)_2POC_6H_4]_2CMe_2$. $b_{0.005}$ 189-91°, n_D^{20} 1.5307.[1094]
1,2-$[(C_6H_5O)_2PO]_2C_6H_4$. Ib. d^{20} 1.2113, n_D^{20} 1.5800.[932]
1,4-$[(C_6H_5O)_2PO]_2C_6H_4$. Ib. d^{20} 1.2457, n_D^{20} 1.6040.[932]
4,4'-$[(C_6H_5O)_2POC_6H_4]_2$. Ib. 1.2417, n_D^{20} 1.6119.[932]
1,4-$[(C_6H_5O)_2PO]_2$-2,5-di-t-BuC_6H_2. Ib. m. 105-7°.[932]
4,4'-$[(C_6H_5O)_2POC_6H_4]_2CMe_2$. Ib.[932] Ie.[1228] As liquid residue, d^{25} 1.158,[1228] d^{20} 1.2110, n_D^{20} 1.6031,[932] n_D^{25} 1.5723.[1228]
[6-$(C_6H_5O)_2PO$-3Me-5-t-$BuC_6H_2]_2CH_2$. Ib. n_D^{20} 1.5820.[932]
$[(C_6H_5O)_2POCH_2CH_2]_2S$. Ib. d^{20} 1.2255, n_D^{20} 1.5829.[932]
4,4-$[(C_6H_5O)_2POC_6H_4]_2S$. Ib. d^{20} 1.2577, n_D^{20} 1.6162.[932]
4,4'-$[(C_6H_5O)(C_{10}H_{21}O)POC_6H_4]_2CMe_2$. Ie. As liquid residue, d^{25} 1.027, n_D^{25} 1.511.[1228]

4,4'-[(C$_6$H$_5$O)(4-C$_9$H$_1$$_9C_6H_4$O)POC$_6H_4$]$_2CMe_2$. Ie. As very
   viscous liquid residue, d^5^{0} 1.047, n$_D^2$^{5} 1.508.[1228]
[(2-MeC$_6$H$_4$O)$_2$POCH$_2$CH$_2$]$_2$S. Ib. d^2^{0} 1.1816, n$_D^2$^{0} 1.5718.[932]
[(3-MeC$_6$H$_4$O)$_2$POCH$_2$CH$_2$]$_2$S. Ib. d^2^{0} 1.1823, n$_D^2$^{0} 1.5712.[932]
4,4'-[(3-MeC$_6$H$_4$O)$_2$POC$_6$H$_4$]$_2$S. Ib. d^2^{0} 1.2024, n$_D^2$^{0}
   1.5990.[932]
[(4-MeC$_6$H$_4$O)$_2$POCH$_2$CH$_2$]$_2$S. Ib. d^2^{0} 1.1786, n$_D^2$^{0} 1.5700.[932]
4,4'-[(4-MeC$_6$H$_4$O)$_2$POC$_6$H$_4$]$_2$S. Ib. d^2^{0} 1.2056, n$_D^2$^{0}
   1.5996.[932]
[(4-t-BuC$_6$H$_4$O)$_2$POCH$_2$CH$_2$]$_2$S. Ib. n$_D^2$^{0} 1.5468.[932]
4,4'-[(4-t-BuC$_6$H$_4$O)$_2$POC$_6$H$_4$]$_2$S. Ib. n$_D^2$^{0} 1.5738.[932]
4,4'-[(4-C$_9$H$_1$$_9C_6H_4$O)(C$_1$$_0H_2$$_1$O)POC$_6H_4$]$_2CMe_2$. Ie. As viscous liquid residue, d^5^{0} 0.973, n$_D^2$^{5} 1.542.[1228]
4,4'-[(4,4'-HOC$_6$H$_4$CMe$_2$C$_6$H$_4$O)POC$_6$H$_4$]$_2$CMe$_2$. Ie. As white
   solid residue, m. about 70°.[1228]

TYPE: [(RO)$_2$PO]$_3$R'.

[(PhO)$_2$POCH$_2$CH$_2$]$_3$N. Ie. From (HOR-)$_3$N, as liquid residue, d^2^{5} 1.2092, n$_D^2$^{5} 1.5820.[543]
[(PhO)$_2$POCH$_2$CH$_2$]$_2$NBu. Ie. From (HOR-)$_2$NBu.[543]

I.1.14. Esters with Phosphorus in a Ring System

<u>Derivatives of 1,3,2-Dioxaphospholane</u>

TYPE:
$$\begin{matrix} (R^1)(R^2)C-O \\ | \quad\quad\quad\quad\quad\; \searrow \\ (R^3)(R^4)C-O \end{matrix} P-OR^5$$

OCH$_2$CH$_2$OPOMe. Ib.[105] Ih.[345] Ie.[345] b$_1$$_4$ 44.5°,[123] b$_2$$_3$
   55-6°,[105] b$_3$$_5$ 60-2°,[1226] b$_4$$_0$ 63-6°,[345] d$_0^2$^{0} 1.2159,[105]
   d$_2^2$$_0^0$ 1.2067,[1226] d$_4^2$^{0} 1.2044,[1226] d$_0^2$^{0} 1.2159,[123] n$_D^2$^{0}
   1.4460,[105] n$_D^2$^{0} 1.4448,[1226] n$_D^2$^{0} 1.4440,[123] n$_D^2$^{5}
   1.4423,[345] ^{3}^{1}P -132.4, -131.6 ppm,[569,826,1182] ^{1}H
   NMR.[432] CuI salt, m. 132-3°.[105]

OCH$_2$CH$_2$OPOCH$_2$C$_6$H$_5$. ^{1}H NMR.[432]

OCH$_2$CH$_2$OPOCH$_2$C$_6$H$_4$OMe. ^{3}^{1}P -135.4 ppm.[826]

OCH$_2$CH$_2$OPOEt. Ib. b$_1$$_5$ 50.5-1°,[105] b$_1$$_9$$_.$$_5$ 61-2.5°,[1226]
   b$_1$$_6$ 54-5°,[270] d$_0^2$^{0} 1.1317,[105] d$_4^2$^{0} 1.1191,[1226] n$_D^2$^{0}
   1.4395,[105] n$_D^2$^{0} 1.4397,[1226] n$_D^2$^{0} 1.4411,[270] ^{3}^{1}P -131,
   -134, -132.8 ppm.[197,569,628] CuI salt, m. 90°.[105]
   On treatment with water this ester gives what appears
   to be an open-chain ester (EtO)(HOCH$_2$CH$_2$O)POH.[105]

OCH$_2$CH$_2$OPOCH:CH$_2$. Ib. b$_1$$_5$ 58-9°, d$_4^2$^{0} 1.1672, n$_D^2$^{0} 1.4577,
   MR$_D$ 31.34 (calc. 31.07).[446]

OCH$_2$CH$_2$OPOCH$_2$CH$_2$Cl. Ia [(RO)$_2$PCl + ROH in dioxane].[351]
   Ib.[105] Ic. b$_1$ 57°,[944] b$_6$$_.$$_5$ 78.5-9.5°, d^2^{0} 1.3206,

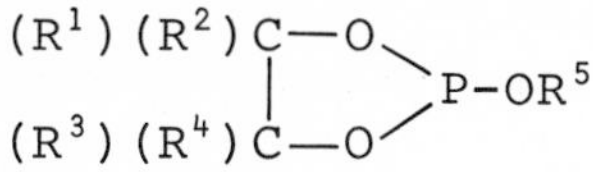

n_D^{20} 1.4755,[105] n_D^{20} 1.4810.[944]

$\overline{OCH_2CH_2OP}OCH_2CH_2OCH{:}CH_2$. Ib. b_8 98-9°, d^{20} 1.1881, n_D^{20} 1.4706,[692] IR.[1271]

$\overline{OCH_2CH_2OP}OCH_2CH_2O_2C(CH_3){:}CH_2$. Ib. n_D^{20} 1.4467, b_{15} 97-104°.[1076]

$\overline{OCH_2CH_2OP}OCH_2CH_2NEt_2$. Ib. b_4 95-7°, n_D^{25} 1.4620.[270]

$\overline{OCH_2CH_2OP}OCClPhMe$. Ic, from styrene oxide. b_1 135-7°, d^{20} 1.2960, n_D^{20} 1.5400.[1112]

$\overline{OCH_2CH_2OP}OCH_2CH_2OH$. Ia (dioxane) low yield.[351]

$\overline{OCH_2CH_2OP}OPr$. Ib. b_{12} 64-6°, d_0^{20} 1.1026, n_D^{20} 1.4445,[123] ^{31}P -134.4, -132.0 ppm.[826]

$\overline{OCH_2CH_2OP}OCH_2CH{:}CH_2$. Ib. b_{10-11} 69.5-70°, d_0^{20} 1.1553, n_D^{20} 1.4635,[123] IR.[1271]

$\overline{OCH_2CH_2OP}OCH_2\overline{CHCH_2O}$. Ib. $b_{0.05}$ 69-70°, d^{20} 1.2967, n_D^{20} 1.4723.[1293]

$\overline{OCH_2CH_2OP}O$-i-Pr. $b_{10-10.5}$ 53.5-4°, d_0^{20} 1.0829, n_D^{20} 1.4348.[123]

$\overline{OCH_2CH_2OP}OCHMeCH_2Cl$. Ic. b_{10} 85-6°, d^{20} 1.2718, n_D^{20} 1.4725.[1116]

$\overline{OCH_2CH_2OP}OCH(CH_2Cl)_2$. Ic. b_2 110-1°, d^{20} 1.4112, n_D^{20} 1.4960.[1111]

$\overline{OCH_2CH_2OP}OCH(CH_2Cl)CH_2OMe$. Ic. b_5 111-3°, d^{20} 1.2831, n_D^{20} 1.4721.[1111]

$\overline{OCH_2CH_2OP}OCH(CH_2Cl)CH_2OEt$. Ic. b_5 116-8°, d^{20} 1.2386, n_D^{20} 1.4677.[1111]

$\overline{OCH_2CH_2OP}OCH(CH_2Cl)CH_2OCH_2CH_2Cl$. Ib. $b_{0.22}$ 110-2°, n_D^{20} 1.4893, d_4^{20} 1.4420.[687]

$\overline{OCH_2CH_2OP}OCMe_2CN$. Ib. b_{10} 103-5°, d_4^{20} 1.1642, n_D^{20} 1.4515.[774]

$\overline{OCH_2CH_2OP}OBu$. Ib.[105] Ie.[345,430] Ie from acetate.[345] Ih.[345] $b_{0.2}$ 30°,[345] $b_{2.7}$ 57.5-8°,[345] $b_{3.3-2.5}$ 55-65°,[345] b_8 71-2°,[430] $b_{8.5}$ 71-2°,[105] d_0^{20} 1.0819,[105] n_D^{20} 1.4470,[105] n_D^{20} 1.4465,[430] n_D^{25} 1.4439,[345] n_D^{25} 1.444-1.4446,[345] ^{31}P -132, -133.1, -132.7 ppm.[826]

$\overline{OCH_2CH_2OP}OCH_2CHClCH{:}CH_2$. $(CH_2O)_2PCl$ + butadiene oxide. b_2 76-7°, d^{20} 1.2600, n_D^{20} 1.4889.[1112]

$\overline{OCH_2CH_2OP}O$-i-Bu. Ib. $b_{4-4.5}$ 54°, d_0^{20} 1.0652, n_D^{20} 1.4420,[123] ^{31}P -134.1 ppm.[826]

$\overline{OCH_2CH_2OP}OC_{10}H_{21}$. Ie. n_D^{25} 1.4563.[541]

$\overline{OCH_2CH_2OP}O$-i-$C_{10}H_{21}$. Ie, from formate. $b_{0.05}$ 70-1°.[345]

$\overline{OCH_2CH_2OP}OC_6H_5$. ^1H NMR,[432] ^{31}P -120, -124, -128.1 ppm.[826]

$\overline{OCH_2CH_2OP}OC_6H_4OH$-1,2. Ie, from monoacetate. b_1 91-3°, n_D^{20} 1.5325.[431]

$\overline{OCH_2CH_2OP}OCH_2CH_2O\overline{POCH_2CH_2O}$. Ib.[1096] Ie, from acetate.[345] b_1 122.5-3.5°,[1096] $b_{0.04-0.1}$ 99-106°,[345] d^{20} 1.3712,[1096] n_D^{20} 1.4889-1.4893,[345] n_D^{20} 1.4908, MR_D 51.09.[1096]

$\overline{OCH_2CH_2OP}OCHMeCH_2O\overline{POCH_2CH_2O}$. Ib. b_3 136-7°, d^{20} 1.3098, n_D^{20} 1.4835, MR_D 55.93.[1096]

$\overline{OCH_2CH_2OP}O(CH_2)_3O\overline{POCH_2CH_2O}$. Ib. b_3 144-5°, d^{20} 1.3259, n_D^{20} 1.4870, MR_D 55.53.[1096]

$\overline{OCH_2CH_2OP}O(CH_2)_3CHMeO\overline{POCH_2CH_2O}$. Ib. b_3 137-8°, d^{20} 1.2779, n_D^{20} 1.4823, MR_D 60.32.[1096]

$\overline{OCH_2CH_2OP}OCH_2CH_2OCH_2CH_2O\overline{POCH_2CH_2O}$. Ib. b_1 142-5°, d^{20} 1.3364, n_D^{20} 1.4902, MR_D 61.92.[1096]

$(\overline{OCH_2CH_2OP}O)_2C_6H_4$-o. Ib. b. 162-3°, d^{20} 1.3741, n_D^{20} 1.5400, MR_D 66.22.[1096]

$\overline{OCHMeCH_2OP}OMe$. ^1H NMR.[493a]

$\overline{OCHMeCH_2OP}OEt$. ^{31}P -135.4 ppm.[826]

$\overline{OCHMeCH_2OP}OCH_2CH_2Cl$. Ic.[1116] b_2 64-6°, d^{20} 1.2631, n_D^{20} 1.4670.[1116]

$\overline{OCHMeCH_2OP}OCH_2C{\equiv}CH$. Ib. d_{20} 1.2498, n_D^{20} 1.4962.[918]

$\overline{OCHMeCH_2OP}OCH_2\overline{CHCH_2O}$. Ib. $b_{0.005}$ 64-5°, d^{20} 1.2188, n_D^{20} 1.4651.[1293]

$\overline{OCHMeCH_2OP}OCHMeCH_2Cl$. Ic. b_3 70-2°, d^{20} 1.2007, n_D^{20} 1.4602.[1116]

$\overline{OCHMeCH_2OP}OCH(CH_2Cl)_2$. Ic. b_2 90-1°, d^{20} 1.3362, n_D^{20} 1.4820.[1116]

$\overline{OCHMeCH_2OP}OCH(CH_2Cl)CH_2OCH_2CH_2Cl$. Ib. $b_{0.075}$ 107-8°, n_D^{20} 1.4820, d_4^{20} 1.2890.[687]

$\overline{OCHMeCH_2OP}OCH_2CH_2O\overline{POCH_2CHMeO}$. Ib. b_3 126.5-7°, d^{20} 1.2535, n_D^{20} 1.4758, MR_D 60.80.[1096]

$\overline{OCHMeCH_2OP}OCHMeCH_2O\overline{POCH_2CHMeO}$. Ib. $b_{2.5}$ 128-9°, d^{20} 1.2159, n_D^{20} 1.4703, MR_D 65.51.[1096]

$\overline{OCH_2CH(CH_2Cl)OP}OCH_2\overline{CHCH_2O}$. Ib. $b_{0.005}$ 88-9°, d^{20} 1.3697, n_D^{20} 1.4900.[1293]

$\overline{OCH_2CH(CH_2Cl)OP}OCH(CH_2Cl)CH_2OCH_2CH_2Cl$. Ib. $b_{0.11}$ 128°, d_4^{20} 1.3900, n_D^{20} 1.5000.[687]

$\overline{OCH(CH_2Cl)CH_2OP}OBu$. Ib. b_8 108.5-10°, d_0^{20} 1.1629, n_D^{20} 1.4601.[105]

$\overline{OCH(CH_2OMe)CH_2OP}OMe$. Ib. b_9 77-8°, d_0^{20} 1.1798, n_D^{20} 1.4459. On treatment with water the ring system appears to be retained.[105]

$\overline{OCH(CH_2OMe)CH_2OPO}Et$. Ib. b_{10} 84-5°, d_0^{20} 1.1415, n_D^{20} 1.4498.[105]

$\overline{OCH(CH_2OMe)CH_2OPO}CH_2CH_2Cl$. Ic. b_3 106-7°, d^{20} 1.2977, n_D^{20} 1.4722.[1116]

$\overline{OCH(CH_2OMe)CH_2OPO}CH_2\overline{CHCH_2O}$. Ib. $b_{0.005}$ 86-7°, d^{20} 1.2443, n_D^{20} 1.4678.[1293]

$\overline{OCH(CH_2OMe)CH_2OPO}CHMeCH_2Cl$. Ic. b_1 88-9°, d^{20} 1.2425, n_D^{20} 1.4670.[1116]

$\overline{OCH(CH_2OMe)CH_2OPO}CH(CH_2Cl)_2$. Ic. b_6 143-4°, d^{20} 1.3562, n_D^{20} 1.4877.[1116]

$\overline{OCH(CH_2OMe)CH_2OPO}Bu$. Ib. b_9 107-7.5°, d_0^{20} 1.0713, n_D^{20} 1.4450.[105]

$\overline{OCH(CH_2OMe)CH_2OPO}C_6H_5$. Ib. b. 145.5-6°, d_0^{20} 1.2130, n_D^{20} 1.4768.[105]

$\overline{OCH(CH_2OEt)CH_2OPO}Et$. Ib. b_7 93-4°, d_0^{20} 1.0937, n_D^{20} 1.4401.[105]

$\overline{OCHMeCHMeOPO}CH_2CH_2OCH{:}CH_2$. Ib. b_5 96.5°, d^{20} 1.090, n_D^{20} 1.4615.[147]

$\overline{OCHMeCHMeOPO}CH_2CH_2Cl$. Ic. b_2 70-2°, d^{20} 1.1982, n_D^{20} 1.4596.[1116]

$\overline{OCHMeCHMeOPO}CH_2\overline{CHCH_2O}$. Ib. $b_{0.006}$ 74-6°, d^{20} 1.1784, n_D^{20} 1.4630.[1293]

$\overline{OCHMeCHMeOPO}CHMeCH_2Cl$. Ic. b_3 78-80°, d^{20} 1.1527, n_D^{20} 1.4546.[1116]

$\overline{OCHMeCHMeOPO}CH(CH_2Cl)_2$. Ic. b_2 96-8°, d^{20} 1.2743, n_D^{20} 1.4718.[1116]

$\overline{OCMe_2CH_2OPO}\text{-}i\text{-}C_{10}H_{21}$. Ie, from acetate. $b_{0.08}$ 98-9°.[345]

$\overline{OCMe_2CMe_2OPO}Me$. Ib. Liquid b_{48} 91-2.5°, d_0^0 1.0622, d_0^{90} 1.0449, d_{20}^{20} 1.0469, n_D^{20} 1.4417.[82]

$\overline{OCMe_2CMe_2OPO}Et$. Ib. Liquid b_{14} 75-6°, d_0^0 1.0322, d_0^{20} 1.0136, d_{20}^{20} 1.0156, n_D^{20} 1.4392.[82]

$\overline{OCMe_2CMe_2OPO}Pr$. Ib. Liquid $b_{11.5}$ 84.5-6.0°, d_0^0 1.0138, d_0^{20} 0.9961, d_{20}^{20} 0.9981, n_D^{20} 1.4392.[82]

$\overline{OCMe_2CMe_2OPO}CH(CH_2Cl)CH_2OCH_2CH_2Cl$. Ib. $b_{0.18}$ 134-83°, n_D^{20} 1.4725, d_4^{20} 1.2000.[687]

$\overline{OCMe_2CMe_2OPO}Bu$. Ib. Liquid $b_{14.5}$ 105-6.5°, d_0^0 1.0076, d_0^{20} 0.9901, d_{20}^{20} not cited, d_{25}^{25} 0.9780, n_D^{25} 1.4413.[82]

$\overline{OCMe_2CMe_2OPO}CMeEtCCl_3$. Ib. $b_{0.2}$ 102°, d^{20} 1.2471, n_D^{20} 1.4878.[368]

$\overline{OCHPhCHPhOPO}\text{-}i\text{-}C_{10}H_{21}$. Ie, from acetate. $b_{0.04}$ 114-5°.[345]

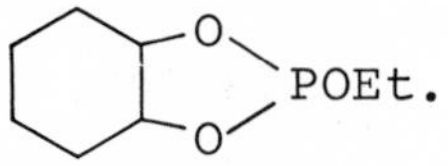

^{31}P -136, -139 ppm.[197]

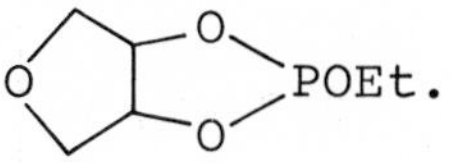

^{31}P -134 ppm.[197]

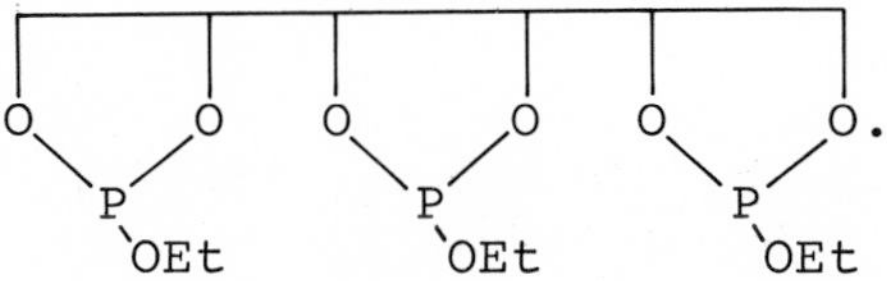

Ie, from D-mannitol. $b_{0.06}$ 152-4°, n_D^{20} 1.4799. Extremely hygroscopic, hydrolyzed in air. IR.[1360,1361]

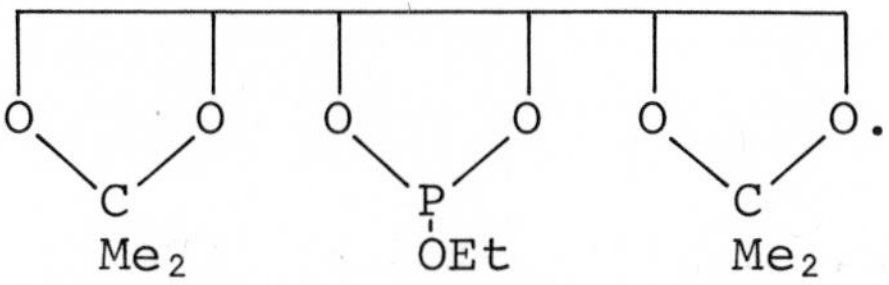

Ie, from di-O-isopropylidene-D-mannitol. $b_{0.8}$ 73-5°, n_D^{20} 1.4562, IR.[1360]

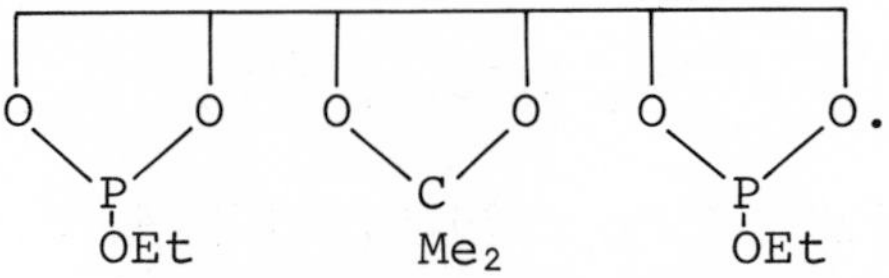

Ie, from mono-O-isopropylidene-D-mannitol. $b_{0.8}$ 146-7°, n_D^{20} 1.4720. (For this and preceding 2 compounds) strong bands at 615-22, 775-6, and 924-30 cm^{-1} were assigned to the cyclophosphite groupings.[1360]

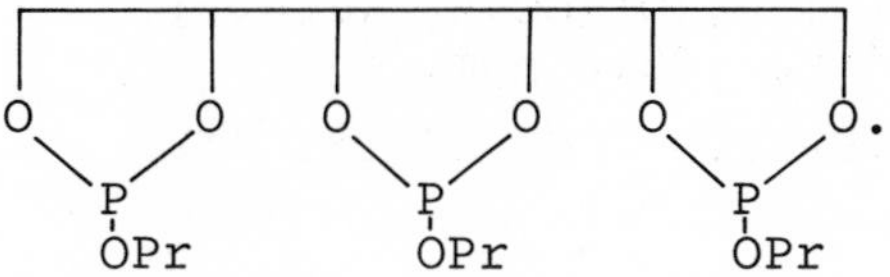

Ib. Ie. $b_{0.08}$ 169-71°, n_D^{20} 1.4760.[1361]

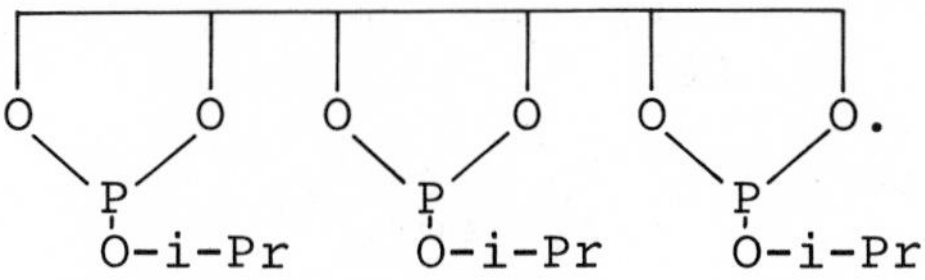

Ib. Ie. $b_{0.05}$ 145°, n_D^{20} 1.4588.[1361]

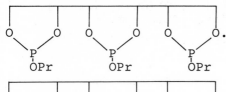

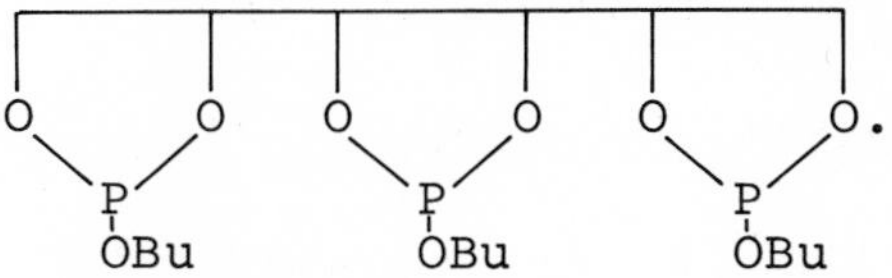

Ib. Ie. $b_{0.05}$ 178-80°, n_D^{20} 1.4752.[1361]

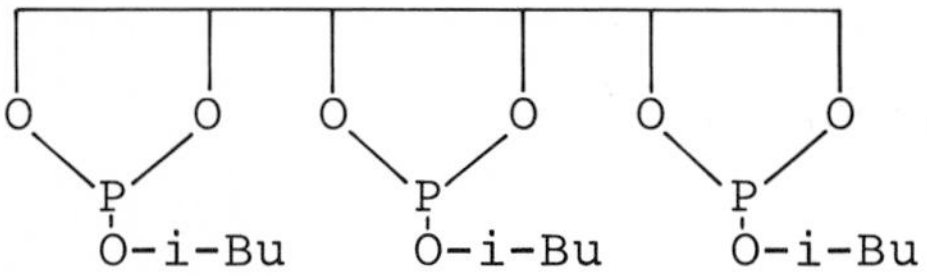

. Ib. Ie. $b_{0.8}$ 178-80°, n_D^{20} 1.4631.[1361]

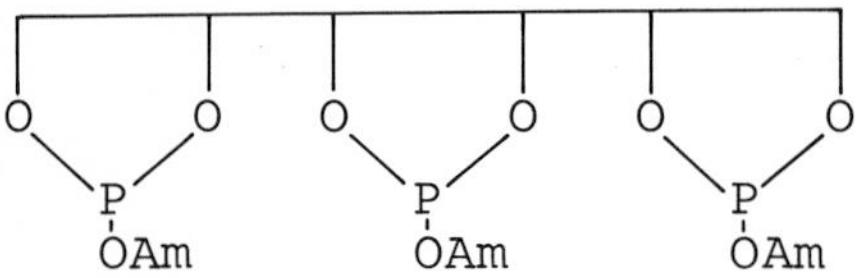

. Ib. Ie. $b_{0.05}$ 202-4°, n_D^{20} 1.4702.[1361]

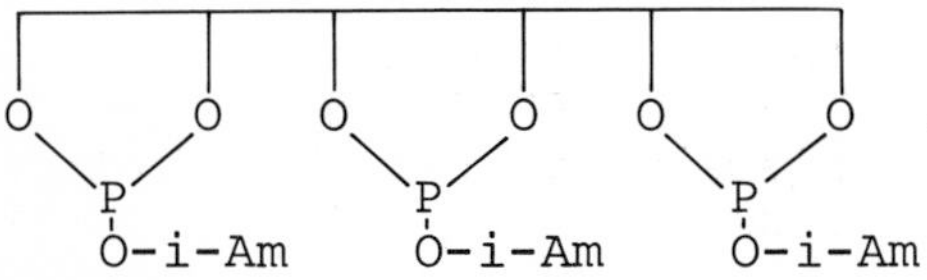

. Ib. Ie. $b_{0.05}$ 182-4°, n_D^{20} 1.4685.[1361]

TYPE: $\overline{OCH_2CH_2OP}OR'P(O)(OR)_2$

$\overline{OCH_2CH_2OP_\alpha}OCH[CH{=}CH_2]P_\beta(O)(OCH_3)_2$. $^{31}P_\alpha$ -130.8, $^{31}P_\beta$ -18.0 ppm.[826]

$\overline{OCH_2CH_2OP_\alpha}OC(CH_3)_2P_\beta(O)(OC_2H_5)_2$. $^{31}P_\alpha$ -126.5, $^{31}P_\beta$ -19.5 ppm.[826]

$\overline{OCH_2CH_2OP_\alpha}OCH(CH_3)P_\beta(O)(OC_2H_5)_2$. $^{31}P_\alpha$ -130.6, -139.4, $^{31}P_\beta$ -20.8, -22.4 ppm.[826]

$\overline{OCH_2CH_2OP_\alpha}OC(CH_3)[P_\beta(O)(OC_2H_5)_2]_2$. $^{31}P_\alpha$ -128.8, $^{31}P_\beta$ -16.0 ppm.[826]

$\overline{OCH_2CH_2OP_\alpha}OC(CH_3)[P_\beta(O)(OC_2H_5)_2][P_\gamma(O)(OC_2H_5)(C_6H_5)]$. $^{31}P_\alpha$ -136.1, $^{31}P_\beta$ -18.3, $^{31}P_\gamma$ -33.6 ppm.[826]

$\overline{OCH_2CH_2OP_\alpha}OC(CN)(CH_3)P_\beta(O)(OC_2H_5)_2$. $^{31}P_\alpha$-141.8, $^{31}P_\beta$ -13.5 ppm.[826]

$\overline{OCH_2CH_2OP_\alpha}OC$  $P_\beta(O)(OC_2H_5)_2$. $^{31}P_\alpha$ -127.0, $^{31}P_\beta$ -21.6 ppm.[826]

$\overline{OCH_2CH_2OP_\alpha}OC$ 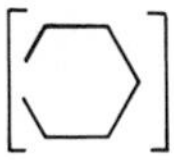 $P_\beta(O)(OC_2H_5)_2$. $^{31}P_\alpha$ -126.8, $^{31}P_\beta$ -21.6 ppm.[826]

$\overline{OCH_2CH_2OP_\alpha}OC(CH_3)(CH_2CH_2CH{=}CH_2)P_\beta(O)(OC_2H_5)_2$. $^{31}P_\alpha$ -130.0, $^{31}P_\beta$ -24.3 ppm.[826]

$\overline{OCH_2CH_2OP_\alpha}OC(CH_3)[CH_2C(O)(OC_2H_5)]P_\beta(O)(OC_2H_5)_2$. $^{31}P_\alpha$ -129.0, $^{31}P_\beta$ -20.9 ppm.[826]

$\overline{OCH_2CH_2OP_\alpha}OCH(p\text{-}C_6H_4OH)P_\beta(O)(OC_2H_5)_2$. $^{31}P_\alpha$ -137.3, $^{31}P_\beta$ -18.6 ppm.[826]

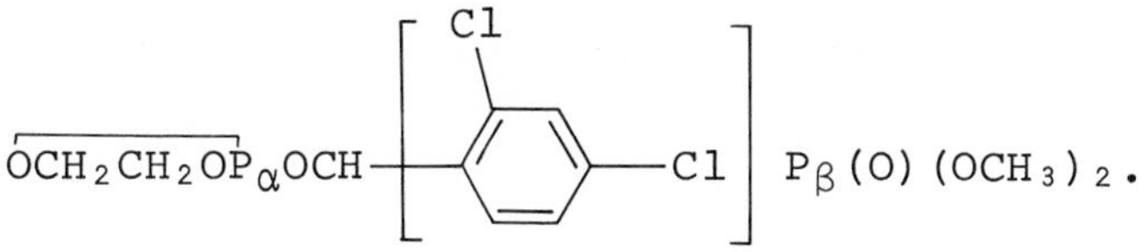

$\overline{OCH_2CH_2OP_\alpha OCH}$ $\left[\begin{array}{c} Cl \\ \end{array}\right]$ $-Cl$ $P_\beta(O)(OCH_3)_2$. $^{31}P_\alpha$ -131.5, $^{31}P_\beta$ -18.4 ppm.[826]

$\overline{OCH_2CH_2OP_\alpha OCMe:CMeOP_\beta}(O)(OCH_2CHClMe)_2$. $^{31}P_\alpha$ -137.9, $^{31}P_\beta$ $+7.9$ ppm.[826]

2-Acyloxy-1,3,2-Dioxaphospholanes

$\overline{OCH_2CH_2OPOAc}$. $(RO)_2PCl$ + acetate.[345,430,955] $b_{0.5}$ 55°,[345] b_6 80-1°,[430] $b_{8.5}$ 83.5-5°,[955] d^{20} 1.2880,[430] d_4^{20} 1.2958,[955] n_D^{20} 1.4533,[955] n_D^{20} 1.4532,[430] ^1H NMR,[432] ^{31}P -127.2 ppm.[826]

$\overline{OCHMeCHMeOPOAc}$. $(RO)_2PCl$ + NaOAc. $b_{9.5}$ 99.5-100°, d_4^{20} 1.1787, n_D^{20} 1.4883.[955]

$\overline{OCMe_2CMe_2OPOAc}$. $(RO)_2PCl$ + NaOAc. b_8 106.5°, d_4^{20} 1.1168, n_D^{20} 1.4504.[955]

Derivatives of 1,3,2-Dioxaphosphorinane

TYPE:

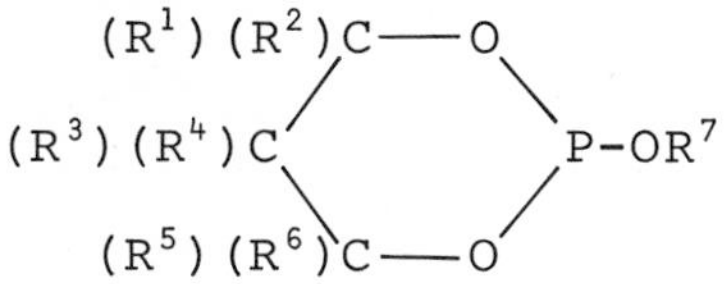

$$\begin{array}{c} (R^1)(R^2)C-O \\ (R^3)(R^4)C \qquad P-OR^7 \\ (R^5)(R^6)C-O \end{array}$$

$\overline{OCH_2CH_2CH_2OPOMe}$. b_{11} 50-1°,[143] d_0^{20} 1.1758,[143] n_D^{20} 1.4465.[143] Thermogravimetry.[143]

$\overline{OCH_2CH_2CH_2OPOCH_2Ph}$. b_{13-14} 147-8°, d_0^{20} 1.1873, n_D^{20} 1.5270.[143] Thermogravimetry.[143]

$\overline{OCH_2CH_2CH_2OPOEt}$. ^{31}P -128, -132 ppm.[197,628]

$\overline{OCH_2CH_2CH_2OPOCH_2CH_2NEt_2}$. Ib. $b_{3.5-4.0}$ 107-10°, n_D^{25} 1.4609.[270]

$\overline{OCH_2CH_2CH_2OPOCH_2CHCH_2O}$. Ib. $b_{0.005}$ 83-4°, d^{20} 1.2333, n_D^{20} 1.4707.[1293]

$\overline{OCH_2CH_2CH_2OPOC_6H_5}$. $b_{1.5-2}$ 105-8°, m. 45-8°.[143] Thermogravimetry.[143]

$\overline{OCHMeCH_2CH_2OPOMe}$. Ib.[105] b_{13} 62°, d_0^{20} 1.1092, n_D^{20} 1.4420,[105,143] ^1H NMR.[338] Thermogravimetry.[143]

$\overline{OCHMeCH_2CH_2OPOEt}$. Ib.[105] b_8 63-4°, d_0^{20} 1.0696, n_D^{20} 1.4410.[105,143] ^{31}P -131 ppm.[197] Thermogravimetry.[143]

$\overline{OCHMeCH_2CH_2OPOPr}$. $b_{6.5-7}$ 72-2.5°, d_0^{20} 1.0425, n_D^{20} 1.4435.[143] Thermogravimetry.[143]

$OCHMeCH_2CH_2OPOCH_2CHCH_2O$. Ib. $b_{0.005}$ 80-1°, d^{20} 1.1922, n_D^{20} 1.4691.[1293]

$OCHMeCH_2CH_2OPO$-i-Pr. Ie, from acetate. b_6 62-4°, n_D^{20} 1.4431.[1061]

$OCHMeCH_2CH_2OPOBu$. Ie.[430,960,961,1061] b_7 90-0.5°,[961] b_7 90-1°,[430,960] b_9 94-5°,[1061] n_D^{20} 1.4472,[960,961] n_D^{20} 1.4474,[1061] d^{20} 1.0250.[961]

$OCHMeCH_2CH_2OPO$-t-Bu. Ie, from acetate. b_6 68-70°, n_D^{20} 1.4458.[1061]

$OCHMeCH_2CH_2OPOC_6H_{13}$. Ie.[430,960,961] b_7 102-4°,[430] b_7 102-3°,[961] b_7 114-5°,[960] n_D^{20} 1.4482,[430,961] n_D^{20} 1.4495,[960] d^{20} 0.9936.[961]

$OCHMeCH_2CH_2OPO\text{-}2\text{-}C_8H_{17}$. Ie, from acetate. b_7 129-30°, d^{20} 1.0743, n_D^{20} 1.4500.[961]

$OCHMeCH_2CH_2OPOC_8H_{17}$. Ie, from FCH_2CH_2 ester. b_2 105-6°, d^{20} 0.9690, n_D^{20} 1.4510.[957]

$OCHMeCH_2CH_2OPOCH_2CH_2OPOCH_2CH_2CHMeO$. Ib. b_2 138-9°, d^{20} 1.2168, n_D^{20} 1.4780, MR_D 69.36.[1096]

$OCHMeCH_2CH_2OPO(CH_2)_3OPOCH_2CH_2CHMeO$. Ie. b_3 152-3°, d^{20} 1.1800, n_D^{20} 1.4728.[957]

$OCHMeCH_2CH_2OPOC_6H_5$. Ie, from acetate. b_3 132-3°, d^{20} 1.1560, n_D^{20} 1.5145,[961] d_{15}^{25} 1.178, n_D^{25} 1.515130.[541]

$OCH_2CHMeCH_2OPOMe$. ^1H NMR.[338]

$OCH_2CMe_2CH_2OPOMe$. ^1H NMR.[338]

$OCH_2CMe_2CH_2OPOCH_2CH_2Cl$. Ic. $b_{0.9}$ 82°, n_D^{20} 1.4747.[944] ^{31}P -122.6 ppm.[826]

$OCH_2CMe_2CH_2OPOPr$. ^{31}P -138 ppm.[826]

$OCH_2CMe_2CH_2OPOCH_2CHClCH_3$. ^{31}P -121.3 ppm.[826]

$OCH_2CMe_2CH_2OPOCH_2CHClC_5H_{11}$. ^{31}P -123.2 ppm.[826]

$OCH_2CMe_2CH_2OPOC_6H_5$. Ie.[541,548] d^{29} 1.135, n_D^{25} 1.5157,[548] n_D^{25} 1.50375,[541] ^{31}P -114.8 ppm.[826]

$OCH_2C(CHMe_2)[CH(Me)OMe]CH_2OPOEt$. Ib. $b_{0.03}$ 76°, d_4^{20} 1.0711, n_D^{20} 1.4596.[210,225]

$OCH_2C(CHMe_2)[CH(Me)OCHMe_2]CH_2OPOMe$. Ib. $b_{0.02}$ 67°, d_4^{20} 1.0837, n_D^{20} 1.4668.[210,225]

$OCH_2C(CHMe_2)[CH(Me)OCHMe_2]_2CH_2OPOEt$. Ib. $b_{0.02}$ 70°, d_4^{20} 1.0388, n_D^{20} 1.4578.[210,225]

$OCH_2C(Et)(CH_2OCH_2CH{:}CH_2)CH_2OPOC_6H_5$. Ie.[426] $b_{0.5}$ 110-2°,[425] $b_{0.5}$ 110-25°.[426]

$OCH_2C(Et)(CH_2OCH_2CH{:}CH_2)CH_2OPOCH_2CH{:}CH_2$. Ie. b_{10} 90°.[425]

$\overline{OCHMeCH_2CHMeOP}OMe$. ^1H NMR.[338]

$\overline{OCHMeCH_2CMe_2OP}OC_6H_5$. Ie. b_1 95°, d_{15}^{25} 1.0934, n_D^{25} 1.51245.[541]

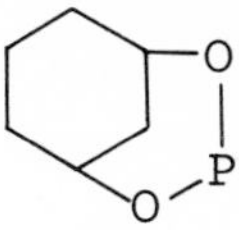
POEt. ^{31}P -124 ppm.[197]

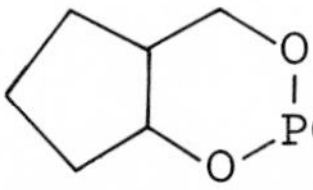
POEt. ^{31}P -125 ppm.[197]

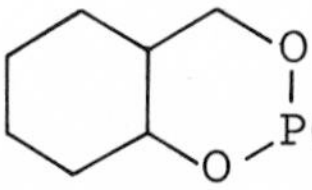
POEt. ^{31}P -125 ppm.[197]

TYPE: $\overline{OCH_2CMe_2CH_2OP}OR'P(O)(OR)_2$

$\overline{OCH_2CMe_2CH_2OP_\beta}OCMe{:}CMeOP_\alpha(O)(OEt)_2$. $^{31}P_\alpha$ +6.0, $^{31}P_\beta$ -117.8 ppm.[826]

$\overline{OCH_2CMe_2CH_2OP_\beta}OCMe{:}CMeOP_\alpha(O)(OC_6H_{13})_2$. $^{31}P_\alpha$ +5.8, $^{31}P_\beta$ -118.6 ppm.[826]

$\overline{OCH_2CMe_2CH_2OP_\alpha}OCH$ 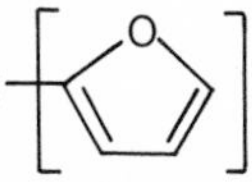 $P_\beta(O)(OC_2H_5)_2$. $^{31}P_\alpha$ -124.1, $^{31}P_\beta$ -16.4 ppm.[826]

2-Acyloxy-1,3,2-Dioxaphosphorinanes

$\overline{OCHMeCH_2CH_2OP}OC(O)H$. $(RO)_2PCl$ + HCOOK. b_3 90° (some decomp.), d^{20} 1.2210, n_D^{20} 1.4570.[430]

$\overline{OCHMeCH_2CH_2OP}OAc$. $(RO)_2PCl$ + NaOAc.[955,960] $b_{9.5}$ 111-2.5°,[955] b_{12} 114-5°, d^{20} 1.1836,[1061] d^{20} 1.2160,[960] d_4^{20} 1.1886,[955] n_D^{20} 1.4560,[960] n_D^{20} 1.4560,[1061] n_D^{20} 1.4530.[955]

$\overline{OCHMeCH_2CH_2OP}OC(O)Et$. $(RO)_2PCl$ + NaO_2CEt. d^{20} 1.2400, n_D^{20} 1.4555.[960]

$\overline{OCHMeCH_2CH_2OP}OC(O)i\text{-}Pr$. $(RO)_2PCl$ + i-PrCOOK. b_2 104°, d^{20} 1.1050, n_D^{20} 1.4550.[430]

$\overline{OCHMeCH_2CH_2OP}OC(O)C_6H_5$. $(RO)_2PCl$ + PhCOOK. b_2 138-40°, m. 52°.[430]

TYPE: $\overline{O(CH_2)_4OP}OR$

$\overline{OCH_2CH_2CH_2CH_2OP}OMe$. Ib. $b_{4.5-5}$ 54-5°, d_0^{20} 1.1640, n_D^{20} 1.4642. CuI salt, m. 142-4°.[105]

$\overline{OCH_2CH_2CH_2CH_2OP}OBu$. Ib. b_{9-10} 100-2°, d_0^{20} 1.0557, n_D^{20} 1.4540.[105]

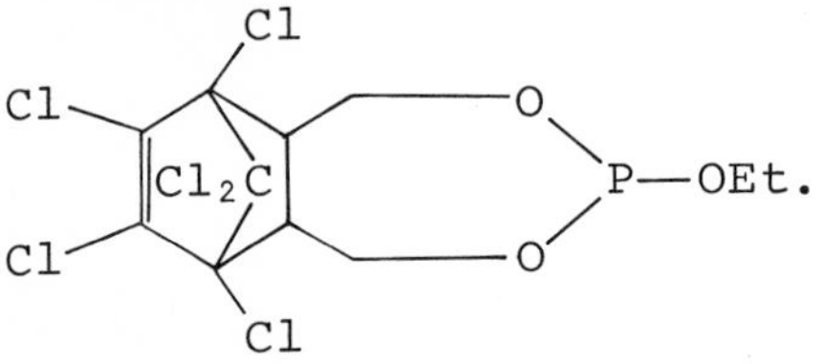
Ib. b_8 110°.[509]

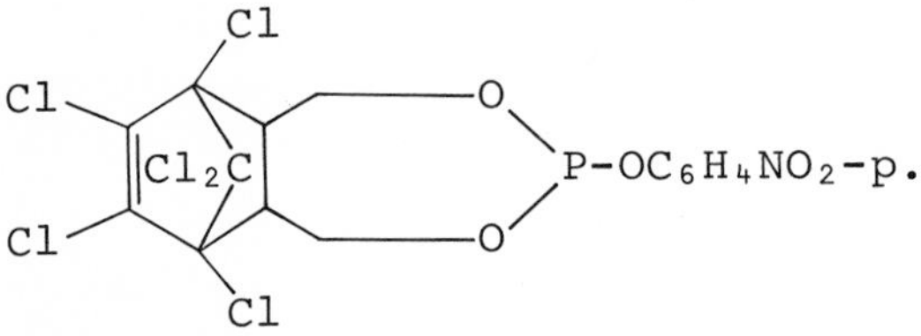
Ib. m. 158-62°.[509]

Derivatives of Pentaerythritol

TYPE: 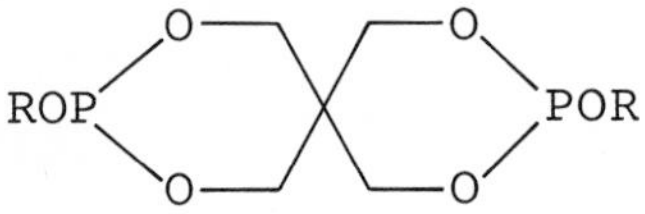

$MeOP(OCH_2)_2C(CH_2O)_2POMe$. Ib.[675,705] Ie.[705] m. 124-7°.[675]

$EtOP(OCH_2)_2C(CH_2O)_2POEt$. Ib.[705] Ie.[675,705] b_1 121-4°, m. 76-9° (m. 80-2° after sublimation in vacuo). Very hygroscopic. IR.[675]

$PrOP(OCH_2)_2C(CH_2O)_2POPr$. Ib. Ie. $b_{0.5}$ 135-7°, m. 70-2°, m. 71-3°.[675]

$MeCHClCH_2OP(OCH_2)_2C(CH_2O)_2POCH_2CHClMe$. Ie. d^{20} 1.355, n_D^{25} 1.5012.[1381]

i-$PrOP(OCH_2)_2C(CH_2O)_2PO$-i-Pr. Ie. $b_{0.5}$ 128-30°, m. 51-3°.[675]

$MeCH(CH_2Cl)OP(OCH_2)_2C(CH_2O)_2POCH(CH_2Cl)Me$. Ie. d^{35} 1.318, n_D^{20} 1.4372.[541]

$BuOP(OCH_2)_2C(CH_2O)_2POBu$. Ib. $b_{0.5}$ 139-40°, m. 25-7°.[675]

$BuEtCHCH_2OP(OCH_2)_2C(CH_2O)_2POCH_2CHEtBu$. Ie. d^{25} 1.04, n_D^{25} 1.4688.[541]

$C_8H_{17}OP(OCH_2)_2C(CH_2O)_2POC_8H_{17}$. Ib. Ie. n_D^{20} 1.4745.[705]

$C_9H_{19}OP(OCH_2)_2C(CH_2O)_2POC_9H_{19}$. Ib. Ie. n_D^{20} 1.4730.[705]

$C_{10}H_{21}OP(OCH_2)_2C(CH_2O)_2POC_{10}H_{21}$. Ib. Ie. m. 24-5°, n_D^{25} 1.4700,[705] n_D^{25} 1.4723.[428,1150]

n-$C_{10}H_{21}OP(OCH_2)_2C(CH_2O)_2PO$-n-$C_{10}H_{21}$. Ie. d^{25} 1.030, n_D^{25} 1.4731.[1381]

i-$C_{10}H_{21}OP(OCH_2)_2C(CH_2O)_2PO$-i-$C_{10}H_{21}$. Ie. d^{25} 1.00, n_D^{25} 1.4731.[541]

$C_{11}H_{23}OP(OCH_2)_2C(CH_2O)_2POC_{11}H_{23}$. Ib. Ie. m. 30-5°.[705]

$C_{12}H_{25}OP(OCH_2)_2C(CH_2O)_2POC_{12}H_{25}$. Ie. d^{25} 1.015, n_D^{25} 1.4725, m. approx. 35°.[541]

$C_{14}H_{29}OP(OCH_2)_2C(CH_2O)_2POC_{14}H_{29}$. Ib. Ie. m. 50-5°.[705]

$C_{18}H_{37}OP(OCH_2)_2C(CH_2O)POC_{18}H_{37}$. Ib.[705] Ie.[428,541,705] m. 60-4°,[705] m. 40°, d^{25} 0.949, n_D^{25} 1.4602, n_D^{60} 1.4582.[541]

$C_6H_5OP(OCH_2)_2C(CH_2O)_2POC_6H_5$. Ie. m. 123°.[541]

TYPE: 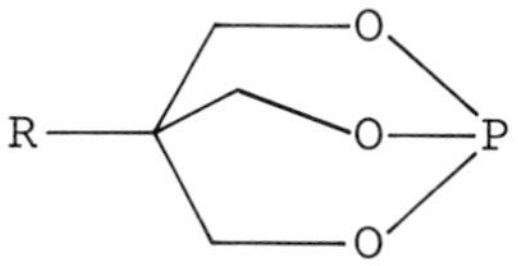

$MeC(CH_2O)_3P$ (metriol phosphite). Ia.[1348] By evolvement of $PhCH_2OH$ from

$CH_3\cdot C(CH_2O)_2PHO$
  |
CH_2OCH_2Ph

at 250-60°/170 mm and by heating 2-(hydroxymethyl)-2-methyltrimethylene hydrogen phosphite at 160-70°/45 mm.[970] m. 97-8°,[1348] m. 96-7°,[970] m. 90°, b_{174} 155°,[28] IR,[680] ^{31}P -91.5, -91.8 ppm, J_{POCH} 2 Hz.[826] Oxidized to metriol phosphate m. 249-50°. Chromatography.[970]

$HOCH_2C(CH_2O)_3P$. Ie.[674] m. 58-9°,[674] m. 62.4-4.4°,[28] b_1 117-18°,[674] $b_{2.5}$ 143.4°.[28]

$(EtO)_2POCH_2C(CH_2O)_3P$. Ie. $b_{0.5}$ 113-5°, n_D^{20} 1.4822, d^{20} 1.2469, IR.[675]

$P(OCH_2)_3CCH_2OCH_2C(CH_2O)_3P$. ^{31}P -93.4 ppm.[826]

$EtOP[OCH_2C(CH_2O)_3P]_2$. Ie. m. 95-105°. Very hygroscopic. IR.[675]

$\overline{OCH_2CH_2CH_2OP}OCH_2C(CH_2O)_3P$. Ib. b_1 155-6.5°.[705]

$\overline{OCMe_2CH_2CH_2OP}OCH_2C(CH_2O)_3P$. Ib. $b_{1.5}$ 158-9.5°, m. 90-1°.[705]

$\overline{OCH_2CH_2OP}OCH_2C(CH_2O)_3P$. Ib. b_1 151-2°.[705]

1,2-$C_6H_4O_2POCH_2C(CH_2O)_3P$. Ib. Ie. $b_{1.5}$ 160-1°, m. 92-3°, n_D^{20} 1.4745.[705]

$EtC(CH_2O)_3P$. Ie.[566] m. 44-5.5°,[28] m. 56.5°, b_8 100°,[566] b_{14} 114°, b_{30} 135°,[28] ^{31}P -94 ppm.[826]

$BuC(CH_2O)_3P$. Ie. From $P(CH_2OH)_3$ + $(MeO)_3P$. m. 75-6°, sublimes ca. 50°/0.1 mm. b_9 95°, m. 79-80°.[566] ^1H NMR.[310a]

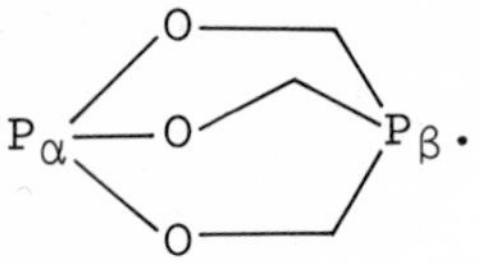

$^{31}P_{\alpha}$ -90.0 ppm, J_{POCH} 2.6 Hz, $^{31}P_{\beta}$ +67.0 ppm, J_{PCH} 8.9 Hz.

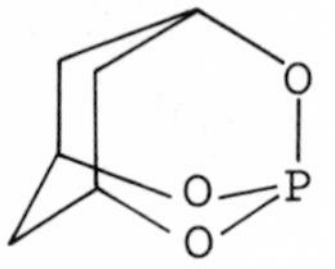

Ia. m. 207°,[1294] IR,[680] ^{31}P -137 ppm, J_{POCH} 6 Hz.[826]

TYPE: 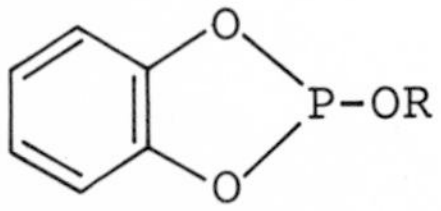

1,2-$C_6H_4O_2$POMe. Ia.[61] Ih.[96] b_8 73°,[96] b_{15} 76-7°,[61] d_0^{15} 1.2568, n_D^{19} 1.5209. CuBr salt, m. 130-5°.[96]
1,2-$C_6H_4O_2POCH_2C_6H_5$. Ib. b_1 138-9°, d^{20} 1.2427, n_D^{20} 1.5782.[704]
1,2-$C_6H_4O_2$POEt. Ia.[61] Ib.[301] Ih.[96] b_{11} 83-4°,[61] b_{11} 86°,[96,143] b_{19} 99-100°,[301] d_0^0 1.2420,[143] n_D^{17} 1.5085.[143] CuBr salt, m. 142-5°.[96]
1,2-$C_6H_4O_2POCH_2CH_2Cl$. Ic. $b_{2.5}$ 107-8°, d_0^{20} 1.3455, d_4^{20} 1.3444, n_D^{20} 1.5430. CuCl salt, m. 135-7°.[629]
1,2-$C_6H_4O_2POCH$:CH_2. Ib. $b_{2-2.5}$ 71°, d_4^{20} 1.2306, n_D^{20} 1.5357, MR_D 46.13 (calc. 45.94).[446]
1,2-$C_6H_4O_2POCH_2CH_2O_2CC(CH_3)$:$CH_2$. Ib. n_D^{20} 1.5285.[1076]
1,2-$C_6H_4O_2POCH_2CH_2OCH{=}CH_2$. Ib. $b_{0.8}$ 101-2°, d^{20} 1.2204, n_D^{20} 1.5272,[692] IR.[1271]
1,2-$C_6H_4O_2$POPr. Ia.[61] Ie.[430] Ih.[96] b_8 94-5°,[430] b_9 97°,[96] b_{13} 100-2°,[61] d_0^0 1.1120, d_0^{17} 1.1296, n_D^{17} 1.4841, n_D^{20} 1.5025.[430] CuI salt, m. 138°.[96]
1,2-$C_6H_4O_2POCH_2C{\equiv}CH$. Ib.[598,723] m. 44-5°, IR. White or colorless complex with CuCl.[598,723]
1,2-$C_6H_4O_2$POPr-iso. Ih. b_3 73-4°, d_0^{17} 1.1171, n_D^{17} 1.4724. CuCl salt, m. 143°; CuI salt, m. 178-9° (with decomposition).[96]
1,2-$C_6H_4O_2POCMe_2CN$. Ib. m. 54-5°, b_{10} 139-42°.[704]
1,2-$C_6H_4O_2POCH(CH_2Cl)_2$. Ib. b_{11} 154°, d^{20} 1.3930, n_D^{20} 1.5469.[701]
1,2-$C_6H_4O_2$POBu. Ia.[61] Ie.[430] Ih.[96] b_5 104°,[430] b_8 116°,[96] b_{12} 116-7°,[61] d_0^0 1.1457, d_0^{17} 1.1255, n_D^{18} 1.5053, n_D^{20} 1.5053.[430] CuCl salt sinters at 150°, m. 202°.[96]
1,2-$C_6H_4O_2POCH_2CH_2CH$:CH_2. Ib. b_{13} 117-20°, d^{20} 1.1510, n_D^{20} 1.5190, MR_D 55.40.[703]
1,2-$C_6H_4O_2$POBu-iso. Ih. b_8 105°, d_0^0 1.1208, n_D^{15} 1.4950 d_0^{19} 1.0997. CuCl salt sinters at 158°, m. 208-10°.[9]

$1,2-C_6H_4O_2PO$-i-Am. Ib. b_{10} 120-2°, d^{20} 1.0928, n_D^{20} 1.4991.[704]

$1,2-C_6H_4O_2POCH_2C(CH_2OH)_2Me$. Ie. From $1,2-C_6H_4O_2POAc$ and metriol. m. 63-4°.[430]

$1,2-C_6H_4O_2POC_6H_{13}$. Ib. b_{11} 139-40°, d^{20} 1.0930, n_D^{20} 1.5010.[704]

$1,2-C_6H_4O_2POC_8H_{17}$. Ib. b_2 127-8°, d^{20} 1.0570, n_D^{20} 1.4956.[704]

$1,2-C_6H_4O_2POC_{10}H_{21}$. Ib. b_3 164-5°, d^{20} 1.0360, n_D^{20} 1.4945.[704]

$1,2-C_6H_4O_2POC_{12}H_{25}$. Ib. b_1 183-9°, d^{20} 1.0254, n_D^{20} 1.4930.[704]

$1,2-C_6H_4O_2POC_{14}H_{29}$. Ib. m. 25°, b_1 189-90°, d^{20} 1.0044, n_D^{20} 1.4922.[704]

$1,2-C_6H_4O_2POC_{18}H_{37}$. Ib. m. 43-3.5°, b_1 219-23°.[704]

$1,2-C_6H_4O_2PO$-cyclo-C_6H_{11}. Ib. b_{10} 144-5°, d^{20} 1.1637, n_D^{20} 1.5274.[704]

$1,2-C_6H_4O_2PO$-2-isobornyl-4-MeC_6H_3. Ib. $b_{1.5}$ 189-90°.[707]

$1,2-C_6H_4O_2PO$-2,6-diisobornyl-4-MeC_6H_2. Ib. b_2 239-43°.[707]

$1,2-C_6H_4O_2POPh$. Ia. b_{12} 150°.[61]

$1,2-C_6H_4O_2POC_6H_4OMe$-o. Ia. $b_{0.13}$ 137°, b_{13} 184°.[61]

$1,2-C_6H_4O_2POC_6H_4OH$-o. Ia. Original preparation[54] is not satisfactory for synthetic purpose. Use of dry ether in reaction of pyrocatechol with 1.5 mole of PCl_3 gives an 85% yield of this substance,[55] m. 112-3°,[54] m. 117°, d_4^{137} 1.256,[61] b_{12} 155°.[54] Its acetyl derivative said to be undistillable, actually $b_{0.02}$ 135°.[54] This ester does not add sulfur,[61] contrary to expectations.

$1,2-C_6H_4O_2PO-C_6H_4NO_2$-4. Ib. m. 62-4°.[595]

$1,2-C_6H_4O_2PO$-4-ClC_6H_4. Ib. m. 32-3°, b_2 130-2°.[704]

$1,2-C_6H_4O_2PO$-2,3,4,5-Cl_4C_6H. Ib. m. 75-7°, b_1 180-7°.[704]

$1,2-C_6H_4O_2POC_6H_4Me$-o. Ia. b_{13} 159-60°.[61]

$1,2-C_6H_4O_2POC_6H_4Me$-m. Ia. b_{11} 158-9°.[61]

$1,2-C_6H_4O_2PO$-4-MeC_6H_4. Ia.[61] Ib.[704] b_{12} 164°,[61] b_3 108-10°,[704] m. 25°,[61] d^{20} 1.2362, n_D^{20} 1.5625.[704]

$1,2-C_6H_4O_2PO$-2,4,6-$Me_3C_6H_2$. Ib. b. 170-1°, n_D^{20} 1.5645, d^{20} 1.1830.[707]

$1,2-C_6H_4O_2PO$-3,4,6-$Me_3C_6H_2$. Ib. b_1 124-5°, n_D^{20} 1.5661.[707]

$1,2-C_6H_4O_2PO$-3,4,5-$Me_3C_6H_2$. Ib. b_2 131-2°, n_D^{20} 1.5641, d^{20} 1.1523.[707]

$1,2-C_6H_4O_2PO$-4-BuC_6H_4. Ib. b_4 154-5°, d^{20} 1.1447, n_D^{20} 1.5523.[704]

$1,2-C_6H_4O_2PO$-4-PhC_6H_4. Ib. m. 152°.[704]

$1,2-C_6H_4O_2PO$-2-PhC_6H_4. Ib. b_3 200-3°.[704]

$1,2-C_6H_4O_2PO$-4-$C_8H_{17}C_6H_4$. Ib. b_2 174-7°, d^{20} 1.1120, n_D^{20} 1.5450.[704]

$1,2-C_6H_4O_2PO$-4-$(PhCHMe)C_6H_4$. Ib. b_1 193-4°, d^{20} 1.2034,

n_D^{20} 1.6005.[707]
1,2-$C_6H_4O_2$PO-2,6-di-t-Bu-4-MeC_6H_2. Ib. m. 78-9°, b_5 205-6°.[704]
1,2-$C_6H_4O_2$PO-2-Me-4,6-di-t-BuC_6H_2. Ib. $b_{1.5}$ 152-4°, m. 72-4°.[707]
1,2-$C_6H_4O_2$PO-4-($PhMe_2C$)C_6H_4. Ib. m. 88-90°.[704]
1,2-$C_6H_4O_2$PO-2-(PhCHMe)-4-MeC_6H_3. Ib. b_1 183-4°, d^{20} 1.1947, n_D^{20} 1.5940.[707]
1,2-$C_6H_4O_2$PO-2,4,6-tri-t-BuC_6H_2. Ib. m. 101-2°, b_1 194-6°.[704]
1,2-$C_6H_4O_2$PO-4-$C_{12}H_{25}C_6H_4$. Ib. b_2 209-16°, d^{20} 1.0681, n_D^{20} 1.5331.[704]
1,2-$C_6H_4O_2$PO-2,6-bis(PhCHMe)-4-MeC_6H_2. Ib. b_1 221-3°.[707]
1,2-$C_6H_4O_2$PO-1-$C_{10}H_7$. Ib. m. 86-7°.[49,704]
1,2-$C_6H_4O_2$PO-2-$C_{10}H_7$. Ib. m. 72-4°.[704]
4-Me-1,2-$C_6H_3O_2$PO-2,4,6-tri-t-BuC_6H_2. Ib. b_2 188-8.5°.[708]
4-Me-1,2-$C_6H_3O_2$PO-1-$C_{10}H_7$. Ib. $b_{1.5}$ 171-2°.[708]
4-t-Bu-1,2-$C_6H_3O_2POC_6H_5$. Ib. b_9 172-3.5°.[708]
4-t-Bu-1,2-$C_6H_3O_2$PO-2,6-di-t-Bu-4-MeC_6H_2. Ib. b_2 188-9°.[708]
4-t-Bu-1,2-$C_6H_3O_2$PO-2,4,6-tri-t-BuC_6H_2. Ib. b_2 200-1°.[708]
4-t-Bu-1,2-$C_6H_3O_2$PO-1-$C_{10}H_7$. Ib. b_2 192-4°.[708]
4-t-Am-1,2-$C_6H_3O_2POC_6H_5$. Ib. b_2 148-9°, d^{20} 1.1443, n_D^{20} 1.5490.[708]
4-t-Am-1,2-$C_6H_3O_2$PO-2,4,6-tri-t-BuC_6H_2. Ib. b_1 197-8°.[708]
4-t-Am-1,2-$C_6H_3O_2$PO-1-$C_{10}H_7$. Ib. b_1 184-5°.[708]
3,5-di-t-Bu-1,2-$C_6H_2O_2POC_6H_5$. Ib. b_2 124°.[708]
3,5-di-t-Bu-1,2-$C_6H_2O_2$PO-2,6-di-t-Bu-4-MeC_6H_2. Ib. b_2 199-200°.[708]
3,5-di-t-Bu-1,2-$C_6H_2O_2$PO-2,4,6-tri-t-BuC_6H_2. Ib. $b_{1.5}$ 210-4°.[708]
3,5-di-t-Bu-1,2-$C_6H_2O_2$PO-1-$C_{10}H_7$. Ib. b_2 202-4°.[708]
3,5-di-t-Bu-1,2-$C_6H_2O_2$PO-2-$C_{10}H_7$. Ib. b_3 190-3°.[708]

TYPE: 1,2-$C_6H_4O_2$POR'P(O)$(OR)_2$

1,2-$C_6H_4O_2P_\alpha$OCH(CH_3)P_β(O)$(OCH_3)_2$. $^{31}P_\alpha$ -132.5, $^{31}P_\beta$ -24.6 ppm.[826]

1,2-$C_6H_4O_2P_\alpha$OC  P_β(O)$(OC_6H_{13})_2$. $^{31}P_\alpha$ -140.0, $^{31}P_\beta$ -21.8 ppm.[826]

1,2-$C_6H_4O_2P_\alpha$OC 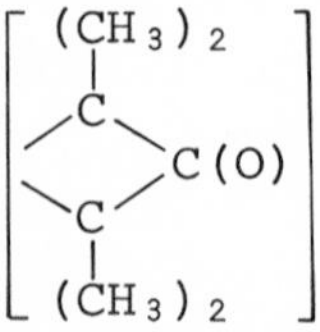

P_β(O)$(OC_2H_5)_2$. $^{31}P_\alpha$ -141.9, $^{31}P_\beta$ -19.3 ppm.[826]

1,2-$C_6H_4O_2P_{\alpha}OCH(p\text{-}C_6H_4OCH_3)P_{\beta}(O)(OCH_3)_2$. $^{31}P_{\alpha}$ -133.5, $^{31}P_{\beta}$ -18.8 ppm.[826]
1,2-$C_6H_4O_2P_{\alpha}OCH(C_6H_5)P_{\beta}(O)(OCH_3)_2$. $^{31}P_{\alpha}$ -127.5, $^{31}P_{\beta}$ -16.2 ppm.[826]

TYPE: 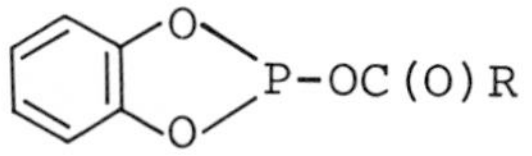

1,2-$C_6H_4O_2POCOMe$. >P-Cl + $(RCO)_2O$ or RCO_2M.[430,442] $b_{0.2}$ 77°,[442] b_2 90-1°,[430] m. 31-2°,[430] m. 31°,[442] d^{20} 1.3174,[442] n_D^{20} 1.5313.[442]
1,2-$C_6H_4O_2POCOEt$. >P-Cl + $(RCO)_2O$ or RCO_2M. $b_{0.07}$ 83°, d^{20} 1.2635, n_D^{20} 1.5222.[442]

I.1.15. Diphosphites and Triphosphites with P in Ring System

TYPE: 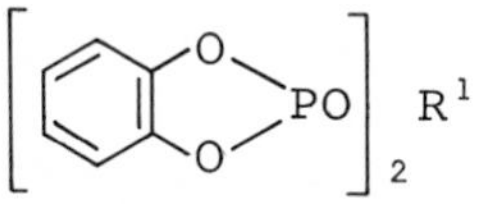

$(1,2\text{-}C_6H_4O_2POCH_2)_2$. Ib. $b_{3.5}$ 184-5°.[931]
1,2-$C_6H_4O_2POCH_2CHMeOPO_2C_6H_4$-1,2. Ib. b_3 187-9°, d^{20} 1.3302, n_D^{20} 1.5621.[931]
1,2-$C_6H_4O_2PO(CH_2)_3OPC_6H_4$-1,2. Ib. $b_{0.5}$ 162-3°, d^{20} 1.3420, n_D^{20} 1.5700.[705]
1,2-$C_6H_4O_2POCMe_2(CH_2)_2OPC_6H_4$-1,2. Ib. $b_{1.5}$ 185-6°, d^{20} 1.2740, n_D^{20} 1.5540.[705]
$(1,2\text{-}C_6H_4O_2POCH_2CH_2)_2O$. Ib. b_3 210.5-11.5°, d^{20} 1.7728, n_D^{20} 1.5759.[931]
$(1,2\text{-}C_6H_4O_2POCH_2CH_2)_2S$. Ib. d^{20} 1.3811, n_D^{20} 1.5800.[932]
$(1,2\text{-}C_6H_4O_2POCH_2CH_2OCH_2)_2$. Ib. b_4 246-7°, d^{20} 1.3162, n_D^{20} 1.5500.[931]
$(1,2\text{-}C_6H_4O_2PO)_2C_6H_4$-1,2. Ib. $b_{3.5}$ 214-6°, d^{20} 1.6018, n_D^{20} 1.6018.[931]
$(1,2\text{-}C_6H_4O_2PO)_2C_6H_4$-1,3. Ib. $b_{2.5}$ 212°, d^{20} 1.3659, n_D^{20} 1.6007.[931]
$(1,2\text{-}C_6H_4O_2PO)_2C_6H_4$-1,4. Ib. b_4 216-8°. A viscous glasslike substance.[931]
$[4\text{-}(1,2\text{-}C_6H_4O_2PO)C_6H_4]_2$. Ib. $b_{1.5}$ 281-3°. A viscous glasslike substance.[931]
$[4\text{-}(1,2\text{-}C_6H_4O_2PO)C_6H_4]_2S$. Ib. $b_{3.5}$ 295.5-6°.[931]
1,4-$(1,2\text{-}C_6H_4O_2PO)_2$-2,6-di-t-$BuC_6H_2$. Ib. b_4 223-5°.[931]
$[4\text{-}(1,2\text{-}C_6H_4O_2PO)C_6H_4]_2CMe_2$. Ib. b_3 284.5-6°. A viscous glasslike substance.[931]
$[2\text{-}(1,2\text{-}C_6H_4O_2PO)\text{-}3\text{-}t\text{-}Bu\text{-}5\text{-}MeC_6H_2]_2CH_2$. Ib. m. 49-51°.[931]
$(4\text{-}i\text{-}Pr\text{-}1,2\text{-}C_6H_3O_2POCH_2CH_2)_2S$. Ib. d^{20} 1.1333, n_D^{20} 1.5350.[932]
4,4'-$(4\text{-}i\text{-}Pr\text{-}1,2\text{-}C_6H_3O_2POC_6H_4)_2S$. Ib. n_D^{20} 1.5860.[932]

TYPE: 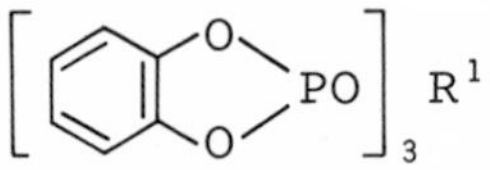

$(1,2\text{-}C_6H_4O_2POCH_2)_3CMe$. Ie, from monoacetate. White mass, 1 mm.[431]

I.1.16. Derivatives of 2,2'-Biphenylene Phosphite

TYPE: 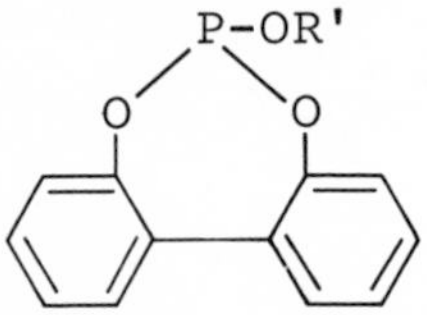

2,2'-$\overline{OC_6H_4C_6H_4O}$POMe. Ib. $b_{0.004}$ 110-1°, d^{20} 1.2587, n_D^{20} 1.6085.[1347]

2,2'-$\overline{OC_6H_4C_6H_4O}$POEt. Ib. $b_{1.5}$ 153.5°, d^{20} 1.2284, n_D^{20} 1.5960.[1347]

2,2'-$\overline{OC_6H_4C_6H_4O}$PO-i-Pr. Ib. $b_{0.007}$ 123-4°, d^{20} 1.2052, n_D^{20} 1.5856.[1347]

2,2'-$\overline{OC_6H_4C_6H_4O}$POBu. Ib. $b_{0.5}$ 174°, d^{20} 1.1750, n_D^{20} 1.5788.[1347]

2,2'-$\overline{OC_6H_4C_6H_4O}$POAm. Ib. $b_{0.003}$ 128-9°, d^{20} 1.1565, n_D^{20} 1.5733.[1347]

2,2'-$\overline{OC_6H_4C_6H_4O}$PO-n-$C_6H_{13}$. Ib. $b_{0.003}$ 138°, d^{20} 1.1401, n_D^{20} 1.5663.[1347]

2,2'-$\overline{OC_6H_4C_6H_4O}POC_7H_{15}$. Ib. $b_{0.002}$ 148°, d^{20} 1.1222, n_D^{20} 1.5603.[1347]

2,2'-$\overline{OC_6H_4C_6H_4O}POC_8H_{18}$. Ib. $b_{0.002}$ 161-1.5°, d^{20} 1.1052, n_D^{20} 1.5545.[1347]

2,2'-$\overline{OC_6H_4C_6H_4O}POC_9H_{19}$. Ib. $b_{0.003}$ 165°, d^{20} 1.0917, n_D^{20} 1.5510.[1347]

2,2'-$\overline{OC_6H_4C_6H_4O}POC_{10}H_{21}$. Ib. $b_{0.001}$ 166°, d^{20} 1.0774, n_D^{20} 1.5450.[1347]

2,2'-$\overline{OC_6H_4C_6H_4O}$PO-4-Me$C_6H_4$. Ib. b_1 240°, m. 76-8°.[1347]

2,2'-$\overline{OC_6H_4C_6H_4O}$PO-4-t-Bu$C_6H_4$. Ib. $b_{0.008}$ 210-1°, m. 91-2°.[1347]

2,2'-$\overline{OC_6H_4C_6H_4O}$PO-1-$C_{10}H_7$. Ib. $b_{0.03}$ 224-5°.[1347]

2,2'-$\overline{OC_6H_4C_6H_4O}$PO-2-$C_{10}H_7$. Ib. b_2 252.5°.[1347] Naphthyl esters by heating with ROH (130°/2 hr).[1347]

I.1.17. Derivatives of Salicyclic Acid

TYPE: 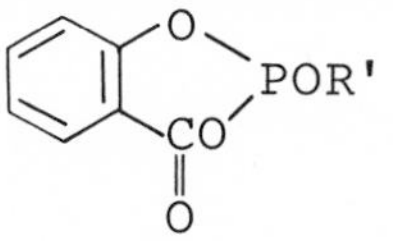

1,2-$\overline{OC_6H_4CO_2P}OCH_2CH_2Cl$. II. $b_{0.2}$ 122-4°, d^{20} 1.4049, n_D^{20} 1.5575.[952,953]

1,2-$\overline{OC_6H_4CO_2P}OCH_2CCl_3$. II. $b_{0.04}$ 132°, d^{20} 1.5292, n_D^{20} 1.5620.[952,953]

1,2-$\overline{OC_6H_4CO_2P}OCH_2CH_2OMe$. II. $b_{0.01}$ 132°, d^{20} 1.2799, n_D^{20} 1.5305.[952,953]

1,2-$\overline{OC_6H_4CO_2P}OCH_2CH_2OEt$. II. $b_{0.05}$ 136°, d^{20} 1.2390, n_D^{20} 1.5227.[952,953]

1,2-$\overline{OC_6H_4CO_2P}OCH_2CH_2OAc$. II. $b_{0.01}$ 142°, d^{20} 1.3258, n_D^{20} 1.5302.[952,953]

1,2-$\overline{OC_6H_4CO_2P}OCH_2CH{:}CH_2$. II. $b_{0.02}$ 142°, d^{20} 1.2641, n_D^{20} 1.5450.[952,953]

1,2-$C_6H_4CO{\cdot}OPOBu$-n (O bridging C_6H_4 and P). II. $b_{0.03}$ 97-9°, $b_{0.03}$ 99-100°, d_4^{20} 1.191, n_D^{20} 1.5250, $[R_L]_D$ 61.80 (calc. $[R_L]_D$ 60.52).[245,247]

1,2-$\overline{OC_6H_4CO_2P}O$-i-Am. II. $b_{0.5}$ 119-20°, d^{20} 1.1610, n_D^{20} 1.5135.[698]

1,2-$\overline{OC_6H_4CO_2P}OC_6H_{13}$. II. b_1 118-9°, d^{20} 1.1320, n^{20} 1.5092.[698]

1,2-$\overline{OC_6H_4CO_2P}OC_8H_{18}$. II. b_1 138-9°, d^{20} 1.0930, n_D^{20} 1.5050.[698]

1,2-$\overline{OC_6H_4CO_2P}OC_{12}H_{25}$. II. b_1 186-8°, d^{20} 1.0210, n_D^{20} 1.4939.[698]

1,2-$\overline{OC_6H_4CO_2P}OC_6H_5$. II. $b_{0.2}$ 135-7°, b_1 155-7°, d^{20} 1.3160, d^{20} 1.3084, n_D^{20} 1.5909, n_D^{20} 1.5881.[952,953]

1,2-$\overline{OC_6H_4CO_2P}O$-4-ClC_6H_4. II. b_2 178-80°, d^{20} 1.3890, n_D^{20} 1.5950.[698]

1,2-$\overline{OC_6H_4CO_2P}O$-2-MeC_6H_4. II. b_2 155-7°, d^{20} 1.2670, n_D^{20} 1.5822.[698]

1,2-$\overline{OC_6H_4CO_2P}O$-3-MeC_6H_4. II. b_3 150-2°, d^{20} 1.2720, n_D^{20} 1.5815.[698]

1,2-$\overline{OC_6H_4CO_2P}O$-4-MeC_6H_4. II. b_4 187-9°, d^{20} 1.2680, n_D^{20} 1.5804.[698]

1,2-$\overline{OC_6H_4CO_2P}O$-4-t-BuC_6H_4. II. b_2 184-6°, d^{20} 1.1959,

n_D^{20} 1.5625.[698]

1,2-$\overline{OC_6H_4CO_2P}$O-2,6-di-t-Bu-4-MeC$_6$H$_2$.   II.   b$_1$ 201-2°, m. 133-4°.[698]

1,2-$\overline{OC_6H_4CO_2P}$O-2,4,6-tri-t-BuC$_6$H$_2$.   II.   b$_2$ 197-9°, m. 147-8°.[698]

1,2-$\overline{OC_6H_4CO_2P}$O-2-C$_{10}$H$_7$.   II.   b$_1$ 195-6°.[698]

1,2-$\overline{OC_6H_4CO_2P}$OAc.   II.   m. 72-4°.[952,953]

1,2-$\overline{OC_6H_4CO_2P}$OCOEt.   II.   m. 55-60°.[952,953]

1,2-$\overline{OC_6H_4CO_2P}$OCOPr.   II.   m. 37-40°.[952,953]

1,2-$\overline{OC_6H_4CO_2P}$OCO-i-Pr.   II.   m. 122-5°.[952,953]

1,2-$\overline{OC_6H_4CO_2P}$OCO-i-Bu.   II.   Undistillable, uncrystal-lizable.[952,953]

1,2-$\overline{C_6H_4CO \cdot O\overset{\displaystyle O}{P}O}$Bz.   II.   m. 107-10°.[247]

I.1.18.   Anhydrides Containing P-O-P Linkages

Pyrophosphites

TYPE: (RO)$_2$POP(OR)$_2$

(EtO)$_2$POP(OEt)$_2$.   (RO)$_2$PONa + (RO)$_2$PCl.[80]   (RO)$_2$PONa + Cl$_2$ or Br$_2$.[79a,80]   b$_{11}$ 102-4°,[80] b$_4$ 87-8°,[80] b$_{2-3}$ 82-3°,[79a,80] d$_0^0$ 1.0748,[80] n$_D^{10}$ 1.4377,[80] n$_D^{20}$ 1.4322.[80] Adducts: 2CuCl, m. 111.5-2.5°; 2CuBr, m. 135-6°; 2CuI, m. 119-20°; 2AgCl, m. 115-6° (decomp.).[80]

(PrO)$_2$POP(OPr)$_2$.   (RO)$_2$PONa + (RO)$_2$PCl.   (RO)$_2$PONa + Br$_2$. b$_6$ 147.5-9°, d$_0^0$ 1.0664, n$^{16.5}$ 1.4408.[94]

(NCCMe$_2$O)$_2$POP(OCMe$_2$CN)$_2$.   (RO)$_2$PCl + (RO)$_2$PHO + C$_5$H$_5$N, in Et$_2$O at 0°.   b$_{0.2}$ 165-71°, d$_4^{20}$ 1.1370, n$_D^{20}$ 1.4555. Slowly solidifies, m. 28°.[673]

(BuO)$_2$POP(OBu)$_2$.   (RO)$_2$PONa + (RO)$_2$PCl.[76]   Also as by-product from preparation of (CH$_2$O)$_2$POP(OBu)$_2$.[117] b$_2$ 136-7°,[117] b$_7$ 175-6°, d$_0^{15}$ 0.9908, n$_D^{20}$ 1.4451,[76] n$_D^{20}$ 1.4466, d^{20} 0.9970.[117]

$\overline{OCH_2CH_2OP}$OP(OEt)$_2$.   (CH$_2$O)$_2$PCl + (EtO)$_2$PONa.   b$_2$ 84-5°, n$_D^{20}$ 1.4557, d^{20} 1.1890.[117]

($\overline{OCH_2CH_2OP}$)$_2$O.   Ig.   b$_4$ 100-1°, n$_D^{20}$ 1.4900, d^{20} 1.4293.[117]

$\overline{OCH_2CH_2OP}$OP(OPr)$_2$.   From (CH$_2$O)$_2$PCl and (PrO)$_2$PONa.   b$_2$ 93-4°, n$_D^{20}$ 1.4600, d^{20} 1.1446.[117]

$\overline{OCH_2CH_2OP}$OP(OCHMe$_2$)$_2$.   From (CH$_2$O)$_2$PCl and (i-PrO)$_2$PONa. b$_2$ 90-1°, d^{20} 1.1392, n$_D^{20}$ 1.4515.[117]

$\overline{OCH_2CH_2OP}$OP(OBu)$_2$.   From (CH$_2$O)$_2$PCl and (BuO)$_2$PONa.   b$_2$ 111-2°, d^{20} 1.110, n$_D^{20}$ 1.4593.[117]

$\overline{O\cdot CHMe\cdot CH_2\cdot O\cdot P}OP(OEt)_2$. From $(CH_2O)_2PCl$ and $(EtO)_2PONa$. b_3 73-4°, n_D^{20} 1.4520, d^{20} 1.1493.[117]

$\overline{O\cdot CHMe\cdot CH_2\cdot O\cdot P}OP(OPr)_2$. From $\overline{OCHMeCH_2OP}Cl$ and $(PrO)_2PONa$. b_2 100°, n_D^{20} 1.4530, d^{20} 1.1090.[117]

$\overline{O\cdot CHMe\cdot CH_2\cdot O\cdot P}OP(OCHMe_2)_2$. From $(RO)_2PCl$ and (i-PrO)$_2$PONa. b_3 86-8°, n_D^{20} 1.4530, d^{20} 1.1070.[117]

$\overline{O\cdot CHMe\cdot CH_2\cdot O\cdot P}OP(OBu)_2$. From $(RO)_2PCl$ and $(BuO)_2PONa$. b_3 120-1°, d^{20} 1.080, n_D^{20} 1.4550.[117]

$\overline{O\cdot CH(CH_2Cl)\cdot CH_2\cdot O\cdot P}OP(OEt)_2$. From $(RO)_2PCl$ and $(EtO)_2ONa$. b_{1-2} 110°, d^{20} 1.2470, n_D^{20} 1.4660.[117]

$\overline{O\cdot CH(CH_2Cl)\cdot CH_2\cdot O\cdot P}OP(OPr)_2$. From $(RO)_2PCl$ and $(PrO)_2ONa$. $b_{1.5}$ 126-9°, d^{20} 1.1990, n_D^{20} 1.4690.[117]

$\overline{O\cdot CH(CH_2Cl)\cdot CH_2\cdot O\cdot P}OP(OBu)_2$. From $(RO)_2PCl$ and $(BuO)_2ONa$. $b_{2.5}$ 146-9°, d^{20} 1.1986, n_D^{20} 1.473.[117]

$(\overline{O\cdot CH_2\cdot CH(CH_2Cl)\cdot O\cdot P})_2O$. Ig. b_3 144-5°, d^{20} 1.5126, n_D^{20} 1.5130.[117]

$(\overline{O\cdot CH_2\cdot CH_2\cdot CH_2\cdot O\cdot P})_2O$. Ig. b_{2-3} 82-3°, n_D^{20} 1.4625, d^{20} 1.2772.[117]

$\overline{O\cdot CHMe\cdot CH_2\cdot CH_2\cdot O\cdot P}OP(OEt)_2$. From $(RO)_2PCl$ and $(EtO)_2PONa$. b_5 113-3.5°, n_D^{20} 1.4563, d^{20} 1.1368.[117]

$\overline{O\cdot CHMe\cdot CH_2\cdot CH_2\cdot O\cdot P}OP(OPr)_2$. From $(RO)_2PCl$ and $(PrO)_2PONa$. b_2 110-4°, n_D^{20} 1.4580, d^{20} 1.1001.[117]

$\overline{O\cdot CHMe\cdot CH_2\cdot CH_2\cdot O\cdot P}OP(OCHMe_2)_2$. From $(RO)_2PCl$ and (i-PrO)$_2$PONa. b_2 98-102°, d^{20} 1.0645, n_D^{20} 1.4460.[117]

$\overline{O\cdot CHMe\cdot CH_2\cdot CH_2\cdot O\cdot P}OP(OBu)_2$. From $(RO)_2PCl$ and $(BuO)_2PONa$. b_7 152-6°, d^{20} 1.0663, n_D^{20} 1.4580.[117]

$(\overline{O\cdot CHMe\cdot CH_2\cdot CH_2\cdot O\cdot P})_2O$. Ig. b_3 118-20°, d^{20} 1.2329, n_D^{20} 1.4745.[117]

1,2-$C_6H_4O_2POP(OEt)_2$. o-$C_6H_4O_2PCl$ + $NaOP(OEt)_2$. b_1 115-6°, d^{20} 1.1816, n_D^{20} 1.4800.[98]

1,2-$C_6H_4O_2POPO_2C_6H_4$-1,2. Byproduct from above. b_1 170-2°, d^{20} 1.3107, n_D^{20} 1.5502.[98]

Mixed Phosphite-Phosphate or Phosphite-Phosphonate Anhydrides

$(EtO)_2P\cdot O\cdot P(O)Et_2$. $(EtO)_2PCl$ + $Et_2P(O)(OH)$ + C_5H_5N. $b_{0.001}$ 69°, n_D^{20} 1.4508.[867]

$(EtO)_2P\cdot O\cdot P(O)(OEt)Et$. $(EtO)_2PCl$ + $Et(EtO)P(O)(OH)$ + C_5H_5N. $b_{0.02}$ 66°, n_D^{20} 1.4361.[867]

$(EtO)_2P\cdot O\cdot P(O)(OEt)_2$. $(EtO)_2PCl$ + $(EtO)_2P(O)OH$ + base.[872,873] $b_{0.01}$ 80°,[872] $b_{0.01}$ 74°,[873] n_D^{20} 1.4289,[872] n_D^{25} 1.4254.[873]

$(EtO)_2P\cdot O\cdot P(O)(OPr)_2$. $(EtO)_2PCl$ + $(PrO)_2P(O)OH$ + base.

$b_{0.001}$ 84-5°, n_D^{25} 1.4277.[873]
$(EtO)_2P{\cdot}O{\cdot}P(O)(OCH_2C_6H_5)_2$. $(EtO)_2PCl + (C_6H_5CH_2O)_2P(O)OH$ + base. n_D^{25} 1.5206 (crude).[873]
$(EtO)_2P{\cdot}O{\cdot}\overline{P(O)OCH_2CMe_2CH_2O}$. $(EtO)_2PCl$ + >P(O)(OH) + C_5H_5N. n_D^{20} 1.4570.[867]
$(PrO)_2P{\cdot}O{\cdot}P(O)(OPr)Me$. $(PrO)_3P + MePO_2$ or $(PrO)_2PCl$ + MeP(O)(OPr)OH + base. $b_{0.03}$ 60-1°, d^{20} 1.0822, n_D^{20} 1.4330, IR.[377]
$(PrO)_2P{\cdot}O{\cdot}P(O)(OPr)C_6H_5$. $(PrO)_3P + C_6H_5PO_2$. $b_{0.001}$ 92-8°, d^{20} 1.1035, n_D^{20} 1.4847, IR.[377]
$(i\text{-}PrO)_2P{\cdot}O{\cdot}P(O)(OEt)_2$. $(i\text{-}PrO)_2PCl + (EtO)_2P(O)(OH) + C_5H_5N$. $b_{0.07}$ 83°, n_D^{20} 1.4224.[867]
$(i\text{-}PrO)_2P{\cdot}O{\cdot}P(O)(O\text{-}i\text{-}Pr)_2$. $(i\text{-}PrO)_2PCl + (i\text{-}PrO)_2P(O)OH$ + base. $b_{0.15}$ 106-8°, n_D^{25} 1.4186.[873]
$(BuO)_2P{\cdot}O{\cdot}P(O)(OPr)_2$. $(BuO)_2PCl + (PrO)_2P(O)OH$ + base. $b_{0.001}$ 114-5°, n_D^{25} 1.4342.[873]
$(BuO)_2P{\cdot}O{\cdot}P(O)(OBu)_2$. $(BuO)_2PCl + (BuO)_2P(O)OH$ + base. $b_{0.005}$ 106-7°, n_D^{25} 1.4341.[873]
$(BuO)_2P{\cdot}O{\cdot}P(O)(OBu)C_6H_5$. $(BuO)_3P + C_6H_5PO_2$. $b_{0.0015}$ 97-100°, d^{20} 1.0883, n_D^{20} 1.4832, IR.[377]
$\overline{O(CH_2)_2OP}{\cdot}OP(O)(OEt)_2$. $\overline{O(CH_2)_2OP}Cl + (EtO)_2P(O)(OH) + C_5H_5N$. $b_{0.01}$ 65°, n_D^{20} 1.4405.[867]
$\overline{O(CH_2)_3OP}{\cdot}OP(O)(OEt)_2$. $\overline{O(CH_2)_2OP}Cl + (EtO)_2P(O)(OH) + C_5H_5N$. $b_{0.05}$ 81-2°, n_D^{20} 1.4480.[867]

<u>Phosphites Containing Two or More P-O-P Linkages</u>

$[(EtO)_2PO]_2PEt$. Ib. $b_{0.5}$ 126-8°, d^{20} 1.1524, n_D^{20} 1.4768, b_1 128-30°, d^{20} 1.1531, n_D^{20} 1.4772.[122]
$[(PrO)_2PO]_2PEt$. Ib. b_1 145-8°, d^{20} 1.0785, n_D^{20} 1.4662.[122]
$[(i\text{-}PrO)_2PO]_2PEt$. Ib. b_1 130-2°, d^{20} 1.0693, n_D^{20} 1.4622.[122]
$[(i\text{-}BuO)_2PO]_2PEt$. Ib. b_2 165-7°, d^{20} 1.0420, n_D^{20} 1.4643.[122]

I.1.19. Miscellaneous Phosphites of Unspecified Structure

The following phosphites have been named:
Isodecyl neohexylene phosphite. Ie. n_D^{25} 1.4602.[428]
Triethylene glycol dineohexylene diphosphite. Ie. n_D^{25} 1.4805, d 1.145.[428,1150]
Tris(methoxycarbowax 350) phosphite. Ie. n_D^{25} 1.4675.[428]
Isodecyl bis(propylene glycol) phosphite. Ie. n_D^{21} 1.4561.[1005]
1,3-propylene di(1,3-butylene) phosphite. Ie. b_9 185°, n_D^{20} 1.4726.[958]
Dipropylene glycol tetrol diphosphite. Ie. n_D^{25} 1.4652,[424] n_D^{25} 1.4654.[423]

Dipropylene glycol pentol triphosphite. Ie. n_D^{25} 1.4660,[423] n_D^{25} 1.4661.[424]
Dipropylene glycol hexol tetraphosphite. Ie. n_D^{25} 1.4662,[423] n_D^{25} 1.4661.[424]
Polypropylene glycol 425 tetrol diphosphite. Ie. n_D^{25} 1.4565.[423,424]
Polypropylene glycol 425 pentol triphosphite. Ie. n_D^{25} 1.4566.[423,424]
Phosphites based on 3,3,5,5-tetrakis(hydroxymethyl)-4-hydroxy-tetrahydropyran (AEH) and substituted cyclohexanols. Ie. A considerable number are named in detail, but are described as pot residues resulting from the observed removal of the volatile ROH. The following are illustrative structures:[521]

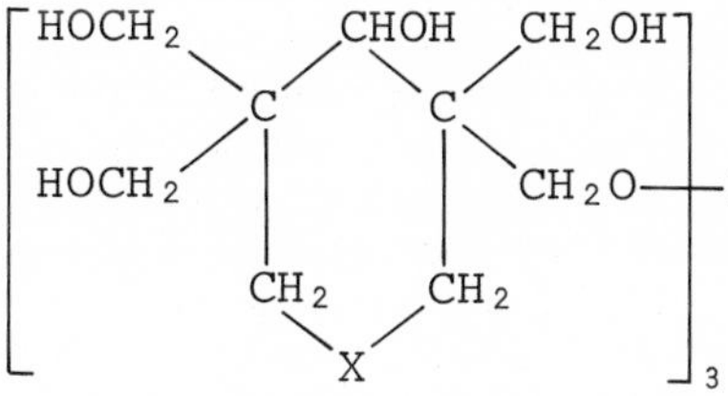

X = O, $-CH_2$

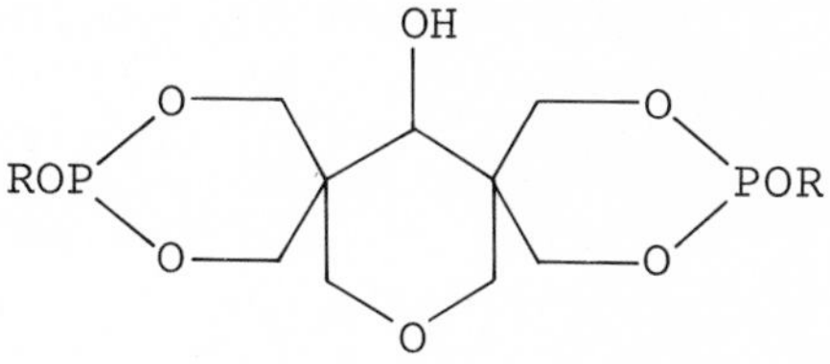

R = alkyl, aryl, or H

I.1.20. Dialkyl or Diaryl Phosphites

TYPE: $(RO)_2PHO$

$(MeO)_2PHO$. Ia.[127,845] Ib.[818,898] Ih.[69,70] Ii.[571] $b_{2.5}$ 38°,[855] b_8 56.5°,[69,70] b_{10} 56-8°,[898] b_{10} 55-5.5°,[127] b_{26} 72-7°,[818] d_0^0 1.2184,[70] d_0^{20} 1.2004,[127] d_0^{25} 1.1909,[69,70] d_4^{20} 1.2008,[571] n_D^{20} 1.4036,[127] n_D^{20} 1.403,[571] n_D^{20} 1.4030,[855] n_D^{20} 1.4029,[1233] n_D^{25} 1.4018.[845] Ag salt: crystals [from (EtOH)],[70] UV,[328a,893a] IR,[855,1003,1263,1264] mass spect.,[527] ^{31}P -11 to -12.8 ppm, J_{PH} 696 to 710 Hz,[381a,396,628,826,913,1342] ^1H NMR,[496a,806] Raman.[855]
$(EtO_2CCH_2O)_2PHO$. Ia.[1196] $b_{0.01}$ 110-2°,[1055] $b_{0.15}$ 150-2°,[1196] $b_{0.2}$ 133-4°,[135] d^{20} 1.2587,[1055] d_0^{20} 1.2485,[1196] d^{20} 1.2507,[135] n_D^{25} 1.4408,[135] n_D^{20} 1.4422,[1196] n_D^{20} 1.4462.[1055]
$(EtO_2C{\cdot}CHPh{\cdot}O)_2PHO$. Ia. Undistilled residue, n_D^{20} 1.5200, α_D^{18} -93.6° (from (-)-mandelate α_D^{18} -131°) (l = 10 cm).[450]
$(C_6H_5CH_2O)_2PHO$. Ib. b. 110-20°, n_D^{18} 1.5521, m. 0-5°, decomp. 120-60°,[150] IR,[176] ^{31}P -7.9 ppm, J_{PH} 713 Hz.[826]
$(4\text{-}O_2NC_6H_4CH_2O)_2PHO$. Ia. m. 75°.[402]
$(4\text{-}BrC_6H_4CH_2O)_2PHO$. Ib + Ig. m. 93-4°.[911]
$(4\text{-}ClC_6H_4CH_2O)_2PHO$. Ib + Ig. m. 75°.[911]

$(4\text{-}MeC_6H_4CH_2O)_2PHO$. Ib + Ig. m. 62°.[911]

$(EtO)_2PHO$. Ia.[77,127,837,845,996,1232] Ib.[818,898] If.[73,893] Ih.[69,70] Ii.[571,1029,1323] From ROH + amidite.[1413] From $(RO)_3P$ + $HO(S)P(OR')_2$.[1136] b_1 49-50°,[1413] b_{8-10} 72°,[69 70] b_9 66-7°,[996] b_9 72-3°, b_{10} 71.0-1.5°,[88] b_{10} 68-70°,[93] b_{11} 72°,[88] b_{13} 77-7.5°,[88] b_{14} 74-5°,[996] b_{14} 74-6°,[1136] b_{15} 80-5°,[858] b_{15} 75°,[127] b_{17} 77°,[855] b_{20} 87°,[70] b_{20} 90°,[1029] b_{33} 94-6°,[817] b_{754} 187-8°,[70] b_{760} 187-8°,[69] d_0^0 1.0912,[69,70] d_4^0 1.0961,[88] d_4^0 1.093,[1232] d_4^{20} 1.07368,[87] d_4^{20} 1.0726,[571] d_4^{20} 1.0756,[93] 1.0685,[1136] d_0^{20} 1.0722,[69,70] d_0^{20} 1.0742,[127] d^{20} 1.0753,[817] d_4^{188} 0.88955,[69,70] n_D^{20} 1.4080,[127] n_D^{20} 1.40823,[87] n_D^{20} 1.4101,[817] n_D^{20} 1.408,[571] n_D^{20} 1.4080,[1136] n_D^{20} 1.4085,[1413] n_D^{25} 1.4112,[845] UV,[328a,893a] IR,[176,324,855,1003] Raman,[855] mass spect.,[527] ^1H NMR,[806] ^{31}P -8 to -6.2 ppm, J_{PH} 670 to 690 Hz.[396,628,826,913,1342] Sodium salt: needles decomp. 142-3° if prepared from metallic sodium in ether.[621,858,895] Obtained in solution when sodium ethoxide is used instead of metallic sodium.[996] Potassium salt: from metallic potassium; crystals.[77] Silver salt: from the ester or its sodium salt with silver nitrate.[74,75,900] Colorless and stable when pure. Magnesium salt: from metallic magnesium powder. Cuprous salt: from cuprous oxide. Insoluble colorless solid that is stable when pure.[74,75,900]

$(FCH_2CH_2O)_2PHO$. Ia. b_2 102°, n_D^{20} 1.4143, d^{20} 1.3738.[634]

$(ClCH_2CH_2O)_2PHO$. Ia.[26,445] Ii.[1374] $b_{0.5}$ 110°,[1374] $b_{3.5-4}$ 119-20°, b_3 119-20°,[26,445] b_4 118-9°,[1054] d^{20} 1.4025, d^{20} 1.4034,[26,445] d^{20} 1.4035,[1054] d_{25}^{25} 1.1057,[1374] n_D^{20} 1.4708,[1054] n_D^{20} 1.4719,[26,445] n_D^{25} 1.4701,[1374] IR.[26,445]

$(BrCH_2CH_2O)_2PHO$. Ia. b_1 160-2°, d^{20} 1.8823, n_D^{20} 1.5112, IR.[26]

$(NCCHMeO)_2PHO$. Ig. $b_{0.15}$ 112-5°, d^{20} 1.1605, n_D^{20} 1.4400 (crude).[673]

$(EtSCH_2CH_2O)_2PHO$. Ib + Ig. $b_{0.0007}$ 110-2°, d^{20} 1.1551, n_D^{20} 1.5048, MR_D 66.33 (calc. 66.19). (Must be freed from all acidic material prior to distillation.[848]) ^{31}P -7.7 ppm, J_{PH} 712 Hz.[826]

$(HO_2C{\cdot}CHMe{\cdot}O)_2PHO$. Modified Ia (reaction of PI_3 with conc. lactic acid, followed by treatment with water; the primary product appears to be an anhydride, $C_6H_9O_6P$, m. 120°). Isolated as water-soluble calcium salt (octahydrate).[440]

$(EtO_2C{\cdot}CHMe{\cdot}O)_2PHO$. Ia (best in Et_2O solution).[302,477] Ig.[477] $b_{0.2}$ 135°,[302] b_2 140-50° (crude),[477] as residue α_D^{16} + 79.4° (l = 1 dm) from ethyl (+)-lactate.[477]

$(PhCH_2CH_2O)_2PHO$. Ie.[132] From $(RO)_3P$ + HCl.[478] $b_{0.05}$ 147°,[478] $b_{0.05}$ 183.5-5.5°, d_0^{20} 1.1333, n_D^{20} 1.5465,[132] n_D^{26} 1.5429.[478]

$(ClCH_2CHPh)_2PHO$. Ii. (Not distilled.) n_D^{25} 1.5625.[1374]
$(C_6H_5OCH_2CH_2O)_2PHO$. Ia. m. 47-8°.[198]
$(4\text{-}Cl\text{-}C_6H_4OCH_2CH_2O)_2PHO$. Ia. As viscous liquid residue,[530,1338] m. 69-70°.[198]
$(2,4\text{-}Cl_2C_6H_3OCH_2CH_2O)_2PHO$. Ia. As viscous liquid residue at 135°/0.53 mm, n_D^{26} 1.5738,[530,1338] n_D^{20} 1.5741.[198]
$(2,4,5\text{-}Cl_3C_6H_2OCH_2CH_2O)_2PHO$. Ia. As viscous liquid residue, n_D^{26} 1.5907,[530,1338] m. 53-5°.[198]
$(4\text{-}FC_6H_4OCH_2CH_2O)_2PHO$. Ia. m. 79-80°.[198]
$(2\text{-}Cl\text{-}4\text{-}FC_6H_3OCH_2CH_2O)_2PHO$. Ia. n_D^{20} 1.5425.[198]
$(2\text{-}F\text{-}4\text{-}ClC_6H_3OCH_2CH_2O)_2PHO$. Ia. n_D^{20} 1.5398.[198]
$(2,4\text{-}Cl_2\text{-}5\text{-}FC_6H_2OCH_2CH_2O)_2PHO$. Ia. m. 46-7°.[198]
$(2,6\text{-}Cl_2\text{-}4\text{-}FC_6H_2OCH_2CH_2O)_2PHO$. Ia. n_D^{20} 1.5525.[198]
$(2,4\text{-}Cl_2\text{-}6\text{-}FC_6H_2OCH_2CH_2O)_2PHO$. Ia. n_D^{20} 1.5462.[198]
$(PrO)_2PHO$. Ia.[311,845] Ib.[818] Ie.[770] If.[73] Ih.[69,70] Ij.[94] From $(RO)_3P$ + HO(S)P(OR')$_2$.[1136] b_2 43-5°,[1136] b_4 70-2°,[770] b_{8-10} 91°,[69,70] b_8 88-90°,[997] b_{11} 91.5°,[127] b_{15} 78-80°,[1157] b_{17} 101-2°,[311] b_{90-100} 143-5°,[817] d_0^0 1.0366,[69,70] d^{20} 1.0184,[817] d_0^{20} 1.0207,[69,70] d_0^{20} 1.0184,[127] d_4^{20} 1.0184,[93] 1.0257,[1136] d_4^{20} 1.1232,[1136] n_D^{20} 1.4163,[997] n_D^{20} 1.4140,[1157] n_D^{20} 1.4175,[311] n_D^{20} 1.4172,[127] n_D^{20} 1.4180,[817] n_D^{20} 1.4183,[93] 1.4230, n_D^{23} 1.4175, n_D^{25} 1.4155,[845] UV,[893a,896a] IR,[571] mass spect.,[527] ^{31}P -6.5, -7.4 ppm, J_{PH} 685 Hz,[826] ^1H NMR.[806] Sodium salt: solid.[621] Silver salt: essentially insoluble in water prepared best from the sodium salt.[74,621,900] SO_2Cl_2 gives ClP(O)(OPr)$_2$.[1157]
$(MeCHClCH_2O)_2PHO$. Ii. b_1 120°, n_D^{25} 1.4609, d_{25}^{25} 1.2640.[1374]
$(ClCH_2CH_2CH_2O)_2PHO$. Ia. $b_{0.28}$ 128-9°, d^{20} 1.2963, n_D^{20} 1.4695, IR.[26]
$(ClCH_2CHClCH_2O)_2PHO$. Ia. $b_{0.035}$ 151-2.5°, d^{20} 1.4831, n_D^{20} 1.5040, IR.[26]
$(BrCH_2CH_2CH_2O)_2PHO$. Ia. $b_{0.045}$ 75°, d^{20} 1.6832, n_D^{20} 1.5023, IR.[26]
$(BrCH_2CHBrCH_2O)_2PHO$. Ia. Undistilled, d^{20} 2.2111, n_D^{20} 1.5742, IR.[26]
$(PhCH_2CH_2CH_2O)_2PHO$. $(RO)_3P$ + HCl. $b_{0.2}$ 186°, n_D^{14} 1.5382.[478]
$(CH_2{:}CHCH_2O)_2PHO$. Ia. b_8 97.5-8.5°, d_0^0 1.1001, d_0^{20} 1.0793, n_D^{20} 1.4430, Na deriv.[676] ^{31}P -7.8 to -4.9 ppm, J_{PH} 708, 700 Hz,[826] mass spect.[527]
$(CH_2{=}CClCH_2O)_2PHO$. Ia. $b_{0.06}$ 84°, d^{20} 1.3239, n_D^{20} 1.4822.[142]
$(CHI{:}CICH_2O)_2PHO$. Modified Ia (propargyl alcohol, iodine, and red phosphorus). Needles m. 48-9° (from EtOH).[549a]
$(i\text{-}PrO)_2PHO$. Ia.[127,311,837,845] Ib.[818] Ih.[69,70] $b_{0.5}$ 41.5°,[855] $b_{6.5}$ 72-3°,[892] b_8 69-71°,[997] b_9 69.5°,[127]

b_{8-10} 69.5°,[69,70] b_{10} 71-3°,[818] b_{10} 76-7°, b_{17} 85-6°,[892] b_{17} 82.5°,[837] b_{53} 106-8°,[311] d_0^0 1.0159, d_0^{18} 0.9972,[69,70] d_0^{20} 0.9981,[127] d_4^{20} 0.9975,[571] n_D^{25} 1.4082,[845] n_D^{20} 1.4008,[127] n_D^{20} 1.4070,[997] n_D^{20} 1.4080,[855] n_D^{20} 1.4089,[311] IR,[32,500,571,855] UV,[893a,1263] mass spect.,[527] Raman,[855] ^{31}P -4.2 to -3.2 ppm, J_{PH} 690 to 670 Hz,[396,628,826,935] ^1H NMR.[496a] Cuprous salt: colorless solid.[900] Silver salt: colorless crystalline solid, soluble in alkali.[74,900]

$(ClCH_2CHMeO)_2PHO$. Ia. $b_{0.25}$ 85-6°, d^{20} 1.2687, n_D^{20} 1.4606, IR.[26]

$[(ClCH_2)_2CHO]_2PHO$. Ia. Unstable, $b_{0.5}$ 168-70°, b_2 180°, d^{20} 1.4894, n_D^{20} 1.5035, IR.[26,302]

$[(BrCH_2)_2CHO]_2PHO$. Ia. Undistilled. d^{20} 2.2143, n_D^{20} 1.5752, IR.[26]

$[ClCH_2(BrCH_2)CHO]_2PHO$. Ia. Undistilled. d^{20} 1.8621, n_D^{20} 1.5437, IR.[26]

$(NCCMe_2O)_2PHO$. Ig. $b_{0.15}$ 118-20°, d^{20} 1.1128, n_D^{20} 1.4420.[673]

$(HO_2C{\cdot}CMe_2O)_2PHO$. Ia. m. 129°.[772]

$[MeOCH_2CH(CH_2Cl)O]_2PHO$. Ia. $b_{0.5}$ 149-51°, d^{20} 1.2948, n_D^{20} 1.4655.[37]

$[EtOCH_2CH(CH_2Cl)O]_2PHO$. Ia. $b_{0.6}$ 160-1°, d^{20} 1.2316, n_D^{20} 1.4616.[37]

$[PrOCH_2CH(CH_2Cl)O]_2PHO$. Ia. $b_{0.5}$ 175-7°, d^{20} 1.1746, n_D^{20} 1.4587.[37]

$[PhCH_2CH(Me)O]_2PHO$. IIIc. $b_{0.02}$ 147°, n_D^{20} 1.5291, α_D^{20} + 21.68° from $(RO)_3P$ α_D^{21} + 21.12° (l = 1).[478]

$[(PhCH_2)_2CHO]_2PHO$. IIIc. Needles, m. 77-9°.[478]

$(BuO)_2PHO$. Ia.[127,311,845] Ib.[898] Ie.[770] $b_{0.6}$ 84-6°,[311] b_8 115-6°,[1054] b_8 116-7°,[997] b_{10} 115°,[127,770] b_{12} 124-5°,[898] b_{12} 125-6°,[1055] b_{27} 126-30°, d_0^{20} 0.9888,[127] d^{20} 0.9972,[1055] d^{20} 0.9950,[1054] d_4^{20} 0.99503,[898] d_4^{20} 0.9883,[571] d^{20} 0.9886,[845] n_D^{20} 1.4240,[845] n_D^{20} 1.4248,[1055] n_D^{20} 1.423,[997] n_D^{20} 1.4240,[127] n_D^{20} 1.4238,[571] n_D^{20} 1.4236,[1054] n_D^{25} 1.4213,[845] IR,[324,568,571] ^1H NMR,[496a] mass spect.,[527] ^{31}P -8 to -5 ppm, J_{PH} 670 to 780 Hz, J_{POCH} ± 9.2 Hz.[628,826,913,1342] Sodium salt is very soluble in hydrocarbons.

$(NCCHPrO)_2PHO$. Ig. $b_{0.15}$ 134-5°, d^{20} 1.0846, n_D^{20} 1.4486.[673]

$(MeEtCHO)_2PHO$. Ia.[311,497] $b_{0.02}$ 48°,[497] $b_{0.7}$ 82-4°,[311] b_{12} 101°,[405] b_{14} 103-4°,[90] n_D^{20} 1.4190,[90] n_D^{20} 1.4183,[497] n_D^{20} 1.4186,[311] d_0^{20} 0.9754,[90] ^{31}P -4.7,[497] -9.9 ppm,[826] J_{PH} 691,[497] 660 Hz,[826] α_D^{20} + 24.92° from ROH α_D^{20} + 11.21° (l = 1).[497]

$(i\text{-}BuO)_2PHO$. Ia.[69,86,127,302,311] Ie.[770] If.[73] b_7 108-9°,[311] b_9 105°,[127] b_{11} 112.5°,[653] b_{12} 105.6-6.5°,[302] b_{13} 111-3°,[770] b_{14} 117.5°,[86] b_{20} 114-6°,[817] b. 235-6°,[86] d_0^0 0.9257,[86] d_0^0 0.9942,[73,86] d_0^{20} 0.9766,[127] d_4^{20} 0.9759,[73,86] d_4^{20} 0.9761,[817] n_D^{20}

1.4195,[311] n_D^{20} 1.4200,[127] n_D^{20} 1.4202,[817] UV.[893a] Sodium salt: amorphous, stable to 200°.[86] Silver salt: colorless needles.[74,900]

[i-PrCH(CN)O]$_2$PHO. Ig. $b_{0.15}$ 123-4°, d^{20} 1.0903, n_D^{20} 1.4460.[673]

(t-BuO)$_2$PHO. Ib.[467,494,827] $b_{0.1}$ 45°, $b_{0.4}$ 48-50°,[467] b_4 62-2.5°, $b_{4.5}$ 78°, $b_{0.5}$ 66-8°,[494] $b_{0.4}$ 42°,[827] n_D^{22} 1.4205, n_D^{25} 1.4168, n_D^{23} 1.4186,[827] d_{19}^{19} 0.9835, d^{25} 0.975,[827] M_D 50.03, IR,[494] ^1H NMR,[827] ^{31}P 3.2 to 3.9 ppm, J_{PH} 678 to 736 Hz.[826,827]

[Me_2CH(CCl_3)O]$_2$PHO. Ig. m. 45-6° (dried in vacuo),[19] ^{31}P 3.1 ppm, J_{PH} 735 Hz.[826]

(AmO)$_2$PHO. Ia.[311,817] $b_{0.05}$ 96-7°,[311] b_{40} 166-8°,[817] d^{20} 0.9635,[817] n_D^{20} 1.4287,[311] n_D^{20} 1.4305,[817] b_2 108-9°,[198] d^{18} 0.9613,[198] n_D^{18} 1.4310.[198]

(i-AmO)$_2$PHO. Ia.[302,405,1398] Ie.[770] $b_{0.15}$ 75°,[302] b_3 99-100°,[770] b_6 131-2°,[1054] b_6 131-3°,[1055] b_{10} 133°,[405] b_{15} 143-4°,[817] $d_0^{19.5}$ 0.967,[1398] d^{20} 0.9564,[817] d^{20} 0.9505,[1054] d^{20} 0.9694,[1055] n_D^{20} 1.4290,[817] n_D^{20} 1.4268,[1054] n_D^{20} 1.4331.[1055]

($MeCH_2CHMe\cdot CH_2O$)$_2$PHO. Ia.[278] b_{15} 142°,[405] b_{15} 140°, n_D^{20} 1.4301.[278]

(Me_3CCH_2O)$_2$PHO. Ia.[475] $b_{0.3}$ 53-6°,[575] b_{10} 109°, b_9 106-8°, b_{13} 116-9°, n_D^9 1.4200, 1.4240, n_D^{13} 1.4200,[475] n_D^{20} 1.4230,[575] ^{31}P -7.7 ppm, J_{PH} 695 Hz,[575] IR.[176]

(PrMeCHO)$_2$PHO. Ia.[279,311] $b_{0.01}$ 51°,[279] $b_{0.1}$ 73°,[311] $b_{0.1}$ 69°, n_D^{20} 1.4240,[279,311] ^{31}P -4.3 ppm, J_{PH} 687 Hz.[279] α_D^{22} + 10.42° or α_D^{25} + 10.08° from ROH α_D^{20} + 4.46° (l = 1).[279]

(Et_2CH)$_2$PHO. Ia.[302,311] $b_{0.2}$ 72°,[302] $b_{0.1}$ 90-1°,[311] n_D^{20} 1.4277.[311]

(i-PrMeCHO)$_2$PHO. Ia. b_{15} 116°, n_D^{20} 1.4265.[278]

(CH≡CMe_2O)PHO. Ia. b_5 100°.[184]

Bis(tetrahydrofurfuryl) phosphite. Ie. $b_{0.06}$ 134-6°, n_D^{20} 1.4735, d^{20} 1.2046.[1325]

(n-C_6H_{13}O)$_2$PHO. Ia.[127] Ie.[132] b_1 121-2°,[198] $b_{1.5}$ 143-4°,[1055] b_2 145-6°,[127] b_2 138.5-9°,[132] b_{13} 163-4°,[817] d^{20} 0.9488,[817] d_0^{20} 0.9486,[127] d^{20} 0.9501,[1055] n_D^{20} 1.4395,[1055] n_D^{20} 1.4350,[817] n_D^{20} 1.4325,[127] n_D^{18} 1.4345,[198] n_D^{20} 1.4332.[132]

($Me_2CH\cdot CH_2\cdot CHMe\cdot O$)$_2$PHO. Ia. $b_{0.2}$ 81°.[302]

[(CH_2CH_2)$_2$C(CCl_3)O]$_2$PHO. Ig. d^{20} 1.4770, n_D^{20} 1.5235 (dried in vacuo).[19]

(C_6H_{11}O)$_2$PHO (cyclohexyl). Ie. b_3 152-3°.[770]

[CH_2(CH_2)$_2$C(CO_2H)O]$_2$PHO. Ia. m. 94-5°.[772]

(n-C_7H_{15}O)$_2$PHO. Ia.[127] $b_{0.4}$ 126-8°,[817] b_1 134-5°,[198] b_1 162-3°,[1054] b_2 166-7°,[127] d_0^{20} 0.9363,[127] d^{20} 0.9366,[817] d^{20} 0.9341,[1054] n_D^{20} 1.4387,[817] n_D^{20} 1.4382,[127] n_D^{20} 1.4364.[1054]

[CH_2(CH_2CH_2)$_2$C(CCl_3)O]$_2$PHO. Ib. m. 138-9° (dried in vacuo).[19]

$(n\text{-}C_8H_{17}O)_2PHO$. Ia.[127,311] Ie.[132] $b_{0.03}$ 151-3°,[311] $b_{0.4}$ 146-8°,[817] b_2 186-8°,[1055] b_3 190-1°,[127] $b_{4.5}$ 191-2°,[132] d_0^{20} 0.9286,[127] d^{20} 0.9285,[1055] d^{20} 0.9284,[817] d_0^{20} 0.9263,[132] n_D^{20} 1.4420,[127,1055] n_D^{20} 1.4413,[311] n_D^{20} 1.4410,[132] ^{31}P -7 ppm.[1342]

$(2\text{-}C_8H_{17}O)_2PHO$. Ia.[311,450,573] Ib.[450] Id.[450] If.[450] $b_{0.6}$ 148-50°,[311] $b_{0.1}$ 116-8°,[573] b_2 138-40°, d_4^{15} 0.9218, d_4^{18} 0.9176, d_4^{21} 0.9133, n_D^{18} 1.4375, n_D^{21} 1.4391, n_D^{23} 1.4370,[450] n_D^{20} 1.4370,[311] α_D^{20} + 14.16° from ROH α_D^{20} + 7.98° (l = 1).[311]

$[Me(CH_2)_3CHEtCH_2O]_2PHO$. Ie.[1374] b_1 148-51°, n_D^{25} 1.4430, d_{25}^{25} 0.940,[1374] n_D^{25} 1.4457.[1376]

$(C_9H_{19}O)_2PHO$. Ia.[817] Ie.[132] $b_{0.04}$ 174.5-5.5°,[132] $b_{0.4}$ 184-6°,[817] b_2 200-1°,[1055] b_2 200-2°,[1054] d^{20} 0.9265,[1054,1055] d_0^{20} 0.9212,[132] d^{20} 0.925,[817] n_D^{20} 1.4458,[132] n_D^{20} 1.4460,[817] n_D^{20} 1.4435,[1055] n_D^{20} 1.4418.[1054]

$(C_{10}H_{21}O)_2PHO$. Ia.[817] Ie.[132] $b_{0.0005}$ 136-40°,[1057] $b_{0.04}$ 190-1°,[132] $b_{0.5}$ 195-6°,[817] b_1 201-2°,[1055] d_0^{20} 0.9157,[132] d_0^{20} 0.9158,[817] d^{20} 0.9187,[1055] d^{20} 0.9200,[1057] n_D^{20} 1.4483,[1057] n_D^{20} 1.4502,[132] n_D^{20} 1.4460,[1055] n_D^{20} 1.4486.[817]

$(C_{12}H_{25}O)_2PHO$. Ib. $b_{0.001}$ 193-5°, m. 30°, n_D^{20} 1.4472,[535] n_D^{25} 1.4521.[1376]

$(C_{14}H_{29}O)_2PHO$. Ib. $b_{0.04}$ 230°, m. 39-40°.[535]

$(C_{16}H_{33}O)_2PHO$. Ib. Ie. $b_{0.02}$ 245-50°, m. 46-7°, m. 51-2°,[132,535] m. 55-7°.[1054]

Distearyl phosphite. Ie. Ii. Solidified at 54-7°.[1376]

$(PhO)_2PHO$. If.[570,899] Ig.[774] Ii.[1374] $b_{0.008}$ 100°,[1374] $b_{0.15}$ 145-8°,[774] b_{25} 218-9°,[899] m. 25°,[570] d^{20} 1.231,[570] d^{20} 1.2159,[774] d_{25}^{25} 1.2268,[1374] n_D^{20} 1.557,[570] n_D^{20} 1.5599,[774] n_D^{25} 1.5570,[1374] n_D^{25} 1.5562.[1375]

$(2\text{-}ClC_6H_4O)_2PHO$. Ii. $b_{0.005}$ 125°, n_D^{25} 1.5750.[1374]

$(4\text{-}ClC_6H_4O)_2PHO$. Ii.[1374,1375] $b_{0.005}$ 125°,[1374] m. 42-3°,[1375] m. 44°,[1374] n_D^{25} 1.5708 (supercooled).[1374,1375]

$(4\text{-}NO_2C_6H_4O)_2PHO$. Ii. m. 98-100°.[1375]

$(2\text{-}MeC_6H_4O)_2PHO$. If.[570] Ii.[1374] $b_{0.005}$ 100°,[1374] d^{20} 1.178,[570] d_{25}^{25} 1.1827,[1374] n_D^{20} 1.546,[570] n_D^{25} 1.5495.[1375]

$(3\text{-}MeC_6H_4O)_2PHO$. If.[570] Ii.[1374] $b_{0.005}$ 90-100°,[1374] d^{20} 1.168,[570] d_{25}^{25} 1.1806,[1374] n_D^{20} 1.546,[570] n_D^{25} 1.5468,[1375] n_D^{20} 1.5471.[1055]

$(4\text{-}MeC_6H_4O)_2PHO$. If.[570] Ii.[1374] $b_{0.005}$ 110°,[1374] m. 30°,[570] d^{20} 1.170,[570] d_{25}^{25} 1.1569,[1374] n_D^{20} 1.547,[570] n_D^{25} 1.5466,[1375] n_D^{20} 1.5468.[1055]

$(t\text{-}BuC_6H_4O)_2PHO$. Ii.[1374,1375] $b_{0.005}$ 160°, m. 30-3°,[1374] n_D^{25} 1.5280.[1374,1375]

$(2,6\text{-di-}t\text{-}BuC_6H_3O)_2PHO$. Ig. m. 147-9°.[1266,1267]

$(2,6\text{-di-}t\text{-Bu-4-}MeC_6H_2O)_2PHO$. Ig. m. 161-3°.[1266]

$(2,6\text{-di-}t\text{-Bu-4-}ClC_6H_2O)_2PHO$. Ig. m. 147-8.5°.[1266]

$(2\text{-Naphthyl-O})_2PHO$. Ia. m. 83-5°.[204]

TYPE: (RO)(RO')PHO

(MeO)(EtO)PHO. Ia.[1262] Ia + H_2O.[816] $b_{1.5}$ 53°,[1262] $b_{0.2}$ 37°,[297] b_{15} 70-3°,[816] n_D^{22} 1.4041,[297] n_D^{20} 1.4070, d^{20} 1.1288.[816]

(MeO)($ClCH_2CH_2O$)PHO. Id (from sym esters). $b_{1.5}$ 87°.[389]

(MeO)(Cl_3CCH_2O)PHO. MeO(OH)ONa + Cl_3CCH_2OCOCl.[297] Id (from sym esters).[389] $b_{1.5}$ 85°,[389] $b_{0.2}$ 87°, n_D^{22} 1.4694.[297]

(MeO)[HO(CH_2)$_2$O]PHO. Ie. d_4^{20} 1.2856, n_D^{20} 1.4382, MR_D 28.66 (28.45), IR.[972]

(MeO)[HO(CH_2)$_2$O(CH_2)$_2$O]PHO. Ie. d_4^{20} 1.2896, n_D^{20} 1.4590, MR_D 39.03 (38.33), IR.[972]

(MeO)(PrO)PHO. Ia + H_2O.[816] Ih.[84] Ie.[84] b_8 64-6°,[816] b_{12} 63°,[84] d^{20} 1.0987,[816] d_4^{20} 1.0686, n_D^{20} 1.4120,[84] n_D^{20} 1.4112,[816] MR_D 32.15 (calc. 31.71).[84]

(MeO)[HO(CH_2)$_3$O]PHO. Ie. d_4^{20} 1.2204, n_D^{20} 1.4436, MR_D 33.50 (33.06), IR.[972]

(MeO)(i-PrO)PHO. Ia.[1262] Ib.[1255] MeO(OH)ONa + i-PrOCOCl.[297] $b_{0.2}$ 42°,[297] b_2 53°,[1262] b_4 52°,[1255] n_D^{22} 1.4056.[297]

(MeO)[($ClCH_2$)$_2$CHO]PHO. Id. b_1 100°.[298]

(MeO)(BuO)PHO. Ia.[1262] Ih.[84] Ie.[84] $b_{2.5}$ 65-8°,[1262] b_{10} 79-81°, d_4^{20} 1.0430, n_D^{20} 1.4174, MR_D 36.71 (calc. 36.33).[84]

(MeO)(cyclo-$C_6H_{11}O$)PHO. Id (from sym esters). b_1 80°.[389]

(MeO)(C_6H_5O)PHO. Ih.[273] Ii.[1392] MeO(OH)ONa + C_6H_5OCOCl.[297] $b_{0.2}$ 94°,[297] b_{13} 85°,[273] d^{20} 1.2184, n_D^{20} 1.5085, MR_D 42.1.[1392]

(EtO)(tetrahydrofurylmethylO)PHO. Id. $b_{0.01}$ 58°.[298]

(EtO)($ClCH_2CH_2O$)PHO. EtO(OH)ONa + $ClCH_2CH_2OCOCl$. $b_{0.2}$ 82°, n_D^{22} 1.4403.[297]

(EtO)[HO(CH_2)$_2$O]PHO. Ie. b_{11} 142-3°,[105] b_{15} 51-1.5°,[105] d_0^{20} 1.1317, d_4^{20} 1.2120,[972] n_D^{20} 1.4380,[972] n_D^{20} 1.4395, n_D^{20} 1.4825, MR_D 33.36 (33.06), IR, ^1H NMR.[972]

(EtO)(PrO)PHO. Ia + H_2O.[816] Ie.[84] Ih.[84] b_{20} 98-102°,[816] b_{24} 97-9°, d_4^{20} 1.0354,[84] d^{20} 1.0389,[816] n_D^{20} 1.4145,[816] n_D^{20} 1.4110, MR_D 36.48 (calc. 36.33).[84]

(EtO)(i-PrO)PHO. Ia + H_2O. b_8 73-5°, n_D^{20} 1.4100, d^{20} 1.0301.[816]

(EtO)(BuO)PHO. Ia.[814] Ia + H_2O.[816] Ie.[732] IV.[1413] $b_{0.5}$ 60-1°,[1413] $b_{1.5}$ 68-9°,[814] b_2 71-3°,[814] b_{13} 99-100°,[732] b_{28} 112-6°,[816] d^{20} 1.0180,[814,816] d^{20} 1.020,[814] d^{20} 1.0120,[732] n_D^{20} 1.4170,[1413] n_D^{27} 1.4139,[732] n_D^{20} 1.4160,[814,816] n_D^{20} 1.4158,[814] MR_D 40.98 (calc. 40.94).[732]

(EtO)(i-BuO)PHO. Ia + H_2O.[816] Ie.[84] Ih.[84] EtOP(OH)ONa + i-BuOCOCl.[297] $b_{0.2}$ 56°,[297] b_{12} 90°,[84] b_{28} 116-20°,[816] d_4^{20} 1.0115,[84] d^{20} 1.0218,[816] n_D^{20} 1.4100,[816] n_D^{20} 1.4171,[84] n_D^{22} 1.4140,[297] MR_D 41.32 (calc. 40.95).[84]

(EtO)(AmO)PHO. Ia + H_2O.[816] Ie.[732] b_9 100-3°,[816] b_{13} 109-10°,[732] d^{20} 0.9979,[816] d^{27} 0.9935,[732] n_D^{20} 1.4220,[816] n_D^{27} 1.4189, MR_D 45.71 (calc. 45.56).[732]

$[HO(CH_2)_5O](EtO)PHO$. Ie. d_4^{20} 1.153, n_D^{20} 1.4585, MR_D 46.44 (46.28), IR.[972]

(EtO)(i-AmO)PHO. Ia.[814a] Ie.[84] Ih.[84] b_1 71-3°,[814a] b_8 93-4°, d_4^{20} 0.9964,[84] d^{20} 1.001,[814] n_D^{20} 1.4201,[814a] n_D^{20} 1.4210, MR_D 45.87 (calc. 45.56).[84]

$(EtO)(C_6H_{13}O)PHO$. Ia.[814a] Ia + H_2O.[816] Ib.[535] Ie.[132] $b_{0.2}$ 82-4°, b_1 78°,[298] $b_{1.5}$ 105-7°,[814a] $b_{3.5}$ 104-5°,[132] b_{12} 118-20°,[535] b_{12-13} 121-5°,[816] d^{20} 0.987,[814a] d^{20} 0.9901,[816] d_0^{20} 0.9883,[132] n_D^{20} 1.4268,[132] n_D^{20} 1.4249,[535] n_D^{20} 1.4226,[814a] n_D^{22} 1.4504, n_D^{20} 1.4243.[816]

$(EtO)(cyclo\text{-}C_6H_{11}O)PHO$. EtOP(OH)ONa + $C_6H_{11}OCOCl$.[297] Ia.[814a] $b_{0.2}$ 82-4°,[297] $b_{0.8}$ 106-8°,[814a] b_1 81°,[298] d^{20} 1.060, n_D^{20} 1.4510,[814a] n_D^{22} 1.4504.[297]

$(EtO)(C_7H_{15}O)PHO$. Ia.[814a] Ia + H_2O.[816] b_1 107-9°,[814a] b_{10} 125-8°,[816] d^{20} 0.9738,[816] d^{20} 0.979, n_D^{20} 1.4299,[814a] n_D^{20} 1.4270.[816]

$(EtO)(C_8H_{17}O)PHO$. Ia.[814] Ib.[535] Ie.[132] $b_{0.02}$ 92-6°,[535] b_1 114-6°,[814] $b_{4.5}$ 126-8°,[132] d_0^{20} 0.9779,[132] d^{20} 0.9634,[814] n_D^{20} 1.4278,[814] n_D^{20} 1.4312,[132] n_D^{20} 1.4290.[535]

$(EtO)(C_6H_5O)PHO$. Ia.[814a] Ih.[273] Ii.[1392] EtO(OH)ONa + C_6H_5OCOCl.[297] $b_{0.2}$ 94-6°,[297] $b_{0.3}$ 88-90°,[814a] b_{13} 95°,[273] d^{20} 1.152,[814a] d^{20} 1.1668,[1392] n_D^{22} 1.4991,[297] n_D^{20} 1.4995,[1392] n_D^{20} 1.4924,[814a] MR_D 46.8.[1392]

$(EtO)(2\text{-}ClC_6H_4O)PHO$. Ih. b_{13} 85-7°.[274]

$(EtO)(4\text{-}ClC_6H_4O)PHO$. Ih. b_{13} 112-3°.[274]

$(EtO)(4\text{-}MeC_6H_4O)PHO$. Ia. $b_{0.3}$ 108-10°, d^{20} 1.129, n_D^{20} 1.5003.[814a]

(PrO)(BuO)PHO. Ia + H_2O.[816] Ie.[84] Ih.[84] b_3 81-3°,[84] b_9 98-100°,[816] d^{20} 0.9988,[84] d^{20} 0.9989,[816] n_D^{20} 1.4200,[816] MR_D 45.75 (calc. 45.56).[84]

(PrO)(AmO)PHO. Ia + H_2O. b_{10} 110-13°, n_D^{20} 1.4222, d^{20} 0.9906.[816]

(PrO)(i-AmO)PHO. Ia. b_1 86-8°, d^{20} 1.003, n_D^{20} 1.4271.[814a]

$(PrO)(C_6H_5O)PHO$. Ii. d^{20} 1.135, n_D^{20} 1.495, MR_D 51.3.[1392]

$(\overline{OCMe_2OCH_2C}HCH_2O)(C_6H_5O)PHO$. $b_{0.0005}$ 127-30°, d^{20} 1.1406, n_D^{20} 1.4980.[1057]

$(i\text{-}PrO)(cyclo\text{-}C_6H_{11}O)PHO$. Ia. $b_{0.4}$ 101-3°, d^{20} 1.040, n_D^{20} 1.4450.[814a]

(i-PrO)(PhO)PHO. Ia.[814a] Ii.[1392] Ih.[273] $b_{0.4}$ 95°,[814a] b_{13} 97°,[273] d^{20} 1.125,[1392] d^{20} 1.013,[814a] n_D^{20} 1.4376,[814a] n_D^{20} 1.4940, MR_D 51.8.[1392]

(BuO)(i-AmO)PHO. Ia.[814,814a] $b_{1.2}$ 98-100°,[814a] b_8 120-1°,[814] d^{20} 0.989,[814a] d^{20} 0.973,[814] n_D^{20} 1.420,[814] n_D^{20} 1.4259.[814a]

$(BuO)(C_7H_{15}O)PHO$. Ie. Ih. b_2 151-2°, d^{20} 0.9481, n_D^{20}

1.4325, MR_D 64.71 (calc. 64.04).[84]

(BuO)($C_8H_{18}O$)PHO. Ia. $b_{0.4}$ 112-4°, d^{20} 0.952, n_D^{20} 1.431.[814]

(BuO)(PhO)PHO. Ia.[814a] Ii.[1392] $b_{0.4}$ 110°,[814a] d^{20} 1.0958,[1392] d^{20} 1.026,[814a] n_D^{20} 1.485,[1392] n_D^{20} 1.4630,[814a] MR_D 55.9.[1392]

(i-BuO)(cyclo-$C_6H_{11}O$)PHO. Ia. $b_{0.4}$ 103-5°, d^{20} 0.974, n_D^{20} 1.4260.[814a]

(i-AmO)($C_6H_{13}O$)PHO. Ia. $b_{0.3}$ 108-10°, d^{20} 0.957, n_D^{20} 1.429.[814]

(i-AmO)($C_7H_{15}O$)PHO. Ia. $b_{0.2}$ 105-7°, d^{20} 0.953, n_D^{20} 1.431.[814]

(i-AmO)($C_8H_{17}O$)PHO. Ia. $b_{0.2}$ 120-1°, d^{20} 0.953, n_D^{20} 1.4315.[814]

(i-AmO)(s-$C_8H_{17}O$)PHO. Ia. $b_{0.2}$ 111-3°, d^{20} 0.942, n_D^{20} 1.429.[814]

(i-$C_8H_{17}O$)(C_6H_5O)PHO. Ia. $b_{0.5}$ 155°, d^{20} 0.994, n_D^{20} 1.4691.[814a]

(4-$Me_3CCH_2CMe_2C_6H_4O$)PH(O)OC_6H_2-2,5-t-Bu(4-OH). Element anal. given. Original paper gives C_8H_{17}, and C_4H_9; C.A. gives as shown.[1025]

Erythromycin A_2'-(methylhydrogen phosphite). $ROPO_2H_2$ + erythromycin + dicyclohexylcarbodiimide. $[\alpha]_D^{25}$ -58.3°.[627]

Erythromycin A_2'-(phenyl hydrogen phosphite). $ROPO_2H_2$ + erythromycin + dicyclohexylcarbodiimide. $[\alpha]_D^{25}$ -48.1°.[627]

Erythromycin A_2'-(benzyl hydrogen phosphite). $ROPO_2H_2$ + erythromycin + dicyclohexylcarbodiimide. $[\alpha]_D^{25}$ -54.2°.[627]

TYPE: (RO)(R'CO_2)PHO

(BuO)(AcO)PHO. BuOP(OAc)$_2$ + AcOH. n_D^{20} 1.4320.[958]

I.1.21. Bis(Hydrogen Phosphites)

TYPE: [(RO)P(O)(H)]$_2$R'

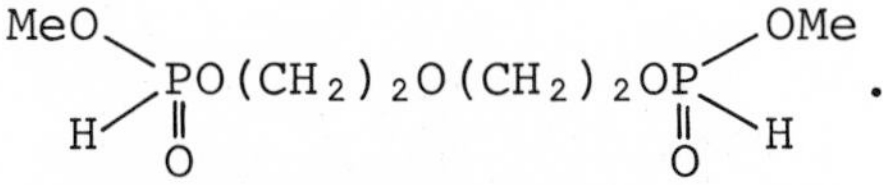

MeO$\diagdown$
$\qquad$PO(CH_2)$_2$O(CH_2)$_2$OP$\diagup$ ^{OMe} . Ie. n_D^{20} 1.4589.[1406]
H$\diagup$ $\overset{\parallel}{O}$ $\qquad\qquad\qquad \overset{\parallel}{O}$ $\diagdown$H

[(MeO)P(O)(H)O]$_2$(CH_2)$_2$. Ie. $b_{0.0001}$ 110-20°, d^{20} 1.4136, n_D^{20} 1.4541.[1406]

[(MeO)P(O)(H)O]$_2$(CH_2)$_3$. Ie. $b_{0.0001}$ 100-10°, d^{20} 1.3236, n_D^{20} 1.4475.[1406]

[(MeO)P(O)(H)O]$_2$(CH_2)$_4$. Ie. $b_{0.0001}$ 120-30°, d^{20} 1.2840, n_D^{20} 1.4450.[1406]

[(MeO)P(O)(H)O]$_2$(CH_2)$_5$. Ie. n_D^{20} 1.4498.[1406]

[(MeO)P(O)(H)O]$_2$(CH_2)$_6$. Ie. $b_{0.0001}$ 130-50°, d^{20} 1.2070, n_D^{20} 1.4481.[1406]

$[(EtO)P(O)(H)O]_2(CH_2)_2$. Ie. $b_{0.001}$ 115-20°, d^{20} 1.2996, n_D^{20} 1.4485.[1406]

I.1.22. Tris(Hydrogen Phosphites)

TYPE: $[(RO)P(O)(H)]_3R'$

$[(EtO)P(O)(H)OCH_2]_3CMe$. Ie. d_4^{20} 1.2754, n_D^{20} 1.4550, MR_D 84.28 (84.11).[1406]

I.1.23. Hydrogen Phosphites (Phosphonates) with Phosphorus in Ring System

Derivatives of 1,3,2-Dioxaphospholane

$\overline{OCHMeCH_2OP}HO$. Ie.[1028] Ig.[1415] $b_{0.01}$ 66-7°,[1415] $b_{2.3}$ 85-6°,[1028] n_D^{20} 1.4410,[1415] n_D^{20} 1.4580,[1028] IR.[1415]

$\overline{OCH(CH_2Cl)CH_2OP}HO$. Ie. $b_{2.3}$ 125-6°, n_D^{20} 1.4910.[1028]

$\overline{OCH(CH_2OMe)CH_2OP}HO$. By H_2O + methyl ester. b_{10} 156-8°, n_D^{20} 1.4719.[105]

$\overline{OCHMeCHMeOP}HO$. Ie.[1028] Ig.[1415] $b_{0.05}$ 67-8°,[1415] $b_{2.3}$ 84-5°,[1028] n_D^{20} 1.4385,[1415] n_D^{20} 1.4616,[1028] IR.[1415]

$\overline{OCMe_2CMe_2OP}HO$. Ia. Ig. m. 104-6°,[1415] m. 106.5-8°,[82] IR.[1415]

$\overline{OCMeEtCMeEtOP}HO$. Ia. Ig. $b_{0.01}$ 76-7°, $b_{0.02}$ 75-6°, n_D^{20} 1.4592, n_D^{20} 1.4578, IR.[1415]

$\overline{OCEt_2CEt_2OP}HO$. Ig. $b_{0.01}$ 88°, n_D^{20} 1.4707, IR.[1415]

$\overline{OCH(C_6H_{11})CH(C_6H_{11})OP}HO$. Ia. Ig. m. 90-2°,[1415] m. 92-2.5°,[83] IR.[1415]

Derivatives of 1,3,2-Dioxaphosphorinane

$\overline{OCH_2CH_2CH_2OP}HO$. Ie.[1028] Ig.[1415] $b_{0.05}$ 92-3°,[1415] $b_{2.5}$ 97-8°,[1028] m. 29-30°, n_D^{20} 1.4579,[1415] n_D^{20} 1.4522,[1028] IR.[1415]

$\overline{OCHMeCH_2CH_2OP}HO$. Ie.[1028] Ig.[1415] $b_{0.05}$ 90-1°,[1415] $b_{2.3}$ 103-4°, b_{11} 150-5°,[101] n_D^{20} 1.4548,[1415] n_D^{20} 1.4547,[1028] m. 53-4°,[1415] m. 48-50°,[101] IR.[1415]

$\overline{OCHPrCHEtCH_2OP}HO$. Ie. $b_{2.5}$ 117-8°, n_D^{20} 1.4600.[1028]

$\overline{OCH_2CMe_2CH_2OP}HO$. Ie.[1028] Ig.[1415] $b_{0.05}$ 93-4°,[1415] $b_{2.3}$ 103-4°,[1028] m. 53-5°,[1028,1415] IR.[1415]

$\overline{OCH_2C(Me)(CH_2OH)CH_2OP}HO$. Gives metriol phosphite on heating (q.v.).[970]

$\overline{OCH_2C(Me)(CH_2OCH_2Ph)CH_2OP}HO$. Gives metriol phosphite on heating (q.v.).[970]

$\overline{OCH_2C(Et)(CH_2OCH_2CH{:}CH_2)CH_2OP}HO$. Ie. b_3 120-5°.[425]

Derivatives of $\overline{O(CH_2)_4OP}HO$

$\overline{O(CH_2)_4OP}HO$. Ig. $b_{0.01}$ 68-9°, m. 44.5-5.5°, IR.[1415]

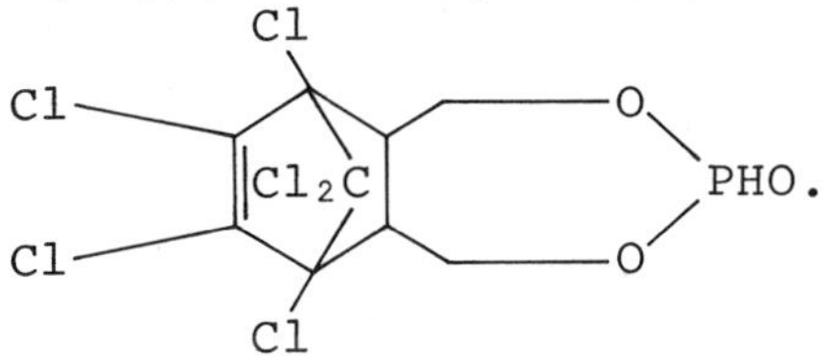

Ig. m. 170-6°.[509]

I.1.24. Monoalkyl or Monoaryl Phosphites

TYPE: ROP(O)(OH)H

$MeOPO_2H_2$. If.[997] Ig.[1246] Na salt: crystals, decomp. 125° (from $EtOH-Et_2O$).[997] Ba salt; Ca salt monohydrate; unstable solids.[1246] Free acid is an unstable oil.

$PhCH_2OPO_2H_2$. From HCl + Na salt.[873] By heating dibenzyl hydrogen phosphite and a tertiary base. Oil isolated as the ammonium salt, needles, m. 154°. Has significance in the stepwise removal of the benzyl group from tribenzyl phosphite.[158] n_D^{25} 1.5278 (undistilled).[873]

$PhCH(CO_2Et)OPO_2H_2$. Ig. The free ester is an oil insoluble in water.[450]

$EtOPO_2H_2$. If.[622,900,996,997] Ig.[1398] From HCl + Na salt.[873] Ba salt: hygroscopic solid (from EtOH).[900,1398] Pb salt: soluble in EtOH.[1398] Na salt: needles, m. 183° (from EtOH).[622,996,997] n_D^{20} 1.4230 (undistilled).[873]

$ClCH_2CH_2OPO_2H_2$. From poorly characterized reaction product of glycol with PCl_3. Isolated as barium salt (from H_2O).[269]

$HOCH_2CH_2OPO_2H_2$. Ii. Isolated as hygroscopic Ca and Ba salts.[269]

$H_2O_2POCH_2CH_2OPO_2H_2$. Combined If-Ig, on $ClP(OCH_2CH_2O)_2PCl$. Isolated as poorly characterized calcium salt.[269]

$4-PhCH_2CH_2C_6H_4CH_2CH_2OPO_2H_2$. Ig. The free ester is an oil that gives a crystalline sodium salt.[1285]

$PrOPO_2H_2$. If. From HCl + Na salt.[873] Ferric salt solid almost insoluble in water.[900] Sodium salt: needles, m. 195-6° (from EtOH).[997] n_D^{20} 1.4257 (undistilled).[873]

1-Chloro-2(or 3)-hydroxypropyl 3(or 2)-phosphite. Ii.[269] Ig, from poorly characterized reaction product of

PCl_3 and glycerol prepared in ether solution. Isolated as poorly described barium and calcium salts.[269]

$CH_2{:}CHCH_2OPO_2H_2$. From HCl + Na salt.[873] n_D^{20} 1.4390 (undistilled).[873]

1(or 2)-glyceryl phosphite. Ii. Isolated as water-soluble barium salt, which forms a trihydrate.[269,802]

iso-$PrOPO_2H_2$. If. From HCl + Na salt.[873] Barium salt: hygroscopic solid.[900] Sodium salt: needles, m. 132-3° (from EtOH-Et_2O).[997] Ferric salt: almost insoluble in water.[900] n_D^{20} 1.4211 (undistilled).[873]

$(CF_3)CHMeOPO_2H_2$. Ig. Isolated as barium salt, which is soluble 1:25 in water.[1305]

$PhOCH_2CH(OPO_2H_2)CH_2OPh$. Ig. Needles, m. 119-20° (from EtOAc). Ammonium salt is an oil. Calcium salt: crystals (from dil. EtOH).[215]

p-$MeC_6H_4OCH_2CH(OPO_2H_2)CH_2OPh$. Ig. Needles, m. 106-7° (from EtOAc). The ammonium salt is a water-soluble solid.[215]

o-$MeC_6H_4OCH_2CH(OPO_2H_2)CH_2OC_6H_4Me$-o. Ig. Prisms, m. 88-9° (from EtOAc). The calcium salt tetrahydrate forms needles (from dil. EtOH).[216]

m-$MeC_6H_4OCH_2CH(OPO_2H_2)CH_2OC_6H_4Me$-m. Ig. Needles, m. 85-7° (from CS_2-petroleum ether).[216]

p-$MeC_6H_4OCH_2CH(OPO_2H_2)CH_2OC_6H_4Me$-p. Ig. Crystals, m. 111-2° (from EtOAc).[215]

$BuOPO_2H_2$. If. From HCl + Na salt.[873] Isolated as sodium salt: needles, m. 177.5-8.5° (from EtOH).[997] n_D^{20} 1.4292 (undistilled).[873]

iso-$BuOPO_2H_2$. If. Isolated as poorly soluble ferric salt.[900]

iso-$AmOPO_2H_2$. If.[900] Ig.[1398] Free ester is an unstable oil. Ferric salt is a poorly soluble solid.[900]

2-$C_8H_{17}OPO_2H_2$. If.[450] Ig.[450] Oil, d_4^{15} 1.0210, d_4^{50} 0.9914, n_D^{16} 1.4400.[450]

Monoerythrityl phosphite. Ii. Obtained as a crude, poorly stable product, readily decomposed by cold water.[269]

l-Menthyl phosphite. Ig. Crystals, m. 29°, decompose at 135° with liberation of menthene. Ca, Ag, and Pb salts isolated.[897]

Cholesteryl phosphite. Ig. Crystals, m. 158° (from benzene-petroleum ether).[376]

4-Hydroxy-tetrahydrofuryl-3-phosphite. If, from tetrahydrofurylene-3,4-phosphite. Free ester is very unstable. Calcium salt monohydrate: needles (from EtOH-Me_2CO).[269]

Erythromycin A_2^1 dihydrogen phosphite. From erythromycin and dicyclohexylcarbodiimide. $[\alpha]_D^{23}$ - 58.3°.[627]

4-$CH_2{:}CHCH_2$-2-$MeOC_6H_3OPO_2H_2$. Ig. Poorly soluble yellow powder, slightly soluble in hot water.[1011]

2,6-di-t-$BuC_6H_3OPO_2H_2$. Ig. m. 136.5-8° (hexane).[1266]

2,6-di-t-Bu-4-Me$C_6H_2OPO_2H_2$. Ig. m. 190-5°.[1266]
2,6-di-t-Bu-4-$C_9H_{19}C_6H_2OPO_2H_2$. Ig. Viscous oil. Na salt has good surface active properties.[1266]
2,6-di-t-Bu-4-Cl$C_6H_2OPO_2H_2$. Ig. m. 140-2°.[1266]
2,4,6-tris(phenethyl)phenyl PHO(OH). Ig. A viscous oil.[1266]
1-$C_{10}H_7OPO_2H_2$. Ig. Powder, m. 82°, Phenylhydrazine salt: m. 83°.[769]
2-$C_{10}H_7OPO_2H_2$. Ig. Crystals, m. 111°. Phenylhydrazine salt: m. 98-9°.[769]
[6,1,3-{2,3-$(HO)_2POC_{10}H_6CONH$}$(HO)C_{10}H_5SO_2]_2NC_6H_5 \cdot 3H_2O$. Ia. m. 254-7° decomp. Other related derivatives, without characterization constants, are given.[1213]
2,5,7-PhNH[$(HO)_2PO]C_{10}H_5SO_2C_6H_4NMe_2$-p·1/2$H_2O$. From PCl_3 + 2,5,7-PhNH(HO)$C_{10}H_5SO_3H$ + $PhNMe_2$, followed by HCl aq. Decomposed 175-80°. Alkali cleavage gave a product approx. $C_{24}H_{21}O_5N_2SPNa_2$1.25H_2O, decomp. 195-202°.[1025]

I.1.25. Monoacyl Phosphites

TYPE: $RCO_2P(O)(OH)H$

$MeCO_2PO_2H_2$. II [by MeCOCl[220] or by $(MeCO)_2O$[1365]]. Hygroscopic plates, decomp. 100°.[220,1365]

I.1.26. Phosphites Containing Both >PHO and -P(O)(OH)H Groups

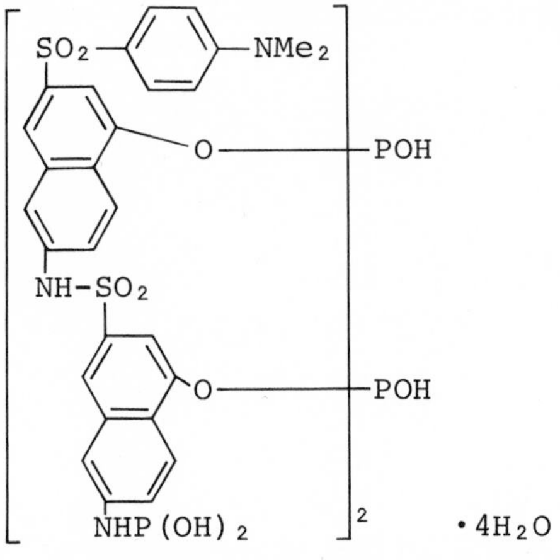

From 2,5,7-$NH_2(HO)C_{10}H_5SO_3H$ + $PhNMe_2$ + PCl_3. Decomposed 279-84°. HCl(15%) at 95-100° gave 2,5,7-$NH_2[(HO)_2PO]C_{10}H_5SO_2C_6H_4NMe_2$-p·1/2EtOH, decomp. 225-30°. NaOH aq. gave $(2,5\text{-}NH_2C_{10}H_6O)_2PONa \cdot 1/2H_2O$ decomp. 271-4°, and [2,7,5-HO_3SNH(p-$Me_2NC_6H_4SO_2)C_{10}H_5O]_2PHO \cdot 2H_2O$ decomp. 230-40°.[1025]

I.1.27. Hydrogen Phosphites (Phosphonates) Containing P-O-P (Anhydride) Linkages

(EtO)P(O)(H)OP(O)$(OEt)_2$. EtOP(O)(H)OH + $(EtO)_2POCl$ + base. $b_{0.005}$ 85-6°, n_D^{25} 1.4225.[873]
(PrO)P(O)(H)OP(O)$(OEt)_2$. PrOP(O)(H)OH + $(EtO)_2POCl$ +

base. $b_{0.005}$ 88-9°, n_D^{25} 1.4235.[873]
(i-PrO)P(O)(H)OP(O)(OEt)$_2$. i-PrOP(O)(H)OH + (EtO)$_2$POCl + base. $b_{0.005}$ 81-2°, n_D^{25} 1.4240.[873]
(BuO)P(O)(H)OP(O)(OEt)$_2$. BuOP(O)(H)OH + (EtO)$_2$POCl + base. $b_{0.005}$ 88-9°. n_D^{25} 1.4230.[873]
($C_6H_5CH_2$O)P(O)(H)OP(O)(OEt)$_2$. $C_6H_5CH_2$OP(O)(H)OH + (EtO)$_2$POCl + base. $b_{0.005}$ 91-2°, n_D^{25} 1.4574.[873]
(CH_2CH:CH_2O)P(O)(H)OP(O)(OEt)$_2$. CH_2CH:CH_2OP(O)(H)OH + (EtO)$_2$POCl + base. $b_{0.005}$ 86-7°, n_D^{25} 1.4248.[873]

I.1.28. Salts and Other Derivatives of Dialkyl Hydrogen Phosphites

TYPE: (RO)$_2$POM

(EtO)$_2$POLi. ^{31}P -145 ppm in diglyme.[913]
(EtO)$_2$PONa. ^{31}P -153 ppm in dyglyme.[913,914a]
(EtO)$_2$POK. ^{31}P -152, -152.3 ppm in diglyme.[913,914a]
(BuO)$_2$POLi. ^{31}P -145 ppm in diglyme.[913,914a]
(BuO)$_2$PONa. ^{31}P -152, -153.5 ppm in diglyme.[913,914a]
(BuO)$_2$POK. ^{31}P -153, -152 ppm in diglyme.[913,914a]
(C_6H_5O)$_2$POLi. ^{31}P -142 ppm in diglyme.[913]
(C_6H_5O)$_2$PONa. ^{31}P -148, -146.7 ppm in diglyme.[913,914a]
(C_6H_5O)$_2$POK. ^{31}P -139 ppm in diglyme.[913]

TYPE: $[(RO)_2PO]^-[R^1R^2NH_2]^+$

$[(EtO)_2PO]^-[n\text{-}C_{16}H_{33}NH_3]^+$. (EtO)$_2$PHO + amine. m. 84-5°.[741]
$[(EtO)_2PO]^-[(n\text{-}C_{16}H_{33})(Me)NH_2]^+$. (EtO)$_2$PHO + amine. m. 87-8°.[741]
$[(i\text{-}PrO)_2PO]^-[n\text{-}C_{16}H_{33}NH_3]^+$. (i-PrO)$_2$PHO + amine. m. 52-3°.[741]
$[(i\text{-}PrO)_2PO]^-[(n\text{-}C_{16}H_{33})(Me)NH_2]^+$. (i-PrO)$_2$PHO + amine. m. 86-7°.[741]
$[(BuO)_2PO]^-[n\text{-}C_{16}H_{33}NH_3]^+$. (BuO)$_2$PHO + amine. m. 54-5°.[741]
$[(BuO)_2PO]^-[(n\text{-}C_{16}H_{33})(Me)NH_2]^+$. (BuO)$_2$PHO + amine. m. 21-2°.[741]
$[(C_6H_5O)_2PO]^-[n\text{-}C_{16}H_{33}NH_3]^+$. ($C_6H_5$O)$_2$PHO + amine. m. 66-7°.[741]
$[(C_6H_5O)_2PO]^-[(n\text{-}C_{16}H_{33})(Me)NH_2]^+$. ($C_6H_5$O)$_2$PHO + amine. m. not given.[741]

TYPE: $[(RO)_2PO]^-[H_3NCONH_2]^+$.

$[(EtO)_2PO]^-[H_3NCONH]^+$. (EtO)$_2$PHO + CO($NH_2$)$_2$. m. 119-21°.[593]
$[(C_9H_{19}O)_2PO]^-[H_3NCONH]^+$. ($C_9H_{19}$O)$_2$PHO + CO($NH_2$)$_2$. m. 114-5.5°.[593]

TYPE: $[(RO)(R'O)PO]^-[H_3NCSNH_2]^+$

$[(MeO)_2PO]^-[H_3NCSNH_2]^+$. $(MeO)_2PHO + CS(NH_2)_2$. m. 116-6.5°, UV.[593]
$[(AmO)_2PO]^-[H_3NCSNH_2]^+$. $(AmO)_2PHO + CS(NH_2)_2$. m. 103-4°, UV.[593]
$[(C_6H_{13}O)_2PO]^-[H_3NCSNH_2]^+$. $(C_6H_{13}O)_2PHO + CS(NH_2)_2$. m. 108-9°, UV.[593]
$[(C_7H_{15}O)_2PO]^-[H_3NCSNH_2]^+$. $(C_7H_{15}O)_2PHO + CS(NH_2)_2$. m. 109-10.5°, UV.[593]
$[(C_8H_{17}O)_2PO]^-[H_3NCSNH_2]^+$. $(C_8H_{17}O)_2PHO + CS(NH_2)_2$. m. 106-7°, UV.[593]
$[(C_9H_{19}O)_2PO]^-[H_3NCSNH_2]^+$. $(C_9H_{19}O)_2PHO + CS(NH_2)_2$. m. 107-8°, UV.[593]
$[(C_{10}H_{21}O)_2PO]^-[H_3NCSNH_2]^+$. $(C_{10}H_{21}O)_2PHO + CS(NH_2)_2$. m. 110-2°, UV.[593]
$[(MeO)(PrO)PO]^-[H_3NCSNH_2]^+$. $(MeO)(PrO)PHO + CS(NH_2)_2$. m. 84-5°, UV.[593]
$[(MeO)(BuO)PO]^-[H_3NCSNH_2]^+$. $(MeO)(BuO)PHO + CS(NH_2)_2$. m. 64-5°, UV.[593]
$[(MeO)(AmO)PO]^-[H_3NCSNH_2]^+$. $(MeO)(AmO)PHO + CS(NH_2)_2$. m. 72-3°, UV.[593]
$[(MeO)(C_6H_{13}O)PO]^-[H_3NCSNH_2]^+$. $(MeO)(C_6H_{13}O)PHO + CS(NH_2)_2$. m. 76-7°, UV.[593]
$[(MeO)(C_7H_{15}O)PO]^-[H_3NCSNH_2]^+$. $(MeO)(C_7H_{15}O)PHO + CS(NH_2)_2$. m. 81.5-2°, UV.[593]
$[(MeO)(C_8H_{17}O)PO]^-[H_3NCSNH_2]^+$. $(MeO)(C_8H_{17}O)PHO + CS(NH_2)_2$. m. 89-90°, UV.[593]
$[(MeO)(C_9H_{19}O)PO]^-[H_3NCSNH_2]^+$. $(MeO)(C_9H_{19}O)PHO + CS(NH_2)_2$. m. 53-7°, UV.[593]
$[(MeO)(C_{10}H_{21}O)PO]^-[H_3NCSNH_2]^+$. $(MeO)(C_{10}H_{21}O)PHO + CS(NH_2)_2$. m. 83-4°, UV.[593]

TYPE: $(RO)_2P(O)HgX$

$(EtO)_2P(O)HgOAc$. $(EtO)_2PHO + Hg(OAc)_2$. m. 106.8-7.6°.[410]
$(PrO)_2P(O)HgOAc$. $(PrO)_2PHO + Hg(OAc)_2$. m. 86.3-7.2°.[410]
$(i\text{-}PrO)_2P(O)HgOAc$. $(i\text{-}PrO)_2PHO + Hg(OAc)_2$. m. 146.2-6.8°.[410]
$(n\text{-}BuO)_2P(O)HgOAc$. $(n\text{-}BuO)_2PHO + Hg(OAc)_2$. m. 80.5-1.0°.[410]
$(i\text{-}BuO)_2P(O)HgOAc$. $(i\text{-}BuO)_2PHO + Hg(OAc)_2$. m. 116-6.6°.[410]
$(AmO)_2P(O)HgOAc$. $(AmO)_2PHO + Hg(OAc)_2$. m. 72.5-3.5°.[410]
$(C_6H_{13}O)_2P(O)HgOAc$. $(C_6H_{13}O)_2PHO + Hg(OAc)_2$. m. 73.9-4.6°.[410]
$(C_7H_{15}O)_2P(O)HgOAc$. $(C_7H_{15}O)_2PHO + Hg(OAc)_2$. m. 83.2-4.0°.[410]
$(EtO)_2P(O)HgCl$. $(EtO)_2PHO + HgCl_2$. m. 104-4.5°.[410]
$(PrO)_2P(O)HgCl$. $(PrO)_2PHO + HgCl_2$. m. 80-1.5°.[410]
$(i\text{-}PrO)_2P(O)HgCl$. $(i\text{-}PrO)_2PHO + HgCl_2$. m. 114.6-5.2°.[410]

$(BuO)_2P(O)HgCl$. $(BuO)_2PHO + HgCl_2$. m. 91-2.0°.[410]
$(i\text{-}BuO)_2P(O)HgCl$. $(i\text{-}BuO)_2PHO + HgCl_2$. m. 118-9.2°.[410]
$(EtO)_2P(O)HgBr$. $(EtO)_2PHO + HgBr_2$. m. 86-7.0°.[410]
$(PrO)_2P(O)HgBr$. $(PrO)_2PHO + HgBr_2$. m. 70-0.4°.[410]
$(BuO)_2P(O)HgBr$. $(BuO)_2PHO + HgBr_2$. m. 78-8.4°.[410]
$(EtO)_2P(O)HgI$. $(EtO)_2PHO + HgI_2$. m. 102-2.5°.[410]
$(PrO)_2P(O)HgI$. $(PrO)_2PHO + HgI_2$. m. 75.2-6.0°.[410]
$(BuO)_2P(O)HgI$. $(BuO)_2PHO + HgI_2$. m. 56-7.2°.[410]
$(EtO)_2P(O)HgSCN$. $(EtO)_2PHO + Hg(SCN)_2$. m. 124.2-4.6°.[410]

TYPE: $[(RO)_2P(O)]_2Hg$

$[(MeO)_2P(O)]_2Hg$. $(MeO)_2PHO + HgO$. m. 121.6-3.0°.[1346]
$[(EtO)_2P(O)]_2Hg$. $(EtO)_2PHO + HgO$. m. 56.8-8.2°.[1346]
$[(PrO)_2P(O)]_2Hg$. $(PrO)_2PHO + HgO$. m. 35.6-6.8°, n_D^{20} 1.5062 (supercooled).[1346]
$[(i\text{-}PrO)_2P(O)]_2Hg$. $(i\text{-}PrO)_2PHO + HgO$. m. 124.4-5.0°.[1346]
$[(BuO)_2P(O)]_2Hg$. $(BuO)_2PHO + HgO$. m. 25.0-7.2°, n_D^{20} 1.4991 (supercooled).[1346]
$[(i\text{-}BuO)_2P(O)]_2Hg$. $(i\text{-}BuO)_2PHO + HgO$. m. 132.0-3.5°.[1346]

I.2. Thiophosphites and Their Derivatives

I.2.1. Halogeno and Cyanato Compounds

TYPE: $RSPCl_2$

$MeSPCl_2$. ^{31}P -206 ppm.[914a]
$C_6H_5CH_2SPCl_2$. ^{31}P -205.5 ppm.[914a]
$EtSPCl_2$. VIb. VIe. b. 171°, b_{10} 64°, b_{10} 53°, b. 172-5°, d_0^{12} 1.30,[353,856,1063] ^{31}P -210.7 ppm.[826,914a]
$ClCH_2CH_2SPCl_2$. By $PCl_3 + \overline{SCH_2CH_2}$. b_{15} 110-5°, d_4^{20} 1.5100, n_D^{20} 1.5860.[993]
$PrSPCl_2$. ^{31}P -212.6 ppm.[914a]
$i\text{-}PrSPCl_2$. ^{31}P -211 ppm.[914a]
$BuSPCl_2$. ^{31}P -209.7 ppm.[826]
$CH_2{:}CHCH_2SPCl_2$. ^{31}P -210.1 ppm.[914a]
$AmSPCl_2$. ^{31}P -210.4 ppm.[914a]
$C_7H_{15}SPCl_2$. ^{31}P -211 ppm.[914a]
$C_6H_5SPCl_2$. VIa. b. 213-7°, b_{10} 125°, d_{15}^{15} 1.2560,[299,860] ^{31}P -204.2 ppm.[914a]

TYPE: $(RS)_2PCl$

$(MeS)_2PCl$. ^{31}P -188.2 ppm, J_{PH} 11 Hz.[914a]
$(C_6H_5CH_2S)_2PCl$. ^{31}P -180.3 ppm, J_{PH} ca. 11 Hz.[914a]
$(EtS)_2PCl$. VIb. VIe. b_5 84-5°, b_4 83°, $b_{0.4-8.8}$ 39°, d_4^{20} 1.2050, n_D^{20} 1.5850,[353,820] ^{31}P -186.2 ppm, J_{PH} ca. 11 Hz.[914a]
$(PrS)_2PCl$. b_5 ca. 14°,[853] ^{31}P -189.4 ppm, J_{PH} ca. 13 Hz.[914a]

$(i\text{-}PrS)_2PCl$. ^{31}P -181.6 ppm.[914a]
$(CH_2{:}CHCH_2S)_2PCl$. ^{31}P -185.1 ppm.[914a]
$(BuS)_2PCl$. ^{31}P -185.2, -185.5, -187.4 ppm.[826]
$(t\text{-}BuS)_2PCl$. ^{31}P -173.2 ppm.[826]
$(AmS)_2PCl$. ^{31}P -187.2 ppm, J_{PH} ca. 11 Hz.[914a]
$(C_7H_{15}S)_2PCl$. ^{31}P -187.7 ppm.[914a]
$(C_6H_5S)_2PCl$. ^{31}P -182.7, -182.9 ppm.[826,914a]

Chloro Derivatives with P in a Ring System

$\overline{SCH_2CH_2SP}Cl$. From $(CH_2SH)_2 + PCl_3$. b_{10} 102-3°, d_0^{20} 1.5297, n_D^{20} 1.6700.[102]

$\overline{SCH_2CH_2SP}SCH_2CH_2S\overline{PSCH_2CH_2S}$. From $(CH_2SH)_2 + PCl_3$. m. 130°.[102]

TYPE: (RS)(R'O)PCl

(MeS)(MeO)PCl. VIb.[1254]
(MeS)(EtO)PCl. VIb. b_5 47°.[1254]
(EtS)(EtO)PCl. VIb. b_1 46°, b_{12} 62°.[1254]
$(EtSCH_2CH_2S)(EtO)PCl$. VIb. b_1 111°.[1254]
(i-PrS)(EtO)PCl. VIb. b_1 52°.[1254]
(i-BuS)(EtO)PCl. VIb. b_1 57°.[1254]

TYPE: $RSPBr_2$

$MeSPBr_2$. ^{31}P -203.5 ppm, J_{PH} ca. 7 Hz.[914a]
$C_6H_5SPBr_2$. ^{31}P -203.5 ppm.[914a]

TYPE: $(RS)_2PBr$

$(C_6H_5S)_2PBr$. ^{31}P -184.2 ppm.[914a]

TYPE: $RSP(NCO)_2$

$EtSP(NCO)_2$. $EtSPCl_2$ + NaOCN. $b_{0.1}$ 38-40°, d^{20} 1.3039, n_D^{20} 1.5420.[1273]

TYPE: $(RS)_2PNCO$

$(EtS)_2PNCO$. $(EtS)_2PCl$ + NaOCN. b_8 110-2°, d^{20} 1.1919, n_D^{20} 1.5614.[1273]

I.2.2. Trialkyl or Triaryl Esters

TYPE: $(RS)_3P$

$(MeS)_3P$. VIc. $b_{0.2}$ 85-90°, n_D^{30} 1.6397,[1396] ^{31}P -124.1 to -125.6 ppm, J_{PH} 10, 12 Hz.[826,914a]
$(PhCH_2S)_3P$. VIc.[1295]

$(EtS)_3P$. VIa (very poor).[856] $b_{0.2}$ 85°,[1396] b_{18} 140-3°, m. -32°, d_4^0 1.1883, d_4^{25} 1.1585,[856] n_D^{30} 1.5850,[1396] n_D^{25} 1.5689,[856] ^{31}P -114.1 to -115.6 ppm.[628,826,935] MeI adduct, m. 191°. EtI adduct, m. 125-7°. $HgBr_2$ derivative, m. 184°. HgI_2 derivative, m. 187°. $AuCl_3$ derivative, m. 225°.[353,794]
$(PrS)_3P$. VIb. b_{15} 164-9°, m. -65°, d_4^0 1.1277, d_4^{25} 1.0932, n_D^{25} 1.5350,[794] ^{31}P -118.2, -117.5 ppm.[628,826,935] MeI adduct, m. 191°. $HgBr_2$ derivative, m. 176°. HgI_2 derivative. m. 182°. $AuCl_3$ derivative, m. 208°.[794]
$(BuS)_3P$. VIb. VId.[794] VIc.[1295] $b_{1.5}$ 160-70°,[1295] b_{15} 174-80°,[794] $b_{0.3}$ 144°, n_D^{30} 1.5460,[1396] m. -100°, d_4^0 1.0773, d_4^{25} 1.0421, n_D^{25} 1.5305,[794] ^{31}P -116 to -117.0 ppm.[826,914a] MeI adduct, m. 198°. $HgBr_2$ derivative, m. 148°. HgI_2 derivative, m. 162°. $AuCl_3$ derivative, m. 182°.[794]
$(C_6H_{11}S)_3P$. VIc.[1295]
$(C_{12}H_{25}S)_3P$. VId.[1397]
$(C_6H_5S)_3P$. VIa. VId. Scales or needles, m. 76-7°,[860,1313,1396] ^{31}P -130.5, -133.3 ppm.[914a,1249]
$(4\text{-}MeC_6H_4S)_3P$. VIa. n_D^{20} 1.6810.[910]
$(4\text{-}EtC_6H_4S)_3P$. VIa. n_D^{20} 1.6569.[910]
$(4\text{-}i\text{-}PrC_6H_4S)_3P$. VIa. n_D^{20} 1.6328.[910]
$(4\text{-}t\text{-}BuC_6H_4S)_3P$. VIa. n_D^{20} 1.6216.[910]
$(4\text{-}s\text{-}BuC_6H_4S)_3P$. VIa. n_D^{20} 1.6208.[910]
$(4\text{-}t\text{-}AmC_6H_4S)_3P$. VIa. n_D^{20} 1.6117.[910]
$(4\text{-}s\text{-}AmC_6H_4S)_3P$. VIa. n_D^{20} 1.6026.[910]
$(4\text{-}s\text{-}C_6H_{13}C_6H_4S)_3P$. VIa. n_D^{20} 1.5914.[910]
$(4\text{-}s\text{-}C_7H_{15}C_6H_4S)_3P$. VIa. n_D^{20} 1.5836.[910]
$(4\text{-}s\text{-}C_8H_{18}C_6H_4S)_3P$. VIa. n_D^{20} 1.5774.[910]
$(4\text{-}s\text{-}C_9H_{19}C_6H_4S)_3P$. VIa. n_D^{20} 1.5711.[910]
$(4\text{-}C_{10}H_{21}C_6H_4S)_3P$. VIa. n^{20} 1.5670.[910]

TYPE: $(RS)_2(R'S)P$

$(MeS)_2(BuS)P$. ^{31}P -121.7 ppm.[826]
$(BuS)_2(MeS)P$. ^{31}P -118.4 ppm.[826]
$(BuS)_2(C_6H_5S)P$. VIb. $b_{0.1}$ 128°, n_D^{22} 1.5968, d^{20} 1.088.[499]
$(BuS)_2(Me_2C_6H_3S)P$. VIb. $b_{0.1}$ 156-63°, n_D^{21} 1.5828, d^{20} 1.060.[499]

TYPE: $(RS)_2(R'O)P$

$(EtS)_2(EtO)P$. VIg. b_{10} 108-11°, d_0^{20} 1.0681, n_D^{15} 1.5326,[116] ^{31}P -153.5 ppm.[826]
$(EtS)_2(PrO)P$. VIg. b_{15} 128°, d_0^{15} 1.0487, n_D^{20} 1.5278.[116]
$(EtS)_2PO\text{-}4\text{-}MeC_6H_4$. VIb. n_D^{23} 1.120, d^{20} 1.120.[499]
$(PrS)_2POC_6H_5$. VIb. $b_{0.5}$ 122-7°, n_D^{23} 1.5709, d^{20} 1.125.[499]

$(PrS)_2PO$-4-ClC_6H_4. VIb. n_D^{23} 1.5728, d^{20} 1.167.[499]
$(PrS)_2PO$-2,4-$Cl_2C_6H_3$. VIb.[499]
$(PrS)_2PO$-2,4,5-$Cl_3C_6H_2$. VIb.[499]
$(BuS)_2POMe$. ^{31}P -162.1 ppm.[826]
$(BuS)_2POC_6H_5$. VIb. $b_{1.2}$ 164°, n_D^{23} 1.5588, d^{20} 1.069.[499]
$(BuS)_2PO$-4-ClC_6H_4. VIb. n_D^{25} 1.5562, d^{20} 1.116.[499]
$(BuS)_2PO$-2,4-$Cl_2C_6H_3$. VIb. m. 40-5°.[499]
$(BuS)_2POC_6Cl_5$. VIb.[499]
$(BuS)_2PO$-2-MeC_6H_4. VIb. $b_{0.05}$ 158-63°, n_D^{24} 1.5574, d^{20} 1.053.[499]
$(BuS)_2PO$-3-MeC_6H_4. VIb. $b_{0.05}$ 148-52°, n_D^{27} 1.5516, d^{20} 1.164.[499]
$(BuS)_2PO$-4-MeC_6H_4. VIb. n_D^{26} 1.5564, d^{20} 1.125.[499]
$(BuS)_2PO$-4-t-BuC_6H_4. VIb. $b_{0.05}$ 146-56°, n_D^{23} 1.5462, d^{20} 1.026.[499]
$(BuS)_2PO$-2,6-i-$Pr_2C_6H_3$. VIb. $b_{0.01}$ 146°, n_D^{22} 1.5411, d^{20} 1.011.[499]
$(BuS)_2PO$-4-$C_8H_{17}C_6H_4$. VIb. n_D^{23} 1.4379, d^{20} 1.003.[499]
$(BuS)_2PO$-4-$C_6H_5C_6H_4$. VIb. n_D^{22} 1.5930, d^{20} 1.105.[499]
$\overline{SCH_2CH_2SP}OMe$. VIb. $b_{7.5}$ 97-8.5°, d^{20} 1.3426, n_D^{20} 1.6200.[102]
$\overline{SCH_2CH_2SP}OEt$. VIb. b_5 98-9°, d^{20} 1.2629, n_D^{20} 1.5922.[102]

TYPE: $(RS)(R'O)_2P$

$(EtS)(EtO)_2P$. VIg. b_{10} 75-7°, d_0^{20} 1.0213, n_D^{20} 1.4592.[116]
$(EtS)(Me_2Cl_3CCO)_2P$. VIb. $b_{0.004}$ 159-60°, d^{20} 1.4374, n_D^{20} 1.5311.[367]
$(PrS)(PrO)_2P$. VIg. b_{12} 120-4°, d_0^{15} 1.0562, n_D^{17} 1.5241.[116]
$(PrS)(Me_2Cl_3CCO)_2P$. VIb. $b_{0.17}$ 151-2°, d^{20} 1.4199, n_D^{20} 1.5194.[367]
$(PrS)(C_6H_5O)_2P$. VIb. $b_{0.05}$ 153-7°, n_D^{27} 1.5738, d^{20} 1.320.[499]
$(BuS)(Me_2Cl_3CCO)_2P$. VIb. $b_{0.007}$ 160-2°, d^{20} 1.3776, n_D^{20} 1.5240.[367]
$EtS\overline{POCH_2CH_2O}$. VIb. $b_{1.5}$ 53-6°, $n_D^{21.5}$ 1.5222.[366]
$EtS\overline{POCHMeCH_2O}$. VIb. $b_{0.4}$ 47°, $n_D^{21.5}$ 1.5062.[366]
$EtS\overline{POCHMeCHMeO}$. VIb. $b_{1.4}$ 65°, $n_D^{22.5}$ 1.5015.[366]
$EtS\overline{POCMe_2CMe_2O}$. VIb. $b_{2.4-3.0}$ 81-94°, $n_D^{23.5}$ 1.4968.[366]
$\overline{OCH(C_6H_5)CH(C_6H_5)SP}O$-i-$C_{10}H_{21}$. Ie, from acetate. $b_{0.03}$ 138-40°, n_D^{25} 1.531.[345]
$C_6H_5SPO_2C_6H_4$-1,2. VIb. b_3 182-6°, d^{20} 1.2873, n_D^{20} 1.6180.[704]

Pyrothiophosphites

TYPE: $(RO)_2PSP(OR)_2$

$[(EtO)_2P]_2S$. $(EtO)_2PCl + Et_3N + H_2S$. $b_{0.05}$ 66-8°.[865]

TYPE: (RO)(R'S)(R"O)P

(PrS)(BuS)(C_6H_5O)P. VIb. $b_{0.2}$ 144-5°, n_D^{27} 1.5635, d^{20} 1.252.[499]
(PrS)(BuS)(2-MeC_6H_4O)P. VIb. $b_{0.2}$ 147-8°, n_D^{27} 1.5526, d^{20} 1.227.[499]
(PrS)(BuS)(3-MeC_6H_4O)P. VIb. $b_{0.2}$ 159-63°, n_D^{27} 1.5541, d^{20} 1.279.[499]
(PrS)(BuS)(4-MeC_6H_4O)P. VIb. $b_{0.15}$ 153-63°, n_D^{23} 1.5612, d^{20} 1.075.[499]

TYPE: $(ROCS_2)_3P$

$(MeOCS_2)_3P$. $ROCS_2K + PCl_3$. m. 67-9°.[797]
$(EtOCS_2)_3P$. $ROCS_2K + PCl_3$. m. 57-60°.[797]
(i-Pr$OCS_2)_3P$. $ROCS_2K + PCl_3$. m. 92-5°.[797]
$(BuOCS_2)_3P$. $ROCS_2K + PCl_3$. m. 29-30°.[797]
(s-Bu$OCS_2)_3P$. $ROCS_2K + PCl_3$. m. 56-7°.[797]
(i-Am$OCS_2)_3P$. $ROCS_2K + PCl_3$. n_D^{30} 1.5800.[1395]
$(C_6H_{13}OCS_2)_3P$. $ROCS_2K + PCl_3$. n_D^{30} 1.5749.[1395]
$(C_7H_{15}OCS_2)_3P$. $ROCS_2K + PCl_3$. n_D^{30} 1.5548.[1395]
$(C_8H_{18}OCS_2)_3P$. $ROCS_2K + PCl_3$. n_D^{30} 1.5500.[1395]
$(C_{10}H_{21}OCS_2)_3P$. $ROCS_2K + PCl_3$. n_D^{30} 1.5448.[1395]
$(C_{12}H_{25}OCS_2)_3P$. $ROCS_2K + PCl_3$. n_D^{30} 1.5275.[1395]

TYPE: $(RO_2CS)_3P$

$(EtO_2CS)_3P$. $EtO_2CSK + PCl_3$. Oil, m. 68-70°, IR, NMR.[1308]

TYPE: $(R_2NCS_2)_3P$

$[Et_2NC(S)S-]_3P$. $(Et_2N)_3P + CS_2$. By PCl_3 + sodium diethyldithiocarbaminate. m. 131°.[214,1008]
$[Pr_2NC(S)S-]_3P$. $(Pr_2N)_3P + CS_2$. m. 117-9°.[1008]
$[Bu_2NC(S)S-]_3P$. $(Bu_2N)_3P + CS_2$. m. 106-7°.[1008]

TYPE: $(R_2NCS_2)_2(R'O)P$

$[Me_2NC(S)S-]_2POC_6H_4$t-Bu-4. $ROP(NMe_2)_2 + CS_2$. m. 139°.[1008]

I.2.3. Dialkyl Thiophosphites

TYPE: $(RO)_2PHS$

$(MeO)_2PHS$. VIi3.[636] $b_{2.5}$ 25°,[1245] b_8 44-5°, $b_{16.5}$ 52-3.5°, d^{20} 1.1892, n_D^{20} 1.4768.[636]

$(EtO)_2PHS$. VIi1.[864] VIi3.[636] b_{12} 67.5-8.5°,[636] b_{14} 74-6°,[30] b_{15} 73°,[864] b_{15} 73-4°,[936] b_{20} 80-1°,[871] d^{20} 1.0828,[636] d^{20} 1.0798,[30] d_4^{20} 1.0735,[936] d^{20} 1.0758,[864] n_D^{20} 1.4597,[636] n_D^{20} 1.4475,[936] n_D^{20} 1.4632,[864] n_D^{20} 1.4520,[30] n_D^{23} 1.4634.[871] Yields $[(EtO)_2PS]_2Pb$, m. 75-6°.[30] By controlled alkali hydrolysis, yields (EtO)P(S)HONa (colorless plates).[637]

$(PrO)_2PHS$. VIi3. b_3 62-3°, b_4 74.5-5.5°, d^{20} 1.0290, n_D^{20} 1.4581.[636]

$(i\text{-}PrO)_2PHS$. VIi3.[636,936] b_3 49-50°,[636] b_3 47-8°,[636] b_{10} 76-8°,[936] d^{20} 1.0135,[636] d_4^{20} 1.0158,[936] n_D^{20} 1.4541,[636] n_D^{20} 1.4533,[636] n_D^{20} 1.4541.[936]

$(BuO)_2PHS$. VIi1.[864] VIi3.[636] b_4 89-90°,[636] b_3 88-9°,[936] b_{18} 133-4°,[864] d^{20} 0.9974,[636] d_4^{20} 0.9954,[936] n_D^{20} 1.4583,[636] n_D^{20} 1.4535,[936] n_D^{20} 1.4626.[864] Controlled alkali hydrolysis yields (BuO)P(S)HONa (scales).[637]

TYPE: (RO)(R'O)PHS

(EtO)(PrO)PHS. VIi3. b_{15} 86-7°, n_D^{20} 1.4572.[877]

(EtO)(BuO)PHS. VIi3. b_{10} 90-1°, d_4^{20} 1.0395, n_D^{20} 1.4590.[877]

(EtO)(AmO)PHS. VIi3. b_{12} 95-7°, n_D^{20} 1.4610.[877]

Hydrogen Thiophosphites with P in a Ring System

$OCHMeCH_2OPHS$. VIi1. $b_{0.1}$ 52-3°,[1415] b_3 82-3°,[1121a] n_D^{20} 1.5160,[1415] n_D^{20} 1.4710,[1121a] IR.[1415]

OCHMeCHMeOPHS. VIi1. $b_{0.03}$ 58-9°,[1415] $b_{0.5}$ 84-5°,[1121a] n_D^{20} 1.5019,[1415] n_D^{20} 1.4915,[1121a] IR.[1415]

$OCH(C_6H_{11})CH(C_6H_{11})OPHS$. VIi1. Colorless needles from petroleum, m. 76.5-7°, IR.[1415]

$OCH_2CH_2CH_2OPHS$. VIi1. $b_{0.01}$ 68-70°,[1415] $b_{0.02}$ 60°,[743a] m. 34°,[743a] IR.[1415]

$OCHMeCH_2CH_2OPHS$. VIi1. $b_{0.1}$ 72-3°, n_D^{20} 1.5245, IR.[1415]

$OCH_2CMe_2CH_2OPHS$. VIi1. Colorless leaflets from benzene-petroleum, m. 83-4.5°,[1415] m. 81-2.5°,[365] IR.[1415]

I.3. Selenophosphites

I.3.1. Dialkyl Selenophosphites

TYPE: $(RO)_2PHSe$

$(EtO)_2PHSe$. VIj. $b_{1.2}$ 46-7°, d_4^{25} 1.3650, n_D^{25} 1.4965,[774] ^{31}P -71.9,[774] -71 ppm,[826] J_{PH} 630 Hz.[774,826] IR.[774]

$(PrO)_2PHSe$. VIj. $b_{0.5}$ 54-5°, d_4^{25} 1.2646, n_D^{25} 1.4870.[774]

$(BuO)_2PHSe$. VIj. $b_{0.5}$ 76-8°, d_4^{25} 1.1975, n_D^{25} 1.4834.[744]
$(4\text{-}ClC_6H_4CH_2O)_2PHSe$. VIj. m. 56-7° (from petroleum ether).[744]

I.4. Silyl Phosphites

I.4.1. Trisilyl Phosphites

TYPE: $(R_3SiO)_3P$

$(Me_3SiO)_3P$. V.[1020,1022,1356,1359] $b_{2.5}$ 166-7°, b_1 58-60°,[1022] b_{25} 129°,[1359] d_4^{20} 0.9762, d_4^{20} 0.9017,[1022] d_4^{20} 0.9594,[1359] n_D^{20} 1.4460, n_D^{20} 1.4145,[1022] n_D^{20} 1.4120,[1359] IR,[1022] Raman.[1359]
$(Me_2EtSiO)_3P$. V.[1356] b_3 147-9°,[1356] d_4^{20} 0.9723,[1356] n_D^{20} 1.4350,[1356] IR.[1020]
$(Et_2MeSiO)_3P$. V. $b_{1.5}$ 122-4°, d^{20} 0.9221, n_D^{20} 1.4471.[1018]
$(Et_3SiO)_3P$. V.[1018,1020,1359] b_{10} 203-6°, b_1 145-6°,[1020] b_1 135-6°, d_4^{20} 0.9739, d_4^{20} 0.9307,[1020] d_4^{20} 9293, n_D^{20} 1.4480, n_D^{20} 1.4550,[1020] n_D^{20} 1.4448,[1020] IR.[1020]
$(Me_2PhSiO)_3P$. V. $b_{1.5}$ 201-3°, d_4^{20} 1.0526, n_D^{20} 1.5297.[1018]

TYPE: $(R_3SiO)_2(R'_3SiO)P$

$(Et_3SiO)_2(Me_3SiO)P$. V. b_5 144-8°, d_4^{20} 0.9254, n_D^{20} 1.4380.[1018]
$(Et_3SiO)_2(Me_2PhSiO)P$. V. $b_{2.5}$ 165-70°, d_4^{20} 0.9777, n_D^{20} 1.4830.[1018]
$(Me_2PhSiO)_2(Et_3SiO)P$. V. $b_{1.5-2}$ 188-90°, d_4^{20} 1.0208, n_D^{20} 1.5100.[1018]

TYPE: $(R_3SiO)_2(R'O)P$

$(Et_3SiO)_2(EtO)P$. V.[1018,1020] b_2 112-4°,[1020] $b_{1.5}$ 106.5-7°,[1018] d_4^{20} 0.9313,[1020] d_4^{20} 0.9304,[1018] n_D^{20} 1.4440,[1020] n_D^{20} 1.4448.[1018]
$(Et_3SiO)_2(PrO)P$. V. b_3 128-9°, d_4^{20} 0.9233, n_D^{20} 1.4441.[1018]

TYPE: $(R_3SiO)(R'O)_2P$

$(Me_3SiO)(EtO)_2P$. V.[285] b_{11} 60-2°,[225,285] b_{11} 59-61°,[225] d_0^{20} 0.9485,[225,285] d_4^{20} 0.9493,[225] n_D^{20} 1.4116,[225,285] n_D^{20} 1.4115,[225] IR,[225,285] ^{31}P -126.4 ppm.[826]
$(Et_3SiO)(EtO)_2P$. V.[285,1018,1020] b_1 78°, b_2 78°,[1020] b_7 95-7°,[225] b_{10} 105-6°,[225,285] d_4^{20} 0.9421,[1020] d_4^{20} 0.9340,[225,285] d_4^{20} 0.9332,[225] d_4^{20} 0.9338,[225] n_D^{20} 1.4318,[1020] n_D^{20} 1.4332,[225,285] n_D^{20} 1.4333,[225] IR.[225,285]

I.4.2. Disilyl Phosphites

TYPE: $(R_3SiO)_2PHO$

$(Me_3SiO)_2PHO$. V.[1022,1358] b_3 74-5°, b_{763} 196°,[1357] b_6 100-2°,[1022] d_4^{20} 0.9661,[1357] d_4^{20} 1.0020,[1022] n_D^{20} 1.4145,[1357] n_D^{20} 1.4140,[1022] Raman.[1357]
$(Me_2EtSiO)_2PHO$. V. b_3 104-6°, d_4^{20} 0.9657, n_D^{20} 1.4235.[1022]
$(MeEt_2SiO)_2PHO$. V.[1018,1022,1355,1357] b_6 138-45°,[1357] b_3 136-7°,[1022] b_3 136-8°, $b_{1.5-2}$ 93-5°, d_4^{20} 0.9498,[1357] d_4^{20} 0.9670,[1022] d_4^{20} 0.9685, n_D^{20} 1.4299,[1357] n_D^{20} 1.4330,[1022] n_D^{20} 1.4348, Raman.[1357]
$(Et_3SiO)_2PHO$. V.[1018,1022,1355,1357] b_1 118-9°, b_3 152-3°,[1357] b_5 160-2°,[1022] b_5 162-4°, d_4^{20} 0.9625, d_4^{20} 0.9584,[1357] d_4^{20} 0.9688,[1022] d_4^{20} 0.9690, n_D^{20} 1.4402,[1357] n_D^{20} 1.4412,[1022] n_D^{20} 1.4439, Raman.[1357]
$(MePr_2SiO)_2PHO$. V. b_{17} 185-6°, d_4^{20} 0.9373, n_D^{20} 1.4385, Raman.[1357]
$(Me_2PhSiO)_2PHO$. V.[1018,1022] b_4 206-8°,[1022] b_1 169-71°, d_4^{20} 1.0999,[1022] d_4^{20} 1.0899, n_D^{20} 1.5160,[1022] n_D^{20} 1.5208.

TYPE: $(R_3SiO)(R'O)PHO$

$(Et_3SiO)(EtO)PHO$. V. b_2 82-3°, d_4^{20} 0.9883, n_D^{20} 1.4320.[1018]
$(Et_3SiO)(PrO)PHO$. V. b_5 99-101°, d_4^{20} 0.9918, n_D^{20} 1.4320.[1018]
$(Me_2PhSiO)(EtO)PHO$. V. b_2 111-2°, d_4^{20} 1.0817, n_D^{20} 1.4900.[1018]

I.5. Stannyl Phosphites

$(MeO)_2P{\cdot}OSnMe_2Cl$. $(MeO)_3P + Me_2SnCl_2$. m. 120.5-2°.[359]

I.6. Amidophosphites and Their Derivatives

I.6.1. Halogeno, Cyano, and Cyanato Compounds

TYPE: R_2NPF_2

Me_2NPF_2. By $Me_2NPCl_2 + SbF_3$, or + NaF (in tetramethyl sulfone),[1250] or + ZnF_2.[985] By $(Me_2N)_3P + BF_3$ in xylene.[985] Prior claim[1316] acknowledged.[1250] b. 47°,[1250] b_{704} 50°, m. -88°, d_4^{23} 1.095, n_D^{20} 1.3605,[985] n_D^{20} 1.3580, IR, ^{19}F NMR,[1250] ^{31}P -143, -145 ppm, J_{PF} 1200 to 1196 Hz, J_{PH} 9.25 Hz,[826,1250] ^1H NMR.[1350a]
Et_2NPF_2. By $Et_2NPCl_2 + SbF_3$, or + NaF (in tetramethyl sulfone). b. 96°, b_{100} 47°, b_{95} 42°, n_D^{20} 1.3840, $n_D^{26.5}$ 1.3840.[1250] Formerly described as a gas

condensing to a liquid at -78°.[583,1256] ^{31}P -144, -147.2 ppm, J_{PF} 1204 to 1194 Hz, J_{PNCH} 9.0 Hz.[165,1209,1250]

$(CH_2CH_2)_2NPF_2$. From $-PCl_2 + SbF_3$. b_7 32-3°, ^{31}P -146.2 ppm, J_{PF} 1204, 1198 Hz, J_{PNCH} 3.3 Hz, ^1H NMR, ^{19}F NMR, IR.[165]

$C_5H_{10}NPF_2$. By $C_5H_{10}NPCl_2 + SbF_3$, or + NaF (in tetramethyl sulfone). b_{30} 47°, n_D^{26} 1.4252, n_D^{25} 1.4256,[1250] ^1H NMR,[165] ^{19}F NMR,[165] ^{31}P -139.1, -140.5 ppm, J_{PF} 1193 to 1205 Hz, J_{PNCH} 2.9 Hz.[165,1209,1250]

TYPE: $RNHPF_2$

$MeNHPF_2$. From $MeN(PF_2)_2 + MeNH_2$. b. 52-3° (decomp.), ^{31}P -140.5 ppm, J_{PF} 1191 Hz, ^{19}F NMR, ^1H NMR, IR.[164]

TYPE: $RN(PF_2)_2$

$MeN(PF_2)_2$. IV.[980] By $RN(Cl_2)_2 + SbF_5$.[980] b. 40-2°,[980] ^{31}P -141.5 ppm, J_{PF} 1264 Hz, $J_{PP'}$ 437 Hz, $J_{PF'}$ 47 Hz.[981] ^1H NMR, IR, Raman.[980]

$EtN(PF_2)_2$. IV.[980] By $RN(Cl_2)_2 + SbF_3$.[980] b. 62.3°,[980] ^{31}P -145.3 ppm, J_{PF} 1261 Hz, $J_{PP'}$ 446 Hz, $J_{PF'}$ 52 Hz,[981] ^{19}F NMR,[981] ^1H NMR, IR, Raman.[980]

$C_6H_5N(PF_2)_2$. By $RN(Cl_2)_2 + SbF_5$.[980] ^{31}P -132.7 ppm, J_{PF} 1252 Hz, $J_{PP'}$ 371 Hz, $J_{PF'}$ 40 Hz.[981]

$3\text{-}ClC_6H_4N(PF_2)_2$. By $RN(Cl_2)_2 + SbF_5$.[980] ^{31}P -131.3 ppm, J_{PF} 1285 Hz, $J_{PP'}$ 372 Hz, $J_{PF'}$ 41.5 Hz.[981]

TYPE: R_2NPCl_2

Me_2NPCl_2. IVa.[229] and by $[(Me_2N)_2P]_2 + PCl_3$.[983] b. 145-8°,[983] b. 150°,[229] ^{31}P -166 ppm,[914a,1343] ^1H NMR.[1350a]

Et_2NPCl_2. IVa.[857] b. 189°,[857] b_{14} 72-5°,[857] b_{10} 69-70°,[1416] b_{17} 78°,[583] b_{13} 73-4°,[583] d_0^{15} 1.096,[1343] n_D^{20} 1.4679,[1416] ^{31}P -162, -163 ppm, J_{PNCH} 12 Hz.[914a,1343]

Pr_2NPCl_2. IVa. b. 220-3°, b_{11} 95°.[857]

$i\text{-}Pr_2NPCl_2$. IVa.[1344]

$i\text{-}Bu_2NPCl_2$. IVa. m. 37-8°, b_{10} 116-7°.[857]

$i\text{-}Am_2NPCl_2$. IVa. b_8 140°.[857]

$(CH_2CH_2)_2NPCl_2$. IVb. b_{13} 92-3°, $b_{19.5}$ 98.5-102°, ^{31}P -164.5 ppm.[165]

$C_5H_{10}NPCl_2$. IVa. b_{10} 94-5°.[857]

$PhMeNPCl_2$. IVa. b. 251°, b_{10} 138-40°,[857] ^{31}P -158.3 ppm.[826]

$PhEtNPCl_2$. IVa. b_{12} 143°.[857]

TYPE: $(R_2N)_2PCl$

$(Me_2N)_2PCl$. IVa.[984] IVd.[983,1343] b_1 33°,[569] b_1 29°,[983] b_{10} 64°,[984] b_{726} 184°,[984] m. -33°,[984] d_4^{20} 1.070,[1343] d_4^{20} 1.060,[984] n_D^{20} 1.495,[1343] n_D^{20} 1.5005,[569] IR,[569] ^1H NMR,[569,1350a] ^{31}P -158.7 to -160 ppm, J_{PNCH} 12 Hz.[914a,1250a,1343] Flammable on standing in air. Reacts explosively with water.[984]
$(Et_2N)_2PCl$. IVd.[1343] $b_{0.02}$ 60°,[369] $b_{0.6}$ 63-4°,[1414] n_D^{20} 1.4901,[1414] n_D^{20} 1.4900,[369] ^{31}P -154 ppm.[914a,1343] Explodes violently with water.[1414]

TYPE: 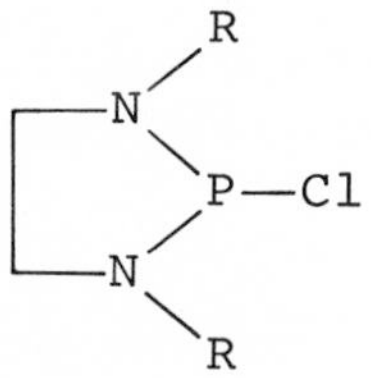

$\overline{MeNCH_2CH_2N(Me)P}Cl$. IVa. $b_{0.2}$ 70°,[1183] $b_{0.1}$ 39-40°,[1340] n_D^{20} 1.5300,[1340] ^{31}P -167.3, -172.8 ppm.[1182,1190] ^1H NMR.[1183] See footnote* page 228.

Hydrazine Derivative

$ClP(NMe\text{-}NMe)_2PCl$, cyclic. PCl_3 + $P(NMe\text{-}NMe)_3P$, quantitative yield.[1036]

TYPE: $RNHPCl_2$

$EtNHPCl_2$. IVb. b_{11} 92°, b. 222-5° (decomp.).[857]
$PrNHPCl_2$. IVb. b_{10} 97°, d_0^{15} 1.226.[857]
$CH_2{:}CHCH_2NHPCl_2$. IVb. Undistillable, d_4^{20} 1.391, n_D 1.5522.[1049]
$i\text{-}BuNHPCl_2$. IVb. b_{10} 101°, d_0^{15} 1.213.[857]
$AmNHPCl_2$. IVb. b_8 101°.[857]

TYPE: $(RN{:}PCl)_n$

$(EtNPCl)_3$. Ring trimer. $BuEtN(SiMe_3)_2$ + PCl_3. $b_{0.2}$ 128-30°, d_4^{20} 1.371, n_D^{20} 1.5732, MW 329 (ebullioscopic in benzene), IR. Stable under N_2; rapidly hydrolyzed and oxidized by water and oxygen.[2]
PhN:PCl, probably a dimer. IVd. By RNH_3Cl + PCl_3. White crystals, m. 136-7° (from benzene).[863]

TYPE: $RN(PCl_2)_2$

$MeN(PCl_2)_2$. IV. By RNH_3Cl + PCl. $b_{0.5}$ 47-52°, ^1H NMR, IR, Raman.[980]
$EtN(PCl_2)_2$. IV. $b_{0.4}$ 62°, ^1H NMR, IR, Raman.[980]

$PhN(PCl_2)_2$. IVd (PCl_3 in excess). m. 46° (from PCl_3).[492]

TYPE: R_2NPBr_2

Me_2NPBr_2. IVa.[984] By $[(Me_2N)_2P]_3$ + 2 Br_2.[983] $b_{0.7}$ 26°,[983] $b_{0.9}$ 29°,[984] b_{712} 214°,[984] m. -29°,[984] d^{20} 1.964,[984] n_D^{20} 1.6012.[983]

TYPE: $(R_2N)_2PBr$

$(Me_2N)_2PBr$. IVa.[984] IVi.[984] $Ph_2NP(NMe_2)_2$ + Br_2.[983,1352] b_2 45°, $b_{0.9}$ 29°,[983,1352] m. -13°,[984] d^{20} 1.312,[984] n_D^{22} 1.5522, n_D^{20} 1.5513.[983,1352]

TYPE: $R_2NP(CN)_2$

$Me_2NP(CN)_2$. $MeNPCl_2$ + AgCN. b_2 59°, m. 11°, d^{21} 1.077, n_D^{20} 1.4780, ^1H NMR.[1350a]

TYPE: $(R_2N)_2PCN$

$(Me_2N)_2PCN$. ^1H NMR.[1350a]

TYPE: 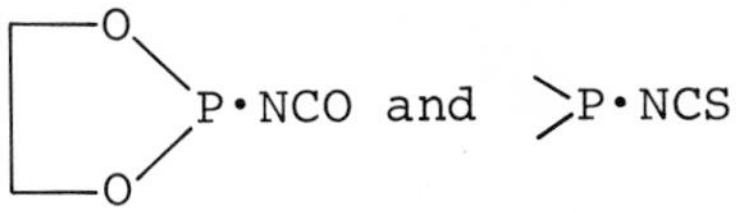

$\overline{OCH_2CH_2OP}NCO$. $(RO)_2PCl$ + sodium cyanate. b_{15} 55-7°, IR.[979]

$\overline{OCHMeCH_2OP}NCO$. As above. b_{50} 84-5°, IR.[979]

$\overline{OCHMeCHMeOP}NCO$. $(RO)_2PCl$ + Na cyanate. b_{15} 93-4°.[979]

$\overline{OCH_2CH_2OP}NCS$. By $(RO)_2PCl$ + NH_4 thiocyanate. b_1 58°, d_4^{20} 1.3830, n_D^{20} 1.5810, IR, ^1H NMR.[384]

TYPE: 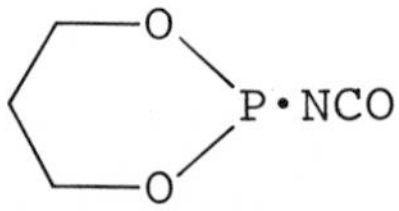

$\overline{OCH_2CH_2CH_2OP}NCO$. $(RO)_2PCl$ + K cyanate.[979]

$OCNP[(OCH_2)_2C(CH_2O)_2]PNCO$. By ClP···PCl + NaCNO in liquid SO_2 at -10°. Used in situ to prepare the OCNP(O)···P(O)NCO, by SO_3/SO_2, OCNP(S)···P(S)NCO, by S in xylene at 140°.[1072] The following compound was prepared from it.

$OCNP(S)[(CH_2)_2C(CH_2O)_2]PNCO$. From the diisocyanate and S in xylene at 140° (48 hr) m. 135-8°. Adds S on further heating with S.[1072]

$\overline{OCH_2(CH_2)_2CH_2OP}NCO$. $(RO)_2PCl$ + Na cyanate. b_{15} 93-4°, IR.[979]

TYPE: $(R_2N)(R'O)PCl$

$Et_2NPCl(OEt)$. IVg.[1416] b_{13} 90-2° (gives EtCl at 170-90°).[857] $b_{1.3}$ 44-5°, n_D^{20} 1.4680.[1416]
$Et_2NPCl(Oi\text{-}Pr)$. IVg. $b_{0.8}$ 39-42°, n_D^{20} 1.4626.[1416]
$Et_2NPCl(OBu)$. IVg. $b_{1.2}$ 62-3°, n_D^{20} 1.4655.[1416]
$Bu_2NPCl(OEt)$. IVg. $b_{0.05}$ 64-5°, n_D^{20} 1.4672.[1416]
$Bu_2NPCl(OBu)$. IVg. $b_{0.08}$ 73-6°, n_D^{20} 1.4675.[1416]
$[O(CH_2CH_2)_2N\text{-}]PCl(OEt)$. IVg. $b_{0.03}$ 51-4°, n_D^{20} 1.4990.[1416]
$C_5H_{10}NPCl(OEt)$. IVf. b_{25} 125°. Gives EtCl at 170-90°.[857]

TYPE: $(R_2N)(R'S)PCl$

$Et_2NPCl(SCH_2CH_2Cl)$. By Et_2NPCl_2 + $\overline{SCH_2CH_2}$, and $RSPCl_2$ + Et_2NH. b_1 110-2°, $b_{0.2}$ 88°, d_4^{20} 1.1984, n_D^{20} 1.5420.[993]

$Et_2NPCl(SCH_2:CHMeCl)$. By Et_2NPCl_2 + $\overline{SCH_2CHMe}$ and $RSPCl_2$ + Et_2NH. b_1 97-101°, d_4^{20} 1.1646, n_D^{20} 1.5291.[993]

TYPE:

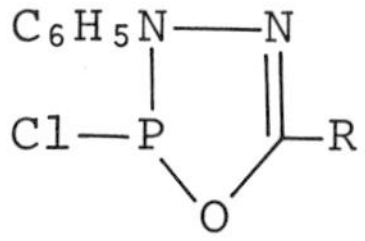

$C_6H_5\overline{NN:C(Me)OP}Cl$. $C_6H_5NHNHCOMe$ + PCl_3 + Et_3N. $b_{0.2}$ 97°.[597]

$C_6H_5\overline{NN:C(Et)OP}Cl$. $C_6H_5NHNHCOEt$ + PCl_3+ Et_3N. $b_{0.3}$ 112-4.5°.[597]

$C_6H_5\overline{NN:C(Pr)OP}Cl$. $C_6H_5NHNHCOPr$ + PCl_3 + Et_3N. $b_{0.2}$ 120-3°.[597]

$C_6H_5\overline{NN:C(i\text{-}Pr)OP}Cl$. $C_6H_5NHNHCO\text{-}i\text{-}Pr$ + PCl_3 + Et_3N. $b_{0.15}$ 106-7.5°.[597]

$C_6H_5\overline{NN:C(C_6H_5)OP}Cl$. $C_6H_5NHNHCOC_6H_5$ + PCl_3 + Et_3N. $b_{0.3}$ 204-10°.[597]

I.6.2. Trisamido Phosphites and Mixed Esters

TYPE: $(R_2N)_3P$

$(Me_2N)_3P$. IVab. b_{760} 163.5°,[229] $b_{736.4}$ 161°,[563] b_{760} 162-4°,[824] b_2 62-3°,[1414] b_{15} 51-2°,[229] b_{10} 48.2°,[985] b_{12} 49-51°,[824] b_{700} 161°,[985] m. -44°,[985] d^0 0.911,[229]

n_D^{20} 1.4656,[1414] n_D^{20} 1.4660,[985] d_4^{20} 0.895,[985] v.p. at 20° 2.2 mm,[563] n_D^{25} 1.4636,[1105] ^{31}P -121.5 to -123 ppm, J_{PNCH} 9 Hz, 8.9 Hz.[826] ^{31}P -122.4 ppm,[569] ^1H NMR.[1350a]

$(Et_2N)_3P$. IVab. b. 245-6°,[857] b_{10} 120-2°,[1302] b_{10} 80-90°,[857] $b_{0.6}$ 72-4°,[1414] n_D^{20} 1.4735,[1414] n_D^{20} 1.4578,[1302] ^{31}P -117.5 to -119 ppm.[826]

$(Pr_2N)_3P$. IVab. b. 310-5°, b_{15} 160-5°,[857] $b_{0.02}$ 83-4°.[1414] Methiodide, m. 83-4°,[857] $b_{0.15}$ 101-3°,[1302] n_D^{20} 1.4718,[1414] n_D^{20} 1.4721,[1302] ^{31}P -121.5 ppm.[826]

$(i\text{-}Pr_2N)_3P$. IVab. b_{18} 190-200°. Methiodide, m. 138°.[857] ^{31}P -133.6 ppm.[826]

$(Bu_2N)_3P$. $b_{0.005}$ 104-5°,[1414] $b_{0.1}$ 140-1°,[1302] n_D^{20} 1.4698,[1414] n_D^{20} 1.4700.[1302]

$(\overline{CH_2\text{-}CH_2N})_3P$. IVab. b_{10} 82-3°, d_4^{20} 1.1028, n_D^{20} 1.5279.[990,991]

$(Me_2\overline{C\text{-}CH_2N})_3P$. IVab. $b_{0.009}$ 62-3°, d_4^{20} 0.9377, n_D^{20} 1.4798, IR.[991]

$(C_4H_8N)_3P$. IVh. $b_{0.1}$ 103-4°.[232] n_D^{19} 1.5268, d_4^{20} 1.0490.

$[O(CH_2CH_2)_2N]_3P$. IVh. m. 154°.[232]

$(C_5H_{10}N)_3P$. IVf. IVh.[232] m. 37-8°.[101,857,861] $b_{0.2}$ 120-30° (crude). Methiodide, m. 242°.[232]

$(C_9H_{10}N)_3P$. From tetrahydroquinoline. m. 202-4° (from benzene).[857]

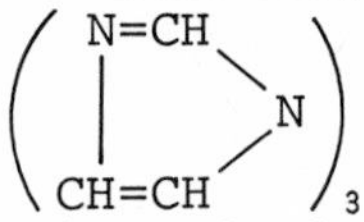
P. By N-(trimethylsilyl) imidazole + PCl_3. m. 154°.[191]

Tri-N,N',N"-(4-pyridyl)phosphorous amide. (Byproduct.) Crystals, m. 305-8°.[719]

Tri-N-indolyl phosphorous amide. Byproduct from the reaction of PCl_3 with indolemagnesium. Crystals, m. 223-5° (from MeOH). Separated from the tri-3-indolylphosphine by insolubility in acetone. Readily decomposed by aqueous alkali.[903]

Tri-N-(2-methylindolyl) phosphorus. Prepared as above from 2-methyl analog. Crystals, m. 180° (from MeOH).[903]

TYPE: $(R_2N)_2(R_2^1N)P$

$\overline{CH_2\text{-}CH_2N}P(NMe_2)_2$. IVe. b_{30} 68-9°, d_4^{20} 0.9448, n_D^{20} 1.4764.[990]

$Me_2\overline{C\text{-}CH_2N}P(NMe)_2$. IVe. b_{12} 66-7°, d_4^{20} 0.9112, n_D^{20} 1.4676, IR.[991]

$(Me_2N)_2PN(CH_2CH_2)_2O$. IVh. $b_{0.2}$ 79-82°, d_4^{20} 1.006, n_D^{20} 1.483, IR, NMR.[569]

$(Me_2N)_2PN(Me)C(S)NMe_2$. $(Me_2N)_3P$ + MeN=C=S. m. 66°.[1008,1010]

$(Me_2N)_2PN(Ph)C(S)NMe_2$. $(Me_2N)_3P$ + PhN=C=S. m. 70°, m. 68°.[1008,1010]

$\overline{CH_2-CH_2N}P(NEt_2)_2$. IVe. b_7 85-6°, d_4^{20} 0.9406, n_D^{20} 1.4823.[990]

$Me_2\overline{C-CH_2N}P(NEt)_2$. IVe. $b_{0.5}$ 75-6°, d_4^{20} 0.9035, n_D^{20} 1.4702, IR.[991]

$(Et_2N)_2PN(Ph)C(S)NEt_2$. $(Et_2N)_3P$ + PhN=C=S. $b_{0.25}$ 70°.[1008]

$(\overline{CH_2-CH_2N})_2PNMe_2$. IVe. b_{70} 105-6°, d_4^{20} 0.9775, n_D^{20} 1.4914.[990]

$(\overline{CH_2-CH_2N})_2PNEt_2$. IVe. b_{28} 113-4°, d_4^{20} 0.9903, n_D^{20} 1.4946.[990]

$(Me_2\overline{C-CH_2N})_2PNMe_2$. IVe. b_{11} 84-5°, d_4^{20} 0.9284, n_D^{20} 1.4765, IR.[991]

$(Me_2\overline{C-CH_2N})_2PNEt_2$. IVe. $b_{0.5}$ 70-1°, d_4^{20} 0.9237, n_D^{20} 1.4762, IR.[991]

$i\text{-}Bu_2N(C_5H_{10}N)_2P$. IV. Undistilled liquid.[857]

TYPE: $(R_2N)(R'NH)_2P$

$Me_2NP(NHPh)_2$. IVh, from $(Me_2N)_3P$. Viscous oil. Loses Me_2NH at 100° in vacuo, and gives following compound in hot benzene:

$(PhN{:}PNHPh)_2$. m. 248-50° (petroleum).[1351]

TYPE: $(R_2N)_2(R'NH)P$

$[(CH_2)_2N]_2PNHCH_2CH{:}CH_2$. IVe. Undistillable, d_4^{20} 1.1434, n_D^{20} 1.5368.[1049]

TYPE: $R_2NP(NH_2)_2$

$Et_2NP(NH_2)_2$. IVk. m. 107-8° (decomp.). White amorphous hygroscopic solid, d^{25} 1.93.[171] HCl gave a ppt. washed with liquid ammonia to give $[HN{:}PNEt_2)_x$.[171]

$MePhNP(NH_2)_2$. IVk. m. 140-50° (decomp.). Yellowish amorphous powder, d^{25} 1.96.[171]

TYPE: $(RNH)_3P$

$(PhCH_2NH)_3P$. IVb. Undistilled oil.[857]

$(CH_2{:}CHCH_2NH)_3P$. IVb. $b_{0.0001}$ 90-100°, d_4^{20} 1.105, n_D^{20} 1.5560.[1049]

$(i\text{-}BuNH)_3P$. IVb. Undistilled oil.[857]

$(C_{12}H_{25}NH)_3P$. IVb. Colorless solid.[397]

TYPE: $(RN{:}PNHR)_2$

$[P(NEt)(NHEt)]_2$. IVc.[562]
$[P(Nt\text{-}Bu)(NHt\text{-}Bu)]_2$. IVa. Sublimes, m. 138-42°. IR, NMR, mass spectrum, soluble in benzene, toluene, pentane, acetone, chloroform, carbon tetrachloride, acetic acid, insoluble in water.[562]
$(PhN{:}PNC_5H_{10})_2$. $(PhNPCl)_2$ + piperidine. Colorless needles, m. 202-3°.[863]
$(PhN{:}PNHPh)_2$. $(PhNPCl)_2$ + aniline. m. 251-3°.[863] Molecular weight in dioxane shows dimer.[516] Given as PhNHP(NHPh)NPhP:NPh. Hydrate, m. 152-3°.[863] HCl in xylene at 70° gave PCl_3 + $PhNH_2{\cdot}HCl$.[863] Heated in alcohol + pyridine gives $(EtO)_3P$. Heated in vacuum at 250° gives $PhNH_2$, and probably $(PhN)_3P_2$. Water in dioxane at 100° gives $(PhNH)_2POH$, m. 156°.[492]
2-$MeOC_6H_4N{:}PNHC_6H_4OMe$-2. IVe. m. 138-40°. Molecular weight in $CHBr_3$ shows dimer.[516]
4-$MeC_6H_4N{:}PNHC_6H_4Me$-4. IVe. m. 197-200°.[516]

TYPE: $(RN{:}P)_2NR$

$(4\text{-}MeOC_6H_4N{:}P)_2NC_6H_4OMe$-4. IVd. Gave uncharacterized 4-$MeOC_6H_4N(PCl_2)_2$, which was heated in vacuum, and then treated with 4-$MeOC_6H_4NH_2$. Octahedrons, m. 315° (from benzene or dioxane). μ 2.05 D.[492]

Hydrazine Derivatives

$P(NMe\text{-}NMe)_3P$, bicyclic hydrazodiphosphine. Heat $P(NMe_2)_3$ and MeNHNHMe,2HCl in benzene 64 hr. Sublimed at 70-80°/1 mm., m. 116-7°. H_2O_2 in alcohol gives O=P(NMe-NMe)P=O, m. 320-5°. Gives 1:1 adduct with MeI under forcing conditions. HCl in benzene gives MeNHNHMe,2HCl quantitatively. IR, NMR. Sulfur in xylene gives disulfide, m. > 360°.[1036]

TYPE: 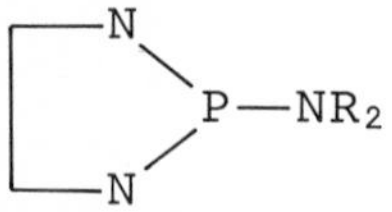

$\overline{MeNCH_2CH_2N(Me)P}NMe_2$.* IVe. $b_{0.5}$ 36°,[1183] n_D^{20} 1.4869,[1340] ^1H NMR.[1183] ^{31}P -114.2, -116.5 ppm,[826] IR.[1183]

$\overline{MeNCH_2CH_2N(Me)P}N(CH_2CH_2)_2$.* IVe. $b_{0.2}$ 55°, n_D^{20} 1.5065, ^1H NMR, IR,[1183] ^{31}P -104.8 ppm.[1182] When the crude

*Warning: The foregoing cyclic aminophosphines should be handled with great care. Exposure to the vapors may result in acute nausea and vomiting.

chlorophosphine was used, the yield was 55%; but the pure reagent gave a yield of 85%.

Polycyclic Compounds

$P_2N_3Me_3$ or $P_4N_6Me_6$, cage structure. Addition of PCl_3 (1 mole) to $MeNH_2$ (8 mole, i.e., excess) at low temperature during 4 days; base hydrochloride separated by petroleum (dry N_2 atmosphere). m. 122.0-2.8°, b_{739} 303-4°; molecular weight in benzene (cryoscopic) corresponds to $P_4N_6Me_6$, conductance in benzene negligible, heat of fusion, 0.2 kcal/mole. IR, ^1H NMR, x-ray, mass spect., ^{31}P -78.4 ppm, J_{PH} 15.7 ± 0.6 Hz. ^{31}P showed equivalence of P atoms. Soluble in benzene, toluene, n-hexane, ethyl ether, acetone, carbon tetrachloride, and ethanol (with reaction). Slowly dissolved by water. MeI gives $P_4N_6Me_7I$, soluble in water. BCl_3 gives empirically $P_2N_3Me_3B_2Cl_8$, oxygen gives ($P_2N_3Me_3O_2$), sulfur gives $P_4S_4N_6Me_6$, m. 240° (decomp.), HCl gives PCl_3 and $MeNH_2HCl$ at -78°.[557, 560,561]

TYPE: $(R_2N)_2(R'O)P$

$(Me_2N)_2POEt$. ^{31}P -135.7 ppm.[569]

$(Me_2N)_2POCH_2CH_2Cl$. IVf. b_8 76-9°, d^{20} 1.0649, n_D^{20} 1.4750.[641]

$(Me_2N)_2POCH_2\overline{CHCH_2OCHMe\cdot O}$. IVg. [From $(Me_2N)_2PCl$]. b_2 78-9°, d^{20} 1.0435, n_D^{20} 1.4650, MR_D 62.52 (calc. 62.30).[138]

$(Me_2N)_2PO\overline{CH_2CH_2\cdot O\cdot CHMe\cdot O\cdot CH}_2$. IVg. [From $(Me_2N)_2PCl$.] b_2 85°, d^{20} 1.0522, n_D^{20} 1.4708, MR_D 62.66 (calc. 62.30).[138]

$(Me_2N)_2P\cdot OCMe{=}CHCO_2Me$. IVg, from $AcCH_2CO_2Me$ (enol). $b_{0.5}$ 85-8°, d^{20} 1.0456, n^{20} 1.4853.[647]

$(Me_2N)_2POCMe_3$. IVg. b_{10} 54°, n_D^{19} 1.444, d_{20}^{20} 0.8968, ^1H NMR, IR.[236]

$(Me_2N)_2POC_6H_{11}$. IVh. b_2 80.5-1°, d_4^{20} 0.958, n_D^{20} 1.472.[569]

$(Et_2N)_2POMe$. By elimn. of EtI from $(Et_2N)_2POEt\cdot MeI$. b_{22} 145-8°.[857]

$(Et_2N)_2POCH_2NEt_2$. IVh. $b_{0.1}$ 127-8°, d^{20} 0.93955, n_D^{20} 1.4721.[429]

$(Et_2N)_2POEt$. IVf. b_{28} 105-8°,[857] ^{31}P -134 ppm.[826]

$(Et_2N)_2POCH_2CH_2Cl$. IVf. b_2 83-5°, d^{20} 1.0229, n_D^{20} 1.4727.[641]

$(Et_2N)_2(MeOCH_2CH_2O)P$. IVh. $b_{0.005}$ 70-3°, d^{20} 0.97355, n_D^{20} 1.4510.[429]

$(Et_2N)_2(EtOCH_2CH_2O)P$. IVh. $b_{0.001}$ 70-2°, d^{20} 0.96410, n_D^{20} 1.4521.[429]

$(Et_2N)_2(EtSCH_2CH_2O)P$. IVh. $b_{0.005}$ 70-2°, d^{20} 0.97683, n_D^{20} 1.4841.[429]

$(Et_2N)_2(Et_2NCH_2CH_2O)P$. IVh. $b_{0.001}$ 72-5°, d^{20} 0.93320, n_D^{20} 1.4611.[429]

$(Et_2N)_2POi\text{-}Pr$. IVh. $b_{0.5}$ 45-6°, d_4^{20} 0.886, n_D^{20} 1.449,[569] ^{31}P -133 ppm.[569]

$(Et_2N)_2POBu$. IVh. $b_{0.5}$ 68-70°, d_4^{20} 0.894, n_D^{20} 1.456,[569] IR, ^{31}P -132.1 ppm.[569]

$(Et_2N)_2(C_6H_5O)P$. IVf. $b_{0.015}$ 84-5°, n_D^{20} 1.5100, d^{20} 0.9886.[226] ^{31}P -131 ppm.[401,628,826]

$(Et_2N)_2(2\text{-}MeC_6H_4O)P$. IVf. $b_{0.03}$ 94-5°, n_D^{20} 1.5117, d^{20} 0.9833.[226]

$(Et_2N)_2(3\text{-}MeC_6H_4O)P$. IVf. $b_{0.027}$ 89-90°, n_D^{20} 1.5095, d^{20} 0.9778.[226]

$(Et_2N)_2(4\text{-}MeC_6H_4O)P$. IVf. $b_{0.02}$ 89-90°, n_D^{20} 1.5090, d^{20} 0.9783.[226]

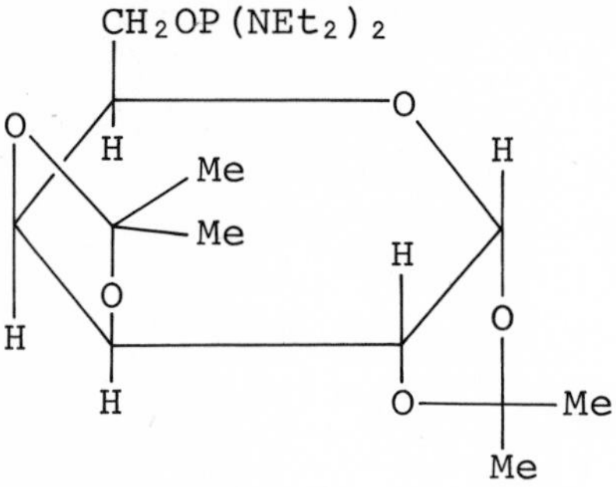

IVh. $b_{0.02}$ 130-6° (bath temp.), n_D^{20} 1.4755, $[\alpha]_D^{20}$ -56.4°.[1060]

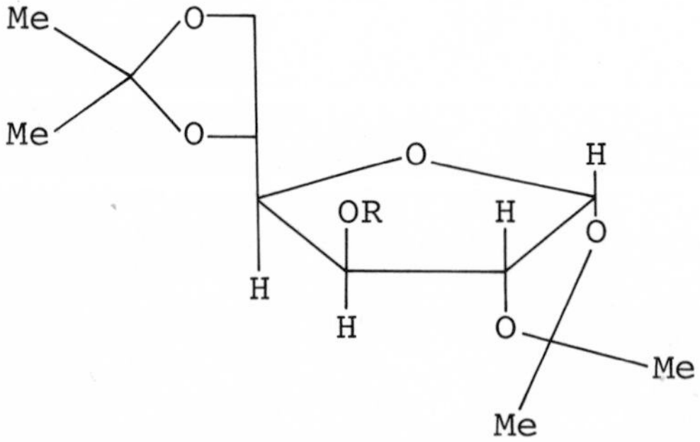

IVh. $b_{0.02}$ 130-6°, n_D^{20} 1.4741, $[\alpha]_D^{20}$ -63°.[1060]

R = $P(NEt_2)_2$

$(Pr_2N)_2POEt$. IVf. b_{29} 143-7°.[857]

$(i\text{-}Bu_2N)_2POPh$. IVf. Undistilled oil. At 100° MeI gives $MeP(N\text{-}i\text{-}Bu_2)_2I_2$, m. 132°.[857]

$(C_5H_{10}N)_2POEt$. IVf. b_{27} 152-4°.[857]

TYPE: 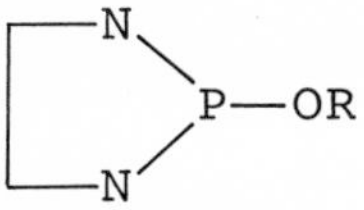

$MeNCH_2CH_2N(Me)POMe$. IVg. b_{20} 56°, n_D^{20} 1.4721, ^1H NMR, IR. Exposure to the vapors may cause acute nausea and vomiting.[1183] ^{31}P -123.2 ppm.[1182]

TYPE: $(R_2N)_2(R_2'NCO_2)P$

$(Me_2N)_2POC(O)NMe_2$. $(Me_2N)_3P + CO_2$. $b_{0.3}$ 75°, n_D^{20} 1.4755.[1008]
$(Et_2N)_2POC(O)NMe_2$. By anhydride + $(R_2N)_3P$. $b_{0.25}$ 122°, n_D^{20} 1.4706.[1009]
$(Et_2N)_2POC(O)NEt_2$. $(Et_2N)_3P + CO_2$. $b_{0.2}$ 120-1°, n_D^{20} 1.4706.[1008]

TYPE: $(R_2N)(R_2'N)(R''O)P$

$(Me_2N)(PhO)PN(Me)C(S)NMe_2$. By MeN:C:S.[1010]
$(Me_2N)(PhO)PN(Ph)C(S)NMe_2$. By $(Me_2N)_2POPh$ + PhN:C:S.[1010]
$(Me_2N)(4\text{-}t\text{-}BuC_6H_4O)PN(Me)C(S)NMe_2$. $ROP(NMe_2)_2$ + MeN=C=S. Undistillable oil.[1008]
$(Et_2N)[O(CH_2CH_2)_2N]POiPr$. IVh. b_1 90°, d_4^{20} 0.983, n_D^{20} 1.470.[569]

$Me_2CCH_2NP(NEt_2)(OPh)$. IVh. $b_{0.02}$ 74-5°, d_4^{20} 1.0089, n_D^{20} 1.5139, IR.[991]

TYPE: 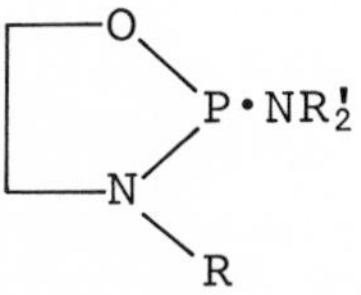

$OCH_2CH_2N(Me)PNMe_2$. IVh. b_{16} 66-8°.[1238]

$OCH_2CH_2N(CH_2C_6H_5)PNMe_2$. IVh. $b_{0.1}$ 95°.[1238]

$OCH_2CH_2\overset{+}{N}Me_2{\cdot}PNMe_2\ Cl^-$. Byproduct from preparation of $(Me_2N)_2POCH_2CHCl$ (q.v.) by IVf. m. 110°.[641]

$OCH_2CH_2N(Et)PNMe_2$. IVh. b_{20} 78-9°.[1238]

$OCH_2CH_2N(C_6H_{11})PNMe_2$. IVh. $b_{0.08}$ 61°.[1238]

$OCH_2CH_2N(2,5\text{-}Me_2C_6H_3)PNMe_2$. IVh. $b_{0.1}$ 85°.[1238]

TYPE:

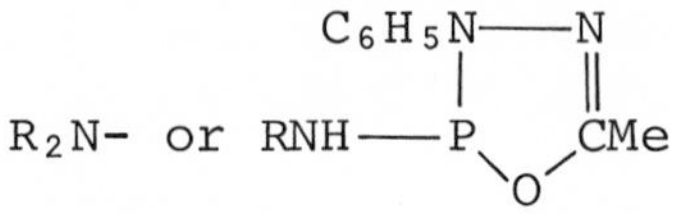

$C_6H_5NN{:}C(Me)OPNEt_2$. $C_6H_5NN{:}C(Me)OPCl + Et_2NH$. $b_{0.1}$ 95-101°.[597]

$C_6H_5NN{:}C(Me)OPN$-i-Pr. $C_6H_5NN{:}C(Me)OPCl$ + i-Pr_2NH. m. 52-5°.[597]

$C_6H_5NN{:}C(Me)OPNC_5H_{10}$. $C_6H_5NN{:}C(Me)OPCl + C_5H_{10}NH$. $b_{0.25}$ 128-9°.[597]

$C_6H_5NN{:}C(Me)OPNHC_8H_{17}$. $C_6H_5NN{:}C(Me)OPCl + C_8H_{17}NH$. m. 115-20°.[597]

$C_6H_5NN{:}C(Me)OPNHC_6H_5$. $C_6H_5NN{:}C(Me)OPCl + C_6H_5NH$. m. 101-3°.[597]

$C_6H_5NN{:}C(Me)OPNHNHC_6H_5$. $C_6H_5NN{:}C(Me)OPCl + C_6H_5NHNH_2$. m. 113-6°.[597]

TYPE: $(H_2N)_2POR$

$MeOP(NH_2)_2$. IVj. m. 130-2° (decomp.). Yellowish hygroscopic solid, d^{25} 1.84.[171]
$EtOP(NH_2)_2$. IVj. m. 140-2° (decomp.). Yellowish amorphous powder, d^{25} 1.975.[171] HCl gave ppt washed with liquid ammonia to give $H_2N[P(OEt)NH]_3P(OEt)NH_2$.
$PhOP(NH_2)_2$. IVj. m. 110-2° (decomp.). Yellowish amorphous solid, d^{25} 2.145.[171]

TYPE: $(R_2N)(R'O)_2P$

$Me_2NP(OEt)_2$. ^{31}P -144.7 ppm.[569]

$Me_2NP(OCH_2CH{\cdot}CH_2{\cdot}O{\cdot}CHMe{\cdot}O)_2$. IVg. [From Me_2NPCl_2]. b_1 132°, d^{20} 1.1340, n_D^{20} 1.4644, MR_D 75.24 (calc. 75.04).[138]

$Me_2NP(OCH{\cdot}CH_2{\cdot}O{\cdot}CHMe{\cdot}O{\cdot}CH_2)_2$. IVg. [From Me_2NPCl_2.] b_1 145°, m. 52-3°.[138]
$Me_2NP(OCMe_3)_2$. IVg. $b_{0.5}$ 76°, $b_{0.1}$ 55-6°, n_D^{19} 1.463, d_{18}^{18} 1.0075, NMR, IR.[236]
$Me_2NP(OC_6H_{11})_2$. IVh. b_2 125-6°, d_4^{20} 1.000, n_D^{20} 1.482.[569]
$(Me_2N)(C_6H_5O)_2P$. IV. $b_{0.15}$ 127-9°, n_D^{20} 1.5670, d^{20} 1.1146.[720]
$Et_2NP(OBu)_2$. ^{31}P -144.0 ppm.[569]

$CH_2CH_2NP(OEt)_2$. IVf. b_{10} 57.5-8.5°, d_4^{20} 1.0070, n_D^{20} 1.4458.[501]
$(Et_2N)(C_6H_5O)_2P$. By $(Me_2C{\cdot}CH_2N)_2PNEt_2$ + 2 PhOH.[991] $b_{0.02}$ 107-8°,[991] $b_{0.45}$ 138-40°, n_D^{20} 1.5568,[720] n_D^{20} 1.5516,[398] n_D^{20} 1.5515,[991] d^{20} 1.0828,[720] IR,[991] ^{31}P -141 ppm.[401,628,826]

$(Pr_2N)(C_6H_5O)_2P$. $b_{0.11}$ 155-6°, n_D^{20} 1.5491, d^{20} 1.0570.[720]
i-$Pr_2NP(OBu)_2$. IVg.[1344]
(i-$Pr_2N)(C_6H_5O)_2P$. IV. $b_{0.03}$ 122-3°, n_D^{20} 1.5406, d^{20} 1.0414.[720]
$(Bu_2N)(BuO)_2P$. IVg. IVh. b_1 128°, b_3 130-2°, d^{20} 0.9575, n_D^{20} 1.4395, n_D^{20} 1.4398.[1048]
$(CH_2CH_2)_2NP(OEt)_2$. IVf. b_1 41-2.5°, d^{20} 0.9941, n_D^{20} 1.4580.[504]
$(CH_2CH_2)_2NP(OPr)_2$. IVf. $b_{0.5}$ 44.5°, d^{20} 0.9648, n_D^{20} 1.4503.[504]
$(CH_2CH_2)_2NP(O$-iPr$)_2$. IVf. b_{11} 85-7°, d^{20} 0.9464, n_D^{20} 1.4483.[504]
$(CH_2CH_2)_2NP(OBu)_2$. IVf. $b_{0.5}$ 63-4°, d^{20} 0.9495, n_D^{20} 1.4560.[504]
$(CH_2CH_2)_2NP(OAm)_2$. IVf. $b_{0.6}$ 88-9°, d^{20} 0.9385, n_D^{20} 1.4609.[504]
$C_5H_{10}NP[OCHPhP(O)(OC_6H_3O)_2]_2$. $C_5H_{10}NPCl_2$ + $(C_6H_{13}O)_3P$ + PhCHO.[192]
$C_5H_{10}NP[OCHEtP(O)(OCH_2CH_2Cl)_2]_2$. $C_5H_{10}NPCl_2$ + $(ClCH_2CH_2O)_3P$ + EtCHO.[192]

TYPE: (RR'N)(R"O)$_2$P

$[Me(EtO_2C)N][EtO]_2P$. $(EtO)_2PCl$ + urethane. b_9 108-9°, n_D^{20} 1.4590, d^{20} 1.059.[65]
$[Et(EtO_2C)N][EtO]_2P$. $(EtO)_2PCl$ + urethane. b_{10} 115-6°, n_D^{20} 1.4350, d^{20} 1.0403.[65]
$[Me(EtO_2C)N][i\text{-}PrO]_2P$. (i-PrO)$_2$PCl + urethane. b_9 112-4°, n_D^{20} 1.4480, d^{20} 1.0091.[65]
$PhMeNP[OCHEtP(O)(OEt)_2]_2$. $PhMeNPCl_2$ + $(EtO)_3P$ + EtCHO.[192]
$PhEtNP(OBu)_2$. IVh. b 129-30°, b_3 121°, d^{20} 0.9725, n_D^{20} 1.4987, n_D^{20} 1.5000.[1048]
$(EtO)_2P(O)N(Me)P(OEt)_2$. $(RO)_2PON(Na)Me$ + $(RO)_2PCl$. b_1 104-6°, n_D^{20} 1.4435, d^{20} 1.1045.[106]
$(PrO)_2P(O)N(Me)P(OPr)_2$. $(RO)_2PON(Na)Me$ + $(RO)_2PCl$. b_1 126-7°, n_D^{20} 1.4405, d^{20} 1.0213.[106]
(i-PrO)$_2$P(O)N(Me)P(O-i-Pr)$_2$. $(RO)_2PON(Na)Me$ + $(RO)_2PCl$. b_1 107-9°, n_D^{20} 1.4330, d^{20} 1.0179.[106]

TYPE: $(R_2N)(R'_2NCO_2)_2P$

$Me_2NP[OC(O)NMe_2]_2$. $(Me_2N)_3P$ + CO_2. $b_{0.1}$ 128°, n_D^{20} 1.4795.[1008]
$Et_2NP[OC(O)NEt_2]_2$. $(Et_2N)_3P$ + CO_2. $b_{0.2}$ 158°, n_D^{20} 1.4700.[1008]

TYPE: (RNH)(R'O)$_2$P

$MeCH(C_6H_5)NHP(OBu)_2$. IVh. b_1 130-1°, d^{20} 0.9516, n_D^{20} 1.4950.[1048]
cyclo-$C_6H_{11}NHP(OEt)_2$. RNH_2 + anhydride. $b_{0.01}$ 50-1°,

$b_{0.01}$ 54.5, n_D^{20} 1.4652, n_D^{20} 1.4659.[867]
$C_6H_5NHP(OEt)_2$. IVf. [From $(EtO)_2PCl$.] b_{17} 144-8°, b_{15} 142-4°.[155]
$C_6H_5NHP(OBu)_2$. IVh. b_3 133°, n_D^{20} 1.5100.[1048]
$4\text{-}NO_2C_6H_4NHP(OEt)_2$. From $(EtO)_2POP(OEt)_2$ + $4\text{-}NO_2C_6H_4NH_2$. m. 40-2°.[632]
$2\text{-}C_{10}H_7NHP(OEt)_2$. [From $(EtO)_2PCl$.] $b_{0.3}$ 126-9°.[155]
$(BuO)_2PNH(CH_2)_6NHP(OBu)_2$. IVh. $b_{0.01}$ 90°, d^{20} 0.9424, n_D^{20} 1.4636.[1048]

TYPE: $(R_2N)(R'O)(R''O)P$

$Me_2NP(OEt)\cdot OCMe{=}CHCO_2Me$. IVg, from $AcCH_2CO_2Me$ (enol). $b_{0.5}$ 76-80°, d^{20} 1.0620, n_D^{20} 1.4705.[647]
$Me_2NP(OEt)OCMe{=}CHAc$. IVg, from $AcCH_2Ac$ (enol). b_2 91-3°, d^{20} 1.0318, n_D^{20} 1.4791.[647]
$(Me_2N)(EtO)(Cl_3CMe_2CO)P$. IVf. b_{14} 129-30°, d_4^{20} 1.2205, n_D^{20} 1.4800.[14]
$(Me_2N)(BuO)(Cl_3CMe_2CO)P$. IVf. b_{14} 153-4°, d_4^{20} 1.1516, n_D^{20} 1.4750.[14]

$CH_2OP(OCH_2C_6H_5)(NEt_2)$

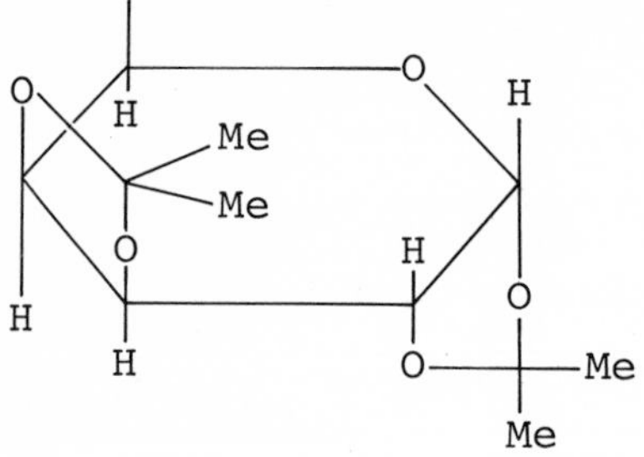

IVh. $b_{0.2}$ 170-85°, n_D^{20} 1.4997.[1060]

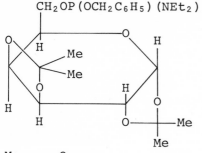

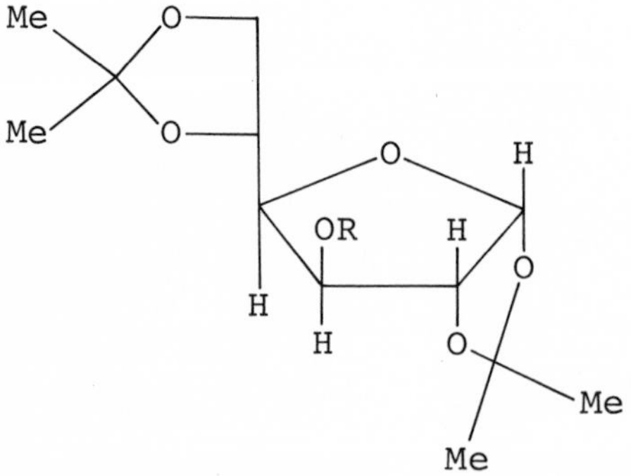

IVh. $b_{0.024}$ 175-82°, n_D^{20} 1.4985.[1060]

R = $P(OCH_2C_6H_5)(NEt_2)$

$(Et_2N)(EtO)(Cl_3CMe_2CO)P$. IVf. b_{11} 139-40°, d_4^{20} 1.1817, n_D^{20} 1.4798.[14]
$(Et_2N)(BuO)(Cl_3CMe_2CO)P$. IVf. b_{10} 158-9°, d_4^{20} 1.1333, n_D^{20} 1.4730.[14]

TYPE: $(R_2N)(R'O)(R''CO_2)P$

$Et_2N(PrO)POC(O)Pr$. $ROP(NEt_2)_2 + (R'CO)_2O$. $b_{1.5}$ 83-5°, d_4^{20} 0.9710, n_D^{20} 1.4430.[381]
$Et_2N(BuO)POC(O)Pr$. $ROP(NEt)_2 + (R'CO)_2O$. b_1 76-8°, d_4^{20} 0.9632, n_D^{20} 1.4448.[381]
$Et_2N(C_6H_{13}O)POC(O)Pr$. $ROP(NEt)_2 + (R'CO)_2O$. b_1 90-2°, d_4^{20} 0.9500, n_D^{20} 1.4468.[381]

TYPE: $(R_2N)(R'O)(R''_2NCO_2)P$

$(Me_2N)(EtO)POC(O)NMe_2$. $EtOP(NMe_2)_2 + CO_2$. $b_{0.15}$ 72°, n_D^{20} 1.4600.[1008]
$(Et_2N)(EtO)POC(O)NEt_2$. $EtOP(NEt_2)_2 + CO_2$.[1008] Also from anhydride + $(R_2N)_3P$.[1009] $b_{0.15}$ 84-5°, n_D^{20} 1.4562.[1008,1009]

TYPE: 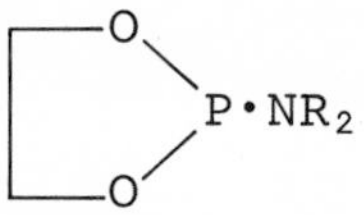

$\overline{OCH_2CH_2OP}NMe_2$. IVf. b_{12} 65-6.5°, b_{11} 61-2°, d_4^{20} 1.1278, n_D^{20} 1.4734,[11] ^{31}P -140.4 ppm.[826]

$\overline{OCH_2CH_2OP}NEt_2$. IVf. b_{10} 82-2.5°, $b_{3.5}$ 66-7°, d_4^{20} 1.10637, n_D^{20} 1.4681.[11]

$\overline{OCH_2CH_2OP}NPr_2$. IVf. b_{12} 80-1°, d_4^{20} 1.0286, n_D^{20} 1.4712.[11]

$\overline{OCH_2CH_2OP}NBu_2$. IVf. b_2 92-3°, d_4^{20} 0.9954, n_D^{20} 1.4681.[11]

$\overline{OCH_2CH_2OP}N(CH_2CH_2)_2O$. IVf.[11] b_7 116-7°,[11] $b_{0.4}$ 79-80°,[232] d_4^{20} 1.2322, n_D^{20} 1.4988.[11]

$\overline{OCH_2CH_2OP}NC_5H_{10}$. IVf. $b_{2.5-3}$ 89-90°.[101]

$\overline{OCH_2CH_2OP}N$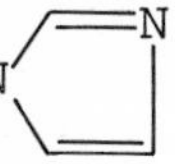. IVf. m. 35°.[44]

$\overline{OCH_2CHMeOP}NMe_2$. IVf.[11,366] $b_{5.5}$ 56°,[366] b_{10} 56-7°,[11] d_4^{20} 1.0734,[11] n_D^{20} 1.4650,[11] $n_D^{22.5}$ 1.4632.[366]

$\overline{OCH_2CHMeOP}NEt_2$. IVf. b_{22} 98-9°, d_4^{20} 1.0215, n_D^{20} 1.4611.[11]

$\overline{OCH_2CHMeOP}NPr_2$. IVf. b_8 92-3°, d_4^{20} 1.0014, n_D^{20} 1.4614.[11]

$\overline{OCH_2CHMeOP}NBu_2$. IVf. b_3 100-1.5°, d_4^{20} 0.9843, n_D^{20} 1.4631.[11]

$\overline{OCH_2CHMeOP}N(CH_2CH_2)_2O$. IVf. b_{10} 117-20°, d_4^{20} 1.1948, n_D^{20} 1.4928.[11]

$\overline{OCH_2CH(CH_2Cl)OP}NEt_2$. IVf. b_{10} 114.5-5°, d_4^{20} 1.1658, n_D^{20} 1.4881.[11]

$OCH_2CH(CH_2Cl)OPNBu_2$. IVf. b_1 121.5-2°, d_4^{20} 1.0805, n_D^{20} 1.4795.[11]

$OCH_2CH(CH_2OMe)OPNEt_2$. IVf. b_1 91-1.5°, d_4^{20} 1.0872, n_D^{20} 1.4708.[11]

$OCH_2CH(CH_2OMe)OPNBu_2$. IVf. b_1 125.5-6°, d_4^{20} 1.0152, n_D^{20} 1.4650.[11]

$OCHMeCHMeOPNMe_2$. IVf.[366] IVh.[1238] $b_{8.5}$ 65.5°, n_D^{24} 1.4619.[366]

$OCMe_2CMe_2OPNMe_2$. IVf.[366] IVh.[1238] b_2 63°,[1238] $b_{5.0-5.5}$ 80-1°, n_D^{22} 1.4655.[366]

$OCH(CO_2Et)CH(CO_2Et)OPNEt_2$. IVg. b_1 132-3°, d^{20} 1.14366, n_D^{20} 1.4643. From d-tartrate: b_1 122-3°, d^{20} 1.20683, n_D^{20} 1.4717.[503]

$OCH(CO_2Me)CH(CO_2Me)OPNEt_2$. IVg. b_1 121-2°, d^{20} 1.20546, n_D^{20} 1.4715. From d-tartrate: $b_{2.5-3}$ 142-4°, d^{20} 1.14555, n_D^{20} 1.4641, $[\alpha]_D^{22}$ -52.36°.[503]

$OCH(CO_2Pr)CH(CO_2Pr)OPNEt_2$. IVg. b_2 163°, d^{20} 1.10627, n_D^{20} 1.46351. From d-tartrate: $b_{1.5}$ 155-6°, d^{20} 1.10921, n_D^{20} 1.4632, $[\alpha]_D^{22}$ -51.81°.[503]

$OCH(CO_2\text{-}i\text{-}Pr)CH(CO_2\text{-}i\text{-}Pr)OPNEt_2$. IVg. b_2 153-4°, d^{20} 1.09212, n_D^{20} 1.4577. From d-tartrate: b_1 138-9°, d^{20} 1.09605, n_D^{20} 1.4572, $[\alpha]_D^{22}$ -49.93°.[503]

$OCH_2(CO_2Bu)CH(CO_2Bu)OPNEt_2$. IVg. b_1 162-3°, d^{20} 1.07974, n_D^{20} 1.4631. From d-tartrate: b_2 174-5°, d^{20} 1.07797, n_D^{20} 1.4630, $[\alpha]_D^{22}$ -39.06°.[503]

$OCH(CO_2\text{-}i\text{-}Bu)CH(CO_2\text{-}i\text{-}Bu)OPNEt_2$. IVg. b_1 156°, d^{20} 1.07445, n_D^{20} 1.4607. From d-tartrate: $b_{1.5}$ 161-2°, d^{20} 1.07268, n_D^{20} 1.4610, $[\alpha]_D^{22}$ -41.09°.[503]

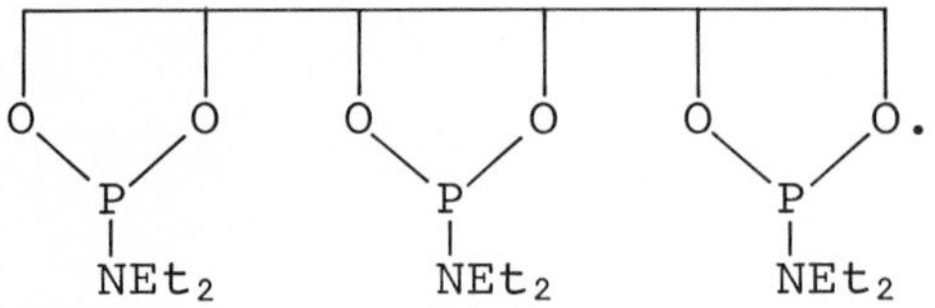

IVh. $b_{0.05}$ 188-90°, n_D^{20} 1.5011.[1361]

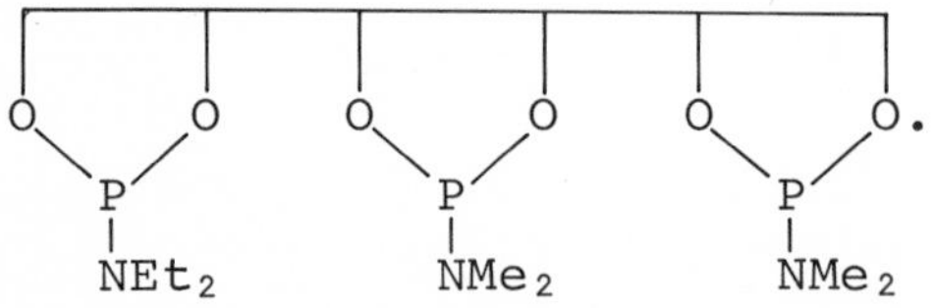

IVh. $b_{0.03}$ 122-4°, n_D^{20} 1.4694.[1361]

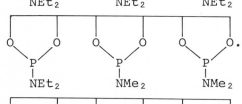

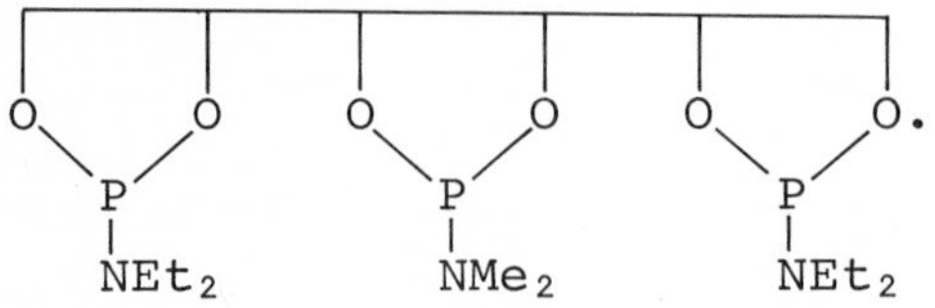

IVh. $b_{0.06}$ 158-9°, n_D^{20} 1.4849.[1361]

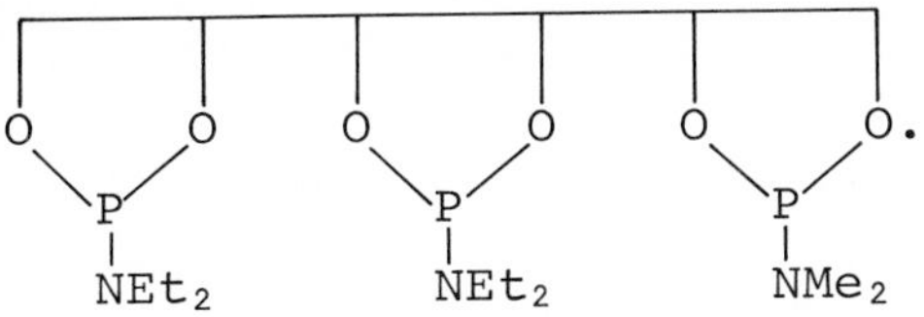

IVh. $b_{0.05}$ 160-2°, n_D^{20} 1.4856.[1361]

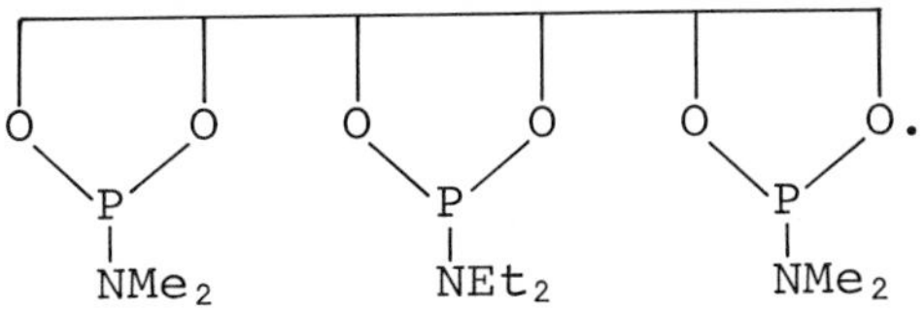

IVh. $b_{0.04}$ 124°, n_D^{20} 1.4691.[1361]

TYPE: 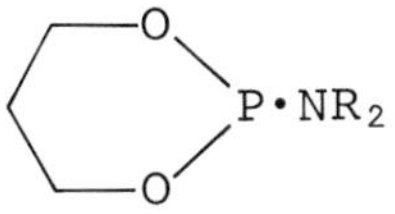

$OCHMeCH_2OPNEt_2$. IVh. $b_{0.5}$ 43-4.5°, d^{20} 1.0426, n_D^{20} 1.4652.[502]

$O(CH_2)_3OPNCHMeCH_2CH_2$. IVh. $b_{1.5}$ 60-2°, d^{20} 1.0873, n_D^{20} 1.4845.[502]

$O(CH_2)_3OPNC_5H_{10}$. IVf. b_4 96-7°.[101]

$OCH(Me)(CH_2)_2OPNMe_2$. IVf. $b_{13.5}$ 71°.[101]

$OCH(Me)(CH_2)_2OPNEt_2$. IVf. b_2 69-70°.[101]

$OCH(Me)(CH_2)_2OPNC_5H_{10}$. IVf. b_3 95-6°.[101]

$OCHMeCH_2CH_2OPN(CH_2CH_2)_2O$. IVh. $b_{0.8}$ 76-7°, b_2 99-100°, d^{20} 1.1439, d^{20} 1.1450, n_D^{20} 1.4932, n_D^{20} 1.4933.[502]

$OCH(Me)(CH_2)_2OPNMePh$. IVf. b_3 133°.[101]

$OCH_2CH_2CHMeOP$-3,5-dimethylpyrazolyl. $(RO)_2PCl$ + pyrazole. $b_{1.5}$ 123-4°, n_D^{25} 1.4623.[956]

$OCH_2CH_2CHMeOP$-4-Et-5-Me-2-pyrazolinyl. $(RO)_2PCl$ + pyrazoline. b_2 132-6°, n_D^{25} 1.4623.[956]

TYPE: 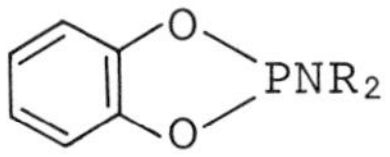

1,2-$C_6H_4O_2P\cdot NMe_2$. IVh. $b_{0.5}$ 60-1°.[1238]

TYPE: 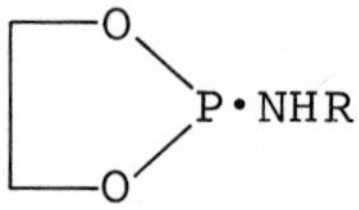

$\overline{O(CH_2)_2OP}\cdot NHC_6H_{11}$. $C_6H_{11}NH_2$ + the anhydride. $b_{0.08}$ 54-5°, n_D^{20} 1.5042.[867]

TYPE: 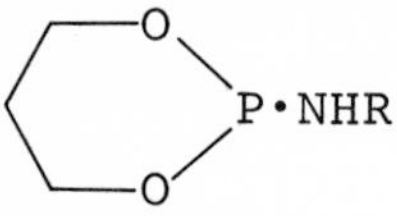

$\overline{O(CH_2)_3OP}\cdot NHC_6H_{11}$. $C_6H_{11}NH_2$ + the anhydride. $b_{0.12}$ 68°.[687]

$\overline{OCH_2CH_2CHMeOP}NHOMe$. $Bu(RO)_2PCl + RONH_2$. b_{12} 152-3°, n_D^{20} 1.4489.[956]

$\overline{OCH_2CH_2CHMeOP}NHOEt$. $Bu(RO_2)_2PCl + RONH_2$. $b_{1.5}$ 108-10°, n_D^{25} 1.4697.[956]

$\overline{OCH_2CH_2CHMeOP}NHNMe_2$. $Bu(RO)_2PCl + H_2NNMe_2$. b_7 99-101°, n_D^{25} 1.4743.[956]

$\overline{OCH(Me)(CH_2)_2OP}NHPh$. IVf. b_{12} 136°.[101]

$BuNHC(O)NHP[(OCH_2)_2C(CH_2O)_2]PNHC(O)NHBu$. From the diisocyanate + $BuNH_2$. m. 196-200° (decomp.).[1072]

TYPE: 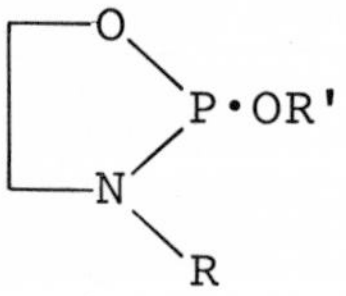

$\overline{OCH_2CH_2N(C_6H_5)P}OEt$. EtOP:NPh + $C_6H_5NH(CH_2)_2OH$.[924] By $PhNHCH_2CH_2OH + ROP(NR_2)_2$.[908] $b_{0.02}$ 91-2°,[924] $b_{0.02}$ 88-91°.[908]

$\overline{OCH_2CH_2N(Me)P}OEt$. EtOP:NPh + $MeNH(CH_2)_2OH$. b_{16} 67-9°.[924]

$\overline{OCH_2CH_2N(C_6H_5)P}OC_6H_5$. C_6H_5OP:NPh + $C_6H_5NH(CH_2)_2OH$. $b_{0.03}$ 130-3°, m. 76-7°.[924]

TYPE: $H_2NP(OR)_2$

$(PhO)_2PNH_2$. IVj. m. 40-2°, decomposed at room temp. forming white phosphorus.[171]

I.6.3. Hydrogen Phosphite Analogs

TYPE: $(R_2N)_2PHO$

$(Me_2N)_2P(O)H$. $(R_2N)_3P + H_3PO_3$.[1377] $(R_2N)_3P + H_2O$.[1414] $b_{1.5}$ 72°,[1377] $b_{1.3}$ 56-7°,[1414] n_D^{25} 1.4518,[1377] n_D^{20} 1.4522,[1414] IR.[1414]
$(Et_2N)_2P(O)H$. $(R_2N)_3P + H_3PO_3$.[1377,1414] $(R_2N)_2PCl + Et_3N + H_2O$.[1414] b_6 120°,[1377] $b_{0.5}$ 76-80°,[1377] $b_{0.03}$ 59-60°,[1414] d_4^{20} 0.960,[569] n_D^{25} 1.4545,[1377] n_D^{20} 1.4552,[569] n_D^{20} 1.4558,[1414] IR,[1414] NMR.[569]
$(Pr_2N)_2P(O)H$. $(R_2N)_3P + H_3PO_3$.[1377] $(R_2N)_2P + H_2O$.[1414] Crude n_D^{20} 1.4586, IR.[1414]
$(Bu_2N)_2P(O)H$. $(R_2N)_3P + H_3PO_3$.[1377] $(R_2N)_3P + H_2O$.[1414] $b_{0.1}$ 140°,[1377] n_D^{25} 1.4560,[1377] n_D^{20} 1.4584 (crude),[1414] IR.[1414]
$[O(CH_2CH_2-)_2N]_2P(O)H$. $(R_2N)_3P + H_3PO_3$.[1377] Low melting solid.[1377]

TYPE: $(RNH)_2PHO$

$(C_6H_5NH)_2PHO$. From $(PhN{:}PNHPh)_2 + H_2O$ in dioxane at 100°. m. 156°.[492]

TYPE: $(R_2N)(R'O)PHO$

$(Et_2N)(EtO)PHO$. $(R_2N)(R'O)PCl + H_2O$.[1416] $R_2NH + ROP(H)(O)OP(O)(OR)_2$.[1413] $b_{0.5}$ 44-5°,[1416] $b_{1.5}$ 57-8°,[1413] n_D^{20} 1.4265,[1413] n_D^{20} 1.4344, IR.[1416] TLC.[1416]
$(Et_2N)(i\text{-}PrO)P(O)H$. $(R_2N)(R'O)PCl + H_2O$.[1416] $(R_2N)(R'O)P + H_2O$.[569] $b_{0.5}$ 52-3°, n_D^{20} 1.4320, IR.[1416] TLC.[1416]
$(Et_2N)(BuO)PHO$. $(R_2N)(R'O)PCl + H_2O$.[1416] $R_2NH + ROP(H)(O)OP(O)(OR)_2$.[1413] $b_{0.05}$ 52-4°,[1416] $b_{0.01}$ 38-9°,[1413] n_D^{20} 1.4276,[1413] n_D^{20} 1.4390,[1416] IR.[1413,1416] TLC.[1416]
$(Bu_2N)(EtO)PHO$. $(R_2N)(R'O)PCl + H_2O$.[1416] $R_2NH + ROP(H)(O)OP(O)(OR)_2$.[1413] $b_{0.03}$ 68-70°,[1416] $b_{0.01}$ 60-1°,[1413] n_D^{20} 1.4360,[1413] n_D^{20} 1.4421,[1416] IR.[1416] TLC.[1416]
$(Bu_2N)(BuO)PHO$. $(R_2N)(R'O)PCl + H_2O$.[1416] $R_2NH + ROP(H)(O)OP(O)(OR)_2$.[1413] $b_{0.05}$ 67°,[1413] $b_{0.05}$ 72-4°,[1416] n_D^{20} 1.4352,[1413] n_D^{20} 1.4421,[1416] IR.[1413,1416] TLC.[1416]
$[O(CH_2CH_2)_2N](EtO)PHO$. $(R_2N)(R'O)PCl + H_2O$.[1416] $R_2NH + ROP(H)(O)OP(O)(OR)_2$.[1413] $b_{0.005}$ 60-1°,[1413] $b_{0.01}$ 66-7°,[1416] n_D^{20} 1.4530,[1413] n_D^{20} 1.4630,[1416] IR.[1413,1416] TLC.[1416]
$(PhNHNH)_2P(O)H$. Hydrolysis of PhNHN=PNHNHPh.[862]

I.6.4. Amidophosphites Having a P-O-P Linkage

$(Et_2N)_2P$-O-P(S)MeEt. $(Et_2N)_3P$ + EtMeP(S)OH,Et_2NH. $b_{0.007}$ 82-3°, d^{20} 1.0368, n_D^{20} 1.4890.[369]

(received March 15, 1971)

GENERAL REFERENCES

1. Abel, E. W., and D. B. Brady, J. Chem. Soc., 1965, 1192.
2. Abel, E. W., and G. Willey, Proc. Chem. Soc., 1962, 308.
3. Abramov, V. S., Dokl. Akad. Nauk, 95, 991 (1954); C.A. 49, 6084d (1955).
4. Abramov, V. S., Khim. Primenenie Fosfororgan. Soedinenii, Akad. Nauk SSSR, Trudy 1-oi Konferents., 1951, 71; C.A. 52, 240b (1958).
5. Abramov, V. S., Khim Primenenie Fosfororgan. Soedinenii Akad. Nauk SSSR, Trudy 1-oi Konferents., 1955, C.A. 52, 297h (1958).
6. Abramov, V. S., Zh. Obshch. Khim., 27, 169 (1957); C.A. 51, 12878e (1957).
7. Abramov, V. S., Dokl. Akad. Sci. SSSR, 117, 811 (1957); C.A. 52, 8038a (1958).
8. Abramov, V. S., V. I. Barabanov, and V. D. Efimova, Zh. Obshch. Khim., 38, 1545 (1968); C.A. 69, 96830u (1968).
9. Abramov, V. S., Yu. A. Bochkova, and A. D. Polyakova, Zh. Obshch. Khim., 23, 1013 (1953); C.A. 48, 8169d (1954).
10. Abramov, V. S., R. Sh. Chenborisov, and A. P. Kirisova., Zh. Obshch. Khim., 39, 350 (1959); C.A. 70, 115242t (1969).
11. Abramov, V. S., and Z. S. Druzhina, Zh. Obshch. Khim., 36, 923 (1966); C.A. 65, 10580g (1966).
12. Abramov, V. S., and M. G. Gubaidullin, Zh. Obshch. Khim., 38, 862 (1968); C.A. 69, 52220r (1968).
13. Abramov, V. S., and N. A. Il'ina, Dokl. Akad. Nauk SSSR, 125, 1027 (1959); C.A. 53, 21747e (1959).
14. Abramov, V. S., and N. A. Il'ina, Khim. Org. Soedin. Fosfora, Akad. Nauk SSSR, Otd. Obshch. Tekh. Khim., 1967, 119; C.A. 69, 43339j (1968).
15. Abramov, V. S., and A. S. Kapustina, Zh. Obshch. Khim., 27, 1012 (1957); C.A. 52, 3666i (1958).
16. Abramov, V. S., and G. Karp, Dokl. Akad. Nauk SSSR, 91, 1095 (1953); C.A. 48, 9906g (1954).
17. Abramov, V. S., and G. Karp, Zh. Obshch. Khim., 24, 1823 (1954); C.A. 49, 13887c (1955).
18. Abramov, V. S., and V. K. Khairullin, Zh. Obshch. Khim., 26, 811 (1956); C.A. 50, 14606b (1956).
19. Abramov, V. S., and V. K. Khairullin, Zh. Obshch.

Khim., 27, 444 (1957); C.A. 51, 15440i (1957).
20. Abramov, V. S., and V. K. Khairullin, Zh. Obshch. Khim., 29, 1222 (1959); C.A. 54, 13473 (1960).
21. Abramov, V. S., and S. N. Kofanov, Trudy Kazansk. Khim. Tekhnol. Inst. S. M. Kirova, 1950, 65; C.A. 51, 5689a (1957).
22. Abramov, V. S., and S. Pall, Zh. Obshch. Khim., 27, 171 (1957); C.A. 51, 12878g (1957).
23. Abramov, V. S., and A. P. Rekhman, Zh. Obshch. Khim., 26, 163 (1956); C.A. 50, 13723h (1956).
24. Abramov, V. S., R. N. Savintseva, and V. E. Ermakova, Zh. Obshch. Khim., 38, 2281 (1968); C.A. 70, 106609r (1969).
25. Abramov, V. S., and N. A. Semenova, Zh. Obshch. Khim., 28, 3056 (1958); C.A. 53, 10091f (1959).
26. Abramov, V. S., and A. L. Shalman, Khim. Org. Soedin. Fosfora, Akad. Nauk SSSR, Otd. Obshch. Tekh. Khim., 1967, 115; C.A. 69, 67468b (1968).
27. Ahmed, W., W. Gerrard, and V. K. Maladkar, J. Appl. Chem. (London), 20, 109 (1970).
28. Ainsworth, B. S., U.S. 3 000 850 (1962); C.A. 56, 2586h (1962).
29. Akhmetzhanov, I. S., Zh. Obshch. Khim., 38, 1090 (1968); C.A. 69, 66829h (1968).
30. Akhmetzhanov, I. S., R. N. Zagidulin, and M. G. Imaev, Dokl.Akad. Nauk SSSR, 163, 362 (1956); C.A. 63, 13060f (1965).
31. Aksnes, G., R. Erksen, and K. Mellingen, Acta Chem. Scand., 21, 10281 (1967).
32. Aksnes, G., and T. Gramstad, Acta Chem. Scand., 14, 1485 (1960).
33. Alimov, P. I., and I. V. Cheplanova, Izv. Kazansk. Filiala Akad. Nauk SSSR, Ser. Khim. Nauk, 1961, 54; C.A. 59, 9858b (1963).
34. Alimova, P. I., and I. V. Cheplanova, Izv. Kazansk. Filiala Akad. Nauk SSSR, Ser. Khim. Nauk., 1961, 61; C.A. 59, 9775e (1963).
35. Allen, J. F., and O. H. Johnson, J. Am. Chem. Soc., 77, 2871 (1955).
36. Allen, J. F., S. K. Reed, O. H. Johnson, and N. J. Brunswold, J. Am. Chem. Soc., 78, 3715 (1956).
37. Alizade, Z. A., and R. K. Velieva, Zh. Obshch. Khim., 39, 599 (1969); C.A. 71, 39091w (1969).
38. Almasi, L., and A. Hantz, Rev. Roumaline Chim., 9, 155 (1964).
39. Almasi, L., A. Hantz, and L. Pakucz, Acad. Rep. Populare Romine, Studii Cercetari Chim., 12, 169 (1961).
40. Almasi, L., A. Hantz, and L. Paskucz, Acad. Rep. Populare Romine, Studii Cercetari Chim., 14, 161 (1963).

41. Almasi, L., and L. Paskucz, Chem. Ber., 96, 2024 (1963).
42. Almasi, L., and L. Paskucz, Chem. Ber., 97, 623 (1964).
43. Anderson, G. W., J. Blodinger, R. W. Young, and P. Welcher, J. Am. Chem. Soc., 74, 5304 (1952).
44. Anderson, G. W., A. McGregor, and R. W. Young, J. Org. Chem., 23, 1236 (1958).
45. Anderson, G. W., A. D. Welcher, and R. W. Young, J. Am. Chem. Soc., 73, 501 (1951).
46. Anderson, G. W., and R. W. Young, J. Am. Chem. Soc., 74, 5307 (1952).
47. Andrews, K. J. M., and F. R. Atherton, J. Chem. Soc., 1960, 4682.
48. Ang, H. G., Chem. Commun., 1968, 1320.
49. Angert, L. G., P. A. Kirpichnikov, A. S. Kuzminskii, and I. E. Saratov, Zh. Prikl. Khim., 36, 2270 (1963); C.A. 61, 621f (1964).
50. Anglamol Ltd, Brit. P. 654 968 (1951); 682 441 (1952); 682 442 (1952); C.A. 46, 3252h (1952); C.A. 47, 6647f (1953); C.A. 47, 6647g (1953).
51. Anschütz, L., Ann. Chem., 1924, 439, 265.
52. Anschütz, L., Ber., 76, 222 (1943).
53. Anschütz, L., H. Boedeker, W. Broecker, and F. Wenger, Ber., 4, 71 (1927).
54. Anschütz, L., and W. Broeker, Ber., 61, 1264 (1928).
55. Anschütz, L., W. Broeker, R. Neher, and A. Ohnheiser, Ber., 76, 218 (1943).
56. Anschütz, L., W. Broeker, and A. Ohnheiser, Ber., 77, 439 (1944).
57. Anschütz, L., F. Koenig, F. Otto, and H. Walbrecht, Ann. Chem., 1936, 525, 297.
58. Anschütz, L., H. Kraft, and K. Schmidt, Ann. Chem., 542, 14 (1939).
59. Anschütz, L., and W. Marquardt, Naturwiss., 42, 644 (1955).
60. Anschütz, L., and W. Marquardt, Chem. Ber., 89, 1119 (1956).
61. Anschütz, L., and H. Walbrecht, J. Prakt. Chem., 133, 65 (1932).
62. Anschütz, L., and F. Wenger., Ann. Chem., 482, 34 (1930).
63. Anschütz, R., and W. D. Emery, Ann. Chem., 239, 301 (1887).
64. Anschütz, R., and W. D. Emery, Ann. Chem., 253, 105 (1889).
65. Antokhina, L. A., and P. I. Alimov, Izv. Akad. Nauk SSSR, Ser. Khim., 1966, 2135; C.A. 66, 95132u (1967).
66. Antokhina, L. A., and P. I. Alimov., Izv. Akad. Nauk SSSR, Ser. Khim., 1968, 180; C.A. 69, 43975p (1968).

67. Arain, R. A., and M. K. Hargreaves, J. Chem. Soc. C, 1970, 67.
68. Arbisman, Ya. S., Yu. A. Kondralev, and S. Z. Ivin, Zh. Obshch. Khim., 37, 509 (1967); C.A. 67, 32284v (1967).
69. Arbuzov, A. E., Chem. Ber., 38, 1171 (1905).
70. Arbuzov, A. E., J. Russ. Phys. Chem. Soc., 38, 687 (1906).
71. Arbuzov, A. E., J. Russ. Phys. Chem. Soc., 38, 293 (1906).
72. Arbuzov, A. E., J. Russ. Phys. Chem. Soc., 45, 79 (1913); C.A. 7, 2225 (1913).
73. Arbuzov, A. E., J. Russ. Phys. Chem. Soc., 46, 291 (1914); C.A. 8, 2551 (1914).
74. Arbuzov, A. E., Trudy Sessii Akad. Nauk Org. Khim., 1939, 211; C.A. 34, 7868 (1940).
75. Arbuzov, A. E., Izv. Akad. Nauk SSSR, Ser. Khim., 1946, 285.
76. Arbuzov, A. E., and V. S. Abramov, Trudy Kazansk. Khim. Tekhnol. Inst., 1, 23 (1935).
77. Arbuzov, A. E., and B. A. Arbuzov, J. Russ. Phys. Chem. Soc., 61, 192 (1929).
78. Arbuzov, A. E., and B. A. Arbuzov, J. Russ. Phys. Chem. Soc., 1929, 61, 217; C.A. 23, 3921 (1929).
79. Arbuzov, A. E., and I. A. Arbuzova, J. Russ. Phys. Chem. Soc., 62, 1533 (1930); C.A. 25, 2414 (1931).
79a. Arbuzov, A. E., and B. A. Arbuzov, J. Prakt. Chem., 131, 337 (1931); C.A. 26, 82 (1932).
80. Arbuzov, A. E., and B. A. Arbuzov, Zh. Obshch. Khim., 2, 348, 368, 371 (1932); Chem. Ber., 65, 195 (1932); C.A. 26, 2168 (1932).
81. Arbuzov, A. E., B. A. Arbuzov, P. T. Alimov, K. V. Nikonorov, N. I. Rizpotozhenskii, and O. N. Fedorova, Trudy Kazansk. Filiala Akad. Nauk SSSR, Ser. Khim. Nauk, 1956, 7; C.A. 51, 10363b (1957).
82. Arbuzov, A. E., and M. M. Azanovskaya, Izv. Akad. Nauk SSSR, Ser. Khim., 1949, 473; C.A. 44, 1905b (1950).
83. Arbuzov, A. E., and M. M. Azanovskaya, Izv. Akad. Nauk SSSR, Ser. Khim., 1951, 544; C.A. 47, 98c (1953).
84. Arbuzov, A. E., and M. G. Imaev, Dokl. Akad. Nauk SSSR, 112, 856 (1957); C.A. 51, 13741g (1957).
85. Arbuzov, A. E., and M. G. Imaev, Izv. Akad. Nauk SSSR, Ser. Khim., 1959, 171; C.A. 53, 16039f (1959).
86. Arbuzov, A. E., and A. A. Ivanov, J. Russ. Phys. Chem. Soc., 45, 681 (1913); C.A. 7, 3598 (1913).
87. Arbuzov, A. E., and A. A. Ivanov, J. Russ. Phys. Chem. Soc., 47, 2015 (1915); C.A. 10, 1342 (1916).
88. Arbuzov, A. E., and G. Kh. Kamai, J. Russ. Phys. Chem. Soc., 61, 619 (1929); C.A. 23, 4443 (1929).
89. Arbuzov, A. E., G. Kamai, and L. V. Nesterov, Trudy

Kazansk. Khim. Tekhnol. Inst., 16, 17 (1951); C.A. 51, 5720f (1957).
90. Arbuzov, A. E., and E. A. Krasil'nilsova, Izv. Akad. Nauk SSSR, Ser. Khim., 1959, 30; C.A. 53, 14982c (1959).
91. Arbuzov, A. E., and P. V. Nesterov, Dokl. Akad. Nauk SSSR, 92, 57 (1953); C.A. 48, 10538b (1954).
92. Arbuzov, A. E., and L. V. Nesterov, Izv. Akad. Nauk SSSR, Ser. Khim., 1954, 361; C.A. 49, 9541b (1955).
93. Arbuzov, A. E., and P. I. Rakov, Izv. Akad. Nauk SSSR, Ser. Khim., 1950, 237; C.A. 44, 8713g (1950).
94. Arbuzov, A. E., and A. I. Razumov, Zh. Obshch. Khim., 7, 1762 (1937); C.A. 32, 4845 (1938).
95. Arbuzov, A. E., and A. I. Razumov, Izv. Akad. Nauk SSSR, Ser. Khim., 1951, 714; C.A. 46, 7517c (1952).
96. Arbuzov, A. E., and F. G. Valitova, Izv. Akad. Nauk SSSR, Ser. Khim. 1940, 529; C.A. 35, 39901 (1941).
97. Arbuzov, A. E., and F. G. Valitova, Trudy Kazansk. Khim. Tekhnol. Inst., 8, 12 (1940); C.A. 35, 2485 (1941).
98. Arbuzov, A. E., and F. G. Valitova, Izv. Akad. Nauk SSSR, Ser. Khim., 1956, 681; C.A. 51, 1877 (1957).
99. Arbuzov, A. E., and V. M. Zoroastrova, Izv. Akad. Nauk SSSR, Ser. Khim., 1950, 357; C.A. 45, 1512d (1951).
100. Arbuzov, A. E., and V. M. Zoroastrova, Izv. Akad. Nauk SSSR, Ser. Khim., 1951, 536; C.A. 47, 97c (1953).
101. Arbuzov, A. E., and V. M. Zoroastrova, Izv. Akad. Nauk SSSR, Ser. Khim., 1952, 770, 779; C.A. 47, 9900e (1953).
102. Arbuzov, A. E., and V. M. Zoroastrova, Izv. Akad. Nauk SSSR, Ser. Khim., 1952, 789; C.A. 47, 10461c (1953).
103. Arbuzov, A. E., and V. M. Zoroastrova, Izv. Akad. Nauk SSSR, Ser. Khim., 1952, 809, 818, 826; C.A. 47, 9898h (1953).
104. Arbuzov, A. E., V. M. Zoroastrova, and T. N. Myasoedova, Izv. Akad. Nauk SSSR, Ser. Khim., 1960, 2127; C.A. 55, 14304i (1961).
105. Arbuzov, A. E., V. M. Zoroastrova, and N. J. Rizpolozhenskii, Izv. Akad. Nauk SSSR, Ser. Khim., 1948, 208; C.A. 42, 4932g (1948).
106. Arbuzov, B. A., P. I. Alimov, and O. N. Fedorova, Izv. Akad. Nauk SSSR, Ser. Khim., 1956, 932; C.A. 51, 4932d (1957).
107. Arbuzov, B. A., P. I. Alimov, M. A. Zvereva, I. D. Neklesova, and M. A. Kudrina, Izv. Akad. Nauk SSSR, Ser. Khim., 1954, 1038; C.A. 50, 161b (1956).
108. Arbuzov, A. E., M. L. Batuev, and V. S. Vinogradova,

Dokl. Akad. Nauk SSSR, 54, 599 (1946); C.A. 41, 5022c (1947).
109. Arbuzov, B. A., and N. P. Bogonostseva, Izv. Akad. Nauk SSSR, Ser. Khim., 1953, 484; C.A. 48, 9905h (1954).
110. Arbuzov, B. A., and N. P. Bogonostseva, Zh. Obshch. Khim., 27, 2356 (1957); C.A. 52, 7129g (1958).
111. Arbuzov, B. A., E. N. Dianova, and V. S. Vinogradova, Izv. Akad. Nauk SSSR, Ser. Khim., 1969, 1109; C.A. 71, 50070g (1969).
112. Arbuzov, B. A., E. N. Dianova, V. S. Vinogradova, and A. K. Shamsutdinova, Dokl. Akad. Nauk SSSR, 160, 99 (1965); C.A. 62, 11848f (1965).
113. Arbuzov, B. A., E. N. Dianova, V. S. Vinogradova, and Yu. Yu. Samitov, Dokl. Akad. Nauk SSSR, 173, 1321 (1967); C.A. 68, 12911q (1968).
114. Arbuzov, B. A., and B. P. Lugovkin, Zh. Obshch. Khim., 22, 1199 (1952); C.A. 47, 4872c (1953).
115. Arbuzov, A. E., and K. N. Nikonorov, Zh. Obshch. Khim., 17, 2139 (1947); C.A. 42, 4546b (1948).
116. Arbuzov, A. E. and K. V. Nikonorov, Dokl. Akad. Nauk SSSR, 62, 75 (1948); C.A. 43, 1004c (1949).
117. Arbuzov, B. A., K. V. Nikonorov, G. M. Vinokurova, O. N. Fedorova, and Z. G. Shishova, Izv. Kazansk. Filiala Akad. Nauk SSSR, Ser. Khim., 1955, 3; C.A. 52, 241i (1958).
118. Arbuzov, B. A., N. A. Polezhaeva, V. S. Vinogradova, and Yu. Yu. Samitov, Dokl. Akad. Nauk SSSR, 173, 93 (1967); C.A. 67, 54221z (1967).
119. Arbuzov, B. A., N. A. Polezhaeva, and V. S. Vinogradova, Izv. Akad. Nauk SSSR, Ser. Khim., 1967, 2281; C.A. 68, 49684u (1968).
120. Arbuzov, B. A., and A. N. Pudovik, Dokl. Akad. Nauk SSSR, 59, 1433 (1948); C.A. 47, 4281g (1953).
121. Arbuzov, B. A., A. O. Vizel, O. A. Raevskii, Yu. F. Tarenko, and L. E. Petrova, Izv. Akad. Nauk SSSR, Ser. Khim., 1969, 460; C.A. 70, 115250u (1969).
122. Arbuzov, B. A., N. I. Rizpozhenskii, and M. A. Zvereva, Izv. Akad. Nauk SSSR, Ser. Khim., 1957, 179; C.A. 51, 11237f (1957).
123. Arbuzov, B. A., M. K. Saikina, and V. M. Zoroastrova, Izv. Akad. Nauk SSSR, Ser. Khim., 1957, 1046; C.A. 52, 7130a (1958).
124. Arbuzov, B. A., and L. A. Shapshinskaya, Izv. Akad. Nauk SSSR, Ser. Khim., 1962, 65; C.A. 57, 13791c (1962).
125. Arbuzov, B. A., T. D. Sorokina, N. P. Bogonostseva, and V. S. Vinogradova, Dokl. Khim. Nauk SSSR, 171, 605 (1966); C.A. 67, 32501p (1967).
126. Arbuzov, B. A., and E. N. Ukhvatova, Izv. Akad. Nauk SSSR, Ser. Khim., 1958, 1395; C.A. 53, 6988g (1959).

127. Arbuzov, B. A., and V. S. Vinogradova, Izv. Akad. Nauk SSSR, Ser. Khim., 617 (1947); C.A. 42, 5844h (1948).
128. Arbuzov, B. A., and V. S. Vinogradova, Izv. Akad. Nauk SSSR, Ser. Khim., 1952, 505; C.A. 47, 4835d (1953).
129. Arbuzov, B. A., and V. S. Vinogradova, Dokl. Akad. Nauk SSSR, 5, 459; 6, 617 (1947); C.A. 42, 3312a, 5844h (1948) [cf. C.A. 45, 1504b (1951)].
130. Arbuzov, B. A., and V. S. Vinogradova, Izv. Akad. Nauk SSSR, Ser. Khim., 1952, 507; C.A. 47, 4834d (1953).
131. Arbuzov, B. A., and V. S. Vinogradova, Izv. Akad. Nauk SSSR, Ser. Khim., 1952, 505; C.A. 47, 4835d (1953).
132. Arbuzov, B. A., and V. S. Vinogradova, Dokl. Akad. Nauk SSSR, 83, 79 (1952); C.A. 47, 2685a (1953).
133. Arbuzov, B. A., and V. S. Vinogradova, Dokl. Akad. Nauk SSSR, Ser. Khim., 99, 85 (1954); C.A. 49, 13925g (1955); 1957, 54; C.A. 51, 10365c (1957).
134. Arbuzov, B. A., and V. S. Vinogradova, Dokl. Akad. Nauk SSSR, 106, 35 (1956); C.A. 51, 3440c (1957).
135. Arbuzov, B. A., and A. O. Vizel, Izv. Akad. Nauk SSSR, Ser. Khim., 1963, 749; C.A. 59, 7362h (1963).
136. Arbuzov, B. A., A. O. Vizel, Yu. Yu. Samitov, and Yu. F. Tarenko, Izv. Akad. Nauk SSSR, Ser. Khim., 1967, 672; C.A. 67, 100193s (1967).
137. Arbuzov, B. A., and D. Kh. Yarmukhametova, Dokl. Akad. Nauk SSSR, 101, 675; C.A. 50, 3214h (1956).
138. Arbuzov, B. A., and D. Kh. Yarmukhametova, Izv. Akad. Nauk SSSR, Ser. Khim., 1957, 292; C.A. 51, 14542b (1957).
139. Arbuzov, B. A., O. D. Zolova, V. S. Vinogradova, and Yu. Yu. Samitov, Dokl. Akad. Nauk SSSR, 173, 335, (1967); C.A. 67, 43886u (1967).
140. Arbuzov, B. A., and V. M. Zoroastrova, Izv. Akad. Nauk SSSR, Ser. Khim., 1959, 1037; C.A. 54, 1498h (1960).
141. Arbuzov, B. A., and V. M. Zoroastrova, Izv. Akad. Nauk SSSR, Ser. Khim., 1960, 1030; C.A. 54, 24627g (1960).
142. Arbuzov, B. A., V. M. Zoroastrova, N. D. Ibragimova, I. D. Neklesova, B. L. Mazur, M. A. Kudrina, N. V. Egorova, and I. S. Iraidova, Izv. Akad. Nauk SSSR, Ser. Khim., 1967, 1278; C.A. 68, 48957 (1968).
143. Arbuzov, B. A., V. M. Zoroastrova, and M. K. Saikiva, Izv. Akad. Nauk SSSR, Ser. Khim., 1959, 1579; C.A. 54, 8619 (1960).
144. Ardis, A. E., and J. A. Wojtowiez, U.S. 3 446 819 (1969); C.A. 71, 38791n (1969).
145. Aronov, Yu. E., Yu. A. Cherburkov, and I. L.

Knunyants, Izv. Akad. Nauk SSSR, Ser. Khim., 1967, 1758; C.A. 69, 2486m (1968).
146. Arutyunyan, E. A., V. I. Gunar, E. P. Gracheva, and S. I. Zav'yalov, Izv. Akad. Nauk SSSR, Ser. Khim., 1968, 445; C.A. 69, 86943u (1968).
147. Atavin, A. S., A. V. Gusarov, B. A. Trofimov, and V. M. Nikitin, Zh. Vses. Khim. Obshch., 11, 594 (1966); C.A. 66, 94653c (1967).
148. Atherton, F. R., V. M. Clark, and A. R. Todd, Rec. Trav. Chim., 69, 295 (1950).
149. Atherton, F. R., H. T. Howard, and A. R. Todd, J. Chem. Soc., 1948, 1106.
150. Atherton, F. R., H. T. Openshaw, and A. R. Todd, J. Chem. Soc., 1945, 382.
151. Atherton, F. R., H. T. Openshaw, and A. R. Todd, J. Chem. Soc., 1945, 660.
152. Atherton, F. R. and A. R. Todd, J. Chem. Soc., 1947, 674.
153. Atkinson, R. E., J. I. G. Cadogan, and J. Dyson, J. Chem. Soc. C, 1967, 2542.
154. Atkinson, R. E., J. I. G. Cadogan, and J. T. Sharp, J. Chem. Soc. B, 1969, 138.
155. Autenrieth, W., and W. Meyer, Chem. Ber., 58, 840, 848 (1925).
156. Ayres, D. C., and H. N. Rydon, J. Chem. Soc., 1957, 1109.
157. Azerbaev, I. N., T. G. Sarbaev, B. D. Abiyurov, and V. S. Bazalitskaya, Izv. Akad. Nauk SSSR, Ser. Khim., 18, 56 (1968); C.A. 70, 57948q (1969).
158. Baddiley, J., V. M. Clark, J. J. Michalski, and A. R. Todd, J. Chem. Soc., 1949, 815.
159. Baker, J. W., and R. E. Stenseth, U.S. 3 457 306 (1969); C.A. 71, 91648x (1969).
160. Ballard, P., Ger. 95 578 (1897).
161. Baranauckas, C. F., and I. Gordon, U.S. 3 310 609 (1967); C.A. 67, 73142r (1967).
162. Baranauckas, C. F., and J. J. Hodan, U.S. 3 331 895 (1967); C.A. 67, 63764f (1967).
163. Barinov, I. V., V. V. Rode, and S. S. Rafikov, Zh. Obshch. Khim., 37, 464 (1967); C.A. 67, 64081t (1967).
164. Barlow, C. G., R. Jefferson, and J. F. Nixon, J. Chem. Soc. A, 1968, 2692.
165. Barlow, C. G., and J. F. Nixon, J. Chem. Soc. A, 1966, 228.
166. Batkowski, T., P. Mastalerz, M. Michalewska, and B. Nitka, Rocz. Chem., 41, 471 (1967).
167. Bauer, S., Oil and Soap, 23, 1 (1946).
168. Bayer, P. W., and J. R. Maugham, U.S. 2 866 807 (1958); C.A. 53, 12240c (1959).
169. Bechamp, A., Compt. Rend., 40, 944 (1855); 42, 244

(1856); Jahr. Fortsch. Chem., 1856, 427.
170. Beck, T. M., and H. Sorstokke, U.S. 3 068 267 (1962); C.A. 58, 10079e (1963).
171. Becke-Goehring, M., and J. Schulze, Chem. Ber., 91, 1188 (1958).
172. Bedell, R., M. J. Frazer, and W. Gerrard, J. Chem. Soc., 1960, 4037.
173. Bell, A., U.S. 2 508 364 (1950); C.A. 44, 7484c (1950).
174. Bellamy, L. J., The Infrared Spectra of Complex Molecules, 2nd. ed., Wiley, New York, 1958.
175. Bellamy, L. J., and L. Beecher, J. Chem. Soc., 1952, 1701.
176. Bellamy, L. J., and L. Beecher, J. Chem. Soc., 1952, 475.
177. Bel'skii, V. E., G. Z. Motygullin, V. N. Eliseenkov, and A. N. Pudovik, Izv. Akad. Nauk SSSR, Ser. Khim., 1969, 1297; C.A. 71, 80416u (1969).
178. Benezra, C., Peintures, Pigments, Vernis, 43, 395 (1967).
179. Bengelsdorf, I. S., J. Org. Chem., 21, 475 (1956).
180. Bengelsdorf, I. S., and L. B. Barron, J. Am. Chem. Soc., 77, 2869 (1955).
181. Bennett, F. W., H. J. Emeleus, and R. N. Haszeldine, J. Chem. Soc., 1954, 3598.
182. Bentrude, W. G., Tetrahedron Letters, 1965, 3543.
183. Bentrude, W. G., and W. D. Johnson, Tetrahedron Letters, 46, 4611 (1967).
184. Bergman, E. D., and D. Herrman, J. Am. Chem. Soc., 73, 4013 (1951).
185. Berlak, M. C., and W. Gerrard, J. Chem. Soc., 1949, 2309.
186. Berlin, K. D., T. H. Austin, M. Peterson, and M. Nagabhushanam, Top. Phosphorus Chem., 1, 17 (1964), Wiley (Interscience).
187. Berlin, K. D., T. H. Austin, and K. Stone, Abstr. Am. Chem. Soc. Meeting, New York, Sept. 1963, p. 690.
188. Berlin, K. D., T. H. Austin, and K. L. Stone, J. Am. Chem. Soc., 86, 1787 (1964).
189. Bilevich, K. A., and V. P. Evdakov, Zh. Obshch. Khim., 35, 365 (1965); C.A. 62, 13030g (1965).
190. Bill, J. C., and B. A. Hunter, U.S. 2 732 365 (1956); C.A. 50, 6830i (1956).
191. Birkofer, L., W. Gilgenberg, and A. Ritter, Angew. Chem., 73, 143 (1961).
192. Birum, G. H., U.S. 3 014 910 (1961); C.A. 57, 4693c (1962).
193. Birum, G. H., U.S. 3 014 944 (1961); C.A. 56, 11622g (1962).
194. Birum, G. H., and J. L. Dever, U.S. 2 961 455

(1961); C.A. 55, 8292g (1961) [cf. U.S. 3 159 017 (1961)].
195. Birum, G. H., and J. L. Dever, U.S. 2 980 722 (1961); C.A. 56, 329c (1962).
196. Blackburn, G. M., and J. S. Cohen, Top. Phosphorus Chem., 1, 187, Wiley (Interscience) (1964).
197. Blackburn, G. M., J. S. Cohen, and A. R. Todd, Tetrahedron Letters, 1964, 2873.
198. Bliznyuk, N. K., A. F. Kolomiets, Z. N. Kvaska, G. S. Kevskaya, and V. V. Antipina, Zh. Obshch. Khim., 36, 375 (1966); C.A. 65, 739a (1966).
199. Bliznyuk, N. K., A. F. Kolomiets, and R. N. Golubeva, U.S.S.R. 188 493 (1966); C.A. 67, 53667n (1967).
200. Bliznyuk, N. K., A. F. Kolomiets, P. S. Khokhlov, and S. G. Zhemchuzhin, Zh. Obshch. Khim., 37, 1353 (1967); C.A. 67, 108126t (1967).
201. Bliznyuk, N. K., Z. N. Kvasha, and A. F. Kolomiets, Zh. Obshch. Khim., 37, 888 (1967); C.A. 67, 90878c (1967).
202. Bliznyuk, N. K., Z. N. Kvasha, L. D. Protasova, and S. L. Varshavskii, U.S.S.R. 250 139 (1969); C.A. 72, 78666c (1970).
203. Bliznyuk, N. K., Z. N. Kvasha, L. M. Solntseva, B. Ya. Libman, A. I. Beim, and I. B. Sevitov, U.S.S.R. 174 624 (1965); C.A. 64, 1961b (1966).
204. Bliznyuk, N. K., N. F. Savenkov, P. S. Khokhlov, and S. L. Varshavskii, U.S.S.R. 189 848 (1966); C.A. 68, 2706n (1968).
205. Bliznyuk, N. K., N. F. Savenkov, P. S. Khokhlov, and S. L. Varshavskii, U.S.S.R. 189 850 (1966); C.A. 68, 21522r (1968).
206. Bliznyuk, N. K., S. G. Zhemchuzhin, and L. D. Protasova, U.S.S.R. 172 328 (1965); C.A. 63, 16261a (1965).
207. Bloch, H. S., U.S. 2 570 512 (1951); C.A. 46, 3555i (1952).
208. Bochwic, B., and J. Michalski, Nature, 167, 1035 (1951).
209. Bochwic, B., and J. Michalski, Roczniki Chem., 25, 338 (1951).
210. Bogatskii, A. V., T. D. Butova, A. A. Kolesnik, and R. A. Sabirova, Khim. Geterotsikl. Soedin., Akad. Nauk Latv. SSR, 1965, 474; C.A. 63, 13261h (1965).
211. Borowitz, I. J., M. Anschel, and S. Firstenberg, J. Org. Chem., 32, 1723 (1967).
212. Bothamley, C. H., and G. R. Thompson, Chem. News, 62, 191 (1890).
213. Boulton, A. J., I. J. Fletcher, and A. R. Katritzky, Chem. Commun., 1968, 62.
214. Bourgeois, L., and J. Bolle, Mem. Services Chim.

Etat (Paris), 34, 411 (1948).
215. Boyd, D. R., J. Chem. Soc., 79, 1224 (1901).
216. Boyd, D. R., J. Chem. Soc., 83, 1138 (1903).
217. Boyer, W. P., J. R. Mangham, and T. M. Melton, U.S. 2 968 670 (1961); C.A. 55, 14383g (1961).
218. Brauns, D. H., Rec. Trav. Chim., 65, 799 (1946); 66, 466 (1947).
219. Broeker, W., J. Prakt. Chem., 1928, 118, 287.
220. Brooks, B. T., J. Am. Chem. Soc., 34, 492 (1912).
221. Brown, D. H., K. D. Crosbie, G. W. Fraser, and D. W. A. Sharp, J. Chem. Soc. A, 1969, 872.
222. Brown, D. M., and P. R. Hammond, J. Chem. Soc., 1960, 4229.
223. Brown, T. L., J. G. Verkade, and T. S. Piper, J. Phys. Chem., 65, 2051 (1961).
224. Buchner, B., and A. F. Jackson, U.S. 3 207 723 (1965); C.A. 69, 8410c (1966).
225. Bugerenko, E. F., E. A. Chernyshev, and E. M. Popov, Izv. Akad. Nauk SSSR, Ser. Khim., 1966, 1391; C.A. 66, 76078q (1967).
226. Buina, N. A., I. A. Nuretdinov, and N. P. Grechkin, Izv. Akad. Nauk SSSR, Ser. Khim., 1967, 1606; C.A. 68, 12594p (1968).
227. Bunyan, P. J., and J. I. G. Cadogan, J. Chem. Soc., 1962, 2953.
228. Burdett, J. L., and L. B. Burger, Can. J. Chem., 44, 111 (1966).
229. Burg, A. B., and P. J. Slota, Jr., J. Am. Chem. Soc., 80, 1107 (1958).
230. Burg, A. B., and P. J. Slota, Jr., J. Am. Chem. Soc., 82, 2148 (1960).
231. Burgada, R., Bull. Soc. Chim. France, 1963, 2335.
232. Burgada, R., Ann. Chim., 8, 374 (1963).
233. Burgada, R., Compt. Rend., 258, 1532 (1964).
234. Burgada, R., Compt. Rend., 258, 4789 (1964).
235. Burgada, R., Bull. Soc. Chim. France, 1967, 347.
236. Burgada, R., G. Martin, and G. Mavel, Bull. Soc. Chim. France, 1963, 2154.
237. Burger, L. L., and R. M. Wagner, Chem. Eng. Data Ser., 3, 310 (1958).
238. Burn, A. J., and J. I. G. Cadogan, J. Chem. Soc., 1963, 5788.
239. Burn, A. J., J. I. G. Cadogan, and P. J. Bunyan, J. Chem. Soc., 1963, 1527.
240. Burn, A. J., J. I. G. Cadogan, and P. J. Bunyan, J. Chem. Soc., 1964, 4369.
241. Butcher, F. K., B. E. Deuters, W. Gerrard, E. F. Mooney, R. A. Rothenbury, and H. A. Willis, Spectrochim. Acta, 20, 759 (1964).
242. Cade, J. A., J. Chem. Soc., 1959, 2266.
243. Cade, J. A., J. Chem. Soc., 1959, 2272.

244. Cade, J. A., and W. Gerrard, Nature (London), 172, 29 (1953).
245. Cade, J. A., and W. Gerrard, Chem. Ind. (London), 1954, 402.
246. Cade, J. A., and W. Gerrard, J. Chem. Soc., 1954, 2030.
247. Cade, J. A., and W. Gerrard, J. Chem. Soc., 1960, 1249.
248. Cadogan, J. I. G., J. Chem. Soc., 1957, 4154.
249. Cadogan, J. I. G., Quart. Rev., 16, 208 (1962).
250. Cadogan, J. I. G., Advan. Free-Radical Chem., 2, 203 (1967).
251. Cadogan, J. I. G., Synthesis, 1969, 11.
252. Cadogan, J. I. G., M. Cameron-Wood, and W. R. Foster, J. Chem. Soc., 1963, 2549.
253. Cadogan, J. I. G., M. Cameron-Wood, R. K. Mackie, and R. J. G. Searle, J. Chem. Soc., 1965, 4831.
254. Cadogan, J. I. G., and W. R. Foster, J. Chem. Soc., 1961, 3071.
255. Cadogan, J. I. G., S. Kulik, and M. J. Todd, Chem. Commun., 1968, 736.
256. Cadogan, J. I. G., R. K. Mackie, and M. J. Todd, Chem. Commun., 1966, 491.
257. Cadogan, J. I. G., R. K. Mackie, and M. J. Todd, Chem. Commun., 1968, 736.
258. Cadogan, J. I. G., and H. N. Moulden, J. Chem. Soc., 1961, 3079.
259. Cadogan, J. I. G., D. J. Sears, and D. M. Smith, J. Chem. Soc. C, 1969, 1314.
260. Cadogan, J. I. G., D. J. Sears, D. M. Smith, and M. J. Todd, J. Chem. Soc. C, 1969, 2813.
261. Cadogan, J. I. G., and J. T. Sharp, Tetrahedron Letters, 1966, 2733.
262. Cadogan, J. I. G., and M. J. Todd, Chem. Commun., 1967, 178.
263. Callis, C. F., J. R. Van Wazer, J. N. Shoolery, and W. A. Anderson, J. Am. Chem. Soc., 79, 2719 (1957).
264. Campbell, C. H., U.S. 2 794 820 (1957); C.A. 51, 15560h (1957).
265. Campbell, C. H., and D. H. Chadwick, J. Am. Chem. Soc., 77, 3379 (1955).
266. Campbell, C. H., D. H. Chadwick, and S. Kaufman, Ind. Eng. Chem., 49, 1871 (1957).
267. Campbell, I. G. M., and J. K. Way, J. Chem. Soc., 1960, 5034.
268. Carlisle Chemical Works, Inc., Fr. Addn. 90 538 (1967); C.A. 70, 19577k (1969).
269. Carre, P., Compt. Rend., 133, 889 (1901); 136, 756 (1903); 136, 1067 (1903); 136, 1457 (1903); 137, 517 (1905); Ann. Chim. (8), 5, 415 (1905); Bull. Soc. Chim. (3), 27 (1902).

270. Cason, J., W. N. Baxter, and W. DeAcetis, J. Org. Chem., 24, 247 (1959).
271. Cassoux, P., and J. F. Labarre, Compt. Rend. Ser. C, 265, 773 (1967).
272. Caubere, P., Bull. Soc. Chim. France, 1967, 3446.
273. Cebrian, G. R., Anales Real Soc. Expan. Fis. Quim., 50B, 673; Pub. Inst. Quim., "Alonso Barba" (Madrid), 8, 191 (1954); C.A. 49, 10873e (1955).
274. Cebrian, G. R., Arch. Inst. Farmacol. Exptl. (Madrid), 8, 61 (1956); C.A. 51, 12020d (1957).
275. Chadwick, D. H., U.S. 2 834 797 (1958); C.A. 52, 14998b (1958).
276. Chadwick, D. H., and T. Reetz, U.S. 2 865 942 (1958); C.A. 53, 7986g (1959).
277. Chambon, E., Jahresber., 1876, 205.
278. Chaudri, B. A., and H. R. Hudson, unpublished work.
279. Chaudri, B. A., D. G. Goodwin, and H. R. Hudson, J. Chem. Soc. B, 1970, 1290.
280. Chaudri, B. A., H. R. Hudson, and W. S. Murphy, J. Chromatog., 29, 218 (1967).
281. Chemische Werke Huls, Brit., 772 486 (1957); C.A. 51, 13919a (1957).
282. Chenborisov, R. Sh., and A. P. Kirisova, Zh. Obshch. Khim., 39, 931 (1969); C.A. 71, 49166p (1969).
283. Cherbuliez, E., F. Hunkeler, G. Weber, and J. Rabinowitz, Helv. Chim. Acta, 47, 1647 (1964).
284. Chernick, C. L., H. A. Skinner, and C. T. Mortimer, J. Chem. Soc., 1955, 3936.
285. Chernyshev, E. A., and E. F. Bugerenko, Izv. Akad. Nauk SSSR, Ser. Khim., 1963, 769; C.A. 59, 10106a (1963).
286. Chernyshev, E. A., and A. D. Petrov, Dokl. Akad. Nauk SSSR, 105, 282 (1955); C.A. 50, 11283g (1956).
287. Chittenden, R. A., and L. C. Thomas, Spectrochim. Acta, 21, 861 (1965).
288. Chopard, P. A., Helv. Chim. Acta, 50, 1021 (1967).
289. Chopard, P. A., V. M. Clark, R. F. Hudson, and A. J. Kirby, Tetrahedron, 21, 1961 (1965).
290. Chrzaszczewska, A., and W. Sobieranski, Rocz. Chem., 7, 470 (1927).
291. Chwalinski, S., and W. Rypinska, Rocz. Chem., 31, 539 (1957).
292. C. I. B. A. Ltd., Brit., 744 360 (1956); C.A. 51, 470i (1957).
293. Clark, R. H., and A. Bell, Trans. Roy. Soc. Canada, 27 III, 97 (1933).
294. Coates, H., D. A. Brown, G. Quesnnel, J. G. C. Givard, and A. Thiot, Brit. 940 697 (1963); C.A. 60, 6747h (1964).
295. Coates, H., and P. C. Crofts, Brit. 713 669 (1954); C.A. 49, 12529d (1955).

296. Coe, D. G., S. R. Landauer, and H. N. Rydon, J. Chem. Soc., 1954, 2281.
297. Coelln, R., Ger. 1 216 278 (1966); C.A. 65, 8762a (1966).
298. Coelln, R., and G. Schrader, Ger. 1 059 425 (1959); C.A. 56, 11446d (1962).
299. Conant, J. B., V. H. Wallingford, and S. S. Gandheker, J. Am. Chem. Soc., 45, 762 (1923).
300. Cook, T. M., E. J. Coulson, W. Gerrard, and H. R. Hudson, Chem. Ind. (London), 1962, 1506.
301. Cook, H. G., J. D. Ilett, B. C. Saunders, G. J. Stracey, H. G. Watson, I. G. E. Wilding, and S. J. Woodcock, J. Chem. Soc., 1949, 2921.
302. Cook, H. G., H. M. McCombie, and B. C. Saunders, J. Chem. Soc., 1945, 873.
303. Cook, H. G., B. C. Saunders, and F. E. Smith, J. Chem. Soc., 1949, 635.
304. Cooke, V. F. G., and W. Gerrard, J. Chem. Soc., 1955, 1978.
305. Coover, H. W., and J. B. Dickey, U.S. 2 789 209 (1957); C.A. 51, 13903d (1957).
306. Coover, H. W., Jr., and M. A. McCall, U.S. 2 921 087 (1960); C.A. 54, 22359f (1960).
307. Corey, E. J., and J. E. Anderson, J. Org. Chem., 32, 4160 (1967).
308. Corey, E. J., F. A. Carey, and R. A. E. Winter, J. Am. Chem. Soc., 87, 934 (1965).
309. Corey, E. J., and G. Markl, Tetrahedron Letters, 1967, 3201.
310. Corey, E. J., and R. A. E. Winter, J. Am. Chem. Soc., 85, 2677 (1963).
310a. Coskran, K. J., and J. G. Verkade, Inorg. Chem., 4, 1655 (1965).
311. Coulson, E. J., W. Gerrard, and H. R. Hudson, J. Chem. Soc., 1965, 2364.
312. Cowdrey, W. A., E. D. Hughes, C. K. Ingold, S. Masterman, and A. D. Scott, J. Chem. Soc., 1937, 1267.
313. Cowley, A. H., and R. P. Pinnell, J. Am. Chem. Soc., 87, 4454 (1965).
314. Cox, J. R., Jr., and F. H. Westheimer, J. Am. Chem. Soc., 80, 5441 (1958).
315. Cramer, F., and K.-G. Gärtner, Angew. Chem., 68, 649 (1956).
316. Cramer, F., and K.-G. Gärtner, Chem. Ber., 91, 704 (1958).
317. Crofts, P. C., and G. M. Kosolapoff, J. Am. Chem. Soc., 75, 3379 (1953).
318. Crofts, P. C., and G. M. Kosolapoff, J. Am. Chem. Soc., 75, 5733 (1953).
319. Crofts, P. C., J. H. H. Markes, and H. N. Rydon,

J. Chem. Soc., 1958, 4250.
320. Crosbie, K. D., G. W. Fraser, and D. W. A. Sharp, Chem. Ind. (London), 1968, 423.
321. Crutchfield, M. M., C. H. Dungan, J. H. Letcher, V. Mark, and J. R. Van Wazer, Topics in Phosphorus Chemistry, Wiley, New York, 1967, Vol. 5.
322. Currell, B. R., and W. Gerrard, Chem. Ind. (London), 1958, 1289.
323. Daasch, L. W., J. Am. Chem. Soc., 80, 5301 (1958).
324. Daasch, L. W., and D. C. Smith, Anal. Chem., 23, 853 (1951).
325. Davidson, R. S., J. Chem. Soc. C, 1967, 2131.
326. Davis, H. E., U.S. 3 283 037 (1966); C.A. 66, 10693t (1967).
327. Deckelmann, E., and H. Werner, Helv. Chim. Acta, 52, 892 (1969).
328. Degen, I. A., D. G. Saunders, and B. P. Woodford, Chem. Ind. (London), 1969, 267.
328a. De Hauss, J. L., Chim. Anal., 34, 248 (1952).
329. Delacre, M., Bull. Soc. Chim. (2), 48, 787 (1887).
330. De Leo, E., and R. Indovina, Ann. Chim. (Rome), 49, 404 (1959).
331. Denney, D. B., and R. R. Dileone, J. Am. Chem. Soc., 84, 4737 (1962).
332. Denney, D. B., W. F. Goodyear, and B. Goldstein, J. Am. Chem. Soc., 82, 1393 (1960).
333. Denney, D. B., and S. T. D. Gough, J. Am. Chem. Soc., 87, 138 (1965).
334. Denney, D. B., and J. W. Hanifin, Jr., Tetrahedron Letters, 1963, 2177.
335. Denney, D. B., and H. M. Relles, J. Am. Chem. Soc., 86, 3897 (1964).
336. Denney, D. B., and S. L. Varga, Tetrahedron Letters, 1966, 4935.
337. Denney, D. Z., G. Y. Chen, and D. B. Denney, J. Am. Chem. Soc., 91, 6838 (1969).
338. Denney, D. Z., and D. B. Denney, J. Am. Chem. Soc., 88, 1830 (1966).
339. Derkach, G. I., E. S. Gubnitskaya, V. A. Shokol, and A. V. Kirsanov, Zh. Obshch. Khim., 32, 1201 (1962); C.A. 58, 5721a (1963).
340. Derkach, G. I., E. S. Gubnitskaya, L. I. Samarai, and V. A. Shokol, Zh. Obshch. Khim., 33, 557 (1963); C.A. 59, 2684c (1963).
341. Derkach, G. I., A. M. Lepesa, and A. V. Kirsanov, Zh. Obshch. Khim., 32, 171 (1962); C.A. 57, 16472e (1962).
342. Derkach, G. I., L. I. Samarai, A. S. Shtepanek, and A. V. Kirsanov, Zh. Obshch. Khim., 32, 3759 (1962); C.A. 58, 13985e (1963).
343. De Rose, A., W. Gerrard, and E. F. Mooney, Chem.

Ind. (London), 1961, 1449.
344. De Selms, R. C., Tetrahedron Letters, 1968, 5545.
345. Dever, J. L., and J. J. Hodan, U.S. 3 459 835 (1969); C.A. 71, 81330y (1969).
346. Dewar, M. J. S., and V. P. Kubba, J. Am. Chem. Soc., 82, 5685 (1960).
347. Dickey, J. B., and H. W. Coover, U.S. 2 721 876 (1955); C.A. 50, 10123i (1956).
348. Diefenbach, E., Ger. 937 956 (1956); C.A. 52, 20106f (1958).
349. Dietrich, M. A., U.S. 2 373 627 (1945); C.A. 40, 1030[4] (1946).
350. Dimroth, K., and B. Lerch, Angew. Chem., 72, 751 (1960).
351. Dimroth, K., and R. Ploch, Chem. Ber., 90, 801 (1957).
352. Dittmer, D. C., and S. M. Kotin, J. Org. Chem., 32, 2009 (1967).
353. Divinskii, A. F., M. I. Kabachnik, and V. V. Siderenko, Dokl. Akad. Nauk SSSR, 60, 999 (1948); C.A. 43, 560g (1949).
354. Doak, G. O., and L. D. Freedman, Chem. Rev., 61, 31 (1961).
355. Drawe, H., Z. Naturforsch., B24, 934 (1969).
356. Drozd, G. I., and S. Z. Ivin, Zh. Obshch. Khim., 38, 1907 (1968); C.A. 69, 95862u (1968).
357. Drozd, G. I., O. G. Strukov, E. P. Sergeeva, S. Z. Ivin, and S. S. Dubov, Zh. Obshch. Khim., 39, 937 (1969); C.A. 71, 49168r (1969).
358. Dumas, J. B. A., and E. M. Peligot, Ann. Chem., 15, 30 (1835).
359. Dunbar, J. E., U.S. 3 100 785 (1963); C.A. 60, 551b (1964).
360. Dupuis, P., Compt. Rend., 150, 622 (1910).
361. Dyatkin, B. L., E. P. Mochalina, Yu. S. Konstantinov, S. R. Sterlin, and I. L. Knunyants, Izv. Akad. Nauk SSSR, Ser. Khim., 1967, 2297; C.A. 68, 77632u (1968).
362. Dye, W. T., Jr., U.S. 2 730 541 (1956); C.A. 50, 11362b (1956).
363. Edmundson, R. S., Chem. Ind. (London), 1963, 784.
364. Edmundson, R. S., Tetrahedron, 20, 2781 (1964).
365. Edmundson, R. S., Chem. Ind. (London), 1965, 1220.
366. Edmundson, R. S., and A. J. Lambie, J. Chem. Soc., 1966, 1997.
367. Eliseenkov, V. N., Zh. Obshch. Khim., 37, 2052 (1967); C.A. 68, 48962w (1968).
368. Eliseenkov, V. N., and V. K. Khairullin, Izv. Akad. Nauk SSSR, Ser. Khim., 1965, 2128; C.A. 64, 11074h (1966).
369. Eliseenkov, V. N., A. N. Pudovik, S. G. Fattakhov,

and N. A. Serkina, Zh. Obshch. Khim., 40, 498 (1970); C.A. 72, 131946x (1970).
370. Emerson, T. R., and C. W. Rees, J. Chem. Soc., 1962, 1921; 1964, 2319.
371. Epshtein, L. M., Z. S. Novikova, L. D. Ashkinadze, and V. N. Kostylev, Dokl. Akad. Nauk SSSR, Ser. Khim., 1969, 184, 1346; C.A. 70, 114410j (1969).
372. Ernsberger, M. L., and J. W. Hill, U.S. 2 661 364 (1953); C.A. 49, 1774b (1955).
373. Ettel, V., and M. Zbirovsky, Collection Czech. Chem. Commun., 21, 1454 (1956).
374. Ettel, V., and M. Zbirovsky, Chem. Listy, 50, 1261 (1956).
375. Ettel, V., and M. Zbirovsky, Chem. Listy, 50, 1265 (1956).
376. Euler, H., and A. Bernton, Chem. Ber., 60B, 1720 (1927).
377. Evdakov, V. P., and E. I. Alipova, Zh. Obshch. Khim., 37, 441 (1967); C.A. 67, 43880n (1967).
378. Evdakov, V. P., K. A. Bilevich, and G. P. Gizova, Zh. Obshch. Khim., 33, 3770 (1963); C.A. 60, 9134d (1964).
379. Evdakov, V. P., L. I. Mizrakh, and L. Yu. Sandalova, Zh. Obshch. Khim., 34, 3124 (1964); C.A. 61, 16089f (1964).
380. Evdakov, V. P., L. I. Mizrakh, and L. Yu. Sandalova, Dokl. Akad. Nauk SSSR, 162, 573 (1965); C.A. 63, 7038f (1965).
381. Evdakov, V. P., and E. K. Shlenkova, Zh. Obshch. Khim., 35, 1587 (1965).
382. Evdakov, V. P., and E. K. Shlenkova, U.S.S.R. 193 507 (1967); C.A. 69, 18599m (1968).
383. Evtikhov, Zh. L., N. A. Razumova, and A. A. Petrov, Dokl. Akad. Nauk SSSR, 181, 877 (1968); C.A. 70, 78084v (1969).
384. Evtikhov, Zh. L., N. A. Razumova, and A. A. Petrov, Zh. Obshch. Khim., 38, 196 (1968); C.A. 69, 77356c (1968).
385. Evtikhov, Zh. L., N. A. Razumova, and A. A. Petrov, Zh. Obshch. Khim., 38, 2341 (1968); C.A. 70, 37879q (1969).
386. Evtikhov, Zh. L., N. A. Razumova, and A. A. Petrov, Zh. Obshch. Khim., 39, 465 (1969); C.A. 70, 115236u (1969).
387. Farbenfabriken Bayer Akt.-Ges., Brit. 691 267 (1953); C.A. 48, 7047e (1954).
388. Farbenfabriken Bayer Akt.-Ges., Brit. 749 550 (1956); C.A. 51, 1243i (1957).
389. Farbenfabriken Bayer Akt.-Ges., Brit. 841 671 (1960); C.A. 55, 3433g (1961).
390. Farbenfabriken Bayer Akt.-Ges., Ger. 1 257 153

(1967); C.A. 68, 95955f (1968).
391. Fearing, R. B., and M. B. McClellan, U.S. 3 391 230 (1968); C.A. 69, 51579c (1968).
392. Feher, F., and A. Blurncke, Chem. Ber., 90, 1934 (1957).
393. Fenton, D. M., U.S. 3 342 909 (1967); C.A. 67, 108236d (1967).
394. Fertig, J., W. Gerrard, and H. Herbst, J. Chem. Soc., 1957, 1488.
395. Filatova, I. M., E. L. Zaitseva, A. P. Simonov, and A. Ya. Yakubovich, Zh. Obshch. Khim., 38, 1304 (1968); C.A. 69, 87091b (1968).
396. Finegold, H., Ann. NY Acad. Sci., 70, 875 (1958).
397. Flint, R. B., and P. L. Salzberg, U.S. 2 151 380 (1937); C.A. 33, 5097[3] (1939).
398. Fluck, E., Z. Anorg. Chem., 307, 38 (1960).
399. Fluck, E., and H. Binder, Z. Anorg. Chem., 354, 139 (1967).
400. Fluck, E., and H. Binder, Z. Naturforsch., b22, 805 (1967).
401. Fluck, E., and J. R. Van Wazer, Z. Anorg. Chem., 307, 113 (1961).
402. Folsch, G., Acta Chem. Scand., 10, 686 (1956).
403. Fontal, B., and H. Goldwhite, Tetrahedron Letters, 22, 3275 (1966).
404. Ford-Moore, A., and J. Williams, J. Chem. Soc., 1947, 1465.
405. Foss, O., Acta Chem. Scand., 1, 8 (1947).
406. Foss, O., Kgl. Norske Videnskab. Selskabs. Forh., 15, 119 (1942).
407. Fossman, J. P., and D. Lipkin, J. Am. Chem. Soc., 75, 3145 (1953).
408. Fox, R. B., "NRL Report," 1959, 5242.
409. Fox. R. B., "NRL Report," 1959, 5335.
410. Fox, R. B., and D. L. Venezky, J. Am. Chem. Soc., 75, 3967 (1953).
411. Fox, R. B., and D. L. Venezky, J. Am. Chem. Soc., 78, 1661 (1956).
412. Francina, A., A. Lamotte, and J. C. Merlin, Compt. Rend. Ser. C, 266, 1050 (1968).
413. Frank, A. W., J. Org. Chem., 30, 3663 (1965).
414. Frank, A. W., and C. F. Baranauckas, J. Org. Chem., 31, 872 (1966).
415. Frankland, E., Ann. Chem., 74, 42 (1850).
416. Frazer, M. J., W. Gerrard, and J. K. Patel, Chem. Ind. (London), 1959, 90.
417. Frazer, M. J., W. Gerrard, and J. K. Patel, Chem. Ind. (London), 1959, 728.
418. Frazer, M. J., W. Gerrard, and J. K. Patel, J. Chem. Soc., 1960, 726.
419. Freeman, K. L., and M. J. Gallagher, Aust. J. Chem.,

21, 145 (1968).
420. French, C. M., and R. C. B. Tomlinson, J. Chem. Soc., 1961, 311.
421. Fridland, S. V., G. Kh. Kamai, and T. M. Odintsova, Zh. Obshch. Khim., 39, 1986 (1969); C.A. 72, 31357a (1970).
422. Friedman, L., U.S. 3 009 939 (1961); C.A. 56, 9965d (1962).
423. Friedman, L., U.S. 3 081 331 (1963); C.A. 59, 6588h (1963).
424. Friedman, L., Fr. 1 356 931 (1964); C.A. 61, 4218h (1964).
425. Friedman, L., Fr. 1 375 860 (1964); C.A. 62, 10614g (1965).
426. Friedman, L., U.S. 3 201 437 (1965); C.A. 63, 130797 (1965).
427. Friedman, L., U.S. 3 442 982 (1969); C.A. 71, 22599v (1969).
428. Friedman, L., and H. Gould, U.S. 3 047 608 (1962); C.A. 57, 16400c (1962).
429. Furdik, M., and J. Masek, Acta Fac. Rerum Nat. Univ., Comenianace Chimica 6, 611 (1961).
430. Fursenko, I. V., G. T. Bakhvalov, and E. E. Nifant'ev, Zh. Obshch. Khim., 38, 1299 (1968); C.A. 70, 11253f (1969).
431. Fursenko, I. V., G. T. Bakhvalov, and E. E. Nifant'ev, Zh. Obshch. Khim., 38, 2528 (1968); C.A. 71, 21790p (1969).
432. Gagnaire, D., J. B. Robert, and J. Verrier, Bull. Soc. Chim., 1966, 3719.
433. Gal, H., Bull. Soc. Chim., 1873 (2), 20, 13.
434. Galbraith, A. R., P. Hale, and J. E. Robertson, J. Am. Oil Chemists' Soc., 41, 104 (1964).
435. Garbuz, N. I., M. D. Balabaeva, and R. G. Zhbankov, Spektrosk. Polim., Sb. Dokl. Vses. Simp., 1965 (Pub. 1968), 121, edited by M. V. Vol'Kenshtein, "Nauk Dumka," Kiev, U.S.S.R.; C.A. 71, 51374y (1969).
436. Garbuz, N. I., R. G. Zhbankov, D. A. Predvoditelev, E. E. Nifant'ev, and Z. A. Rogovin, Zh. Prikl. Spektrosk., 7, 854 (1967); C.A. 68, 60653d (1968).
437. Garbuz, N. I., R. G. Zhbankov, D. A. Predvoditelev, E. E. Nifant'ev, and Z. A. Rogovin, Vysokomolekul. Soedin., 8, 613 (1966); C.A. 65, 4089a (1966).
438. Garner, H. K., and H. J. Lucas, J. Am. Chem. Soc., 72, 5497 (1950).
439. Gash, V. W., J. Org. Chem., 32, 2007 (1967).
440. Gaucher, L., and G. Rollin, Compt. Rend., 172, 390 (1921).
441. Gawron, O., C. Grelecki, W. Reilly, and J. Sands, J. Am. Chem. Soc., 75, 3591 (1953).

442. Gazizov, M. B., T. I. Sobchuk, and A. I. Razumov, Zh. Obshch. Khim., 39, 2595 (1969); C.A. 72, 66527v (1970).
443. Gefter, E. L., Zh. Obshch. Khim., 28, 1908 (1958); C.A. 53, 1120a (1959).
444. Gefter, E. L., V. N. Gorbunov, and D. M. Filippenko, Zh. Obshch. Khim., 38, 2082 (1968); C.A. 70, 28425m (1969).
445. Gefter, E. L., and M. I. Kabachnik, Izv. Akad. Nauk SSSR, Ser. Khim., 1957, 194; C.A. 51, 11238d (1957).
446. Gefter, E. L., and M. I. Kabachnik, Dokl. Akad. Nauk SSSR, 114, 541 (1957); C.A. 52, 295a (1958).
447. Gerhardt, C., Ann. Chim. Phys., 1853 (3), 37, 285.
448. Gerrard, W., Chem. Ind. (London), 1951, 463.
449. Gerrard, W., J. Chem. Soc., 1940, 218, 1464.
450. Gerrard, W., J. Chem. Soc., 1944, 85.
451. Gerrard, W., J. Chem. Soc., 1945, 106, 848; 1946, 741.
452. Gerrard, W., J. Chim. Phys., 1964, 73, and papers cited therein.
453. Gerrard, W., Educ. Chem., 3, 267 (1966).
454. Gerrard, W., Chem. Ind. (London), 1969, 29.
455. Gerrard, W., Chem. Ind. (London), 1969, 146.
456. Gerrard, W., Educ. Chem., 6, 159 (1969).
457. Gerrard, W., Educ. Chem., 7, 167 (1970).
458. Gerrard, W., and W. J. Green, J. Chem. Soc., 1951, 2550.
459. Gerrard, W., W. J. Green, and R. J. Phillips, J. Chem. Soc., 1954, 1148.
460. Gerrard, W., and P. F. Griffey, Chem. Ind. (London), 1959, 55.
461. Gerrard, W., and P. F. Griffey, J. Chem. Soc., 1960, 3170.
462. Gerrard, W., and P. F. Griffey, J. Chem. Soc., 1961, 4095.
463. Gerrard, W., and H. Herbst, J. Chem. Soc., 1955, 277.
464. Gerrard, W., and B. K. Howe, J. Chem. Soc., 1955, 505.
465. Gerrard, W., and H. R. Hudson, J. Chem. Soc., 1964, 2310.
466. Gerrard, W., H. R. Hudson, and F. W. Parrett, Nature (London), 211, 740 (1966).
467. Gerrard, W., M. J. D. Isaacs, M. Machell, K. B. Smith, and P. L. Wyvill, J. Chem. Soc., 1953, 1920.
468. Gerrard, W., and G. J. Jeacocke, Chem. Ind. (London), 1954, 1538.
469. Gerrard, W., and G. J. Jeacocke, J. Chem. Soc., 1954, 3647.
470. Gerrard, W., and W. Lessing, Rec. Trav. Chim., 66, 463 (1947); Nature (London), 160, 467 (1947).

471. Gerrard, W., and M. Lindsay, Chem. Ind. (London), 1960, 152.
472. Gerrard, W., and E. D. Macklen, Chem. Rev., 59, 1105 (1959).
473. Gerrard, W., and V. Maladkar, Chem. Ind. (London), 1970, 925.
474. Gerrard, W., and A. Nechvatal, Nature (London), 159, 812 (1947).
475. Gerrard, W., A. Nechvatal, and B. M. Wilson, J. Chem. Soc., 1950, 2088.
476. Gerrard, W., and N. H. Philip, Research (London), 1, 477 (1948).
477. Gerrard, W., and M. J. Richmond, J. Chem. Soc., 1945, 853.
478. Gerrard, W., and B. D. Shepherd, J. Chem. Soc., 1953, 2069.
479. Gerrard, W., and A. M. Thrush, J. Chem. Soc., 1952, 741; 1953, 2117.
480. Gerrard, W., and E. G. C. Whitbread, J. Chem. Soc., 1952, 914.
481. Gerrard, W., and P. L. Wyvill, Research (London), 2, 536 (1949).
482. Gertsev, V. V., L. A. Vladimirova, and A. V. Karyakin, Zh. Obshch. Khim., 39, 1558 (1969); C.A. 71, 112530v (1969).
483. Gilbert, E. E., and C. J. McGough, U.S. 2 690 450 (1954); C.A. 49, 11682i (1955); 2 690 451 (1954); C.A. 49, 11683c (1955).
484. Gilbert, E. E., and J. A. Otto, U.S. 745 862 (1956); C.A. 51, 1243h (1957).
485. Gilliam, W. F., R. N. Meals, and R. O. Sauer, J. Am. Chem. Soc., 68, 1161 (1946).
486. Gilman, H., and J. Robinson, Rec. Trav. Chim., 48, 328 (1929).
487. Gilman, H., and C. C. Vernon, J. Am. Chem. Soc., 48, 1063 (1926).
488. Gilyarov, V. A., Khim. i Primenenie Fosfororgan Soedinenii, Akad. Nauk SSSR, Trudy I-oi Konferents., 1955, 275; C.A. 52, 243b (1958).
489. Ginsburg, V. A., M. N. Vasil'eva, S. S. Dubov, and A. Ya. Yakubovich, Zh. Obshch. Khim., 30, 2854 (1960); C.A. 55, 17477b (1961).
490. Ginsburg, V. A., and A. Ya. Yakubovich, Zh. Obshch. Khim., 30, 3979 (1960); C.A. 55, 22099c (1961).
491. Ginsburg, V. A., and A. Ya. Yakubovich, Zh. Obshch. Khim., 30, 3987 (1960); C.A. 55, 25743c (1961).
492. Goldschmidt, S., and H. L. Krauss, Ann. Chem., 595, 193 (1955).
493. Goldschmidt, S., and F. Obermeier, Ann. Chem., 588, 24 (1954).
493a. Goldwhite, H., Chem. Ind. (London), 1964, 494.

494. Goldwhite, H., and B. C. Saunders, J. Chem. Soc., 1957, 2409.
495. Gololobov, Yu. G., and L. Z. Soborovskii, Zh. Obshch. Khim., 33, 2955 (1963); C.A. 60, 1572e (1964).
496. Gololobov, Yu. G., I. A. Zaishlova, and A. S. Buntyakov, Zh. Obshch. Khim., 35, 1240 (1965); C.A. 63, 13061g (1965).
496a. Goodell, L. J., and J. T. Yoke, Can. J. Chem., 47, 2461 (1969).
497. Goodwin, D. G., and H. R. Hudson, J. Chem. Soc. B, 1968, 1333.
498. Gottlieb, H. B., J. Am. Chem. Soc., 54, 748 (1932).
499. Goyette, L. E., U.S. 3 037 043 (1962); C.A. 57, 13681f (1962).
500. Gramstad, T., Acta Chem. Scand., 15, 1337 (1961).
501. Grechkin, N. P., Khim. Primenenie Fosfororgan. Soedinenii, Akad. Nauk SSSR, Trudy 1-oi Konferents., 1955, 243; C.A. 52, 241 (1958).
502. Grechkin, N. P., and L. N. Grishina, Izv. Akad. Nauk SSSR, Ser. Khim., 1965, 1502; C.A. 63, 16280f (1965).
503. Grechkin, N. P., and L. N. Grishina, Izv. Akad. Nauk SSSR, Ser. Khim., 1967, 1990; C.A. 68, 114506d (1968).
504. Grechkin, N. P., and L. N. Grishina, Izv. Akad. Nauk SSSR, Ser. Khim., 1969, 1608; C.A. 71, 12721h (1969).
505. Grechkin, N. P., and L. N. Grishina, Izv. Akad. Nauk SSSR, Ser. Khim., 1969, 1608; C.A. 71, 112721h (1969).
506. Grechkin, N. P., I. A. Nuretdinov, and N. A. Buina, Izv. Akad. Nauk SSSR, Ser. Khim., 1969, 168; C.A. 70, 115226r (1969).
507. Grechkin, N. P., R. R. Shagidullin, and L. N. Grishina, Izv. Akad. Nauk SSSR, Ser. Khim., 1968, 854; C.A. 69, 86264y (1968).
508. Green, M., R. I. Hancock, and D. C. Wood, J. Chem. Soc. A, 1968, 2718.
509. Greenbaum, S. B., and N. E. Boyer, U.S. 3 146 253 (1964); C.A. 62, 1678a (1965).
510. Griffin, C. E., Chem. Ind. (London), 1958, 415; Am. Chem. Soc. Abs. of 135th Meeting, 1959.
511. Griffin, C. E., U.S. Dep. Comm. AD 627269 (1965); C.A. 66, 80730p (1967).
512. Griffin, C. E., W. G. Bentrude, and G. M. Johnson, Tetrahedron Letters, 1969, 969.
513. Griffin, C. E., and N. T. Castellucci, J. Org. Chem., 26, 629 (1961).
514. Griffin, C. E., and T. D. Mitchell, J. Org. Chem., 30, 1935 (1965).

515. Griffiths, J. E., and A. B. Burg, J. Am. Chem. Soc., 82, 1508 (1960).
516. Grimmel, H. W., A. Guenther, and J. F. Morgan, J. Am. Chem. Soc., 68, 539 (1946).
517. Gross, H., and B. Costisella, Org. Prep. Proced., 1, 97 (1969).
518. Gryszkiewicz-Trochimowskii, E., Mem. Poudres, 44, 133 (1962).
519. Gubnitskaya, E. S., and G. I. Derkach, Zh. Obshch. Khim., 38, 1530 (1968); C.A. 70, 87914t (1969).
520. Gurvich, Ya. A., P. A. Kirpichnikov, A. I. Karpycheva, Yu. B. Zimin, L. M. Popova, I. N. Serebryakova, and N. V. Tsirul'nikova, Sin. Issled. Eff. Stabil. Polim. Mater. Voronezh, 1964, 196; C.A. 66, 115421t (1967).
521. Guttag, A., U.S. 3 326 939 (1967); C.A. 68, 12859d (1968).
522. Guttag, A., U.S. 3 437 720 (1969); C.A. 71, 21682e (1969).
523. Gzemski, F. C., U.S. 2 353 558 (1944); C.A. 38, 6548[6] (1944).
524. Hall, L. A. R., and C. W. Stephens, J. Am. Chem. Soc., 78, 2565 (1956).
524a. Halman, M., J. Chem. Soc., 1954, 2158.
524b. Halman, M., J. Chem. Soc., 1963, 2853.
525. Hamilton, W. C., S. J. LaPlaca, and F. Ramirez, J. Am. Chem. Soc., 87, 127 (1965).
526. Hanriot, M., Bull. Soc. Chim. (2), 32, 551 (1879).
527. Harless, H. R., Anal. Chem., 33, 1387 (1961).
528. Harman, D., and A. R. Stiles, U.S. 2 630 450 (1953); C.A. 48, 7047h (1954).
529. Harris, G. S., and D. S. Payne, J. Chem. Soc., 1956, 3038.
530. Harris, W. D., and A. W. Feldman, U.S. 2 828 198 (1958); C.A. 53, 1624d (1959).
531. Harvey, R. G., and E. R. De Sombre, in Topics in Phosphorus Chemistry, M. Grayson, and E. J. Griffiths, Eds., Interscience, New York, 1964, Vol. I.
532. Harvey, R. G., H. I. Jacobson, and E. V. Jensen, J. Am. Chem. Soc., 85, 1618 (1963).
533. Harvey, R. G., H. I. Jacobson, and E. V. Jensen, J. Am. Chem. Soc., 85, 1623 (1963).
534. Harvey, R. G., T. C. Myers, H. I. Jacobson, and E. V. Jensen, J. Am. Chem. Soc., 79, 2612 (1957).
535. Hasserodt, U., K. H. G. Pilgram, and F. W. A. G. K. Korte, U.S. 3 247 113 (1966); C.A. 64, 19679c (1966).
536. Hasserodt, U., K. H. G. Pilgram, and F. W. A. G. K. Korte, U.S. 3 318 915 (1967); C.A. 68, 13169r (1968).
537. Hata, T., and T. Mukaiyama, Bull. Chem. Soc. Japan,

35, 1106 (1962).
538. Hata, T., and T. Mukaiyama, Bull. Chem. Soc. Japan, 37, 103 (1964).
539. Haufe, T. B., W. Springs, and J. O. Iverson, U.S. 2 631 161 (1953); C.A. 48, 7047c (1954).
540. Hechenbleikner, I., U.S. 2 852 551 (1958); C.A. 53, 4212i (1959).
541. Hechenbleikner, I., U.S. 3 147 298 (1964); C.A. 64, 8086b (1966).
542. Hechenbleikner, I., and A. T. Gaul, Ger. 1 075 584 (1960); C.A. 55, 15348e (1961).
543. Hechenbleikner, I., and A. T. Gaul, U.S. 2 841 607 (1958); C.A. 53, 770a (1959).
544. Hechenbleikner, I., and A. T. Gaul, U.S. 3 056 823 (1962); C.A. 58, 10081d (1963).
545. Hechenbleikner, I., and F. C. Lanoue, U.S. 2 847 443 (1958); C.A. 52, 19251i (1958).
546. Hechenbleikner, I., C. W. Pause, and F. C. Lanoue, U.S. 2 834 798 (1958); C.A. 52, 17297c (1958).
547. Hecker, A. C., O. H. Knoepke, W. E. Leistner, and M. W. Pollock, Brit. 901 703 (1962); C.A. 58, 1404d (1963).
548. Hecker, A. C., O. H. Knoepke, W. E. Leistner, and M. W. Pollock, U.S. 3 056 824 (1962); C.A. 58, 5575b (1963).
549. Henning, H. G., and M. Morr, Chem. Ber., 101, 3963 (1968).
549a. Henry, L., Chem. Ber., 8, 398 (1875); 17, 1153 (1884).
550. Henry, L., Rec. Trav. Chim., 1905, 24, 332.
551. Hoffmann, F. W., R. J. Ess, T. C. Simmons, and R. S. Hanzel, J. Am. Chem. Soc., 78, 6414 (1956).
552. Hoffmann, F. W., R. J. Ess, and R. P. Usinger, J. Am. Chem. Soc., 78, 5817 (1956).
553. Hoffmann, F. W., T. R. Moore, and B. Kagan, J. Am. Chem. Soc., 78, 6413 (1956).
554. Hoffmann, F. W., and H. D. Weiss, J. Am. Chem. Soc., 79, 4759 (1957).
555. Hoffmann, H., H. Forster, and G. Tor-Poghossian, Montash. Chem., 100, 311 (1969).
556. Holmes, R. R., J. Am. Chem. Soc., 82, 5285 (1960).
557. Holmes, R. R., J. Am. Chem. Soc., 83, 1334 (1961).
558. Holmes, R. R., and E. F. Bertaut, J. Am. Chem. Soc., 80, 2980 (1958).
559. Holmes, R. R., and J. A. Forstner, J. Phys. Chem., 64, 1295 (1960).
560. Holmes, R. R., and J. A. Forstner, J. Am. Chem. Soc., 82, 5509 (1960).
561. Holmes, R. R., and J. A. Forstner, Inorg. Chem., 2, 377 (1963).
562. Holmes, R. R., and J. A. Forstner, Inorg. Chem., 2,

380 (1963).
563. Holmes, R. R., and R. P. Wagner, J. Am. Chem. Soc., 84, 357 (1962).
564. Holmes, R. R., and R. P. Wagner, Inorg. Chem., 2, 384 (1963).
565. Holtschmidt, H., and G. Oertel, Ger. 1 257 153 (1967); C.A. 68, 95955f (1968).
566. Hooker Chemical Corp., Brit. 889 338 (1962); C.A. 58, 8906f (1963).
567. Hooker Chemical Corp., Brit. 937 560 (1963); C.A. 60, 15733g (1964).
568. Horton, C. A., and J. C. White, Talanta, 7, 215 (1961).
569. Houalla, D., M. Sanchez, and R. Wolf, Bull. Soc. Chim. France, 1965, 2368.
570. Houalla, D., and R. Wolf, Compt. Rend., 247, 482 (1958).
571. Houalla, D., and R. Wolf, Bull. Soc. Chim. France, 1960, 129.
572. Houben-Weyl, J., Method. Organisch. Chemie, 3 (2), 15 (1923).
573. Houssa, A. J. H., and H. Phillips, J. Chem. Soc., 1932, 108.
574. Hu, S.-En, U.S. 3 459 662 (1969); C.A. 71, 83328c (1969).
575. Hudson, H. R., J. Chem. Soc. B, 1968, 664.
576. Hudson, H. R., Synthesis, 1969, 112.
577. Hudson, H. R., and (Mrs.) D. Ragoonanan, J. Chem. Soc. B, 1970, 1755.
578. Hudson, R. F., R. J. G. Searle, and F. H. Devitt, J. Chem. Soc. C, 1966, 1001.
579. Hunt, B. B., and B. C. Saunders, J. Chem. Soc., 1957, 2413.
580. Hurwitz, M. I., and A. Carson, U.S. 3 087 958 (1963); C.A. 59, 9794c (1963).
581. Huyser, E. S., and J. A. Dieter, J. Org. Chem., 33, 4205 (1968).
582. I. G. Farbenindustrie, A. G., U.S. 1 936 985 (1933); C.A. 28, 1151[6] (1934).
583. I. G. Farbenindustrie, A. G., Fr. 807 769 (1936); C.A. 31, 5934[3] (1937).
584. Ignat'ev, V. M., B. I. Ionin, and A. A. Petrov, Zh. Obshch. Khim., 36, 1505 (1966); C.A. 66, 11010c (1967).
585. Ignat'ev, V. M., B. I. Ionin, and A. A. Petrov, Zh. Obshch. Khim., 37, 1898 (1967); C.A. 68, 29796d (1968).
586. Ignat'ev, V. M., B. I. Ionin, and A. A. Petrov, Zh. Obshch. Khim., 37, 2135 (1967); C.A. 68, 29787b (1968).
587. Imaev, M. G., Zh. Obshch. Khim., 31, 1762 (1961);

C.A. 55, 24531g (1961).
588. Imaev, M. G., Zh. Obshch. Khim., 31, 1767 (1961); C.A. 55, 24531i (1961).
589. Imaev, M. G., Zh. Obshch. Khim., 31, 1770 (1961); C.A. 55, 24532c (1961).
590. Imaev, M. G., and I. S. Akhmetzhanov, Zh. Obshch. Khim., 36, 85 (1966); C.A. 64, 14082f (1966).
591. Imaev, M. G., and I. S. Akhmetdzhanov, Khim. Seraorg. Soedin., Soderzh. Neftyakh. Niefteprod., 8, 135 (1968); C.A. 71, 83204j (1969).
592. Imaev, M. G., and R. A. Faskhutdinova, Zh. Obshch. Khim., 31, 2934 (1961); C.A. 57, 7090c (1962).
593. Imaev, M. G., V. M. Govina, and V. G. Maslennikov, Zh. Obshch. Khim., 35, 372 (1965); C.A. 62, 13042b (1965).
594. Inukai, K., T. Ueda, and H. Maramatsu, J. Org. Chem., 29, 2224 (1964).
595. Iselin, B. M., W. Rittel, P. Sieber, and R. Schwyzer, Helv. Chim. Acta, 40, 373 (1957).
596. Issleib, K., and W. Seidel, Chem. Ber., 92, 2681 (1959).
597. Italinskaya, T. L., N. N. Mel'nikov, and N. I. Shvetsov-Shilovski, Zh. Obshch. Khim., 38, 2265 (1968); C.A. 70, 37737s (1969).
598. Ivakina, N. M., Yu. A. Kondrat'ev, and S. Z. Ivin, Zh. Obshch. Khim., 37, 1691 (1967); C.A. 68, 39737z (1968).
599. Ivanov, B. E., and A. B. Ageeva, Izv. Akad. Nauk SSSR, Ser. Khim., 1967, 226; C.A. 67, 11538r (1967).
600. Ivanov, B. E., A. B. Ageeva, A. G. Abul'khanov, and T. A. Zyablikov, Izv. Akad. Nauk SSSR, Ser. Khim., 1969, 1912; C.A. 72, 21110n (1970).
601. Ivanov, B. E., A. B. Ageeva, S. V. Pasmanyuk, R. R. Shagidullin, S. G. Salikhov, and E. I. Loginova, Izv. Akad. Nauk SSSR, Ser. Khim., 1969, 1757; C.A. 72, 3531w (1970).
602. Ivanov, B. E., A. B. Ageeva, and Yu. Yu. Samitov, Dokl. Akad. Nauk SSSR, 174, 846 (1967); C.A. 68, 39736y (1968).
603. Ivanov, B. E., A. B. Ageeva, and R. R. Shagidullin, Izv. Akad. Nauk SSSR, Ser. Khim., 1967, 1994; C.A. 68, 69079d (1968).
604. Ivanov, B. E., and L. A. Khismatullina, Izv. Akad. Nauk SSSR, Ser. Khim., 1968, 2150; C.A. 70, 20164e (1969).
605. Ivanov, B. E., L. A. Kudryavtseva, T. G. Bykova, and T. A. Zyablikova, Izv. Akad. Nauk SSSR, Ser. Khim., 1969, 1851; C.A. 72, 31922t (1970).
606. Ivanova, Zh. M., E. S. Levchenko, and A. V. Kirsanov, Zh. Obshch. Khim., 35, 1607 (1965); C.A. 63, 17949f (1965).

607. Ivanov, B. E., and S. V. Pasmanyuk, Izv. Akad. Nauk SSSR, Ser. Khim., 1969, 138; C.A. 70, 115263a (1969).
608. Ivanov, B. E., L. A. Valitova, and T. G. Vavilova, Izv. Akad. Nauk SSSR, Ser. Khim., 1968, 768; C.A. 69, 76330c (1968).
609. Ivanov, B. E., and V. F. Zheltukhin, Izv. Akad. Nauk SSSR, Ser. Khim., 1967, 1878; C.A. 68, 29784y (1968).
610. Ivanov, B. E., and V. F. Zheltukhin, Izv. Akad. Nauk SSSR, Ser. Khim., 1969, 1016, 1022; C.A. 71, 48940z (1969).
611. Ivanov, B. E., V. F. Zheltukhin, and R. R. Shagidullin, Izv. Akad. Nauk SSSR, Ser. Khim., 1968, 2614; C.A. 70, 56892e (1969).
612. Ivanov, B. E., V. F. Zheltukhin, and V. G. Sofronova, Izv. Akad. Nauk SSSR, Ser. Khim., 1967, 940; C.A. 68, 2957v (1968).
613. Ivanov, B. E., V. F. Zheltukhin, and T. G. Vavilov, Izv. Akad. Nauk SSSR, Ser. Khim., 1967, 1285; C.A. 68, 22015q (1968).
614. Ivasyuk, N. V., and I. M. Shermergorn, Izv. Akad. Nauk SSSR, Ser. Khim., 1968, 2388; C.A. 70, 29005m (1969).
615. Ivin, S. Z., V. N. Pastushkov, Yu. A. Kondrat'ev, K. F. Ogloblin, and V. V. Tarasov, Zh. Obshch. Khim., 38, 2069 (1968); C.A. 70, 11759a (1969).
616. Ivin, S. Z., V. K. Promonenkov, A. S. Baberkin, E. V. Volkova, N. F. Sarafanova, and E. A. Fokin, Zh. Prikl. Khim., 42, 472 (1969); C.A. 70, 110484p (1969).
617. Ivin, S. Z., V. K. Promonenkov, and E. A. Fokin, Zh. Obshch. Khim., 37, 1642 (1967); C.A. 68, 12378w (1968).
618. Ivin, S. Z., V. K. Promonenkov, and E. A. Fokin, Zh. Obshch. Khim., 37, 2511 (1967); C.A. 69, 26700c (1968).
619. Jacobson, H. I., R. G. Harvey, and E. V. Jensen, J. Am. Chem. Soc., 77, 6064 (1955).
620. Jahne, O., Ann., 256, 269 (1890).
621. Janczak, M., Rocz. Chem., 4, 180 (1924).
622. Janczak, M., Rocz. Chem., 6, 774 (1926).
623. Jason, E. F., and E. K. Fields, J. Org. Chem., 27, 1402 (1962).
624. Jenker, H., Brit. 1 122 450 (1968); C.A. 69, 108229v (1968).
625. Jonas, H., Ger. 835 145 (1954); C.A. 49, 12529e (1955).
626. Jonas, H., and W. Thraum, Ger. 872 040 (1953); C.A. 52, 16197g (1958).
627. Jones, P. H., and E. A. Rowley, U.S. 3 361 738

(1968); C.A. 69, 44185t (1968).
628. Jones, R. A. Y., and A. R. Katritzky, Angew. Chem. Internat. Edit., 1, 32 (1962).
629. Kabachnik, M. I., Izv. Akad. Nauk SSSR, 1947, 631; C.A. 42, 5845f (1948).
630. Kabachnik, M. I., Izv. Akad. Nauk SSSR, Ser. Khim., 1948, 219; C.A. 42, 5736g (1948).
631. Kabachnik, M. I., C. Y. Chang, and E. N. Tsvelkov, Zh. Obshch. Khim., 32, 3351 (1962); C.A. 58, 9126g (1963).
632. Kabachnik, M. I., V. A. Gilyarov, and E. M. Popov, Zh. Obshch. Khim., 32, 1598 (1962); C.A. 58, 4401b (1963).
633. Kabachnik, M. I., V. A. Gilyarov, and E. M. Popov, Izv. Akad. Nauk SSSR, Ser. Khim., 1961, 1022; C.A. 55, 27014f (1961).
634. Kabachnik, M. I., E. I. Golubeva, D. M. Paikin, M. P. Shabanova, N. M. Gamper, and L. F. Efimova, Zh. Obshch. Khim., 29, 1671 (1959); C.A. 54, 8594a (1960).
635. Kabachnik, M. I., N. I. Kurochkin, T. A. Mastryukova, S. T. Ioffe, E. M. Popov, and N. P. Rodionova, Dokl. Akad. Nauk SSSR, 104, 861 (1955); C.A. 50, 11240a (1956).
636. Kabachnik, M. I., and T. A. Mastryukova, Izv. Akad. Nauk SSSR, Ser. Khim., 1952, 727; C.A. 47, 9909a (1953).
637. Kabachnik, M. I., and T. A. Mastryukova, Izv. Akad. Nauk SSSR, Ser. Khim., 1953, 163; C.A. 48, 3243d (1954).
638. Kabachnik, M. I., T. A. Mastryukova, and A. E. Shilov, Zh. Obshch. Khim., 33, 320 (1963); C.A. 59, 658c (1963).
639. Kabachnik, M. I., T. A. Mastryukova, and A. E. Shipov, Zh. Obshch. Khim., 35, 1574 (1965); C.A. 63, 18145b (1965).
640. Kabachnik, M. I., and T. Ya. Medved, Izv. Akad. Nauk SSSR, Ser. Khim., 1952, 540; C.A. 47, 4848b (1953).
641. Kabachnik, M. I., and T. Ya. Medved, Izv. Akad. Nauk SSSR, Ser. Khim., 1966, 1365; C.A. 66, 54802u (1967).
642. Kabachnik, M. I., and Yu. M. Polikarpov, Dokl. Akad. Nauk SSSR, 115, 512 (1957); C.A. 52, 5327e (1958).
643. Kabachnik, M. I., and P. A. Rossiiskaya, Izv. Akad. Nauk SSSR, Ser. Khim., 1946, 295; C.A. 42, 7241f (1948).
644. Kabachnik, M. I., and P. A. Rossiiskaya, Izv. Akad. Nauk SSSR, Ser. Khim., 1946, 403; C.A. 42, 7242c (1948).
645. Kabachnik, M. I., and P. A. Rossiiskaya, Izv. Akad.

Nauk SSSR, Ser. Khim., 1946, 515; C.A. 42, 5846e (1948).
646. Kabachnik, M. I., and P. A. Rossiiskaya, Izv. Akad. Nauk SSSR, Ser. Khim., 1948, 95; C.A. 42, 7242i (1948).
647. Kabachnik, M. I., P. A. Rossiiskaya, M. P. Shabanova, D. M. Paikin, L. F. Efimova, and N. M. Gamper, Zh. Obshch. Khim., 30, 2218 (1960); C.A. 57, 4531f (1962).
648. Kabachnik, M. I., and E. N. Tsvetkov, Dokl. Akad. Nauk SSSR, 117, 817 (1957); C.A. 52, 8070c (1958).
649. Kabachnik, M. I., and E. N. Tsvetkov, Izv. Akad. Nauk SSSR, Ser. Khim., 1960, 133; C.A. 54, 20822d (1960).
650. Kabachnik, M. I., E. N. Tsvetkov, and C. Y. Chang, Dokl. Akad. Nauk SSSR, 131, 1334 (1960); C.A. 54, 20845a (1960).
651. Kairullin, V. K., A. I. Ledeneva, and V. S. Abramov, Zh. Obshch. Khim., 29, 2551 (1959); C.A. 54, 10835g (1960).
652. Kalibabchuk, N. N., G. V. Sandul, and V. D. Pokhodenko, Zh. Obshch. Khim., 39, 2140 (1969); C.A. 72, 31280g (1970).
653. Kamai, G., Trudy Kazansk. Khim. Tekhnol. Inst., 8, 33 (1940); C.A. 35, 2856[3] (1941).
654. Kamai, G., Trudy Kazansk. Khim. Tekhnol. Inst. S. M. Kirova, 1946, 29; C.A. 51, 6503e (1957).
655. Kamai, G., Dokl. Akad. Nauk SSSR, 79, 795 (1951); C.A. 46, 6081f (1952).
656. Kamai, G., and E. Sh. Basmanov, Zh. Obshch. Khim., 21, 2188 (1951); C.A. 46, 7517h (1952).
657. Kamai, G., and A. P. Bogdonov, Trudy Kazansk. Khim. Tekhnol. Inst., 18, 22 (1953); C.A. 51, 5721b (1957).
658. Kamai, G., and L. P. Egorova, Zh. Obshch. Khim., 16, 1521 (1946); C.A. 41, 5439g (1947); Dokl. Akad. Nauk SSSR, 55, 223 (1947).
659. Kamai, G., and E. A. Gerasimova, Trudy Kazansk. Khim. Tekhnol. Inst., 15, 26 (1950); C.A. 51, 11273i (1957).
660. Kamai, G., and F. M. Kharassova, Trudy Kazansk. Khim. Tekhnol. Inst. S. M. Kirova, 23, 127 (1957); C.A. 52, 9946c (1958).
661. Kamai, G., and F. M. Kharrasova, Trudy Kazansk. Khim. Tekhnol. Inst. S. M. Kirova, 23, 122 (1957); C.A. 52, 9980i (1958).
662. Kamai, G., and F. M. Kharrasova, Izv. Vysshikh. Ucheb. Zavedenii, Khim. Khim. Tekhnol., 4, 229 (1961); C.A. 55, 21762e (1961).
663. Kamai, G., and F. M. Kharrasova, Zh. Obshch. Khim., 27, 953 (1957); C.A. 52, 3666a (1958).

664. Kamai, G., F. M. Kharrasova, G. I. Rakhimova, and R. B. Sultanova, Zh. Obshch. Khim., 39, 625 (1969); C.A. 71, 50104y (1969).
665. Kamai, G., and A. S. Khasanov, Zavedenii, Khim. Khim. Tekhnol., 6, 799 (1963); C.A. 60, 7961c (1964).
666. Kamai, G., and E. S. Koshkina, Kazan. Khim. Tekhnol. Inst., 17, 11 (1953); C.A. 50, 6346i (1956).
667. Kamai, G., and V. A. Kukhtin, Khim. Primenenie Fosfororgan. Soedinenii Akad. Nauk SSSR, Trudy 1-oi Konferents., 1955, 91; C.A. 52, 241b (1958).
668. Kamai, G., and V. A. Kukhtin, Trudy Kazansk. Khim. Tekhnol. Inst. S. M. Kirova, 23, 133 (1957); C.A. 52, 9948f (1958).
669. Kamai, G., and V. A. Kukhtin, Dokl. Akad. Nauk SSSR, 112, 868 (1957); C.A. 51, 13742f (1957).
670. Kamai, G., and V. A. Kukhtin, Zh. Obshch. Khim., 27, 2372, 2376 (1957); C.A. 52, 7127d (1958).
671. Kamai, G., V. A. Kukhtin, and O. A. Strogova, Trudy Kazansk. Khim. Tekhnol. Inst., 21, 155 (1956); C.A. 51, 11994b (1957).
672. Kamai, G., E. V. Kuznetsov, and R. K. Valetdinov, Dokl. Akad. Nauk SSSR, 116, 965 (1957); C.A. 52, 8037c (1958).
673. Kamai, G., E. V. Kuznetsov, and R. Valetdinov, Trudy Kazansk. Khim. Tekhnol. Inst., S. M. Kirova, 23, 190 (1957); C.A. 52, 8940f (1958).
674. Kamai, G., and E. T. Mukmenev, Trudy Kazansk. Khim. Tekhnol. Inst., 30, 296 (1962); C.A. 60, 3993d (1964).
675. Kamai, G., and E. T. Mukmenev, Zh. Obshch. Khim., 33, 3197 (1963); C.A. 60, 3993f (1964); Dokl. Akad. Nauk SSSR, 153, 605 (1963); C.A. 60, 6737k (1964).
676. Kamai, G., and E. I. Shugurova, Dokl. Akad. Nauk SSSR, 72, 301 (1950); C.A. 45, 542f (1951).
677. Kasparek, F., Acta Univ. Palacki. Olomue., Fac. Rerum Natur., 1966, 221; C.A. 70, 47520e (1969).
678. Keay, L., and E. M. Crook, J. Chem. Soc., 1961, 710.
679. Keeber, W. H., and H. W. Post, J. Org. Chem., 21, 509 (1956).
680. Keiter, R. L., and J. G. Verkade, Inorg. Chem., 9, 404 (1970).
681. Kenner, G. W., A. R. Todd, and F. J. Weymouth, J. Chem. Soc., 1952, 3675.
682. Kenyon, J., A. G. Lipscomb, and H. Phillips, J. Chem. Soc., 1931, 2275.
683. Kenyon, J., H. Phillips, and F. M. H. Taylor, J. Chem. Soc., 1931, 382.
684. Kenyon, J., H. Phillips, and V. P. Pittman, J. Chem. Soc., 1935, 1072.
685. Kepler, J. A., F. I. Carroll, R. A. Garner, and

M. E. Wall, J. Org. Chem., 31, 105 (1966).
686. Khairullin, V. K., and V. N. Eliseenkov, Izv. Akad. Nauk SSSR, Ser. Khim., 1967, 94; C.A. 66, 94645b (1967).
687. Khairullin, V. K., and T. I. Sobchuk, Izv. Akad. Nauk SSSR, Ser. Khim., 1963, 320; C.A. 58, 13984c (1963).
688. Khairullin, V. K., and T. I. Sobchuk, Izv. Akad. Nauk SSSR, Ser. Khim., 1965, 2010; C.A. 64, 8062f (1966).
689. Khairullin, V. K., T. I. Sobchuk, and A. N. Pudovik, Zh. Obshch. Khim., 38, 584 (1968); C.A. 69, 59347w (1968).
690. Khairullin, V. K., M. A. Vasyanina, and A. N. Pudovik, Izv. Akad. Nauk SSSR, Ser. Khim., 1967, 1603; C.A. 68, 13073e (1968).
691. Kharash, M. S., and J. S. Bengelsdorf, J. Org. Chem., 20, 1356 (1955).
692. Khasanov, A. S., I. N. Azerbaev, R. D. Kovaleko, Z. A. Navrezova, N. Z. Gabdullina, and M. A. Talasbaeva, Trudy Khim. Met. Inst., Akad. Nauk Kazansk. SSR, 5, 71 (1969); C.A. 72, 121113j (1970).
693. Khaskin, A. N., P. M. Zavlin, and B. I. Ionin, Zh. Obshch. Khim., 39, 176 (1969); C.A. 70, 105878r (1969).
694. Kirillova, K. M., and V. A. Kukhtin, Zh. Obshch. Khim., 32, 2338 (1962); C.A. 58, 9128c (1963).
695. Kirilov, M., and P. Nedkov, Compt. Rend. Acad. Bulgare Sci., 10, 309 (1957).
696. Kirillova, K. M., and V. A. Kukhtin, Zh. Obshch. Khim., 35, 544 (1965); C.A. 63, 523c (1965).
697. Kirillova, K. M., V. A. Kukhtin, and T. M. Sudakova, Dokl. Akad. Nauk 149, 316 (1963); C.A. 59, 7556c (1963).
698. Kirpichnikov, P. A., Ya. A. Gurvich, and G. P. Gren, Zh. Obshch. Khim., 35, 744 (1965); C.A. 63, 4195c (1965).
699. Kirpichnikov, P. A., M. V. Ivanova, and Ya. A. Gurvich, Zh. Obshch. Khim., 36, 1147 (1966); C.A. 65, 10519h (1966).
700. Kirpichnikov, P. A., and V. Ph. Kadyrova, Trudy Kazansk. Khim. Tekhnol. Inst., 33, 193 (1964); C.A. 64, 12815f (1966).
701. Kirpichnikov, P. A., G. Kamai, and R. Sh. Khisamutdinova, Zh. Obshch. Khim., 34, 434 (1964); C.A. 60, 13170b (1964).
702. Kirpichnikov, P. A., N. S. Kolyubakina, K. S. Minsker, N. A. Mukmeneva, and E. G. Chebotareva, Zh. Anal. Khim., 23, 1582 (1968); C.A. 70, 43907p (1969).
703. Kirpichnikov, P. A., E. A. Krasit'nikova, and I. E.

Saratov, Trudy Kazansk. Khim. Tekhnol. Inst., 30, 52 (1962); C.A. 60, 9179c (1964).
704. Kirpichnikov, P. A., A. S. Kuz'minskii, L. M. Popova, and V. N. Spiridonova, Trudy Kazansk. Khim. Tekhnol. Inst., 30, 47 (1962); C.A. 60, 9183e (1964).
705. Kirpichnikov, P. A., E. T. Mukmenev, O. V. Voskresenskaya, and E. I. Vorkunova, Trudy Kazansk. Khim. Tekhnol. Inst. S. M. Kirova, 34, 367 (1965).
706. Kirpichnikov, P. A., N. A. Mukmeneva, A. N. Pudovik, and N. S. Kolyubakina, Dokl. Akad. Nauk SSSR, 164, 1050 (1965); C.A. 64, 1926d (1966).
707. Kirpichnikov, P. A., and L. M. Popova, Zh. Obshch. Khim., 35, 1026 (1965); C.A. 63, 9844g (1965).
708. Kirpichnikov, P. A., L. M. Popova, and G. Ya. Richmond, Zh. Obshch. Khim., 36, 1143 (1966); C.A. 65, 10520c (1966).
709. Kirsanov, A. V., and V. I. Shevchenko, Zh. Obshch. Khim., 26, 74 (1956); C.A. 50, 13786h (1956).
710. Kirsanov, A. V., and V. I. Shevchenko, Zh. Obshch. Khim., 26, 250 (1956); C.A. 50, 13783e (1956).
711. Kirsanov, A. V., and V. I. Shevchenko, Zh. Obshch. Khim., 26, 504 (1956); C.A. 50, 13785g (1956).
712. Kiso, Y., M. Kobayashi, Y. Kitaoka, K. Kawamoto, and J. Takada, Bull. Chem. Soc. Japan, 40, 2779 (1967).
713. Knauer, W., Chem. Ber., 27, 2565 (1894).
714. Knowles, W. S., and Q. E. Thompson, Chem. Ind. (London), 1959, 121.
715. Knunyants, I. L., E. G. Bykhovskaya, B. L. Dyatkin, V. N. Frosin, and A. A. Gevorkyan, Zh. Vses. Khim. Obshchestva D. I. Mendeleeva, 10, 472 (1965); C.A. 63, 14687a (1965).
716. Knunyants, I. L., E. Ya. Pervova, and V. V. Tyuleneva, Dokl. Akad. Nauk SSSR, 129, 576 (1959); C.A. 54, 7536e (1960).
717. Knunyants, I. L., V. V. Tyuleneva, E. Ya. Pervova, and R. N. Sterlin, Izv. Akad. Nauk SSSR, Ser. Khim., 1964, 1797; C.A. 62, 2791d (1965).
718. Kodama, G., and R. W. Parry, Inorg. Chem., 4, 410 (1965).
719. Koenigs, E., and G. Jung, J. Prakt. Chem., 137, 141 (1933).
720. Koketsu, J., S. Sakai, and Y. Ishii, Kogyo Kagaku Zasshi, 72, 2503 (1969).
721. Kolomiets, A. F., L. A. Kalutskii, G. V. Dotsev, and N. K. Bliznyuk, U.S.S.R. 196 827 (1967); C.A. 68, 59099w (1968).
722. Kolotilo, M. V., A. G. Matyusha, and G. I. Derkach, Zh. Obshch. Khim., 39, 188 (1969); C.A. 70, 96866f (1969).

723. Kondrat'ev, Yu. A., V. V. Tarasov, A. S. Vasil'ev, M. M. Ivakina, and S. Z. Ivin, Zh. Obshch. Khim., 38, 1791 (1968); C.A. 69, 106074s (1968).
724. Kondrat'ev, Yu. A., V. V. Tarasov, N. M. Ivakina, S. Z. Ivin, and V. N. Pastushkov, Zh. Obshch. Khim., 38, 2590 (1968); C.A. 70, 47031w (1969).
725. Kondrat'ev, Yu. A., E. S. Vdovina, Ya. S. Arbisman, V. V. Tarasov, O. G. Strukov, S. S. Dubov, and S. Z. Ivin, Zh. Obshch. Khim., 38, 2859 (1968); C.A. 71, 22153v (1969).
726. Koral, M., U.S. 2 974 159 (1961); C.A. 55, 14307h (1961).
727. Kosolapoff, G. M., J. Am. Chem. Soc., 66, 109 (1944).
728. Kosolapoff, G. M., J. Am. Chem. Soc., 66, 1511 (1944).
729. Kosolapoff, G. M., J. Am. Chem. Soc., 69, 1002 (1947).
730. Kosolapoff, G. M., J. Chem. Soc., 1950, 3536.
731. Kosolapoff, G. M., J. Am. Chem. Soc., 73, 4040 (1951).
732. Kosolapoff, G. M., J. Am. Chem. Soc., 73, 4989 (1951).
733. Kosolapoff, G. M., J. Am. Chem. Soc., 74, 4953 (1952).
734. Kosolapoff, G. M., and A. D. Brown, Jr., J. Chem. Soc. D, 1969, 1266.
735. Kosolapoff, G. M., and R. M. Watson, J. Am. Chem. Soc., 73, 4101 (1951).
736. Kosolapoff, G. M., and R. M. Watson, J. Am. Chem. Soc., 73, 5466 (1951).
737. Kovalenko, V. I., Sbornik Statei Obshch. Khim., Akad. Nauk SSSR, 1, 299 (1953); C.A. 49, 868i (1955).
738. Kovalev, L. S., N. A. Razumova, and A. A. Petrov, Zh. Obshch. Khim., 38, 2277 (1968); C.A. 71, 13171x (1969).
739. Kovalev, L. S., N. A. Razumova, and A. A. Petrov, Zh. Obshch. Khim., 39, 869 (1969); C.A. 71, 50077s (1969).
740. Kovalevskii, B. A., J. Russ. Phys. Chem. Soc., 29, 217 (1897).
741. Kozlov, N. S., and M. V. Kosheleva, Vesti Akad. Nauk Belarus SSR, Ser. Khim. Nauk, 1969, 83; C.A. 72, 89685j (1970).
742. Kozlov, N. S., V. D. Pak, and I. N. Levashov, Vestsi Akad. Nauk Belarus SSR, Ser. Khim. Nauk, 1968, 109; C.A. 70, 47545s (1969).
743. Kraft, M. Ya., and V. V. Katyshkina, Zh. Obshch. Khim., 29, 59 (1959); C.A. 53, 21632a (1959).
743a. Krawiecki, C., and J. Michalski, J. Chem. Soc.,

1960, 881.
744. Krawiecki, C., J. Michalski, R. A. Y. Jones, and A. R. Katritzky, Rocz. Chem., 43, 869 (1969).
745. Kreutzkamp, N., Naturwiss., 43, 81 (1956).
746. Kreutzkamp, N., and H. Kayser, Naturwiss. 42, 415 (1955).
747. Kreutzkamp, N., and H. Kayser, Chem. Ber., 89, 1614 (1956).
748. Kreutzkamp, N., and W. Mengel, Ann. Chem., 657, 19 (1962).
749. Krogh, L. C., T. S. Reid, and H. A. Brown, J. Org. Chem., 19, 1124 (1954).
750. Krokhina, S. S., R. I. Pyrkin, Ya. A. Levin, and B. E. Ivanov, Izv. Akad. Nauk SSSR, Ser. Khim., 1968, 1420; C.A. 69, 67478e (1968).
751. Kruglyak, Yu. L., M. A. Landau, G. A. Leibovskaya, I. V. Martynov, L. I. Saltykova, and M. A. Sokalskii, Zh. Obshch. Khim., 39, 215 (1969); C.A. 70, 105743t (1969).
752. Kruglyak, Yu. L., G. A. Leibovskaya, I. I. Svetenskaya, V. V. Sheluchenko, and I. V. Martynov, Zh. Obshch. Khim., 38, 943 (1968); C.A. 69, 58784z (1968).
753. Kruglyak, Yu. L., S. I. Malekin, and I. V. Martynov, Zh. Obshch. Khim., 39, 466 (1969); C.A. 70, 115070k (1969).
754. Kuhn, L. P., J. O. Doali, and C. Wellman, J. Am. Chem. Soc., 82, 4792 (1960).
755. Kuivila, H. G., and W. L. Masterton, J. Am. Chem. Soc., 74, 4953 (1952).
756. Kukhtin, V. A., Dokl. Akad. Nauk SSSR, 121, 466 (1958); C.A. 53, 1105a (1959).
757. Kukhtin, V. A., V. S. Abramov, and K. M. Orekhova, Dokl. Akad. Nauk SSSR, 128, 1198 (1959); C.A. 54, 7536b (1960).
758. Kukhtin, V. A., N. S. Garif'yanov, and K. M. Orekhova, Zh. Obshch. Khim., 31, 1157 (1961); C.A. 55, 23413h (1961).
759. Kukhtin, V. A., and K. M. Kirillova, Zh. Obshch. Khim., 31, 226 (1961); C.A. 56, 3507e (1962).
760. Kukhtin, V. A., and K. M. Orekhova, Dokl. Akad. Nauk SSSR, 124, 819 (1959); C.A. 53, 16037d (1959).
761. Kukhtin, V. A., and K. M. Orekhova, Zh. Obshch. Khim., 30, 1208, 1526 (1960); C.A. 55, 358i, 1567c (1961).
762. Kukhtin, V. A., T. N. Voskoboeva, and K. M. Kirillova, Zh. Obshch. Khim., 32, 2333 (1962); C.A. 58, 9127g (1963).
763. Kuliev, A. M., and Z. A. Alizade, Prisadki Smaz. Maslam., 1967, 80; C.A. 69, 108356j (1968).
764. Kuliev, A. M., and Z. A. Alizade, Prisadki Smaz.

Maslam, 1967, 84; C.A. 70, 11016f (1969).
765. Kuliev, A. M., Z. A. Alizade, and R. K. Veliena, Prisadki Smaz. Maslam, 1969, 57; C.A. 72, 42656k (1970).
766. Kuliev, A. M., and A. B. Kuliev, Prisadki Smaz. Maslam, 1967, 88; C.A. 70, 3411r (1969).
767. Kuliev, A. M., F. A. Mamedov, and F. N. Mamedov, Azerb. Khim. Zh. 1964, 55; C.A. 62, 10357b (1965).
768. Kuliev, E. M., and K. G. Veliev, Sbornik Trudov Azerbaidzhan. Nauch. Issledovatel Inst. po Pererabotke Nefti, 3, 249 (1958); C.A. 55, 6417e (1961).
769. Kunz, P., Chem. Ber., 27, 2559 (1894).
770. Kuskov, V. K., and T. K. Gradis, Dokl. Akad. Nauk SSSR, 92, 323 (1953); C.A. 49, 155g (1955).
771. Kuznetsov, E. V., and M. I. Bakhitov, Dokl. Akad. Nauk SSSR, 141, 1105 (1961); C.A. 57, 4577a (1962).
772. Kuznetsov, E. V., E. B. Kamaeva, R. K. Valetdinov, and A. I. Roikh, Zh. Obshch. Khim., 31, 3013 (1961); C.A. 56, 15342g (1962).
773. Kuznetsov, E. V., A. A. Svishchuk, and A. N. Blakitnyi, Dopov. Akad. Nauk Ukr. RSR, Ser. B., 30, 633 (1968); C.A. 69, 86262w (1968).
774. Kuznetsov, E. V., and R. K. Valetdinov, Trudy Kazansk. Khim. Tekhnol. Inst. S. M. Kirova, 23, 161, 167, 174, 187 (1957); C.A. 52, 8938i (1958).
775. Kuznetsov, E. V., and R. K. Valetdinov, Trudy Kazansk. Khim. Tekhnol. Inst., 21, 167 (1956); C.A. 51, 11985f (1957).
776. Kuznetsov, E. V., and R. K. Valetdinov, Zh. Obshch. Khim., 29, 235 (1959); C.A. 53, 21648d (1959).
777. Kuznetsov, E. V., and L. A. Vlasova, Zh. Obshch. Khim., 39, 698 (1969); C.A. 71, 39092x (1969).
778. Ladd, E. C., and M. P. Harvey, Can. 509 034 (1955); C.A. 50, 10760f (1956).
779. Landauer, S. R., and H. N. Rydon, Chem. Ind. (London), 1951, 313.
780. Landauer, S. R., and H. N. Rydon, J. Chem. Soc., 1953, 2224.
781. Lanham, W. M., Ger. 1 139 478 (1962); C.A. 58, 10081a (1963).
782. Larrison, M. S., Fr. 1 456 954 (1966); C.A. 67, 11209j (1967).
783. Larrison, M. S., U.S. 3 375 304 (1968); C.A. 68, 96559s (1968).
784. Lemper, A. L., Thesis, State University of New York, Buffalo, 1966; C.A. 66, 37117g (1967).
785. Lemper, A. L., and H. Tieckelmann, Tetrahedron Letters, 1964, 3053.
786. Lenton, M. V., and B. Lewis, Chem. Ind. (London), 1965, 946.
787. LeSuer, W. M., U.S. 3 033 890 (1962); C.A. 57,

12385b (1962).
788. Levi-Malvano, M., Atti. Accad. Lincei (5), 17, 1 850 (1908).
789. Levin, Ya. A., I. P. Gozman, and S. G. Salikhov, Izv. Akad. Nauk SSSR, Ser. Khim., 1968, 2609; C.A. 70, 68269v (1969).
790. Levison, J. J., and S. D. Robinson, Chem. Commun., 1968, 140.
791. Lichtenthaler, F. W., Chem. Rev., 61, 607 (1961).
792. Lindet, L., Ann. Chim. Phys. (6), 11, 190 (1887).
793. Linnemann, E., Ann. Chem., 160, 201 (1871).
794. Lippert, A., and E. E. Reid, J. Am. Chem. Soc., 60, 2370 (1938).
795. Lorenz, H. J., V. Franzen, and J. Auer, Ger. 1 244 757 (1967); C.A. 68, 21521q (1968).
796. Lorenz, W., and G. Schrader, Ger. 820 001 (1951); C.A. 47, 3332e (1953).
797. Losse, G., and E. Wottgen, J. Prakt. Chem., 13, 260 (1961).
798. Lowes, F. J., and R. F. Monroe, U.S. 2 809 982 (1957); C.A. 52, 2902f (1958).
799. Lucas, H. J., F. W. Mitchell, Jr., and C. N. Scully, J. Am. Chem. Soc., 72, 5491 (1950).
800. Lucas, H., and D. Pressman, Principles and Practice in Organic Chemistry, Wiley, New York, 1949, p. 267.
801. Lucken, E. A. C., F. Ramirez, V. P. Catto, D. Rhum, and S. Dershowitz, Tetrahedron Letters, 1966, 637.
802. Lumiere, A., L. Lumiere, and F. Perrin, Compt. Rend., 133, 643 (1901).
803. Lutsenko, I. F., R. M. Khomutov, and L. A. Eliseeva, Izv. Akad. Nauk SSSR, Ser. Khim., 1956, 181; C.A. 50, 13730e (1956).
804. Lutsenko, I. F., and V. V. Tyuleneva, Zh. Obshch. Khim., 27, 497 (1957); C.A. 51, 15452h (1957).
805. Lutz, L. J., and H. N. Tatomer, U.S. 3 194 827 (1965); C.A. 63, 9812f (1965).
806. Luz, Z., and B. Silver, J. Am. Chem. Soc., 83, 4513 (1961).
807. Maarsen, J. W., M. C. Smit, and J. Matze, Rec. Trav. Chim., 76, 713 (1957).
808. Machleidt, H., and G. U. Strehlke, Angew. Chem., 76, 494 (1964).
809. Magerlein, B. J., and F. Kagan, U.S. 3 017 423 (1962); C.A. 57, 669a (1962).
810. Maguire, M. H., and G. Shaw, J. Chem. Soc., 1955, 2039.
811. Maier, L., U.S. 3 321 557 (1967); C.A. 67, 73684u (1967).
811a. Maier, L., Chimia, 23, 323 (1969).
812. Maklyaev, F. L., and M. I. Druzin, U.S.S.R. 130 504 (1960); C.A. 55, 7286i (1961).

813. Maklyaev, F. L., and M. I. Druzin, U.S.S.R. 130 886 (1960); C.A. 55, 7286g (1961).
814. Maklyaev, F. L., M. I. Druzin, and I. V. Palagina, Zh. Obshch. Khim., 31, 895 (1961); C.A. 55, 23313e (1961).
814a. Maklyaev, V. K., M. I. Druzin, I. V. Palagina, R. Ya. Aleksandrova, V. K. Prokhodtseva, and R. A. Khamidulina, Zh. Obshch. Khim., 32, 3421 (1962); C.A. 58, 8887g (1963).
815. Mamedova, P. S., Azerb. Neft. Khoz. 42, 34 (1963); C.A. 60, 4037g (1964).
816. Mandel'baum, Ya. A., P. G. Zaks, and N. N. Mel'nikov, Zh. Obshch. Khim., 36, 44 (1966); C.A. 64, 14082h (1966).
817. Mandel'baum, Ya. A., A. L. Itskova, and N. N. Mel'nikov, Khim. Org. Soedin. Fosfora, Akad. Nauk SSSR, Otd. Obshch. Tekh. Khim. 1967, 288; C.A. 69, 43338h (1968).
818. Mangham, J. R., U.S. 2 864 847 (1958); C.A. 53, 7986i (1959).
819. Mann, F. G., and D. Purdie, J. Chem. Soc., 1935, 1549.
820. Mark, V., U.S. 3 106 586 (1963); C.A. 60, 5358b (1964).
821. Mark, V., J. Am. Chem. Soc., 85, 1884 (1963).
822. Mark, V., Tetrahedron Letters, 1964, 3139.
823. Mark, V., Abstracts, 147th meeting, American Chemical Society, Philadelphia, Pa., April 1964, p. 291.
824. Mark, V., Org. Syn., 46, 42 (1966).
825. Mark, V., Mech. Mol. Migr. 2, 319 (1969).
826. Mark, V., C. Dungan, M. Crutchfield, and J. R. Van Wazer, Topics in Phosphorus Chemistry, ed. M. Grayson and E. J. Griffith, Wiley, New York, 1967, Vol. 5.
827. Mark, V., and J. R. Van Wazer, J. Org. Chem., 29, 1006 (1964).
828. Mark, V., and J. R. Van Wazer, J. Org. Chem., 32, 1187 (1967).
829. Markley, F. X., and M. L. Larson, U.S. 2 866 806 (1958); C.A. 53, 19877i (1959).
830. Markowska, A., and J. Michalski, Rocz. Chem., 34, 1675 (1960).
831. Marsden, J. G., U.S. 2 963 503 (1960); C.A. 55, 8292d (1961).
832. Martin, G., and A. Bernard, Compt. Rend., 257, 2463 (1963).
833. Maruszewska-Wieczorkowska, E., and J. Michalski, Bull. Acad. Polon. Sci., Ser. Sci. Chim. Geolet Geograph., 6, 19 (1958).
834. Matousek, J., Czech. 109 145 (1962); C.A. 60, 10549c (1964).
835. Mavel, G., and G. Martin, Compt. Rend., 252, 110

(1961).
836. Mavel, G., in Progress in NMR Spectroscopy, Emsley, J. W., J. Feeney, and L. H. Sutcliffe, Eds., Wiley, New York, 1966, Vol. 2, p. 251.
837. McCombie, H., B. C. Saunders, and G. J. Stacey, J. Chem. Soc., 1945, 380.
838. McCombie, H., B. C. Saunders, and G. I. Stacey, J. Chem. Soc., 1945, 921.
839. McConnell, R. L., and H. W. Coover, Jr., J. Org. Chem., 23, 830 (1958).
840. McConnell, R. L., and H. W. Coover, Jr., J. Org. Chem., 24, 630 (1959).
841. McConnell, R. L., and H. W. Coover, Jr., U.S. 2 952 701 (1960); C.A. 55, 3430d (1961).
842. McConnell, R. L., and H. W. Coover, Jr., U.S. 3 377 409 (1968); C.A. 68, 115204r (1968).
843. McCusker, P. A., and E. L. Reilly, J. Am. Chem. Soc., 75, 1583 (1953).
844. McIvor, R. A., G. A. Grant, and C. E. Hubley, Can. J. Chem., 34, 1611 (1956).
845. McIvor, R. A., G. D. McCarthy, and G. A. Grant, Can. J. Chem., 34, 1819 (1956).
846. McKenzie, A., and G. W. Clough, J. Chem. Soc., 103, 687 (1913).
847. Medved, T. Ya., and M. I. Kabachnik, Izv. Akad. Nauk SSSR, Ser. Khim., 1954, 314; C.A. 48, 10541b (1954).
848. Medved, T. Ya., and M. I. Kabachnik, Izv. Akad. Nauk SSSR, Ser. Khim., 1958, 1212; C.A. 53, 4111c (1959).
849. Mel'nikov, N. N., Ya. A. Mandel'baum, and Z. M. Bakanova, Zh. Obshch. Khim., 28, 2473 (1958); C.A. 53, 3032a (1959).
850. Mel'nikov, N. N., Ya. A. Mandel'baum, and V. I. Lomakina, Zh. Obshch. Khim., 29, 3289 (1959); C.A. 54, 14216a (1960).
851. Mel'nikov, N. N., Ya. A. Mandel'baum, V. I. Lomakina, and V. S. Livshits, Zh. Obshch. Khim., 31, 3949 (1961); C.A. 57, 11072c (1962).
852. Mel'nikov, N. N., K. D. Shvetsova-Shilovskaya, and M. Ya. Kagan, Zh. Obshch. Khim., 30, 2317 (1960); C.A. 55, 9320g (1961).
853. Melton, T. M., U.S. 3 095 439 (1963); C.A. 59, 12645e (1963).
854. Menschutkin, N., Ann. Chem., 139, 343 (1866).
855. Meyrick, C. I., and H. W. Thompson, J. Chem. Soc., 1950, 225.
856. Michaelis, A., Chem. Ber., 5, 4 (1872).
857. Michaelis, A., Ann. Chem., 326, 129 (1903).
858. Michaelis, A., and T. H. Becker, Chem. Ber., 30, 1003 (1897).
859. Michaelis, A., and R. Kahne, Chem. Ber., 31, 1048

(1898).
860. Michaelis, A., and G. L. Linke, Chem. Ber., 40, 3419 (1907).
861. Michaelis, A., and K. Luxembourg, Chem. Ber., 28, 2205 (1895).
862. Michaelis, A., and F. Oster, Ann. Chem., 270, 123 (1892).
863. Michaelis, A., and G. Schroeter, Chem. Ber., 27, 490 (1894).
864. Michalski, J., Rocz. Chem., 29, 960 (1955).
865. Michalski, J., and C. Krawiecki, Chem. Ind. (London), 1957, 1323.
866. Michalski, J., and C. Krawiecki, Rocz. Chem., 31, 715 (1957).
867. Michalski, J., and J. Mikolajczyk, Bull. Acad. Pol. Sci., Ser. Sci. Chim., 14, 829 (1966).
868. Michalski, J., M. Mikolajczyk, and A. Skowronska, Chem. Ind. (London), 1962, 1053.
869. Michalski, J., and T. Modro, Chem. Ind. (London), 1960, 1570.
870. Michalski, J., and T. Modro, Bull. Acad. Polon. Sci., Ser. Sci. Chim., 10, 327 (1962).
871. Michalski, J., and T. Modro, Rocz. Chem., 38, 123 (1964).
872. Michalski, J., and T. Modro, Bull. Acad. Pol. Sci., Ser. Sci. Chim., 15, 201 (1967).
873. Michalski, J., T. Modro, and A. Zwierzak, J. Chem. Soc., 1961, 4904.
874. Michalski, J., and B. Pliszka, Chem. Ind. (London), 1962, 1052.
875. Michalski, J., and B. Pliszka-Krawiecka, J. Chem. Soc. C, 1966, 2249.
876. Michalski, J., W. Stec, and A. Zwierzak, Chem. Ind. (London), 1965, 347.
877. Michalski, J., Z. Tulimowski, and R. Wolf, Chem. Ber., 98, 3006 (1965).
878. Michalski, J., and J. Wasiak, J. Chem. Soc., 1962, 5056.
879. Michalski, J., and J. Wieczorkowski, Rocz. Chem., 31, 585 (1957).
880. Michalski, J., and J. Wieczorkowski, Bull. Acad. Polon. Sci. III, 5, 917 (1957).
881. Michalski, J., and J. Wieczorkowski, J. Chem. Soc., 1960, 885.
882. Michalski, J., J. Wieczorkowski, J. Wasiak, and B. Pliszka, Rocz. Chem., 33, 247 (1959).
883. Michalski, J., and A. Zwierzak, Chem. Ind. (London), 1960, 376.
884. Michalski, J., and A. Zwierzak, Rocz. Chem., 35, 619 (1961).
885. Michalski, J., and A. Zwierzak, Rocz. Chem., 36,

489 (1962).
886. Michalski, J., and A. Zwierzak, Proc. Chem. Soc., 1964, 80.
887. Michalski, J., and A. Zwierzak, Bull. Acad. Polon. Sci., Ser. Sci. Chim., 13, 253 (1965).
888. Miller, B., Topics in Phos. Chemistry, ed. M. Grayson and E. J. Griffith, Wiley, New York, 1965, Vol. 2, p. 133.
889. Miller, B., J. Am. Chem. Soc., 88, 1841 (1966).
890. Miller, C. D., R. C. Miller, and W. Rogers, Jr., J. Am. Chem. Soc., 80, 1562 (1958).
891. Miller, R. C., J. S. Bradley, and L. A. Hamilton, J. Am. Chem. Soc., 78, 5299 (1956).
892. Milobendzki, T., J. Russ. Phys. Chem. Soc., 30, 730 (1898).
893. Milobendzki, T., Chem. Ber., 45, 298 (1912); Chem. Polsk., 15, 89 (1917); C.A. 13, 2867 (1919).
893a. Milobedzki, T., and W. Borowski, Rocz. Chem., 18, 725 (1938).
893b. Milobedzki, T., and W. Borowski, Rocz. Chem., 19, 507 (1939).
894. Milobendzki, T., and J. H. Kolitowska, Rocz. Chem., 6, 67 (1926).
895. Milobendzki, T., and T. Knoll, Chem. Polsk., 15, 79 (1917).
896. Milobendzki, T., and S. Krakoviecki, Bull. Soc. Chim. (4), 41, 932 (1927).
896a. Milobedzki, T., and M. Lewandowski, Rocz. Chem., 19, 509 (1939).
897. Milobendzki, T., and A. Renc, Rocz. Chem., 11, 834 (1931).
898. Milobendzki, T., and A. Sachnowski, Chem. Polsk., 15, 34 (1917).
899. Milobendzki, T., and K. Szulgin, Chem. Polsk., 15, 66 (1917).
900. Milobendzki, T., and M. Szwejkowska, Chem. Polsk., 15, 56 (1917).
901. Milsunobu, O., and J. Mukaiyama, J. Org. Chem., 29, 3005 (1964).
902. Minami, T., H. Miki, and T. Agawa, Kogyo Kagaku Zasshi, 70, 1829 (1967).
903. Mingoia, Q., Gazz. Chim. Ital., 60, 144 (1930).
904. Minich, D., N. I. Rizpolozhenskii, and V. D. Akamsin, Izv. Akad. Nauk SSSR, Ser. Khim., 1968, 1792; C.A. 70, 37096g (1969).
905. Minich, D., N. I. Rizpolozhenskii, V. D. Akamsin, and O. A. Raevskii, Izv. Akad. Nauk SSSR, Ser. Khim., 1969, 876; C.A. 71, 49447f (1969).
906. Mitin, Yu. V., and G. P. Vlasov, Probl. Organ. Sinteza, Akad. Nauk SSSR, Otd. Obshch. Tekhn. Khim., 1965, 297; C.A. 64, 11122h (1966).
907. Mitsunobu, O., K. Kodera, and T. Mukaiyama, Bull.

Chem. Soc. Japan, 41, 461 (1968).
908. Mitsunobu, O., T. Ōashi, M. Kikuchi, and T. Mukaiyama, Bull. Chem. Soc. Japan, 39, 214 (1966).
909. Mitsunobu, O., T. Obata, and T. Mukaiyama, J. Org. Chem., 30, 1071 (1965).
910. Mitsunobu, O., M. Yamada, and T. Mukaiyama, Bull. Chem. Soc. Japan, 40, 935 (1967).
911. Miyano, M., and F. Saburo, J. Am. Chem. Soc., 77, 3522 (1955).
912. Mizrath, L. I., L. Yu. Sandalova, and V. P. Evdakav, Zh. Obshch. Khim., 38, 1107 (1968); C.A. 69, 77367g (1968).
913. Moedritzer, K., J. Inorg. Nucl. Chem., 22, 19 (1962).
914. Moedritzer, K., G. M. Burch, R. N. Van Wazer, and H. K. Hofmeister, Inorg. Chem., 2, 1152 (1963).
914a. Moedritzer, K., L. Maier, and L. C. D. Groenweghe, J. Chem. Eng. Data, 7, 307 (1962).
915. Montrose Chem. Co., Brit. 894 169 (1962); C.A. 57, 11020b (1962).
916. Mooney, E. F., and B. S. Thornhill, J. Inorg. Nucl. Chem., 28, 2225 (1966).
917. Morgan, P. W., and B. C. Herr, J. Am. Chem. Soc., 74, 4526 (1952).
918. Morris, R. C., and J. L. Van Winkle, U.S. 2 728 789 (1955); C.A. 50, 10759c (1956).
919. Mortimer, C. T., Pure Appl. Chem., 2, 71 (1961).
920. Morrison, D. C., J. Am. Chem. Soc., 77, 181 (1955).
921. Moskva, V. V., A. I. Maikova, and A. I. Razumov, Zh. Obshch. Khim., 38, 198 (1968); C.A. 69, 52223u (1968).
922. Moskva, V. V., A. I. Maikova, and A. I. Razumov, Zh. Obshch. Khim., 39, 595 (1969); C.A. 71, 50076r (1969).
923. Moyle, C. L., U.S. 2 170 833 (1939); C.A. 34, 1035[9] (1940). U.S. 2 220 113 (1940); C.A. 35, 1898[1] (1941). U.S. 2 220 845 (1940); C.A. 35, 1897[3] (1941).
924. Mukaiyama, M., K. Fujiwara, H. Mitsunobu, T. Ohashi, and G. Kikuchi, Japan. 6 930 484 (1969); C.A. 72, 66948h (1970).
925. Mukaiyama, T., T. Hata, and T. Tasaka, J. Org. Chem., 28, 481 (1963).
926. Mukaiyama, T., O. Mitsunobu, and T. Obata, J. Org. Chem., 30, 101 (1965).
927. Mukaiyama, T., T. Nagaoka, and S. Fukuyama, Tetrahedron Letters, 1967, 2461.
928. Mukaiyama, T., T. Obata, and O. Mitsunobu, Bull. Chem. Soc. (Japan), 38, 1088 (1965).
929. Mukaiyama, T., and M. Veki, Tetrahedron Letters, 1967, 3429.

930. Mukmeneev, E. T., and G. Kamai, Dokl. Akad. Nauk SSSR, 153, 605 (1963); C.A. 60, 6737k (1964).
931. Mukmeneva, N. A., P. A. Kirpichnikov, and A. N. Pudovik, Zh. Obshch. Khim., 32, 2193 (1962); C.A. 58, 8943a (1963).
932. Mukmeneva, N. A., P. A. Kirpichnikov, and A. N. Pudovik, Zh. Obshch. Khim., 33, 3192 (1963); C.A. 60, 5375e (1964).
933. Mukhamedova, L. A., T. M. Malyshko, and R. R. Shagidullin, Khim. Geterotsibl. Soedin., Akad. Nauk Latv. SSR, 4, 483 (1965); C.A. 64, 3445e (1966).
934. Muller, P., Swiss 326 948 (1958); C.A. 52, 16197e (1958).
935. Muller, N., P. C. Lauterbur, and J. Goldenson, J. Am. Chem. Soc., 78, 3557 (1956).
936. Murav'ev, I. V., N. I. Zemlyanskii, and E. P. Panov, Zh. Obshch. Khim., 38, 133 (1968); C.A. 69, 58789e (1968).
937. Murdock, L. L., and T. L. Hopkins, J. Org. Chem., 33, 907 (1968).
938. Murray, R. W., and M. L. Kaplan, J. Am. Chem. Soc., 90, 537 (1968).
939. Murray, R. W., and M. L. Kaplan, J. Am. Chem. Soc., 91, 5358 (1969).
940. Muskat, I. E., J. Am. Chem. Soc., 56, 2449 (1934).
941. Mustafa, A., M. M. Sidky, and F. M. A. Soliman, Tetrahedron, 23, 99, 107 (1967).
942. Mustafa, A., M. M. Sidky, S. M. A. D. Zayed, and W. M. Abdo, Tetrahedron, 24, 4725 (1968).
943. Mustafa, A., M. M. Sidky, S. M. A. D. Zayed, and M. R. Mahran, Ann. Chem., 712, 116 (1968).
944. Nagy, G., Fr. 1 384 809 (1965); C.A. 62, 9061h (1965).
945. Nakanishi, M., T. Kuriyama, and H. Yugi, Japan. 6 822 092 (1968); C.A. 70, 57396h (1969).
946. National Distillers and Chemical Corp., Brit. 1 167 918 (1969); C.A. 72, 12365x (1970).
947. Neale, E., and L. T. D. Williams, J. Chem. Soc., 1955, 2485.
948. Nesmeyanov, A. N., I. F. Lutsenko, Z. S. Kraits, and A. P. Bokovoi, Dokl. Akad. Nauk SSSR, 124, 1251 (1959); C.A. 53, 16931g (1959).
949. Nesterov, L. V., and R. A. Sabirova, Zh. Obshch. Khim., 31, 897 (1961); C.A. 55, 23424h (1961).
950. Nesterov, L. V., A. Ya. Kessel, and L. I. Maklakov, Zh. Obshch. Khim., 38, 1278 (1968); C.A. 69, 106827q (1968).
951. Nesterov, L. V., N. E. Krepysheva, and R. I. Mutalapova, Zh. Obshch. Khim., 35, 2050 (1965); C.A. 64, 8019h (1966).
952. Nesterov, L. V., and R. A. Sabirova, Zh. Obshch.

Khim., 35, 1976 (1965); C.A. 64, 8072f (1966).
953. Nesterov, L. V., and R. A. Sabirova, Zh. Obshch. Khim., 35, 2006 (1965); C.A. 64, 8072c (1966).
954. Nesterov, L. V., R. A. Sabirova, N. E. Krepysheva, and R. I. Mutalapova, Dokl. Akad. Nauk SSSR, 148, 1085 (1963); C.A. 59, 3757a (1963).
955. Nesterov, L. V., R. A. Sabirova, and N. E. Krepysheva, Zh. Obshch. Khim., 39, 1943 (1969); C.A. 72, 31765u (1970).
956. Nifant'ev, E. E., Zh. Obshch. Khim., 34, 3850 (1964); C.A. 62, 6470f (1965).
957. Nifant'ev, E. E., and I. V. Fursenko, Zh. Obshch. Khim., 35, 1882 (1965); C.A. 64, 3589 (1966).
958. Nifant'ev, E. E., and I. V. Fursenko, Zh. Obshch. Khim., 37, 511 (1967); C.A. 67, 32306d (1967).
959. Nifant'ev, E. E., and I. V. Fursenko, Zh. Obshch. Khim., 38, 1295 (1968); C.A. 69, 58786b (1968).
960. Nifant'ev, E. E., and I. V. Fursenko, Zh. Obshch. Khim., 37, 1134 (1967); C.A. 68, 104510m (1968).
961. Nifant'ev, E. E., and I. V. Fursenko, Zh. Obshch. Khim., 38, 1295 (1968); C.A. 69, 58786b (1968).
962. Nifant'ev, E. E., and I. V. Fursenko, Zh. Obshch. Khim., 39, 1028 (1969); C.A. 71, 60598b (1969).
963. Nifant'ev, E. E., I. V. Fursenko, and A. M. Sokurenko, Zh. Obshch. Khim., 38, 1909 (1968); C.A. 69, 105804t (1968).
964. Nifant'ev, E. E., N. L. Ivanova, and N. K. Bliznyuk, Zh. Obshch. Khim., 36, 765 (1966); C.A. 65, 8698b (1966).
965. Nifant'ev, E. E., N. L. Ivanova, and I. V. Fursenko, Zh. Obshch. Khim., 39, 854 (1969); C.A. 71, 60907b (1969).
966. Nifant'ev, E. E., and I. V. Komlev, U.S.S.R. 186 469 (1966); C.A. 66, 76150g (1967).
967. Nifant'ev, E. E., and L. P. Levitan, Probl. Organ. Sinteza, Akad. Nauk SSSR, Otd. Obshch. Tekhnol. Khim., 1965, 293; C.A. 64, 11074e (1966).
968. Nifant'ev, E. E., and L. V. Matveeva, Zh. Obshch. Khim., 37, 1692 (1967); C.A. 68, 48965z (1968).
969. Nifant'ev, E. E., and I. S. Nasonovskii, Zh. Obshch. Khim., 39, 1948 (1969); C.A. 72, 31142p (1970).
970. Nifant'ev, E. E., and I. M. Petrova, Zh. Obshch. Khim., 38, 2341 (1968); C.A. 70, 19505k (1969).
971. Nifant'ev, E. E., and A. P. Tuseev, Sintez Prirodn. Soedin. Analogov Fragmentov Akad. Nauk SSSR, Otd. Obshch. Tekhn. Khim., 1965, 34; C.A. 65, 5512a (1966).
972. Nifant'ev, E. E., A. I. Zavalishina, I. S. Nasonoskii, and I. V. Komlev, Zh. Obshch. Khim., 38, 2538 (1968); C.A. 70, 106615q (1969).
973. Nifant'ev, E. E., A. I. Zavalishina, and M. R. Ter-

Ovanesyan, Zh. Obshch. Khim., 39, 360 (1969); C.A. 70, 114890r (1969).
974. Nikolaev, A. V., Yu. A. Alfanas'ev, and A. D. Starostin, Izv. Sib. Otd. Akad. Nauk SSSR, Ser. Khim. Nauk, 1968, 3; C.A. 71, 7169m (1969).
975. Nikolenko, L. N., and E. V. Degterev, Zh. Obshch. Khim., 37, 1350 (1967); C.A. 68, 59490s (1968).
976. Nikonorov, K. V., Izv. Kazansk. Filiala Akad. Nauk SSSR, Ser. Khim. Nauk, 1961, 68; C.A. 59, 10112e (1963).
977. Nikonorov, K. V., E. A. Gurylev, R. R. Shagidullin, and A. V. Chernova, Izv. Akad. Nauk SSSR, Ser. Khim., 1968, 593; C.A. 69, 95861t (1968).
978. Nikonorov, K. V., G. M. Vinokurova, and Z. G. Speranskaya, Khim. Primenenie Fosfororgan. Soedinenii, Akad. Nauk SSSR, Trudy 1-oi Konferents., 1955, 223 (1957); C.A. 52, 240 (1958).
979. Nitto Chemical Industry Co., Brit. 1 110 917 (1968); C.A. 69, 19162a (1968).
980. Nixon, J. F., J. Chem. Soc. A, 1968, 2689.
981. Nixon, J. F., J. Chem. Soc. A, 1969, 1087.
982. Noack, E., Ann. Chem. 1883, 218, 85.
983. Noth, H., and H. J. Vetter, Chem. Ber., 94, 1505 (1961).
984. Noth, H., and H. J. Vetter, Chem. Ber., 96, 1109 (1963).
985. Noth, H., and H. J. Vetter, Chem. Ber., 96, 1298 (1963).
986. Noth, H., and H. J. Vetter, Chem. Ber., 96, 1479 (1963).
987. Noth, H., and H. J. Vetter, Chem. Ber., 98, 1981 (1965).
988. Novikova, Z. S., M. A. Krasnovskaya, and I. F. Lutsenko, Zh. Obshch. Khim., 39, 1060 (1969); C.A. 71, 60597a (1969).
989. Novitskii, K. I., N. A. Razumova, and A. A. Petrov, Khim. Org. Soedinenii Fosfora, Akad. Nauk SSSR, Otd. Obshch. Tekh. Khim., 1967, 248; C.A. 69, 43981n (1968).
990. Nuretdinov, I. A., and N. P. Grechkin, Izv. Akad. Nauk SSSR, Ser. Khim., 1964, 1883; C.A. 62, 2747d (1965).
991. Nuretdinov, I. A., and N. P. Grechkin, Izv. Akad. Nauk SSSR, Ser. Khim., 1967, 439; C.A. 67, 21748y (1967).
992. Nuretdinov, I. A., and N. P. Grechkin, Izv. Akad. Nauk SSSR, Ser. Khim., 1968, 2831; C.A. 70, 77241g (1969).
993. Nuretdinova, O. N., and L. Z. Nikonova, Izv. Akad. Nauk SSSR, Ser. Khim., 1969, 1125; C.A. 71, 50105z (1969).

994. Nylen, P., Chem. Ber., 57, 1023 (1924).
995. Nylen, P., Chem. Ber., 59, 1119 (1926).
996. Nylen, P., Dissertation, Uppsala, 1930; Chem. Ber., 57, 1023 (1924); 59, 1119 (1926).
997. Nylen, P., Svensk. Kem. Tid., 48, 2 (1936).
998. Nylen, P., Z. Anorg. Chem., 230, 385 (1937).
999. Nylen, P., Tids. Kjemi. Bergvesen, 18, 59 (1938).
1000. Nylen, P., Z. Anorg. Chem., 235, 161 (1938).
1001. Nyquist, R. A., App. Spectrosc., 11, 161 (1957).
1002. Nyquist, R. A., Spectrochim. Acta, 19, 713 (1963).
1003. Nyquist, R. A., Spectrochim. Acta, 25, 47 (1969).
1004. Nyquist, R. A., and W. W. Muelder, Spectrochim. Acta, 22, 1563 (1966).
1005. Oakes, V., and D. F. W. Cross, Brit. 1 173 763 (1969); C.A. 72, 56501t (1970).
1006. Oakes, V., D. F. W. Cross, and T. Taylor, Brit. 1 180 398 (1970); C.A. 72, 101644e (1970).
1006a. Occolowitz, J. L., and G. L. White, Anal. Chem., 35, 1179 (1963).
1007. Oertel, G., and H. Holtschmidt, Fr 1 332 901 (1963); C.A. 59, 15445g (1963).
1008. Oertel, G., H. Malz, and H. Holtschmidt, Chem. Ber., 97, 891 (1964).
1009. Oertel, G., H. Malz, and H. Holtschmidt, Ger. 1 172 260 (1964); C.A. 63, 17916a (1965).
1010. Oertel, G., H. Malz, H. Holtschmidt, and E. Degener, Ger. 1 155 433 (1963); C.A. 60, 2823d (1964).
1011. Oeser, C., Ann. Chem., 131, 277 (1864).
1012. Olechowski, J. R., G. C. McAlister, and R. F. Clark, J. Inorg. Chem., 4, 246 (1965).
1013. Organic Syntheses, Wiley, New York, 1933, Vol. 13, p. 20.
1014. Orlov, N. F., and B. L. Kaufman, U.S.S.R. 182 146 (1966); C.A. 65, 17002c (1966).
1015. Orlov, N. F., and B. L. Kaufman, Zh. Obshch. Khim., 36, 1155 (1966); C.A. 65, 10608g (1966).
1016. Orlov, N. F., and B. L. Kaufman, U.S.S.R. 193 509 (1967); C.A. 68, 95936a (1968).
1017. Orlov, N. F., and B. L. Kaufman, Zh. Obshch. Khim., 38, 1842 (1968); C.A. 70, 4193h (1969).
1018. Orlov, N. F., B. L. Kaufman, L. Sukhi, L. N. Slesar, and E. V. Sudakova, Khim. Prakt. Primen, Kremniiorg. Soedin. Tr. Sovesch., 1966, 111; C.A. 72, 21738y (1970).
1019. Orlov, N. F., and E. V. Sudakova, U.S.S.R. 229 511 (1968); C.A. 70, 58016w (1969).
1020. Orlov, N. F., and E. V. Sudakova, Zh. Obshch. Khim., 39, 222 (1969); C.A. 70, 87881e (1969).
1021. Orlov, N. F., and E. V. Sudakova, Khim. Prakt. Primen, Kremniiorg. Soedin. Tr. Sovesch., 1966, 123; C.A. 72, 12807t (1970).

1022. Orlov, N. F., and L. N. Volodina, Zh. Obshch. Khim., 36, 920 (1966); C.A. 65, 10608c (1966).
1023. Orudzheva, I. M., and Sh. M. Novruzov, Prisadki Smaz. Maslam., 1967, 67; C.A. 70, 28438t (1969).
1024. Orudzheva, I. M., and Sh. M. Novruzov, Prisadki Smaz. Maslam., 1967, 73; C.A. 69, 86515f (1968).
1025. Orudzheva, I. M., Sh. M. Novruzov, A. A. Vagabova, and P. S. Mamedova, Neftepererab. Neftekhim., 1967, 28; C.A. 68, 115380v (1968).
1026. Orudzheva, M., and A. A. Vagabova, Azerb. Khim. Zh. 1966, 21; C.A. 66, 2300s (1967).
1027. Oswald, A. A., Can. J. Chem., 37, 1498 (1959).
1028. Oswald, A. A., U.S. 3 152 164 (1964); C.A. 61, 16082c (1964).
1029. Palazzo, F. C., and F. Maggiacomo, Atti. Accad. Lincei (5), 17, 1, 432 (1908).
1030. Panteleeva, A. R., and I. M. Shermergorn, Izv. Akad. Nauk SSSR, Ser. Khim., 1968, 1652; C.A. 69, 87106k (1968).
1031. Parry, R. W., and T. C. Bissot, J. Am. Chem. Soc., 78, 1524 (1956).
1032. Pascal, P., D. Voigt, M. C. Labarre, and L. Fournes, Compt. Rend. Ser. C., 262, 1481 (1966).
1033. Pastushkov, V. N., Ya. S. Arbisma, Yu. A. Kondrat'-ev, S. Z. Ivin, and A. S. Vasil'eva, Zh. Obshch. Khim., 38, 1405 (1968); C.A. 69, 87102f (1968).
1034. Pastushkov, V. N., Yu. A. Kondrat'ev, S. Z. Ivin, E. S. Vdovina, and A. S. Vasilev, Zh. Obshch. Khim., 38, 1407 (1968); C.A. 69, 87087e (1968).
1035. Pastushkhov, V. N., E. S. Vdovina, Yu. A. Kondrat'-ev, S. Z. Ivin, and V. V. Tarasov, Zh. Obshch. Khim., 38, 1408 (1968); C.A. 69, 87101e (1968).
1036. Payne, D. S., H. Nöth, and G. Henniger, Chem. Commun., 1965, 327.
1037. Peiffer, G., A. Guillemonat, and G. Buono, Bull. Soc. Chim. Fr., 1969, 946.
1038. Perkow, W., Chem. Ber., 87, 755 (1954).
1039. Perkow, W., E. W. Krockow, and K. Kisoevenagel, Chem. Ber., 88, 662 (1955).
1040. Perkow, W., K. Ullerich, and F. Meyer, Naturwiss., 39, 353 (1952).
1041. Petragnani, N., and M. Moura Campos, Tetrahedron, 21, 13 (1965).
1042. Petragnani, N., V. G. Toscano, and M. Moura Campos, Chem. Ber., 101, 3070 (1968).
1043. Petrov, K. A., and N. K. Bliznyuk, U.S.S.R. 127 649 (1960); C.A. 54, 22359h (1960).
1044. Petrov, K. A., N. K. Bliznyuk, and T. N. Lysenko, Zh. Obshch. Khim., 30, 1964 (1960); C.A. 55, 6362g (1961).
1045. Petrov, K. A., N. K. Bliznyuk, and I. Yu. Mansurov,

Zh. Obshch. Khim., 31, 176 (1961); C.A. 55, 22097i (1961).
1046. Petrov, K. A., V. P. Evdakov, K. A. Bilevich, and V. I. Chernykh, Zh. Obshch. Khim., 32, 3065 (1962); C.A. 58, 11395e (1963).
1047. Petrov, K. A., V. P. Evdakov, K. A. Bilevich, V. P. Radchenko, and E. E. Nifant'ev, Zh. Obshch. Khim., 32, 920 (1962); C.A. 58, 2391h (1963).
1048. Petrov, K. A., V. P. Evdakov, L. I. Mizrakh, and V. P. Romodin, Zh. Obshch. Khim., 32, 3062 (1962); C.A. 58, 11395b (1963).
1049. Petrov, K. A., A. I. Garrilova, and V. P. Korotkova, Zh. Obshch. Khim., 32, 915 (1962); C.A. 58, 1485g (1963).
1050. Petrov, K. A., A. A. Neimysheva, M. G. Fomenko, L. M. Chernushevich, and A. D. Kuntsevich, Zh. Obshch. Khim., 31, 516 (1961); C.A. 55, 22111a (1961).
1051. Petrov, K. A., and E. E. Nifant'ev, U.S.S.R. 136 347 (1961); C.A. 55, 19239i (1961).
1052. Petrov, K. A., and E. E. Nifant'ev, Vysokomolekul. Soedin., 4, 242 (1962); C.A. 57, 3665e (1965).
1053. Petrov, K. A., E. E. Nifant'ev, and R. G. Gol'tsova, Zh. Obshch. Khim., 33, 1485 (1963); C.A. 59, 12840e (1963).
1054. Petrov, K. A., E. E. Nifant'ev, R. G. Gol'tsova, M. A. Belaventsev, and S. M. Korneev, Zh. Obshch. Khim., 32, 1277 (1962); C.A. 58, 6856e (1963).
1055. Petrov, K. A., E. E. Nifant'ev, R. G. Gol'tsova, A. A. Shchegolev, and B. V. Bushmin, Zh. Obshch. Khim., 32, 3723 (1962); C.A. 58, 12594c (1963).
1056. Petrov, K. A., E. E. Nifant'ev, and L. V. Khorkhoyanu, Zh. Obshch. Khim., 31, 2889 (1961); C.A. 57, 860g (1962).
1057. Petrov, K. A., E. E. Nifant'ev, L. V. Khorkhoyanu, and I. G. Shcherba, Zh. Obshch. Khim., 34, 70 (1964); C.A. 60, 10530e (1964).
1058. Petrov, K. A., E. E. Nifant'ev, and T. N. Lysenko, Zh. Obshch. Khim., 31, 1709 (1961); C.A. 55, 23312h (1961).
1059. Petrov, K. A., E. E. Nifant'ev, T. N. Lysenko, and V. P. Evdakov, Zh. Obshch. Khim., 31, 2377 (1961); C.A. 56, 3511i (1962).
1059a. Petrov, K. A., E. E. Nifant'ev, A. A. Shchegolev, M. M. Butilov, and I. F. Rebus, Zh. Obshch. Khim., 33, 899 (1963); C.A. 59, 7555b (1963).
1060. Petrov, K. A., E. E. Nifant'ev, A. A. Shchegolev, and A. P. Tuseev, Zh. Obshch. Khim., 34, 4096 (1964); C.A. 62, 9215g (1965).
1061. Petrov, K. A., E. E. Nifant'ev, and I. I. Sopikova, Dokl. Akad. Nauk SSSR, 151, 859 (1968); C.A. 59,

12627h (1963).
1062. Petrov, K. A., E. E. Nifant'ev, I. I. Sopikova, and V. M. Budanov, U.S.S.R. 135 487 (1961); C.A. 55, 15348c (1961).
1063. Petrov, K. A., E. E. Nifant'ev, I. I. Sopikova, and V. M. Budanov, Zh. Obshch. Khim., 31, 2373 (1961); C.A. 56, 8553a (1962).
1064. Petrov, A. A., N. A. Razumova, and Zh. L. Evtikhov, Zh. Obshch. Khim., 37, 1410 (1967); C.A. 68, 2949u (1968).
1065. Petrov, A. A., N. A. Razumova, A. Kh. Voznesenskaya, Zh. L. Evtikhov, and L. S. Kovalev, Zh. Obshch. Khim., 38, 1201 (1968); C.A. 69, 77357d (1968).
1066. Petrov, K. A., and G. A. Sokol'skii, Zh. Obshch. Khim., 26, 3377 (1956); C.A. 51, 8029a (1957).
1067. Petrov, K. A., and G. A. Sokol'skii, Zh. Obshch. Khim., 26, 3378 (1956); C.A. 51, 8029b (1957).
1068. Petrov, K. A., G. A. Sokol'skii, and B. M. Polees, Zh. Obshch. Khim., 26, 3381 (1956); C.A. 51, 9473i (1957).
1069. Petschik, H., and E. Steger, Angew. Chem., 76, 344 (1964).
1070. Pilgram, K., and H. Ohse, J. Org. Chem., 34, 1592 (1969).
1071. Pistschimuka, P., J. Russ. Phys. Chem. Soc., 44, 1406 (1912).
1072. Pivawer, P. M., A. D. Bliss, and R. Ratz, J. Heterocyclo. Chem., 4, 599 (1967).
1073. Platt, A. E., and B. Tottle, J. Chem. Soc. C, 1967, 1150.
1074. Plets, V. M., Zh. Obshch. Khim., 6, 1198 (1936); C.A. 31, 1355[5] (1937).
1075. Pobedimskii, D. G., P. I. Levin, and Z. B. Chelnokova, Izv. Akad. Nauk SSSR, Ser. Khim., 1969, 2066; C.A. 72, 30938r (1970).
1076. Podashova, G. M., A. N. Smirnov, and A. P. Khardin, Khim. Khim. Tekhnol., 1968, 18; C.A. 72, 67309n (1970).
1077. Podkladchikov, M., J. Russ. Phys. Chem. Soc., 1899, 31, 30.
1078. Poshkus, A. C., and J. E. Herweh, J. Am. Chem. Soc., 79, 4245 (1957).
1079. Poshkus, A. C., and J. E. Herweh, J. Am. Chem. Soc., 79, 6127 (1957).
1080. Poshkus, A. C., and J. E. Herweh, Chem. Ind. (London), 1961, 1316.
1081. Poshkus, A. C., and J. E. Herweh, J. Am. Chem. Soc., 84, 555 (1962).
1082. Poshkus, A. C., J. E. Herweh, and L. F. Hass, J. Am. Chem. Soc., 80, 5022 (1958).

1083. Postnikova, G. B., and I. F. Lutsenko, Zh. Obshch. Khim., 37, 233 (1967); C.A. 66, 95134w (1967).
1084. Pudovik, A. N., Zh. Obshch. Khim., 20, 92 (1950); C.A. 44, 5800d (1950).
1085. Pudovik, A. N., Dokl. Akad. Nauk SSSR, 73, 499 (1950); C.A. 45, 2856d (1951).
1086. Pudovik, A. N., Zh. Obshch. Khim., 22, 109 (1952); C.A. 46, 11099c (1952).
1087. Pudovik, A. N., Zh. Obshch. Khim., 22, 462 (1952); C.A. 47, 2686f (1953); C.A. 46, 6082e (1952).
1088. Pudovik, A. N., Zh. Obshch. Khim., 22, 473 (1952); C.A. 47, 2687h (1953).
1089. Pudovik, A. N., Zh. Obshch. Khim., 22, 1143 (1952); C.A. 47, 4836h (1953).
1090. Pudovik, A. N., Zh. Obshch. Khim., 22, 2047 (1952); C.A. 47, 9910b (1953).
1091. Pudovik, A. N., Zh. Obshch. Khim., 25, 2173 (1955); C.A. 50, 8486i (1956).
1092. Pudovik, A. N., Zh. Obshch. Khim., 27, 2755 (1957); C.A. 52, 7129c (1958).
1093. Pudovik, A. N., and I. M. Aladzheva, Zh. Obshch. Khim., 31, 2052 (1961); C.A. 55, 27013d (1961).
1094. Pudovik, A. N., and I. M. Aladzheva, Zh. Obshch. Khim., 32, 2005 (1962); C.A. 58, 4594h (1963).
1095. Pudovik, A. N., and I. M. Aladzheva, Zh. Obshch. Khim., 33, 707, 708 (1963); C.A. 59, 2851g, 2852h (1963).
1096. Pudovik, A. N., and I. M. Aladzheva, Zh. Obshch. Khim., 33, 1816 (1963); C.A. 59, 8580h (1963).
1097. Pudovik, A. N., and I. M. Aladzheva, Zh. Obshch. Khim., 33, 3096 (1963); C.A. 60, 1788c (1964).
1098. Pudovik, A. N., and I. M. Aladzheva, Zh. Obshch. Khim., 37, 2715 (1967); C.A. 69, 59348x (1968).
1099. Pudovik, A. N., I. M. Aladzheva, V. G. Kotova, and A. F. Zinkovskii, Zh. Obshch. Khim., 39, 1528 (1969).
1100. Pudovik, A. N., I. M. Aladzheva, and L. N. Yakovenko, Zh. Obshch. Khim., 33, 3443 (1963); C.A. 60, 5543d (1964).
1101. Pudovik, A. N., I. M. Aladzheva, and L. N. Yakovenko, Zh. Obshch. Khim., 35, 1210 (1965); C.A. 63, 11609f (1965).
1102. Pudovik, A. N., and B. A. Arbuzov, Izv. Akad. Nauk SSSR, Ser. Khim., 1949, 522; C.A. 44, 1893g (1950).
1103. Pudovik, A. N., and V. P. Aver'yanova, Zh. Obshch. Khim., 26, 1426 (1956); C.A. 50, 14512f (1956).
1104. Pudovik, A. N., and E. S. Batyeva, Zh. Obshch. Khim., 38, 285 (1968); C.A. 69, 52231v (1968).
1105. Pudovik, A. N., and E. S. Batyeva, Zh. Obshch. Khim., 39, 334 (1969); C.A. 70, 115239x (1969).
1106. Pudovik, A. N., and L. G. Biktimirova, Zh. Obshch.

Khim., 27, 1708 (1957); C.A. 52, 3714b (1958).
1107. Pudovik, A. N., and L. G. Biktimirova, Zh. Obshch. Khim., 28, 1496 (1958); C.A. 53, 216g (1959).
1108. Pudovik, A. N., and E. G. Chebotareva, Zh. Obshch. Khim., 28, 2492 (1958); C.A. 53, 3117c (1959).
1109. Pudovik, A. N., and O. S. Durova, Zh. Obshch. Khim., 36, 1460 (1966); C.A. 66, 11009g (1967).
1110. Pudovik, A. N., and G. I. Evstaf'ev, Dokl. Akad. Nauk SSSR, 183, 842 (1968); C.A. 70, 67280e (1969).
1111. Pudovik, A. N., and E. M. Faizullin, Zh. Obshch. Khim., 32, 231 (1962); C.A. 57, 12299h (1962).
1112. Pudovik, A. N., and E. M. Faizullin, Zh. Obshch. Khim., 34, 882 (1964); C.A. 60, 15903g (1964).
1113. Pudovik, A. N., and E. M. Faizullin, Zh. Obshch. Khim., 38, 1908 (1968); C.A. 69, 96828z (1968).
1114. Pudovik, A. N., E. M. Faizullin, and E. Kh. Mukhametzyanova, Zh. Obshch. Khim., 34, 2471 (1964); C.A. 61, 9449h (1964).
1115. Pudovik, A. N., E. M. Faizullin, and S. V. Yakovleva, Zh. Obshch. Khim., 37, 460 (1967); C.A. 67, 43884s (1967).
1116. Pudovik, A. N., E. M. Faizullin, and V. P. Zhukov, Zh. Obshch. Khim., 36, 310 (1966); C.A. 64, 15916b (1966).
1117. Pudovik, A. N., E. M. Faizullin, and G. I. Zhuravlev, Dokl. Akad. Nauk SSSR, 165, 586 (1965); C.A. 64, 6481e (1966).
1118. Pudovik, A. N., and T. Kh. Gazizov, Zh. Obshch. Khim., 38, 140 (1968); C.A. 69, 96836a (1968).
1119. Pudovik, A. N., T. Kh. Gazizov, and A. P. Pashinkin, Zh. Obshch. Khim., 36, 951 (1966); C.A. 65, 8951h (1966).
1120. Pudovik, A. N., T. Kh. Gazizov, Yu. Yu. Samitov, and T. V. Zykova, Dokl. Akad. Nauk SSSR, 166, 615 (1966); C.A. 64, 12716b (1966).
1121. Pudovik, A. N., T. Kh. Gazuzov, Yu. Yu. Samitov, and T. V. Zykova, Zh. Obshch. Khim., 37, 706 (1967); C.A. 67, 54225d (1967).
1121a. Pudovik, A. N., and G. A. Golicyna, Zh. Obshch. Khim., 34, 876 (1964).
1122. Pudovik, A. N., and I. V. Gur'yanova, Zh. Obshch. Khim., 37, 1649 (1967); C.A. 68, 13093m (1968).
1123. Pudovik, A. N., I. V. Gur'yanova, and S. P. Perevezentseva, Zh. Obshch. Khim., 38, 942 (1968); C.A. 69, 10400u (1968).
1124. Pudovik, A. N., I. V. Gur'yanova, S. P. Perevezentseva, and S. A. Terent'eva, Zh. Obshch. Khim., 39, 337 (1969); C.A. 70, 115071m (1969).
1125. Pudovik, A. N., I. V. Gur'yanova, S. P. Perevezentseva, and T. V. Zykova, Zh. Obshch. Khim., 37, 1317 (1967); C.A. 68, 22013n (1968).

1126. Pudovik, A. N., E. I. Kashevarova, and L. I. Goloven'kina, Vysokomolekul. Soedin., 7, 1248 (1965); C.A. 63, 13420g (1965).
1127. Pudovik, A. N., E. I. Kashevarova, and V. M. Gorchakova, Zh. Obshch. Khim., 34, 2213 (1964); C.A. 61, 10704e (1964).
1128. Pudovik, A. N., and N. I. Khlyupina, Zh. Obshch. Khim., 26, 1672 (1956); C.A. 51, 3439f (1957).
1129. Pudovik, A. N., and N. G. Khusainova, Zh. Obshch. Khim., 36, 1236 (1966); C.A. 65, 16994h (1966).
1130. Pudovik, A. N., N. G. Khusainova, and I. M. Aladzheva, Zh. Obshch. Khim., 33, 1045 (1963); C.A. 59, 10115a (1963).
1131. Pudovik, A. N., and Yu. P. Kitaev, Zh. Obshch. Khim., 22, 467 (1952); C.A. 47, 2687a (1953).
1132. Pudovik, A. N., and I. V. Konovalova, Zh. Obshch. Khim., 27, 1617 (1957); C.A. 52, 3713d (1958).
1133. Pudovik, A. N., and I. V. Konovalova, Dokl. Akad. Nauk SSSR, 149, 1091 (1963); C.A. 59, 6434g (1963).
1134. Pudovik, A. N., and I. V. Konovalova, Zh. Obshch. Khim., 35, 1591 (1965); C.A. 63, 17887b (1965).
1135. Pudovik, A. N., I. V. Konovalova, and A. A. Guryleva, Zh. Obshch. Khim., 33, 2924 (1963); C.A. 60, 5541g (1964).
1136. Pudovik, A. N., and V. P. Krupnov, Zh. Obshch. Khim., 38, 194 (1968); C.A. 69, 77345y (1968).
1137. Pudovik, A. N., and V. K. Krupnov, Zh. Obshch. Khim., 38, 304 (1968); C.A. 69, 77369j (1968).
1138. Pudovik, A. N., and V. K. Krupnov, Zh. Obshch. Khim., 38, 1406 (1968); C.A. 69, 66816b (1968).
1139. Pudovik, A. N., and V. K. Krupnov, Zh. Obshch. Khim., 38, 1652 (1968); C.A. 70, 86945s (1969).
1140. Pudovik, A. N., and T. M. Moshkina, Zh. Obshch. Khim., 27, 1611 (1957); C.A. 52, 3712e (1958).
1141. Pudovik, A. N., T. M. Moshkina, and I. V. Konovalova, Zh. Obshch. Khim., 29, 3338 (1959); C.A. 54, 15223g (1960).
1142. Pudovik, A. N., and M. A. Pudovik, Zh. Obshch. Khim., 39, 1645 (1969); C.A. 71, 91592z (1969).
1143. Pudovik, A. N., M. A. Pudovik, and R. R. Shagidullin, Zh. Obshch. Khim., 39, 973 (1969); C.A. 72, 31937b (1970).
1144. Pudovik, A. N., and A. P. Rakov, Dokl. Akad. Nauk SSSR, 161, 1352 (1965); C.A. 63, 4151d (1965).
1145. Pudovik, A. N., and O. S. Shulyndina, Zh. Obshch. Khim., 39, 1014 (1969); C.A. 71, 614937a (1969).
1146. Pudovik, A. N., and R. I. Tarasova, Zh. Obshch. Khim., 34, 293 (1964); C.A. 60, 10579a (1964).
1147. Pudovik, A. N., R. I. Tarasova, and R. A. Bulgakova, Zh. Obshch. Khim., 33, 2560 (1963); C.A. 60,

543g (1964).
1148. Pudovik, A. N., S. A. Terent'eva, and E. S. Batyeva, Dokl. Akad. Nauk SSSR, 175, 616 (1967); C.A. 68, 2961s (1968).
1149. Pudovik, A. N., G. E. Yastrebova, and V. I. Nikitina, Zh. Obshch. Khim., 36, 1232 (1966); C.A. 65, 15418g (1966). CA 58 2392a (1963)
1150. Pure Chemicals Ltd., Brit. 1 002 321 (1960); C.A. 63, 16216c (1965).
1151. Pure Chemicals Ltd., Fr. 1 321 898 (1960); C.A. 59, 12985f (1963).
1152. Quesnel, G., and G. Mavel, Compt. Rend., 248, 295 (1959).
1153. Quesnel, G., A. Thiot, H. Coates, and R. Burgada, Compt. Rend., 256, 717 (1963).
1154. Quin, L. D., and C. H. Rolston, J. Org. Chem., 23, 1693 (1958).
1155. Rabinowitz, R., J. Org. Chem., 28, 2975 (1963).
1156. Railton, R., J. Chem. Soc., 7, 216 (1854); Ann. Chem., 92, 348 (1854).
1157. Ramaswami, D., and E. R. Kirch, J. Am. Chem. Soc., 75, 1763 (1953).
1158. Ramirez, F., Pure Appl. Chem., 9, 337 (1964).
1159. Ramirez, F., Bull. Soc. Chim. Fr., 1966, 2443.
1160. Ramirez, F., Trans. N.Y. Acad. Sci., 30, 410 (1968).
1161. Ramirez, F., Accounts Chem. Res., 1, 168 (1968).
1162. Ramirez, F., S. B. Bhatia, A. V. Patwardhan, E. H. Chen, and C. P. Smith, J. Org. Chem., 33, 20 (1968).
1163. Ramirez, F., S. B. Bhatia, A. V. Patwardhan, and C. P. Smith, J. Org. Chem., 32, 2194 (1967).
1164. Ramirez, F., S. B. Bhatia, and C. P. Smith, Tetrahedron, 23, 2067 (1967).
1165. Ramirez, F., E. H. Chen, and S. Dershowitz, J. Am. Chem. Soc., 81, 4338 (1959).
1166. Ramirez, F., and S. Dershowitz, J. Org. Chem., 22, 1282 (1957).
1167. Ramirez, F., and S. Dershowitz, J. Org. Chem., 22, 856 (1957); 23, 778 (1958).
1168. Ramirez, F., and S. Dershowitz, J. Am. Chem. Soc., 81, 587 (1959).
1169. Ramirez, F., and N. B. Desai, J. Am. Chem. Soc., 82, 2652 (1960).
1170. Ramirez, F., M. B. Desai, and N. Ramanathan, Tetrahedron Letters, 1963, 232.
1171. Ramirez, F., S. R. Glaser, A. J. Bigler, and J. F. Pilot, J. Am. Chem. Soc., 91, 496 (1969).
1172. Ramirez, F., A. S. Gulati, and C. P. Smith, J. Org. Chem., 33, 13 (1968).
1173. Ramirez, F., O. P. Madan, and S. R. Heller, J. Am.

Chem. Soc., 87, 731 (1965).
1174. Ramirez, F., O. P. Madan, and C. P. Smith, J. Am. Chem. Soc., 86, 5339 (1964).
1175. Ramirez, F., O. Madan, and C. P. Smith, J. Am. Chem. Soc., 87, 670 (1965).
1176. Ramirez, F., O. P. Madan, and C. P. Smith, J. Org. Chem., 30, 2284 (1965).
1177. Ramirez, F., O. P. Madan, and C. P. Smith, Tetrahedron, 22, 567 (1966).
1178. Ramirez, F., R. B. Mitra, and N. B. Desai, J. Am. Chem. Soc., 82, 2651 (1960).
1179. Ramirez, F., R. B. Mitra, and N. B. Desai, J. Am. Chem. Soc., 82, 5763 (1960).
1180. Ramirez, F., M. Nagabhushanam, and C. P. Smith, Tetrahedron, 24, 1785 (1968).
1181. Ramirez, F., A. V. Patwardhan, and S. R. Heller, J. Am. Chem. Soc., 86, 514 (1964).
1182. Ramirez, F., A. V. Patwardhan, H. J. Kugler, and C. P. Smith, Tetrahedron Letters, 1966, 3053.
1183. Ramirez, F., A. V. Patwardhan, H. J. Kugler, and C. P. Smith, J. Am. Chem. Soc., 89, 6276 (1967).
1184. Ramirez, F., A. V. Patwardhan, H. J. Kugler, and C. P. Smith, Tetrahedron, 24, 2275 (1968).
1185. Ramirez, F., A. V. Patwardhan, and C. P. Smith, J. Am. Chem. Soc., 87, 4973 (1965).
1186. Ramirez, F., A. V. Patwardhan, and C. P. Smith, J. Org. Chem., 30, 2575 (1965).
1187. Ramirez, F., J. F. Pilot, C. P. Smith, S. B. Bhatia, and S. A. Gulati, J. Org. Chem., 34, 3385 (1969).
1188. Ramirez, F., and N. Ramanthan, J. Org. Chem., 26, 3041 (1961).
1189. Ramirez, F., D. Rhum, and C. P. Smith, Tetrahedron, 21, 1941 (1965).
1190. Ramirez, F., C. P. Smith, and S. Meyerson, Tetrahedron Letters, 1966, 3651.
1191. Ramirez, F., K. Tasaka, N. B. Desai, and C. P. Smith, J. Org. Chem., 33, 25 (1968).
1192. Ramirez, F., K. Tasaka, N. B. Desai, and C. P. Smith, J. Am. Chem. Soc., 90, 751 (1968).
1193. Razumov, A. I., Zh. Obshch. Khim., 14, 464 (1944); C.A. 39, 4586[9] (1945).
1194. Razumov, A. I., Trudy Kazansk. Khim. Tekhnol. Inst. S. M. Kirova, 23, 201 (1957); C.A. 52, 9947d (1958).
1195. Razumov, A. I., B. G. Kiorber, V. V. Moskva, and Z. M. Khammatova, Trudy Kazansk. Khim. Tekhnol. Inst., 30, 265 (1962); C.A. 60, 1571g (1964).
1196. Razumov, A. I., and E. I. Korobkova, Trudy Kazansk. Khim. Tekhnol. Inst. S. M. Kirova, 23, 215 (1957); C.A. 52, 9948d (1958).

1197. Razumov, A. I., A. I. Kukhtin, and N. Sazonova, Zh. Obshch. Khim., 22, 920 (1952); C.A. 47, 4836a (1953).
1198. Razumov, A. I., and B. G. Liorber, Zh. Obshch. Khim., 32, 4063 (1962); C.A. 58, 13984a (1963).
1199. Razumov, A. I., and V. V. Moskva, U.S.S.R. 175 961 (1965); C.A. 64, 9596h (1966).
1200. Razumov, A. I., and N. I. Rizpolozhenskii, Trudy Kazansk. Khim. Tekhnol. Inst., 8, 42 (1940); C.A. 35, 2485[5] (1941).
1201. Razumova, N. A., Zh. L. Evtikhov, and A. A. Petrov, Zh. Obshch. Khim., 38, 1117 (1968); C.A. 69, 96602w (1968).
1202. Razumova, N. A., Zh. L. Evtikhov, A. Kh. Voznesenskaya, and A. A. Petrov, Zh. Obshch. Khim., 39, 176 (1969); C.A. 70, 106608q (1969).
1203. Razumova, N. A., Zh. L. Evtikhov, L. I. Zubtsova, and A. A. Petrov, Zh. Obshch. Khim., 38, 2342 (1968); C.A. 70, 47546t (1969).
1204. Razumova, N. A., L. S. Kovalev, and A. A. Petrov, Zh. Obshch. Khim., 38, 126 (1968); C.A. 69, 5222t (1968).
1205. Razumova, N. A., L. S. Kovalev, and A. A. Petrov, Zh. Obshch. Khim., 38, 323 (1968); C.A. 69, 59332n (1968).
1206. Razumova, N. A., and A. A. Petrov, Zh. Obshch. Khim., 33, 783 (1963); C.A. 59, 8783g (1963).
1207. Razumova, N. A., and A. A. Petrov, Zh. Obshch. Khim., 33, 3858 (1963); C.A. 60, 10711h (1964).
1208. Razumova, N. A., A. A. Petrov, A. Kh. Voznesenskaya, and K. I. Novitskii, Zh. Obshch. Khim., 36, 244 (1966); C.A. 64, 15913g (1966).
1209. Reddy, G. S., and R. Schmutzler, Z. Naturforsch., 20b, 104 (1965).
1210. Reetz, T., U.S. 2 863 905 (1958); C.A. 53, 9058i (1959).
1211. Reetz, T., W. A. Busch, and D. H. Chadwick, U.S. 3 057 904 (1962); C.A. 58, 3316e (1963).
1212. Reetz, T., and B. Katlafsky, J. Am. Chem. Soc., 82, 5036 (1960).
1213. Reichel, I., and D. Purdela, Acad. Rep. Populare Romine, Baza Cercetari Stünt Timisoara, Studii Cercetari Stünt. Chim., 6, 95 (1959); C.A. 55, 4446a (1961).
1214. Reuter, R., U.S. 2 280 450 (1942); C.A. 36, 6007[7] (1942).
1215. Reynolds, R. B., and H. Atkins, J. Am. Chem. Soc., 51, 280 (1929).
1216. Richter, V. von, Organic Chemistry (Translated by E. Newmarche Abbot), Elsevier, New York, 1947, Vol. 1, p. 315.

1217. Rizpolozhenskii, N. I., Nekotorye Vopr. Org. Khim., 1964, 122; C.A. 65, 2290g (1966).
1218. Rizpolozhenskii, M. I., L. V. Boiko, and M. A. Zvereva., Dokl. Akad. Nauk SSSR, 155, 1137 (1964); C.A. 61, 1817f (1964).
1219. Rizpolozhenskii, N. I., and F. S. Mukhametov, Izv. Akad. Nauk SSSR, Ser. Khim., 1968, 2755; C.A. 70, 77229j (1969).
1220. Rizpolozhenskii, N. I., F. S. Mukhametov, and R. T. Shagidullin, Izv. Akad. Nauk SSSR, Ser. Khim., 1969, 1121; C.A. 71, 50107b (1969).
1221. Rizpolozhenskii, N. I., A. A. Mushinkin, Izv. Akad. Nauk SSSR, Ser. Khim., 1961, 1600; C.A. 56, 4791d (1962).
1222. Robinson, S. D., Chem. Commun., 1968, 521.
1223. Rodd, E. H., The Chemistry of Carbon Compounds, Elsevier, Amsterdam, 1951, Vol. la, p. 588.
1224. Rosenbaum, A., J. Prakt. Chem., 37, 200 (1968).
1225. Rosenheim, L., Z. Anorg. Chem., 43, 34 (1905).
1226. Rossiiskaya, P. A., and M. I. Kabachnik, Izv. Akad. Nauk SSSR, Ser. Khim. Otd. Khim. Nauk, 1947, 509; C.A. 42, 2924b (1948).
1227. Rossiiskaya, P. A., and M. I. Kabachnik, Izv. Akad. Nauk SSSR, Ser. Khim. Otd. Khim. Nauk, 1947, 389; C.A. 42, 1558d (1948).
1228. Roussos, M., and Y. Bourgeois, Fr. 1 500 821 (1967); C.A. 69, 86604j (1968).
1229. Rumpf, P., Bull. Soc. Chim. France, 18, 128 (1951).
1230. Rydon, H. N., and B. L. Tonge, J. Chem. Soc., 1956, 3043.
1231. Sakiyama, M., K. Fujiwara, O. Mitsunobu, T. Ohashi, and M. Kikuchi, Japan 67 18 608 (1967); C.A. 69, 96871h (1968).
1232. Saks, Levitskii, J. Russ. Phys. Chem. Soc., 35, 211 (1903).
1233. Samigulin, F. K., I. M. Kafengauz, and A. P. Kafengauz, Kinet. Katal., 9, 898 (1968); C.A. 70, 76960d (1969).
1234. Samuel, A. A., R. Bouvet, S. Hittner, and M. de Beaulieu, Fr. 1 157 174 (1958); C.A. 54, 18360 (1960).
1235. Samuel, D., and B. L. Silver, Chem. Ind. (London), 1961, 556.
1236. Samuel, D., and B. L. Silver, J. Org. Chem., 28, 2089 (1963).
1237. Samuel, D., and B. L. Silver, J. Chem. Soc., 1963, 3582.
1238. Sanchez, M., R. Wolf, R. Burgada, and F. Mathis, Bull. Soc. Chim. France, 1968, 773.
1239. Sander, M., Chem. Ber., 93, 1220 (1960).
1240. Sander, M., Chem. Ber., 95, 473 (1962).

1241. Sanin, P. I., M. G. Voronkov, E. S. Shepeleva, and B. I. Ionin, Dokl. Akad. Nauk SSSR, 132, 145 (1960); C.A. 54, 20940a (1960).
1242. Sass, S., and J. Cassidy, Anal. Chem., 28, 1968 (1956).
1243. Schay, G., Gy. Szekely, Gy. Racz, and G. Traply, Periodica Polytech. 2, 1 (1958).
1244. Scheid, B., Ann. Chem., 218, 207 (1883).
1245. Schiebs, R., S. African 69 02 779 (1969); C.A. 72, 132037p (1970).
1246. Schiff, H., Ann. Chem., 102, 334 (1857); 103, 164 (1857).
1247. Schliebs, R., Ger. 1 079 022 (1960); C.A. 55, 14307i (1961).
1248. Schmadebeck, J. H., U.S. 3 071 442 (1963); C.A. 58, 7640h (1963).
1249. Schmutzler, R., Chem. Ber., 96, 2435 (1963).
1250. Schmutzler, R., Inorg. Chem., 3, 415 (1964).
1250a. Schmutzler, R., J. Chem. Soc., 1965, 5630.
1251. Schofield, J. A., and A. Todd, J. Chem. Soc., 1961, 2316.
1252. Scholler, C., Fr. 1 271 964 (1962); C.A. 57, 2146 (1962).
1253. Schrader, G., Ger. 1 161 556 (1964); C.A. 60, 10548h (1964).
1254. Schrader, G., Ger. 1 112 732 (1960); C.A. 56, 2331c (1962).
1255. Schrader, G., Ger. 1 175 659 (1964); C.A. 62, 1566q (1965).
1256. Schrader, G., and O. Bayer, Ger. 664 438 (1938); C.A. 33, 803[3] (1939); C.A. U.S. 2 146 356 (1939); C.A. 33, 3500[6] (1939).
1257. Schulz, A. R., Fr. 1 395 477 (1965); C.A. 63, 11361f (1965).
1258. Schulze, W. A., G. H. Short, and W. W. Crouch, Ind. Eng. Chem., 42, 916 (1950).
1259. Schwarz, R., and H. Geulen, Chem. Ber., 90, 952 (1957).
1260. Schwyzer, R., B. M. Iselin, and W. Rittel, U.S. 2 938 915 (1960); C.A. 54, 20897g (1960).
1261. Scott, C. B., J. Org. Chem., 22, 1118 (1957).
1262. Sehring, R., and K. Zeile, Ger. 1 078 558 (1960); C.A. 55, 14308d (1961).
1263. Serra, M., and P. Malatesta, Ann. Chim. (Rome), 43, 568 (1953).
1264. Serra, M., and P. Malatesta, Ann. Chim. (Rome), 45, 911 (1955).
1265. Shakirova, A. M., and M. G. Imaev, Zh. Obshch. Khim., 37, 468 (1967); C.A. 67, 43874p (1967).
1266. Shepard, A. F., B. F. Dannels, and F. M. Kujawa, Fr. 1 366 579 (1964); C.A. 62, 2738g (1965).

1267. Shepard, A. F., and B. F. Dannels, U.S. 3 281 506 (1966); C.A. 66, 10739s (1967).
1268. Shepard, A. F., and B. F. Dannels, U.S. 3 415 906 (1968); C.A. 70, 37449z (1969).
1269. Shitov, L. N., and B. M. Gladshtein, Zh. Obshch. Khim., 38, 2340 (1968); C.A. 70, 29010j (1969).
1270. Shrer, S. M., L. P. Bocharova, and I. K. Rubtsova, Zh. Obshch. Khim., 37, 418 (1967); C.A. 67, 43875q (1967).
1271. Shokanov, A. K., and Z. M. Muldakhmetov, Trudy Khim. Met. Inst. Akad. Nauk Kazansk. SSR., 2, 169 (1968); C.A. 70, 7680a (1969).
1272. Shokol, V. A., G. A. Golik, and G. I. Derlach, Khim. Org. Soedin. Fosfora, Akad. Nauk SSSR, Otd. Obshch. Tekh. Khim., 1967, 96; C.A. 69, 10511f (1968).
1273. Shokol, V. A., A. G. Matyusha, L. I. Molyarko, N. K. Mikhailyuchenko, and G. I. Derkach, Zh. Obshch. Khim., 39, 2137 (1969); C.A. 72, 31139t (1970).
1274. Shokol, V. A., L. I. Molyavko, and G. I. Derkach, Zh. Obshch. Khim., 36, 930 (1966); C.A. 65, 12229f (1966).
1275. Short, J. N., U.S. 2 928 861 (1960); C.A. 54, 18360h (1960).
1276. Shuikin, N. I., I. F. Bel'skii, and I. E. Grushko, Izv. Akad. Nauk SSSR, Ser. Khim., 1963, 557; C.A. 59, 2750h (1963).
1277. Shuler, W. E., and R. C. Axtmann, U.S. At. Energy Comm. DP 474 (1960).
1278. Simon, A., and W. Schulze, Chem. Ber., 94, 3251 (1961).
1279. Smeykal, K., H. Baltz, and H. Fischa, J. Prakt. Chem., 22, 186 (1963).
1280. Smith, P. V., Jr., F. Knoth, Jr., and W. E. Wadday, U.S. 2 722 517 (1955); C.A. 50, 3751i (1956).
1281. Sollett, G. P., and W. R. Peterson, J. Organometal. Chem., 19, 143 (1969).
1282. Sommer, L. H., H. D. Blankman, and P. C. Miller, J. Am. Chem. Soc., 73, 3542 (1951).
1283. Songstad, J., Acta Chem. Scand., 21, 1681 (1967).
1284. Sorochkin, I. N., A. P. Tuseev, V. B. Luk'yanov, and V. T. Tsarynov, U.S.S.R. 193 499 (1967); C.A. 69, 43430g (1968).
1285. Speer, J. H., and A. J. Hill, J. Org. Chem., 2, 139 (1937).
1286. Spencer, E. Y., A. R. Todd, and R. F. Webb, J. Chem. Soc., 1958, 2968.
1287. Speziale, A. J., and R. C. Freeman, J. Org. Chem., 23, 1883 (1958).
1288. Speziale, A. J., and R. C. Freeman, J. Am. Chem.

Soc., 82, 903 (1960).
1289. Srivastava, K. C., Chem. Age, India, 18, 561 (1967).
1290. Stauffer Chemical Company, Neth. Appl. 6 612 966 (1967); C.A. 67 53665k (1967).
1291. Stec. W., A. Zwierzak, and J. Michalski, Tetrahedron Letters, 1968, 5873.
1292. Stelling, O., Z. Physik. Chem., 117, 161, 194 (1925).
1293. Stepashkina, L. V., and N. I. Rizpolozhenskii, Izv. Akad. Nauk SSSR, Ser. Khim., 1967, 607; C.A. 67, 108491h (1967).
1294. Stetter, H., and K. H. Steinacker, Chem. Ber., 85, 451 (1952).
1295. Stevens, D. R., and R. S. Spindt, U.S. 2 542 370 (1951); C.A. 45, 5712h (1951).
1296. Steyermark, P. R., J. Org. Chem., 28, 3570 (1963).
1297. Stiles, A. R., U.S. 2 895 982 (1959); C.A. 53, 20681a (1959).
1298. Stiles, A. R., and F. F. Rust. U.S. 2 724 718 (1955); C.A. 50, 10124d (1956).
1299. Strecker, W., and C. Grossmann, Chem. Ber., 49, 63 (1916).
1300. Streich, J. A., W. E. Jones, and R. S. Long, Belg. 613 241 (1962); C.A. 58, 8907a (1963).
1301. Strecker, W., and R. Spitaler, Chem. Ber., 59, 1754 (1926).
1302. Stuebe, C., and H. P. Lankelma, J. Am. Chem. Soc., 78, 976 (1956).
1303. Sundberg, R. J., J. Org. Chem., 30, 3604 (1965).
1304. Sundberg, R. J., J. Am. Chem. Soc., 88, 3781 (1966).
1305. Swarts, F., Bull. Soc. Chim. Belges, 38, 105 (1929).
1306. Takamizawa, A., K. Hirai, and Y. Hamashima, Tetrahedron Letters, 1967, 5081.
1307. Takamizawa, A., K. Hirai, Y. Hamashima, Y. Matsumoto, and S. Tanaka, Chem. Pharm. Bull. (Tokyo), 15, 1764 (1968).
1308. Takamizawa, A., and Y. Sato, Japan 2 335 (1965); C.A. 62, 14501d (1965).
1309. Takemura, K. H., and D. J. Tuma, J. Org. Chem., 34, 252 (1969).
1310. Tanaka, T. Yakugaku Zasshi, 80, 439 (1960).
1310a. Tanaka, T. Yakugaku Zasshi, 79, 1301 (1959).
1311. Tanaka, T. Yakugaku Zasshi, 80, 1053 (1960).
1312. Taniuchi, A., and M. Sonoda, Japan. 19 576 (1967); C.A. 69, 18601f (1968).
1313. Tasker, H. S., and H. O. Jones, J. Chem. Soc., 95, 1910 (1909).
1314. Taylor, G. W., and D. H. Wood, Brit. 1 066 404

(1967); C.A. 67, 12147z (1967).
1315. Teichmann, H., Ann. Chem., 703, 31 (1967).
1316. TerHaar, G., Sr., M. A. Fleming, and R. W. Parry, J. Am. Chem. Soc., 84, 1767 (1962).
1317. Thomas, L. C., Chem. Ind. (London), 1957, 198.
1318. Thomas, L. C., and R. A. Chittenden, Spectrochim. Acta, 20, 467, 489 (1964).
1319. Thomas, L. C., and R. A. Chittenden, Spectrochim. Acta, 22, 1449 (1966).
1320. Thompson, Q. E., J. Am. Chem. Soc., 83, 845 (1961)
1321. Thorpe, T. E., J. Chem. Soc., 1880, 186.
1322. Thorpe, T. E., J. Chem. Soc., 1880, 327.
1323. Thorpe, T. E., and B. North, J. Chem. Soc., 1890, 634.
1324. Thorpe, T. E., and A. Tutton, J. Chem. Soc., 1891, 1019.
1325. Tichy, V., and S. Truchlik, Chem. Zvesti, 12, 345 (1958).
1326. Tolkmith, H., J. Org. Chem., 23, 1682 (1958).
1327. Tolocho, A. F., N. I. Ganushchak, and A. V. Dombrovskii, Zh. Obshch. Khim., 38, 1112 (1968); C.A. 69, 106825n (1968).
1328. Traufel, W., and F. Kunkele, Fettchem. Umschau., 42, 27 (1935).
1329. Trippett, S., J. Chem. Soc., 1962, 2337.
1330. Trofimov, B. A., A. S. Atavin, G. M. Gavrilova, and G. A. Kalabin, Zh. Obshch. Khim., 38, 2344 (1968); C.A. 70, 29102m (1969).
1331. Trofimov, B. A., A. S. Atavin, A. Gusarov, S. V. Amosova, and S. E. Korostova, Zh. Org. Khim., 5, 816 (1969); C.A. 71, 38252n (1969).
1332. Tsai, C., and K. Chen, K'o Hsueh T'ung Pao, 1964, 1003.
1333. Tseng, C.-L., and C.-S. Hou, J. Chin. Chem. Soc., 2, 57 (1934).
1334. Tsivunin, V. S., R. G. Ivanova, and G. Kamai, Zh. Obshch. Khim., 38, 1062 (1968); C.A. 69, 106076u (1968).
1335. Tsvetkov, E. N., and M. I. Kabachnik, "Reactions and Methods of Investigating Organic Compounds," Izd. Khimiya, 13, 295, 377 (1964).
1336. Ueda, T., K. Inukai, and H. Muramatsu, Bull. Chem. Soc. Japan, 42, 1684 (1969).
1337. Ulrich, H., U.S. 3 413 382 (1968); C.A. 70, 58015v (1969).
1338. United States Rubber Company, Brit. 771 866 (1955); C.A. 51, 15575e (1957).
1339. United States Rubber Company, Belg. 643 973 (1964); C.A. 63, 14763g (1965).
1340. Utvary, K., V. Gutman, and C. Kemanater, Inorg. Nucl. Chem. Letters, 1, 75 (1965).

1341. Van Druten, A., Rec. Trav. Chim., 48, 312 (1929).
1342. Van Wazer, J. R., C. F. Callis, J. N. Shoolery, and R. C. Jones, J. Am. Chem. Soc., 78, 5715 (1956).
1343. Van Wazer, J. R., and L. Maier, J. Am. Chem. Soc., 86, 811 (1964).
1344. Van Winkle, J. L., E. R. Bell, and R. C. Morris, U.S. 2 712 029 (1955); C.A. 51, 470d (1957).
1345. Varga, S. L., Diss. Abs., 1968, 2370B.
1346. Venezky, D. L., and R. B. Fox, J. Am. Chem. Soc., 78, 1664 (1956).
1347. Verizhnikov, L. V., and P. A. Kirpichnikov, Zh. Obshch. Khim., 37, 1355 (1967); C.A. 68, 12597s (1968).
1348. Verkade, J. G., and L. T. Reynolds, J. Org. Chem., 25, 663 (1960).
1349. Vetter, H. J., Z. Naturforsch., 19b, 72 (1964).
1350. Vetter, H. J., Z. Naturforsch., 19b, 167 (1964).
1350a. Vetter, H. J., Naturwiss., 51, 240 (1964).
1351. Vetter, H. J., and H. Noth, Chem. Ber., 96, 1308 (1963).
1352. Vetter, H. J., and H. Noth, Chem. Ber., 96, 1816 (1963).
1353. Virginia-Carolina Chemical Corporation, U.S. 2 678 940 (1954); C.A. 49, 4704h (1955); Brit. 853 982 (1960); C.A. 55, 18593e (1961).
1354. Voigt, D., and M. C. Labarre, Compt. Rend., 259, 4632 (1964).
1355. Volodina, L. N., Zh. Obshch. Khim., 37, 513 (1967); C.A. 67, 32729u (1967).
1356. Volodina, L. N., Zh. Obshch. Khim., 38, 200 (1968); C.A. 69, 67460t (1968).
1357. Voronkov, M. G., V. A. Kolesova, and V. N. Zgonnik, Izv. Akad. Nauk SSSR, Ser. Khim., 1957, 1363; C.A. 52, 7128f (1958).
1358. Voronkov, M. G., and Yu. I. Skorik, Izv. Akad. Nauk SSSR, Ser. Khim., 1958, 119; C.A. 52, 11735a (1958).
1359. Voronkov, M. G., and Yu. I. Skorik, Zh. Obshch. Khim., 35, 106 (1965); C.A. 62, 13173d (1965).
1360. Voskresenskaya, O. V., N. A. Makarova, P. A. Kirpichnikov, and E. T. Mukmenev, Izv. Akad. Nauk SSSR, Ser. Khim., 1968, 1393; C.A. 69, 97038d (1968).
1361. Voskresenskaya, O. V., N. A. Makarova, P. A. Kirpichnikov, and E. T. Mukmenev, Izv. Akad. Nauk SSSR, Ser. Khim., 1969, 1626; C.A. 71, 113200z (1969).
1362. Voznesenskaya, A. Kh., and N. A. Razumova, Zh. Obshch. Khim., 38, 1553 (1968); C.A. 69, 96824v (1968).

1363. Voznesenskaya, A. Kh., and N. A. Razumova, Zh. Obshch. Khim., 39, 387 (1969); C.A. 70, 115231p (1969).
1364. Voznesenskaya, A. Kh., N. A. Razumova, and A. A. Petrov, Zh. Obshch. Khim., 39, 1033 (1969); C.A. 71, 61486a (1969).
1365. Vuedenskii, V., J. Russ. Phys. Chem. Soc., 20, 29 (1888).
1366. Wadsworth, W. S., Jr., J. Org. Chem., 32, 1603 (1967).
1367. Wagner-Jauregg, T., Helv. Chim. Acta, 12, 61 (1929).
1368. Walker, J. W., and F. M. G. Johnson, J. Chem. Soc., 87, 1529 (1905).
1369. Walling, C., O. H. Basedow, and E. S. Savas, J. Am. Chem. Soc., 82, 2181 (1960).
1370. Walling, C., and M. S. Pearson, J. Am. Chem. Soc., 86, 2262 (1964).
1371. Walling, C., and R. Rabinowitz, J. Am. Chem. Soc., 79, 5326 (1957).
1372. Walling, C., and R. Rabinowitz, J. Am. Chem. Soc., 81, 1243 (1959).
1373. Walling, C., F. R. Stacey, S. E. Jamison, and E. S. Huyser, J. Am. Chem. Soc., 80, 4546 (1958).
1374. Walsh, E. N., J. Am. Chem. Soc., 81, 3023 (1959).
1375. Walsh, E. N., Ger. 1 078 136 (1960); C.A. 56, 3413c (1962).
1376. Walsh, E. N., U.S. 3 036 109 (1962); C.A. 57, 13612g (1962).
1377. Walsh, E. N., Ger. 1 125 425 (1962); C.A. 57, 7104h (1962).
1378. Warner, P. F., and J. R. Siagle, U.S. 3 136 807 (1964); C.A. 61, 6920h (1964).
1379. Wasserman, E., R. W. Murray, M. L. Kaplan, and W. A. Yager, J. Am. Chem. Soc., 90, 4160 (1968).
1380. Wende, A., H. Priebe, and A. Conrad, Plaste Kautschuk, 11, 515 (1964).
1381. Weston Chemical Corporation, Brit. 943 731 (1963); C.A. 60, 14387h (1964).
1382. Whetstone, R. R., and A. R. Stiles, U.S. 2 802 855 (1957); C.A. 52, 2067b (1958).
1383. White, D. W., G. K. McEwen, and J. E. Verkade, Tetrahedron Letters, 1968, 5369.
1384. Wichelhaus, K. H., Ann. Chem., Suppl., 6, 257 (1868); Chem. Ber., 1, 80 (1868).
1385. Wiley, R. H., U.S. 2 478 441 (1949); C.A. 44, 2010b (1950).
1386. Wiley, D. W., and H. E. Simmons, J. Org. Chem., 29, 1876 (1964).
1387. Williams, J. L., Chem. Ind. (London), 1957, 235.
1388. Williams, R. H., and L. A. Hamilton, J. Am. Chem.

Soc., 74, 5418 (1952).
1389. Williams, R. H., and L. A. Hamilton, J. Am. Chem. Soc., 77, 3411 (1955).
1390. Williamson, A. W., Proc. Roy. Soc., 1854, 131.
1391. Wilms, H., Ger. 924 385 (1955); C.A. 52, 16197i (1958).
1392. Wolf, R., R. Mathis-Noel, and F. Mathis, Bull. Soc. Chim. Fr., 1960, 1240.
1393. Wolf, R., J. R. Miquel, and F. Mathis, Bull. Soc. Chim. Fr., 1963, 825.
1394. Wottgen, E., and G. Gosse, Ger. 1 156 067 (1963); C.A. 60, 10555d (1964).
1395. Wottgen, E., and D. Luft, J. Prakt. Chem., 30, 93 (1965).
1396. Wu, C., U.S. 3 341 632 (1967); C.A. 68, 86840n (1968).
1397. Wu, C., and F. J. Welch, U.S. 3 351 683 (1967); C.A. 68, 21685w (1968).
1398. Wurtz, A., Ann. Chem., 58, 72 (1846).
1399. Yaroshenko, A. A., J. Russ. Phys. Chem. Soc., 29, 233 (1897).
1400. Young, R. W., J. Am. Chem. Soc., 74, 1672 (1952).
1401. Young, R. W., J. Am. Chem. Soc., 75, 4620 (1953).
1402. Young, R. W., K. H. Wood, R. J. Joyce, and G. W. Anderson, J. Am. Chem. Soc., 78, 2126 (1956).
1403. Zakharkin, L. I., A. V. Kazantsev, and M. N. Zhubekova, Izv. Akad. Nauk SSSR, Ser. Khim., 1969, 2056; C.A. 72, 12829b (1970).
1404. Zaripov, R. K., and V. S. Abramov, Trudy Khim. Met. Inst. Akad. Nauk Kazansk. SSR, 5, 50 (1969); C.A. 72, 21744x (1970).
1405. Zaripov, R. K., I. N. Azerbaev, and G. Sh. Shamgunov, Trudy Khim. Met. Inst. Akad. Nauk Kazansk. SSR, 8, 48 (1969); C.A. 72, 31930u (1970).
1406. Zavalishina, A. I., S. F. Sorokina, and E. E. Nifantev, Zh. Obshch. Khim., 38, 2271 (1968); C.A. 71, 13172y (1969).
1407. Zavlin, P. M., and B. I. Ionin, Zhur. Priklad. Khim., 33, 2376 (1960); C.A. 55, 6360g (1961).
1408. Zavlin, P. M., and B. I. Ionin, U.S.S.R. 130 889 (1960); C.A. 55, 7286e (1961).
1409. Zbirovsky, M., and V. Ettel, Collection Czech. Chem. Communs., 21, 1607 (1956).
1410. Zecchini, F., Gazz. Chim. Ital., 24I, 34 (1894).
1411. Zech, J. D., U.S. 2 824 113 (1958); C.A. 52, 11110f (1958).
1412. Zimmermann, C., Ann. Chem., 175, 1 (1875).
1413. Zwierzak, A., Bull. Acad. Polon. Sci. Ser. Sci. Chim., 12, 235 (1964).
1414. Zwierzak, A., Bull. Acad. Polon. Sci. Ser. Sci. Chim., 13, 609 (1965).

1415. Zwierzak, A., Can. J. Chem., 45, 2501 (1967).
1416. Zwierzak, A., and A. Koziara, Tetrahedron, 23, 2243 (1967).

TECHNICAL REFERENCES*

1. Ainsworth, B. S., U.S. 3 000 850 (1962); C.A. 56, 2586h (1962).
2. Akhmedzade, D. A., V. D. Yasnopol'skii, D. A. Mamedova, and E. N. Gerovkova, Azerb. Khim. Zh., 1966, 70; C.A. 65, 17133b (1966).
3. Akhmedzade, D. A., V. D. Yasnopol'skii, and A. S. Zakharyan, Azerb. Khim. Zh., 1968, 80; C.A. 70, 57326k (1969).
4. Aktiebolaget, L., Brit. 778 142 (1957); C.A. 52, 2104g (1958).
5. Albright and Wilson Ltd., Brit. 931 146 (1963); C.A. 59, 11728c (1963).
6. Albright and Wilson Ltd., Fr. 1 391 765 (1965); C.A. 63, 13525f (1965).
7. Albright and Wilson Ltd., Fr. 1 435 890 (1966); C.A. 65, 18404b (1966).
8. Ando, T., H. Osuga, H. Kodera, H. Matsubara, K. Imarura, and T. Kikuchi, Japan 23 089 (1967); C.A. 68, 87855h (1968).
9. Andres, K., G. Oertel, H. Schafer, and H. Holdtsmidt, Brit. 1 119 030 (1968); C.A. 69, 44375e (1968).
10. Angert, L. G., P. A. Kirpichnikov, A. S. Kuz'minskii, V. K. Khairullin, and V. N. Borisova, U.S.S.R. 151 688 (1963); C.A. 60, 4319h (1964).
11. Anglamol Ltd., Brit. 654 968 (1951); C.A. 46, 3252i (1952).
12. Anglamol Ltd., Brit. 682 441, 682 442 (1952); C.A. 47, 6647f (1953).
13. Anon., Federal Register, 28, 1796, 5047 (1963); C.A. 59, 4474d (1963); C.A. 59, 8039h (1963).
14. Anon., Federal Register, 30, 1937 (1965); C.A. 62, 11055c (1965); C.A. 62, 7028b (1965).
15. Anon., Federal Register, 33, 569 (1968) [cf. C.A. 68, 117526 (1968)]; C.A. 68, 48330p (1968).
16. Ardis, A. E., and J. A. Wojtowicz, U.S. 3 446 819 (1969); C.A. 71, 38791n (1969).
17. Argus Chemical Corporation, Brit. 841 890 (1960); C.A. 55, 4054a (1961).
18. Argus Chemical Corporation, Brit. 945 441 (1963); C.A. 60, 9437e (1964).
19. Argus Chemical N.V., Neth. Appl. 6 612 825 (1967); C.A. 67, 11767z (1967).

*For Section H only.

20. Argus Chemical N.V., Fr. 1 519 200 (1968); C.A. 71, 4755p (1969).
21. Asseff, P. A., U.S. 3 033 789 (1962); C.A. 57, 3708i (1962).
22. Ayers, G. W., U.S. 2 866 753 (1958); C.A. 53, 8614h (1959).
23. Badische, Anilin, and A. G. Soda-Fabrik, Brit. 944 440 (1963); C.A. 60, 13164b (1964).
24. Badische, Anilin., and A. G. Soda-Fabrik, Neth. Appl. 6 413 754 (1965); C.A. 63, 18378h (1965).
25. Baer, M., U.S. 2 750 351 (1956); C.A. 51, 1657h (1957).
26. Bahr, U., K. H. Andres, and G. Broun, Ger. 1 098 707 (1961); C.A. 55, 22929i (1961).
27. Bahr, U., K. Andres. and G. Broun, U.S. 3 027 349 (1962); C.A. 57, 16901f (1962).
28. Bahr, U., G. Oertel, G. Nischk, and M. Dahm, Belg. 617 413 (1962); C.A. 58, 9302f (1963).
29. Bahr, U. G., G. Broun, and G. Nischk, Ger. 1 170 636 (1964); C.A. 61, 4560d (1964).
30. Balas, J. G., and L. M. Porter, U.S. 3 040 016 (1962); C.A. 57, 10014g (1962).
31. Bamford, C. H., and K. Hargreaves, Proc. Roy. Soc., Ser. A, 297, 425 (1967).
32. Baranauckas, C. F., and I. Gordon, Belg. 623 965 (1963); C.A. 59, 9894c (1963).
33. Baranauckas, C. F., and I. Gordon, Fr. 1 408 166 (1965); C.A. 64, 2011e (1966).
34. Baranauckas, C. F., and I. Gordon, U.S. 3 320 337 (1967); C.A. 67, 73141q (1967).
35. Baranauckas, C. F., and I. Gordon, U.S. 3 310 609 (1967); C.A. 67, 73142r (1967).
36. Baranauckas, C. F., and J. J. Hodan, U.S. 3 290 264 (1966); C.A. 66, 39056k (1967).
37. Baranauckas, C. F., and G. Irving, U.S. 3 412 051 (1968); C.A. 70, 38509f (1969).
38. N. V. de Bataafsche Petroleum Maatschappij, Dutch. 69 357 (1952); C.A. 47, 143h (1953). See also Brit. 692 261 (1953); C.A. 48, 10052i (1954).
39. Baum, B. O., U.S. 3 119 783 (1964); C.A. 60, 10884d (1964).
40. Bayer, O., H. Malz, E. Roos, and W. Goebel, Ger. 1 218 721 (1966); C.A. 65, 17152c (1966).
41. Beach, L. K., and R. Drogin, U.S. 2 863 900 (1958); C.A. 53, 9059g (1959).
42. Bean, C. T., Jr., U.S. 3 257 355 (1966); C.A. 65, 13884f (1966).
43. Becher, U., Plaste Kaut, 15, 197 (1968); C.A. 69, 3407y (1968).
44. Beindorff, A. B., U.S. 2 806 831 (1957); C.A. 52, 743g (1958).

45. Bellus, D., Z. Manasek, and J. Holcik, Chem. Prum., 15, 217 (1965); C.A. 63, 3123g (1965).
46. Bellus, D., J. Holcik, and Z. Manasek, Czech. 120 154 (1966); C.A. 68, 69763x (1968).
47. Benedetti, L., Ger. 1 150 485 (1963); C.A. 59, 11172h (1963).
48. Beriger, E., and R. Sallmann, U.S. 2 908 605 (1959); C.A. 54, 15819a (1960).
49. Berth, P., B. Blaser, G. Germschied, and K. H. Worms, Ger. 1 119 814 (1961); C.A. 57, 13902i (1962).
50. Billett, M. G., and J. Gibson, Brit. 1 052 751 (1966); C.A. 66, 48094y (1967).
51. Birum, G. H., U.S. 2 818 364 (1957); C.A. 52, 15821h (1958).
52. Birum, G. H., U.S. 3 042 701 (1962); C.A. 58, 3459b (1963).
53. Birum, G. H., and W. E. Weesner, U.S. 2 841 525 (1958); C.A. 52, 15820i (1958).
54. Birum, G. H., and G. A. Richardson, U.S. 3 160 650 (1964); C.A. 62, 7797h (1965).
55. Birum, G. H., U.S. 3 243 370 (1966); C.A. 64, 19292d (1966). (Numbers and C.A.'s for 16 related patents are given in this specification.)
56. Birum, G. H., R. B. Clampitt, and R. M. Anderson, U.S. 3 332 893 (1967); C.A. 68, 30600s (1968).
57. Birum, G. H., U.S. 3 363 031 (1968); C.A. 68, 50605a (1968).
58. Bliznynk, N. K., Z. N. Kvasha, and A. F. Kolomiets, Zh. Obshch. Khim., 37, 181 (1967); C.A. 69, 2985y (1968).
59. Boiron, P. G., Brit. 1 043 832 (1966); C.A. 65, 18794c (1966).
60. Boothman, E., Brit. 1 086 644 (1967); C.A. 68, 77945y (1968).
61. Borisov, G., P. Nikolinski, M. Grigorova, and M. Mikhailov, C. R. Acad. Bulg. Sci., 19, 733 (1966); C.A. 66, 11591g (1967).
62. Borisov, G., P. Nikolinski, M. Grigorova, and M. Mikhailov, C. R. Acad. Bulg. Sci., 19, 799 (1966); C.A. 66, 11228u (1967).
63. Bottenbruch, L., G. Fritz, and H. Schnell, Ger. 1 128 653 (1962); C.A. 57, 2408g (1962).
64. Bown, D. E., and R. I. McDougall, U.S. 3 297 631 (1967); C.A. 66, 38486p (1967).
65. Boyer, W. P., J. R. Maugham, and T. M. Melton, U.S. 3 027 348 (1962); C.A. 57, 4813b (1962).
66. Boyer, N. E., R. R. Huidersinn, and C. T. Bean, Jr., U.S. 3 321 553 (1967); C.A. 67, 33289n (1967). See also U.S. 3 278 464; C.A. 66, 19030u (1967).
67. Brachel, H. V., Ger. 1 103 328 (Appl. 1959); C.A. 56, 7176a (1962).

68. Brandt, H. J., Anal. Chem., 33, 1390 (1961).
69. Brauchesi, M., Fr. 1 561 511 (1969); C.A. 71, 102702p (1969).
70. Braus, H., and J. R. Woltermann, U.S. 3 413 258 (1968); C.A. 70, 29764h (1969).
71. Braus, H., J. E. Hager, and L. A. Hill, Jr., U.S. 3 462 375 (1969); C.A. 71, 92294r (1969).
72. Brennan, J. A., and J. W. Schick, U.S. 3 280 031 (1966); C.A. 66, 12770b (1967).
73. Brindell, G. D., Fr. 1 496 563 (1967); C.A. 69, 20065c (1968). See also Fr. 1 460 949 (1966); C.A. 67, 22498d (1967).
74. British Petroleum Company Ltd., Brit. 734 403 (1955); C.A. 50, 2156h (1956).
75. Brod, G., Ger. 1 065 617 (1959); C.A. 55, 7909c (1961).
76. Brown, E. S., and E. A. Rick, Chem. Commun., 1969, 112.
77. Buckley, R. A., U.S. 3 342 767 (1967); C.A. 67, 100718s (1967).
78. Budnick, E. G., U.S. 3 463 835 (1969); C.A. 71, 91654u (1969).
79. Burk, G. A., and D. N. De Mott, U.S. 3 456 041 (1969); C.A. 71, 61545u (1969).
80. Burpitt, R. D., U.S. 3 454 677 (1969); C.A. 71, 61544t (1969).
81. Camacho, V. G., J. J. Anderson, and W. M. Byrd, Jr., Fr. 1 515 615 (1968); C.A. 70, 114816w (1969).
82. Campbell, C. H., and L. T. Jenkins, U.S. 2 932 661 (1960); 14782g (1960).
83. Canarios, C. M., Belg. 651 024 (1965); C.A. 64, 11402a (1966).
84. Cannon, J. A., and A. Y. Coran, Brit. 909 480 (1962); C.A. 58, 5856d (1963).
85. Cannon, J. A., and J. R. Darby, U.S. 3 240 751 (1966); C.A. 64, 17808c (1966).
86. Carmody, D. R., and A. Zletz, U.S. 3 066 478 (1962); C.A. 58, 7781b (1963).
87. Casey, J. A., U.S. 3 256 237 (1966); C.A. 65, 10754f (1966).
88. Casida, J. E., R. L. Baron, M. Eto, and J. L. Engel, Biochem. Pharmacol. 12, 73 (1963); C.A. 58, 9542f (1963).
89. Chelnokova, Z. B., Yu. B. Zimin, and P. I. Levin, Vysokomol. Soedin. Ser. B, 10, 126 (1968); C.A. 68, 96468m (1968).
90. Chemische Werke Albert, Fr. 1 549 411 (1968); C.A. 71, 39804f (1969).
91. CIBA Ltd., Brit. 940 033 (1962); C.A. 60, 5340d (1964).
92. CIBA Ltd., Belg. 620 538 (1963); C.A. 58, 14229d

(1963).
93. Clark, R. F., C. D. Storrs, and G. B. Barnes, U.S. 3 364 273 (1968); C.A. 68, 95391u (1968).
94. Coates, H., Brit. 796 446 (1958); C.A. 52, 17799f (1958).
95. Coates, H., D. A. Brown, G. Quesnnel, J. G. C. Girard, and A. Thiot, Brit. 940 697 (1963); C.A. 60, 6747h (1964).
96. Coates, H., and M. V. Cooksley, Brit. 1 069 524 (1967); C.A. 67, 33277g (1967).
97. Coler, M. A., Brit. 894 123 (1962); C.A. 58, 633e (1963).
98. Condit, P. C., U.S. 3 099 132 (1963); C.A. 59, 12588g (1963).
99. Conrad, W. R., U.S. 3 330 887 (1967); C.A. 68, 40574a (1968).
100. Cook, W. S., U.S. 3 355 419 (1967); C.A. 68, 22625p (1968).
101. Coover, H. W., Jr., and J. B. Dickey, U.S. 2 784 209 (1957); C.A. 51, 13903d (1957).
102. Coover, H. W., Jr., and J. B. Dickey, U.S. 2 811 543 (1957); C.A. 52, 4922d (1958).
103. Coover, H. W., Jr., and R. L. McConnell, U.S. 2 899 455 (1959); C.A. 54, 1571i (1960).
104. Coover, H. W., Jr., and R. L. McConnell, U.S. 3 053 795 (1962); C.A. 58, 632c (1963).
105. Coover, H. W., Jr., and F. B. Joyner, U.S. 3 184 443 (1965); C.A. 63, 5772e (1965).
106. Corbett, W. M., and J. L. Carrington, Brit. 1 118 876 (1968); C.A. 69, 44540e (1968).
107. Cornish, E. H., U.S. 3 445 439 (1969); C.A. 71, 39819q (1969).
108. Crosby, G. W., and E. W. Brennan, Am. Chem. Soc. Div. Petrol. Chem., Preprints 3, No. 4A, 171 (1958); C.A. 54, 23293g (1960).
109. Cummings, W., U.S. 2 824 085 (1958); C.A. 52, 9656h (1958).
110. Dachselt, E., E. Grimm, J. Wieduwilt, and R. Wilms, Ger. (East) 55 819 (1967); C.A. 67, 82703b (1967).
111. Darby, J. R., U.S. 2 951 052 (1960); C.A. 55, 1086f (1961).
112. Davy, G. S., S. African 67 00 978 (1968); C.A. 70, 69225q (1969).
113. DeGray, R. J., U.S. 2 956 869 (1960); C.A. 55, 3047i (1961).
114. Deinet, A. J., U.S. 3 351 681 (1967); C.A. 68, 95966k (1968).
115. DeLap, R. A., U.S. 2 914 416 (1959); C.A. 54, 3952f (1960).
116. Deligny, J., and R. Zagdoun, Fr. 1 205 729 (1960); C.A. 55, 13916h (1961).

117. Demarcq, M., and G. Perot, Fr. 1 277 936 (1962); C.A. 57, 6128i (1962).
118. Dershowitz, S., U.S. 3 104 257 (1963); C.A. 60, 2848f (1964).
119. Deutsche Gold-und Silber- Scheideanstalt Vorm Roessler, Brit. 1 014 924 (1965); C.A. 64, 9914c (1966).
120. Dever, J. L., and J. J. Hodan, U.S. 3 459 835 (1969); C.A. 71, 81330y (1969).
121. De Young, E. L., U.S. 3 201 348 (1965); C.A. 63, 11228e (1965).
122. Distillers Company Ltd., Neth. Appl. 6 613 062 (1967); C.A. 67, 108439x (1967).
123. Dombrow, B. A., U.S. 3 100 752 (1963); C.A. 59, 12999f (1963).
124. Dorer, C. J., Jr., U.S. 3 416 900 (1968); C.A. 70, 77970u (1969).
125. Dorer, C. J., Jr., U.S. 3 440 247 (1969); C.A. 71, 38965x (1969).
126. Duke, R., H. P. Mayo, and L. Molinario, Fr. 1 470 166 (1967); C.A. 67, 117741s (1967).
127. Dynamit-Nobel, A. G., Belg. 617 536 (1962); C.A. 58, 9251c (1963).
128. Edl, W., H. W. Meyer, and D. Schmidt, Ger. Offen. 1 800 327 (1969); C.A. 71, 125186n (1969).
129. Eimers, E., Ger. 1 242 226 (1967); C.A. 68, 78421m (1968).
130. Elam, E. U., and J. B. Dickey, U.S. 3 336 348 (1967); C.A. 67, 108205t (1967).
131. Elliot, J. S., and E. D. Edwards, U.S. 2 820 766 (1958); C.A. 52, 8540i (1958).
132. Elliott, J. S., and E. D. Edwards, U.S. 2 971 912 (1961); C.A. 55, 11828e (1961).
133. Eliseeva, L. A., E. V. Kuznetsov, T. V. Razumovskaya, and I. M. Shermergorn, U.S.S.R. 216 722 (1968); C.A. 69, 77497z (1968).
134. Ethyl Corporation, Brit. 791 526 (1958); C.A. 53, 1253e (1959).
135. Farbenfabriken Bayer, A. G., Brit. 813 966 (1959); C.A. 53, 19878c (1959).
136. Farbenfabriken Bayer A. G., Brit. 804 080 (1958); C.A. 53, 11223i (1959).
137. Farbenfabriken Bayer A. G., Neth. Appl. 6 404 924 (1964); C.A. 63, 1893h (1965).
138. Farbenfabriken Bayer, A. G., Brit. 998 859 (1965); C.A. 63, 13498b (1965).
139. Farbenfabriken Bayer A. G., Neth. Appl. 6 508 414 (1966); C.A. 65, 901c (1966).
140. Farbenfabriken Bayer A. G., Fr. 1 486 607 (1967); C.A. 68, 50530x (1968).
141. Farbenfabriken Bayer A. G., Fr. 1 497 955 (1967);

C.A. 69, 86606m (1968).
142. Farbenfabriken Bayer A. G., Fr. 1 550 045 (1968); C.A. 71, 40314c (1969).
143. Farbwerke Hoechst A. G., Ger. 1 130 162 (1962); C.A. 57, 6131i (1962).
144. Farbwerke Hoechst A. G., Belg. 611 661 (1962); C.A. 57, 16898f (1962).
145. Farbwerke Hoechst A. G., Neth. Appl. 297 534 (1964); C.A. 62, 4171c (1965).
146. Farbwerke Hoechst A. G., Belg. 659 428 (1965); C.A. 64, 2250g (1966).
147. Farbwerke Hoechst A. G., Neth. Appl. 6 410 154 (1965); C.A. 64, 16083f (1966).
148. Fedorov, S. G., and E. E. Nifant'ev, Zh. Prikl. Khim., 39, 1881 (1966); C.A. 65, 18763c (1966).
149. Ferrari, R. J., and B. A. Hunter, Belg. 613 463 (1962); C.A. 57, 11355a (1962).
150. Ferro Corporation, Brit. 905 074 (1962); C.A. 57, 16888a (1962).
151. Fest, C., Belg. 654 748 (1965); C.A. 64, 19679f (1966).
152. Firestone Tyre and Rubber Company, Ger. Offen. 1 812 034 (1969); C.A. 71, 71700e (1969).
153. Fitch, S. J., U.S. 3 363 032 (1968); C.A. 69, 3005r (1968).
154. Ford, J. F., and P. M. Blanchard, U.S. 3 321 401 (1967); C.A. 67, 45790g (1967).
155. Freeman Chemical Corporation, Brit. 832 999 (1960); C.A. 54, 18308f (1960).
156. Friedman, L., U.S. 3 039 993 (1962); C.A. 57, 8745c (1962).
157. Friedman, L., and H. Gould, Am. Chem. Soc., Div. Polymer Chem., Preprints 4, 98 (1963); C.A. 63, 5827f (1965).
158. Friedman, L., U.S. 3 142 650 (1964); C.A. 61, 10845d (1964).
159. Friedman, L., U.S. 3 142 651 (1964); C.A. 61, 10845e (1964).
160. Friedman, L., U.S. 3 205 269 (1965); C.A. 63, 16557c (1965).
161. Friedman, L., Fr. 1 359 075 (1964); C.A. 62, 4180h (1965).
162. Friedman, L., U.S. 3 246 051 (1966); C.A. 64, 19927g (1966).
163. Friedman, L., U.S. 3 220 961 (1965); C.A. 64, 8420h (1966).
164. Friedman, L., U.S. 3 225 010 (1965); C.A. 64, 12913f (1966) [cf. C.A. 59, 6589a (1963)]; C.A. 61, 10845e (1964).
165. Friedman, L., U.S. 3 318 855 (1967); C.A. 22497c (1967).

166. Friedman, L., U.S. 3 382 191 (1968); C.A. 69, 11128y (1968).
167. Friedman, L., U.S. 3 442 982 (1969); C.A. 71, 22599v (1969).
168. Friedman, L., U.S. 3 471 592 (1969); C.A. 71, 124647b (1969).
169. Frey, H. H., Kunststoffe 52, 667 (1962); C.A. 58, 10354g (1963).
170. Fritz, G., L. Bottenbruch, and H. Schnell, U.S. 3 305 520 (1967); C.A. 66, 86268r (1967).
171. Fuchsman, C. H., and A. M. Nicholson, U.S. 3 082 187 (1963); C.A. 59, 4123c (1963).
172. Funck, D. L., Fr. 1 377 034 (1964); C.A. 62, 11974h (1965).
173. Gagliani, J., U.S. 3 192 243 (1965); C.A. 63, 9812e (1965).
174. Garner, B. L., P. J. Papillo, and H. W. Zussmann, Belg. 633 390 (1963); C.A. 61, 2017c (1964).
175. Gazieva, N. I., A. I. Shchekotikhin, and V. A. Ginsburg, U.S.S.R. 225 190 (1968); C.A. 70, 20219u (1969).
176. Gefter, E. L., and I. K. Rubtsova, U.S.S.R. 132 404 (1960); C.A. 55, 8933i (1961).
177. General Electric Company, Neth. Appl. 6 609 262 (1967); C.A. 67, 54803r (1967).
178. Gerrard, W., and E. F. Mooney, Brit. 1 015 782 (1966); C.A. 64, 9595c (1966).
179. Gevaert Photo-Production N. V., Belg. 609 497 (1962); C.A. 58, 7542g (1963).
180. Gilbert, L. F., U.S. 3 113 005 (1963); C.A. 60, 6684b (1964).
181. Girard, J. G. C., A. Thiot, and and G. Quesnnel, U.S. 3 121 697 (1964); C.A. 61, 5861d (1964).
182. Gleim, C. E., and J. E. Hrivnak, U.S. 3 361 846 (1968); C.A. 68, 50634j (1968).
183. Gmitter, G. T., U.S. 3 208 959 (1965); C.A. 63, 18394g (1965).
184. Goldsmith, H. A., and R. B. Roessler, U.S. 3 138 556 (1964); C.A. 61, 7231f (1964).
185. Goldstein, I. S., and W. J. Oberley, U.S. 3 160 515 (1964); C.A. 62, 9348h (1965).
186. Goodhue, L. D., U.S. 3 006 752 (Appl. 1958); C.A. 56, 3845d (1962).
187. Goodyear Tyre and Rubber Company, Neth. Appl. 6 510 295 (1967); C.A. 67, 33661c (1967).
188. Gotch, M., Japan 5995 (1956); C.A. 52, 19164i (1958).
189. Gould, H., and J. P. Rakus, U.S. 2 961 454 (1960); C.A. 55, 25761i (1961).
190. Goyette, L. E., U.S. 3 112 244 (1963); C.A. 60, 13816d (1964).

191. Gross, H., and B. Costisella, Agnew. Chem. Int. Ed. Engl., 7, 391 (1968).
192. Guest, H. R., and B. W. Kiff, U.S. 2 951 826 (1960); C.A. 55, 1072a (1961).
193. Guest, R., and B. W. Kiff, U.S. 2 975 856 (1960); C.A. 55, 21673b (1961).
194. Guest, H. R., and B. W. Kiff, U.S. 3 010 941 (1961); C.A. 58, 5841d (1963).
195. Guest, H. R., and B. W. Kiff, U.S. 3 012 988 (1961); C.A. 56, 13094e (1962).
196. Gumlick, W., and G. Sprock, Ger. 1 046 874 (1958); C.A. 55, 2185h (1961).
197. Gunderloy, F. C., Jr., and R. F. Neblett, J. Chem. Eng. Data, 7, 142 (1962).
198. Gurvick, Ya. A., P. A. Kirpichnikov, N. V. Tsirul'-nikova, Yu. B. Zimin, A. I. Karpycheva, and L. M. Popova, Khim. Prom., 41, 100 (1965); C.A. 62, 14558f (1965).
199. Guttag, A., U.S. 3 120 849 (1964); C.A. 60, 14703h (1964).
200. Guttag, A., U.S. 3 144 419 (1964); C.A. 61, 12163h (1964).
201. Guttag, A., U.S. 3 305 516 (1967); C.A. 66, 86196r (1967).
202. Guttag, A., U.S. 3 382 236 (1968); C.A. 69, 19481d (1968).
203. Hadert, H., Seifen-Oele-Fette-Wachse, 94, 305 (1968); C.A. 69, 88403s (1968).
204. Hagemeyer, H. J., Jr., U.S. 2 811 514 (1957); C.A. 10608g (1958).
205. Haggis, G. A., D. Williams, and A. M. Wooler, Brit. 927 175 (1963); C.A. 59, 5327b (1963).
206. Halpern, B. D., U.S. 3 069 400 (1962); C.A. 58, 8105g (1963).
207. Hans, J., Fr. 1 445 046 (1966); C.A. 66, 6634d (1967).
208. Hansen, F. R., and B. Zaremsky, U.S. 2 867 594 (1959); C.A. 53, 7671h (1959).
209. Harrison, A., U.S. 2 729 660 (1956); C.A. 50, 11361f (1956).
210. Hasserodt, U., K. H. G. Pilgram, and F. Korte, U.S. 3 318 915 (1967); C.A. 68, 13169r (1968) [cf. U.S. 3 247 113 (1966)]; C.A. 64, 19679c (1966).
211. Hatakeyama, H., and O. Fukumoto, Japan. 6257 (1967); C.A. 68, 70106s (1968).
212. Haward, R. N., and T. H. Boultbee, Brit. 803 557 (1958); C.A. 53, 5750d (1959).
213. Haward, R. N., and C. C. Gosselin, Brit. 846 684 (1960); C.A. 55, 8941h (1961).
214. Haward, R. N., and C. C. Gosselin, Brit. 866 883 (1961); C.A. 55, 24116f (1961).

215. Haward, R. N., and C. C. Gosselin, Ger. 1 169 662 (1964); C.A. 61, 9638e (1964).
216. Hechenbleikner, I., U.S. 2 839 563 (1958); C.A. 53, 292f (1959).
217. Hechenbleikner, I., U.S. 3 147 298 (1964); C.A. 64, 8086b (1966).
218. Hechenbleikner, I., and F. C. Lanoue, U.S. 2 847 443 (1958); C.A. 52, 19251i (1958).
219. Hechenbleikner, I., and F. C. Lanoue, U.S. 3 281 381 (1966); C.A. 66, 29612z (1967).
220. Hechenbleikner, I., and C. W. Pause, U.S. 2 960 463 (1960); C.A. 57, 3362c (1962).
221. Hechenbleikner, I., and K. R. Molt, Belg. 622 418 (1962); C.A. 59, 11424c (1963).
222. Hechenbleikner, I., and K. R. Molt, U.S. 3 096 345 (1963); C.A. 59, 12646g (1963).
223. Hecker, A. C., and W. E. Deistner, U.S. 2 860 115 (1958); C.A. 53, 5748f (1959).
224. Heckner, A. C., and W. E. Leistner, Brit. 884 145 (1961); C.A. 57, 1081h (1962).
225. Hecker, A. C., and N. L. Perry, Belg. 626 323 (1963); C.A. 60, 16062g (1964).
226. Hecker, A. C., M. W. Pollock, and S. Cohen, Ger. 1 150 202 (1963); C.A. 59, 8941h (1963).
227. Hecker, A. C., O. S. Kauder, and N. L. Perry, U.S. 3 255 136 (1966); C.A. 65, 12366g (1966).
228. Hecker, A. C., M. W. Pollock, and S. Cohen, U.S. 3 364 159 (1968); C.A. 68, 50604z (1968).
229. Hemphill, D. D., Proc. North Central Weed Control Conf. 14th, 47 (1957); C.A. 55, 882i (1961).
230. Henkel & Cie. G.m. b.H., Brit. 810 642 (1959); C.A. 53, 15979h (1959).
231. Heufer, G., O. Mauz, and E. Prinz, S. African 68 07 199 (1969); C.A. 71, 82170h (1969).
232. Heyden Newport Chemical Corporation, Brit. 907 877 (1962); C.A. 58, 14227c (1963).
233. Hindersinn, R. R., M. Worsley, and B. O. Schoepfle, Brit. 919 067 (1963); C.A. 58, 14246g (1963).
234. Hindersinn, R. R., and N. E. Boyer, Ind. Eng. Chem., Prod. Res. Develop., 3, 141 (1964).
235. Hodan, J. J., and W. L. Schall, Ger. Offen. 1 800 675 (1969); C.A. 71, 13823m (1969).
236. Hofinger, G., Austrian, 259 232 (1968); C.A. 68, 50570k (1968).
237. Hoke, D. I., U.S. 3 108 119 (1963); C.A. 60, 3014f (1964).
238. Holcik, J., D. Bellus, Z. Manasek, and A. Rybar, Czech. 117 976 (1966); C.A. 66, 116289t (1967).
239. Holock, K. E., U.S. 3 414 536 (1968); C.A. 70, 38510z (1969).
240. Holtschmidt, H., and G. Braun, Ger. 1 143 638

(1963); C.A. 58, 12739d (1963).
241. Hook, E. O., and D. P. Tate, U.S. 3 023 092 (1962); C.A. 56, 13162g (1962).
242. Hooker Chemical Corporation, Brit. 937 560 (1963); C.A. 60, 15733g (1964).
243. Hooker Chemical Corporation, Belg. 621 145 (1962); C.A. 60, 2768e (1964).
244. Hooker Chemical Corporation, Brit. 965 785 (1964); C.A. 63, 495e (1965).
245. Hooker Chemical Corporation, Neth. Appl. 6 400 517 (1964); C.A. 62, 4174b (1965).
246. Hooker Chemical Corporation, Neth. Appl. 300 379 (1965); C.A. 64, 11401e (1966).
247. Hooker Chemical Corporation, Neth. Appl. 6 611 208 (1967); C.A. 67, 109830e (1967).
248. Hopff, H., and W. Zimmerli, Swiss 436 732 (1967); C.A. 68, 115394c (1968).
249. Hort, E. V., U.S. 2 950 290 (1960); C.A. 55, 10385f (1961).
250. Horun, S., Material Plast. 2, 385 (1965); C.A. 65, 5593a (1966).
251. Hu, S., U.S. 3 459 662 (1969); C.A. 71, 83328c (1969).
252. Huhn, J. S. F., and F. J. Sheets, Brit. 898 000 (1962); C.A. 57, 16885i (1962).
253. Huhn, J. S. F., and F. J. Sheets, U.S. 3 061 583 (1962); C.A. 58, 10362f (1963).
254. Hunter, B. A., Ger. 924 532 (1955); C.A. 51, 18681g (1957).
255. Hunter, B. A., Fr. 1 496 562 (1967); C.A. 69, 20232e (1968).
256. Huwyler, R., R. Aenishaenslin, and D. Porrett, Swiss. 456 949 (1968); C.A. 69, 78116m (1968).
257. Ichikawa, A., T. Ozeki, M. Musha, and H. Otani, Japan. 9266 (1967); C.A. 67, 91649r (1967).
258. Illing, G., Belg. 634 559 (1964); C.A. 61, 8482f (1964).
259. Illing, G., Fr. 1 368 372 (1964); C.A. 62, 6633h (1965).
260. Illing, G., Ger. 1 164 077 (1964); C.A. 61, 8486d (1964).
261. Illing, G., Brit. 1 019 348 (1966); C.A. 64, 11400c (1966).
262. Imperial Chemical Industries Ltd., Brit. 791 283 (1958); C.A. 52, 14184a (1958).
263. Imperial Chemical Industries Ltd., Belg. 671 112 (1966); C.A. 66, 46926x (1967).
264. Imperial Chemical Industries Ltd., Neth. Appl. 6 602 632 (1966); C.A. 66, 56140a (1967).
265. Imperial Chemical Industries Ltd., Neth. Appl. 6 605 060 (1966); C.A. 66, 56194w (1967).

266. Imperial Chemical Industries Ltd., Fr. Addn. 90 054 (1967); C.A. 69, 20059d (1968).
267. Imperial Chemical Industries Ltd., Fr. 1 549 277 (1968); C.A. 71, 39866c (1969).
268. Imperial Chemical Industries Ltd., Fr. 1 558 350 (1969); C.A. 71, 82541e (1969).
269. Ishida, S., H. Ohama, and H. Fukuda, Japan. 69 07 947 (1969); C.A. 71, 31050m (1969).
270. Japanese Geon Co Ltd., Brit. 1 145 715 (1969); C.A. 70, 97806y (1969).
271. Jasinski, A. L., U.S. 3 414 463 (1968); C.A. 70, 29731v (1969).
272. Jason, E. F., and E. K. Fields, U.S. 3 217 030 (1965); C.A. 64, 2018h (1966).
273. Jenkins, R. L., U.S. 2 426 691 (1947); C.A. 42, 586 (1948).
274. Jonas, H., P. Hoppe, and E. Walaschewski, Ger. 1 138 219 (1962); C.A. 58, 3568c (1963).
275. Jones, R. W., Fr. 1 379 528 (1964); C.A. 63, 4495e (1965).
276. Jordon, D. P., and D. H. Antonsen, Ind. Eng. Chem., Prod. Res. Develop., 8, 208 (1969).
277. Juredine, G. M., U.S. 3 284 386 (1966); C.A. 66, 19236r (1967).
278. Juredine, G. M., U.S. 3 356 617 (1967); C.A. 68, 30700z (1968).
279. Jureviciene, R., and Y. A. Shlyapnikov, Vysokomolekul. Soedin. Ser. B, 11, 444 (1969); C.A. 71, 81753p (1969).
280. Kadyrova, V. Kh., P. A. Kirpichnikov, and L. G. Tokareva, Trudy Kazansk. Khim. Tekhnol. Inst. No. 30, 58 (1962); C.A. 60, 3109d (1964).
281. Kagan, F., and R. D. Birkenmeyer, U.S. 2 995 576 (1961); C.A. 56, 3355f (1962).
282. Kalle, A. G., Fr. 1 381 924 (1964); C.A. 63, 752c (1965).
283. Kaplan, M., and M. Koral, U.S. 3 007 884 (Appl. 1959); C.A. 56, 4978a (1962).
284. Kauder, O. S., and W. E. Leistner, Belg. 626 102 (1963); C.A. 59, 5325b (1963).
285. Kauder, O. S., Fr. 1 519 200 (1968); C.A. 71, 4755p (1969).
286. Kautsky, G. J., E. G. Lindstrom, and M. R. Barusch, Fr. 1 315 358 (1963); C.A. 59, 1426c (1963).
287. Kazimirchik, I. V., G. F. Bebikh, F. S. Denisov, and M. I. Kabachnik, Zh. Obshch. Khim., 36, 1226 (1966); C.A. 65, 16889h (1966).
288. Kearney, J. J., U.S. 3 026 308 (1962); C.A. 57, 3628c (1962).
289. Keihei, K., and H. Sato, Japan. 9470 (1956); C.A. 52, 17691g (1958).

290. Keil, H., Ger. 1 271 874 (1968); C.A. 69, 53402p (1968).
291. Keil, H., Ger. 1 286 675 (1969); C.A. 70, 69978n (1969).
292. Kellerman, D., R. Wardi, and H. Bernstein, Brit. 1 112 976 (1968); C.A. 69, 59365a (1968).
293. Kennedy, J., E. S. Lane, and B. K. Robinson, J. Appl. Chem. (London), 8, 459 (1958); C.A. 53, 3032h (1959).
294. Kennedy, J., and G. E. Ficken, J. Appl. Chem. (London), 8, 465 (1958); C.A. 53, 3033g (1959).
295. Kikuchi, M., and M. Matsukane, Japan. 3596 (1961); C.A. 55, 21661h (1961).
296. Kirpichnikov, P. A., P. I. Levin, A. G. Popov, L. M. Popova, and L. G. Tokareva, Trudy Kazansk. Khim. Tekhnol. Inst. No. 34, 374 (1965); C.A. 68, 79013s (1968).
297. Kirpichnikov, P. A., and N. A. Mukmeneva, Starenie Stabil. Polim 1966, 168; C.A. 68, 78794k (1968).
298. Kirpichnikov, P. A., N. A. Mukmeneva, A. N. Pudovik, and L. M. Yartseva, U.S.S.R. 162 959 (1964); C.A. 62, 1599c (1965).
299. Kirpichnikov, P. A., N. A. Mukmeneva, and A. N. Pudovik, Sin. Issled. Eff. Stabil. Polim. Mater., Voronezh 1964, 61; C.A. 66, 46882e (1967).
300. Kirpichnikov, P. A., and L. D. Romanova, Plast. Massy, 1965, 8; C.A. 62, 9298e (1965).
301. Klapovskaya, A. O., N. V. Andrianova, B. M. Kovarskaya, I. K. Rubstova, and V. I. Kirilovich, Plast. Massy, 1967, 69; C.A. 67, 117664u (1967).
302. Kline, R. H., Fr. 1 319 836 (1963); C.A. 58, 14273a (1963).
303. Kodolov, V. I., and S. S. Spasskii, Trudy Inst. Khim. Akad. Nauk SSSR, Ural. Filial, 1968, No. 15, 15; C.A. 70, 38391m (1969).
304. Kolodny, E. R., M. Grayson, and R. L. Myers, Fr. 1 487 270 (1967); C.A. 68, 704552e (1968).
305. Kolomiets, A. F., G. S. Levskaya, and N. K. Bliznyuk, U.S.S.R. 192 203 (1967); C.A. 69, 3008u (1968).
306. Kondrat'eva, L. V., B. V. Kotov., E. L. Geffer, and V. I. Serenkov, Vysokomolekul. Soedin. Ser. B, 9, 751 (1967); C.A. 68, 13549q (1968).
307. Kormatsu, H., M. Tanaka, and E. Hayashi, Japan. 69 02 479 (1969); C.A. 70, 107003a (1969).
308. Koral, J. N., U.S. 3 412 055 (1968); C.A. 70, 30190z (1966).
309. Kotrelev, V. N., B. M. Kovarskaya, Yu. V. Tsvetkova, and I. I. Levantovskaya, Plast. Massy, 1969, 30; C.A. 72, 22218r (1970).
310. Kovarskaya, B. M., P. M. Tanunina, I. I. Levantovskaya, L. P. Litvak, P. A. Kirpichnikov, and Ya. A.

Gurvich, Plast. Massy, 1951, 7; C.A. 63, 11792f (1965).
311. Krajewski, R. M., and H. C. Kelly, U.S. 2 887 396 (1959); C.A. 53, 15600d (1959).
312. Krasilova, O. I., A. V. Pavlov, L. V. Krondrat'eva, and V. I. Serenkov, Vysokomolekul. Soedin. Ser. B, 11, 678 (1969); C.A. 72, 32571q (1970).
313. Kubo, M., H. Hatakeyama, and O. Fukomoto, Japan 16 304 (1966); C.A. 66, 56533f (1967).
314. Kuliev, A. M., I. M. Orudzheva, G. A. Zeinalova, A. A. Atal'yan, D. A. Akhmed-Zade, A. M. Levshina, K. I. Sadykhov, and A. B. Abdinova, Trudy 1-oi (Pervoi) Konf., Zakavkazsk. Univ., 1959, 111; C.A. 56, 7583g (1962).
315. Kuliev, A. M., P. S. Mamedova, and I. M. Orudzheva, Azerb. Khim. Zh., 1963, 63; C.A. 61, 504b (1964).
316. Kuliev, A. M., I. M. Orudzheva, Sh. M. Novruzov, and P. S. Mamedova, Neftepererab. Neftekhim 1967, 21; C.A. 67, 34607b (1967).
317. Kuliev, A. M., I. M. Orudzheva, P. S. Mamedova, and Sh. A. Ragimova, Prisadki Smaz. Maslam., 1967, 161; C.A. 69, 108348h (1968).
318. Kuliev, A. M., and A. B. Kuliev, Prisadki Smaz. Maslam., 1967, 88; C.A. 70, 3411r (1969).
319. Kumnick, M. C., and F. K. Watson, Ger. 1 045 090 (1958); C.A. 55, 2049a (1961).
320. Kunoshirua Kagaku Kogyo Kabushiki Kaisha, Brit. 902 580 (1962); C.A. 57, 11418i (1962).
321. Kurashiki Rayon Company, Ltd., Brit. 1 060 401 (1967); C.A. 67, 12432p (1967).
322. Kutepow, N. V., H. Seibt, and F. Meier, Ger. 1 144 268 (1963); C.A. 59, 3790g (1963).
323. Kuznetsov, E. V., and I. N. Fiazullin, U.S.S.R. 146 040 (1962); C.A. 57, 10050e (1962).
324. Kuznetsov, E. V., V. I. Gusev, L. A. Shurygina, and L. S. Semenova, Trudy Kazansk. Khim. Tekhnol. Inst. No. 34, 316 (1965); C.A. 68, 87804r (1968).
325. Kuznetsov, E. V., V. P. Arkhireev, and M. V. Batalina, U.S.S.R. 174 356 (1965); C.A. 64, 643b (1966).
326. Ladd, E. C., and M. P. Harvey, U.S. 2 651 656 (1953); C.A. 48, 10052h (1954).
327. Ladd, E. C., and M. P. Harvey, Can. 509 034 (1955); C.A. 50, 10760g (1956).
328. Lal, J., and J. E. McGrath, Fr. 1 461 530 (1966); C.A. 67, 22737f (1967).
329. Lane, C. A., U.S. 2 934 554 (1960); C.A. 54, 18360b (1960).
330. Lane, E. S., J. Appl. Chem., 8, 687 (1958); see also C.A. 53, 3032h (1959); C.A. 55, 13899a (1961); C.A. 55, 18650h (1961).
331. Lanham, W. M., Brit. 922 198 (1963); C.A. 60, 6749d

(1964).
332. Lanham, W. M., and P. L. Smith, U.S. 3 270 096 (1966); C.A. 65, 18500f (1966).
333. Lankro Chemicals Ltd., Neth. Appl. 6 411 230 (1965); C.A. 63, 5862e (1965).
334. Larrison, M. S., U.S. 3 285 998 (1966); C.A. 66, 39650z (1967).
335. Larrison, M. S., Fr. 1 455 466 (1966); C.A. 66, 95813s (1967).
336. Larrison, M. S., U.S. 3 328 493 (1967); C.A. 67, 64990p (1967).
337. Larrison, M. S., U.S. 3 341 629 (1967); C.A. 67, 100668a (1967).
338. Larrison, M. S., U.S. 3 354 241 (1967); C.A. 68, 22549s (1968).
339. Larrison, M. S., U.S. 3 356 770 (1967); C.A. 68, 30703c (1968).
340. Larrison, M. S., U.S. 3 364 153 (1968); C.A. 68, 50527b (1968).
341. Larrison, M. S., U.S. 3 375 304 (1968); C.A. 68, 96559s (1968).
342. Larrison, M. S., U.S. 3 376 364 (1968); C.A. 68, 105789w (1968).
343. Larrison, M. S., U.S. 3 382 299 (1968); C.A. 69, 20066d (1968).
344. Larrison, M. S., U.S. 3 383 436 (1968); C.A. 69, 18830e (1968).
345. Larrison, M. S., Fr. 1 456 954 (1966); C.A. 67, 11209j (1967).
346. Larrison, M. S., Fr. 1 566 352 (1969); C.A. 71, 113740a (1969).
347. Lazcano, C. S., Span. 274 113 (1962); C.A. 59, 11729h (1963).
348. Leistner, W. E., A. C. Hecker, and O. H. Knoepke, U.S. 2 997 454 (Appl. 1959); C.A. 56, 594d (1962).
349. Leistner, W. E., and A. C. Hecker, Belg. 640 568 (1964); C.A. 62, 14897g (1965).
350. Levin, P. I., Zh. Fiz. Khim., 38, 672 (1964); C.A. 61, 1728a (1964).
351. Levin, P. I., and T. A. Bulgakova, Vysokomolekul. Soedin., 6, 700 (1964); C.A. 61, 9634e (1964).
352. Levin, P. I., P. A. Kirpichnikov, A. F. Lukovnikov, and M. S. Khloplyankina, Vysokomolekul. Soedin., 5, 1152 (1963); C.A. 59, 12981b (1963).
353. Levin, P. I., and L. V. Ruban, Kauch. Rezina, 26, 23 (1967); C.A. 66, 116434m (1967).
354. Levin, P. I., Yu. B. Zimin., L. S. Solodar., and M. B. Neiman, U.S.S.R. 218 420 (1968); C.A. 69, 78170z (1968).
355. Liepins, R., and C. S. Marvel, J. Polymer Sci., 5, 2453 (1967).

356. Lines, W., U.S. 3 392 032 (1968); C.A. 69, 44442z (1968).
357. Lines, E. W., Brit. 1 153 966 (1969); C.A. 71, 71636p (1969).
358. Lipke, P. H., Jr., and R. S. Montgomery, U.S. 2 752 319 (1956); C.A. 51, 14324a (1957).
359. Lorenz, W., Ger. 930 212 (1955); C.A. 52, 14700g (1958).
360. Lorenz, W., and R. Coelln, U.S. 2 962 518 (1960); C.A. 59, 2860f (1963).
361. Lorenz, W., and G. Schrader, Ger. 1 050 331 (1959); C.A. 55, 3430b (1961).
362. Louthan, R. P., U.S. 3 050 452 (1962); C.A. 57, 16404a (1962).
363. Lowenstein-Lom, W., U.S. 2 769 013 (1956); C.A. 51, 3988f (1957).
364. Lowes, F. J., and R. F. Monroe, U.S. 2 809 982 (1957); C.A. 52, 2902f (1958).
365. Lusebrink, R., and W. M. Sawyer, U.S. 2 765 221 (1956); C.A. 51, 3130c (1957).
366. Luttinger, L. B., U.S. 2 951 055 (1960); C.A. 55, 1086h (1961).
367. Mack, G. P., U.S. 2 970 980, 2 970 981 (1961); C.A. 55, 10977a (1961).
368. Mack, P., and E. Parker, U.S. 2 938 877 (1960); C.A. 55, 5035h (1961).
369. Mack, G. P., and E. Parker, Ger. 1 175 874 (1964); C.A. 61, 14857f (1964).
370. Maier, L., U.S. 3 320 251 (1967); C.A. 67, 54262p (1967)[Cf. Brit. 1 068 364 (1967); C.A. 67, 54260m (1967)].
371. Malz, H., O. Bayer, and H. Groene, Ger. 1 144 914 (1963); C.A. 58, 14272g (1963).
372. Mamedor, F. A., F. N. Mamedov, and R. S. Khalilov, Azerb. Khim. Zh., 1965, 64; C.A. 64, 4829b (1966).
373. Mark, V., U.S. 3 328 472 (1967); C.A. 67, 118098z (1967).
374. Mather, J., Brit. 1 102 710 (1968); C.A. 68, 70128a (1968).
375. Matson, H. J., and J. W. Nelson, U.S. 3 239 464 (1966); C.A. 64, 17334b (1966).
376. Matsui, T., I. Suyama, T. Nishizaki, and T. Fujita, Japan. 69 02 115 (1969); C.A. 71, 92289t (1969).
377. Matveeva, E. N., P. A. Kirpichnikov, M. Z. Kreman, N. A. Obol'yaninova, N. A. Lazareva, and L. M. Popova, Plast. Massy, 1964, 37; C.A. 60, 14681b (1964); cf. 60, 9183e (1964).
378. McCaee, M., and H. W. Coover, Jr., Fr. 1 478 047 (1967); C.A. 68, 12557d (1968).
379. McConnell, R. L., and H. W. Coover, U.S. 2 952 701 (1960); C.A. 55, 3430d (1961).
380. McConnell, R. L., and H. W. Coover, U.S. 3 030 340

(1962); C.A. 57, 2416b (1962).
381. McConnell, R. L., and H. W. Coover, Jr., U.S. 3 377 409 (1968); C.A. 68, 115204r (1968).
382. McGrath, H., Brit. 1 002 929 (1965); C.A. 63, 15054d (1965).
383. McGrath, J. J., and H. O. Strange, U.S. 3 318 810 (1967); C.A. 67, 23806q (1967).
384. McMaster, E. L., and H. Tolkmith, U.S. 2 764 560 (1956); C.A. 51, 3874e (1957).
385. McMaster, E. L., and W. K. Glesner, U.S. 2 764 561 (1956); C.A. 51, 3874i (1957).
386. McWhorter, W. F., J. E. Kuhn, and H. S. Klassen, Ger. 1 106 493 (Appl. 1959); C.A. 56, 4955b (1962).
387. McWhorter, W. F., J. E. Kuhn, and H. C. Klassen, U.S. 3 352 826 (1967); C.A. 68, 87895w (1968).
388. Megna, I. S., and F. A. Sullivan, Ger. Offen. 1 912 179 (1969); C.A. 72, 4302j (1970).
389. Meiller, F., Fr. 1 355 621 (1964); C.A. 61, 4513g (1964).
390. Mel'nikov, N. N., B. A. Khaskin, and N. B. Petruchenko, Khim. Org. Soedin. Fosfora, Akad. Nauk SSSR, Otd. Obshch. Tekh. Khim., 1967, 283; C.A. 69, 43990q (1968).
391. Merten, R., O. Bayer, G. Braun, and H. Kaiser, Ger. 1 206 152 (1965); C.A. 64, 12913c (1966).
392. Messina, V., and D. R. Senior, S. African 67 07 230 (1968); C.A. 71, 51983w (1969).
393. Mikhailov, N. V., L. G. Tokareva, K. K. Buravchenko, G. M. Terekhova, and P. A. Kirpichnikov, Vysokomolekul. Soedin., 4, 1186 (1962); C.A. 59, 813c (1963).
394. Mikhailov, N. V., L. G. Tokareva, and A. G. Popov, Vysokomolekul. Soedin., 5, 188 (1963); C.A. 59, 2979h (1963).
395. Mikhailov, N. V., L. G. Tokareva, G. M. Terekhova, B. V. Petukhov, K. K. Buravchenko, and P. A. Kirpichnikov, U.S.S.R. 141 584 (1961); C.A. 56, 11845h (1962).
396. Mills, K. R., and A. A. Harban, Fr. 1 391 241 (1965); C.A. 63, 8581f (1965).
397. Mills, K. R., U.S. 3 409 587 (1968); C.A. 70, 20693b (1969).
398. Minekawa, S., K. Yamaguchi, and K. Toyomoto, Japan. 69 08 747 (1969); C.A. 71, 113949a (1969).
399. Minsker, K. S., P. A. Kirpichnikov, G. T. Fedoseeva, N. S. Kolyubakina, I. L. Benderskii, and L. V. Verizhnikov, Plast. Massy, 1968, 35; C.A. 69, 19940c (1968).
400. Minsker, K. S., P. A. Kirpichnikov, N. S. Kolyubakina, I. L. Benderskii, G. T. Fedoseeva, N. A. Mukmeneva, and L. V. Verizhnikov, Vysokomolekul.

Soedin, Ser. A, 10, 2500 (1968); C.A. 70, 38423y (1969).
401. Mitin, Yu. V., and G. P. Vlasov, Dokl. Akad. Nauk SSSR, 179, 353 (1968); C.A. 69, 77719y (1968).
402. Mitsubishi Rayon Company Limited, Fr. 1 533 474 (1968); C.A. 71, 71856k (1969).
403. Monsanto Chemicals Company, Brit. 876 443 (1961); C.A. 56, 11798f (1962).
404. Monsanto Company, Brit. 941 706 (1963); C.A. 67, 100240e (1967).
405. Monsanto Company, Brit. 961 797 (1964); C.A. 61, 8505e (1964).
406. Morris, E. K., U.S. 2 489 225 (1949); C.A. 44, 1703e (1950).
407. Morris, R. C., and J. L. Van Winkle, U.S. 2 728 806 (1955); C.A. 50, 10760e (1956).
408. Morris, R. E., and R. D. Taylor, Belg. 622 653 (1963); C.A. 59, 7430f (1963).
409. Muller, P., Swiss 329 037 (1958); C.A. 53, 7012a (1959).
410. Murai, K., T. Ito, and S. Kagami, Japan 13 775 (1962); C.A. 58, 11541a (1963).
411. Myers, H., U.S. 3 329 742 (1967); C.A. 67, 81937u (1967).
412. Myers, H., and W. F. Olszewskii, U.S. 3 351 554 (1967); C.A. 68, 59104u (1968).
413. Nagano, M., and S. Ogura, Brit. 1 153 161 (1969); C.A. 71, 41033x (1969).
414. Nagelschmidt, R. (Mrs.), M. Goecke, and H. Kopsch, Belg. 622 690 (1963); C.A. 59, 10319b (1963).
415. Nakaguchi, K., and M. Hirooka, Japan. 9432 (1963); 9433 (1963); C.A. 59, 8895g (1963).
416. Nakamura, T., K. Eto, and K. Take, Japan. 68 01 503 (1968); C.A. 69, 28530w (1968).
417. Nakamura, T., and T. Ueda, Agr. Biol. Chem., 31, 1288 (1967); C.A. 68, 21151u (1968) [cf. C.A. 64, 14502a (1966)].
418. Nakamura, Y., and M. Manabe, Japan. 68 21 527 (1968); C.A. 70, 79428x (1969).
419. Nakaoka, T., H. Matsubara, K. Imamura, and T. Kikuchi, Japan. 10 488 (1967); C.A. 67, 90937w (1967).
420. National Polychemicals, Inc., Neth. Appl. 6 516 646 (1965); C.A. 66, 19032w (1967).
421. National Polychemicals, Inc., Neth. Appl. 6 410 226 (1965); C.A. 63, 5840e (1965).
422. National Polychemicals, Inc., Brit. 1 136 578 (1968); C.A. 70, 48447y (1969).
423. Neiman, M. B., B. M. Kovarskaya, I. I. Levantovskaya, G. V. Dralyuk, M. P. Yazvikova, V. A. Sidorov, V. N. Kochetkov, G. M. Trossman, G. O.

Tatevos'yan, and I. B. Kuznetsova, Plast. Massy, 1962, 6; C.A. 58, 8097h (1963).
424. Nelson, J. W., U.S. 3 159 533 (1964); C.A. 62, 3978f (1965).
425. Nelson, J. W., H. J. Matson, and D. W. Young, U.S. 3 245 979 (1966); C.A. 64, 17334d (1966).
426. Nifant'ev, E. E., and L. P. Levitan, U.S.S.R. 166 694 (1964); C.A. 62, 7895h (1965).
427. Nifant'ev, E. E., and S. G. Fedorov, Dokl. Akad. Nauk SSSR, 164, 1327 (1965); C.A. 64, 3768e (1966) [cf. C.A. 61, 14847b (1964)].
428. Nifant'ev, E. E., I. V. Fursenko, and D. A. Predvoditelev, U.S.S.R. 216 725 (1968); C.A. 69, 76628f (1968).
429. Nifant'ev, E. E., E. A. Kirichenko, T. G. Shestakova, and G. N. Kashirina, U.S.S.R. 220 495 (1968); C.A. 69, 77983e (1968).
430. Nifant'ev, E. E., S. M. Markov, and A. P. Tuseev, Vysokomol. Soedin. 7, 1020 (1965); C.A. 63, 10084a (1965).
431. Nischk, G., and G. Braun, Ger. 1 106 067 (1961); C.A. 55, 27998c (1961).
432. Nischk, G., U. Bahr, and K. Andres, Ger. 1 122 252 (1962); C.A. 56, 11799i (1962).
433. Nischk, G., and G. Braun, Ger. 1 129 686 (1962); C.A. 57, 6154b (1962).
434. Nischk, G., Ger. 1 144 275 (1963); C.A. 59, 7561c (1963).
435. Nishizawa, Y., M. Nakagawa, and Y. Suzuki, Japan 21 321 (1961); C.A. 58, 4425e (1963).
436. Nottes, G., H. J. Sasse, and H. Lemme, Ger. 1 093 168 (1960); C.A. 55, 22806a (1961).
437. Nowach, R., and W. Herwig, Ger. 1 232 142 (1967); C.A. 67, 32776g (1967).
438. Nukushina, K., I. Nakamura, and I. Tamai, Japan. 16 794 (1963); C.A. 59, 14133 (1963).
439. Nunn, L. G., U.S. 3 150 161 (1964); C.A. 62, 737f (1965).
440. Nudenberg, W., and D. B. Merifield, U.S. 3 080 338 (1963); C.A. 58, 14272h (1963).
441. Oertel, G., H. Holtschmidt, and G. Braun, Belg. 617 051 (1962); C.A. 58, 3458f (1963).
442. Oertel, G., H. Holtschmidt, and R. Merten, Fr. 1 332 901, 1 332 902 (1963); C.A. 59, 15445g (1963).
443. Oertel, G., H. Holtschmidt, G. Nischk, and G. Braun, Ger. 1 127 583 (1962); C.A. 57, 10035i (1962).
444. O'Halloran, R., U.S. 3 388 066 (1968); C.A. 69, 37692c (1968); see also 68, 31833g (1968).
445. Oldsberg, L. O., U.S. 3 355 418 (1967); C.A. 68, 13756e (1968).
446. Ono, M., and M. Okamoto, Japan, 68 05 395 (1968);

C.A. 69, 76679y (1968).
447. Orloff, H. D., and G. G. Knapp, U.S. 3 115 463 (1963); C.A. 60, 6685d (1965).
448. Orloff, H. D., and G. G. Knapp, U.S. 3 115 464 (1963); C.A. 60, 10458e (1964).
449. Orlov, N. F., and M. V. Androsova, Izv. Vyssh. Ucheb. Zaved., Tekhnol. Tekst. Prom., 1967, 113; C.A. 67, 44680j (1967).
450. Orudzheva, I. M., Sh. M. Novruzov, A. A. Vagabora, and P. S. Mamedova, Neftepererab. Neftekhim., 1967, 28, C.A. 68, 115380v (1968).
451. Orudzheva, I. M., Sh. M. Novruzov, A. A. Vagabova, and S. A. Rasulova, Azerb. Khim. Zh., 1967, 47; C.A. 69, 37621d (1968).
452. Orudzheva, I. M., and S. A. Rasulova, Azerb. Neft. Khoz., 47, 36 (1968); C.A. 69, 78980p (1968).
453. Osborn, H., and D. V. Haren, Brit. 1 086 396 (1967); C.A. 68, 3691r (1968).
454. Oswald, H. J., and E. Turi, U.S. 3 364 169 (1968); C.A. 68, 50623e (1968).
455. Oswald, J., E. Turi, and R. B. Lund, U.S. 3 425 987 (1969); C.A. 70, 69058n (1969).
456. Ozawa, S., and K. Maekawa, Kobunshi Kagaku, 20, 357 (1963); C.A. 61, 14854c (1964).
457. Palm, C., Ger. 1 230 010 (1966); C.A. 66, 104696k (1967).
458. Pan, H., and T. L. Fletcher, J. Org. Chem., 27, 3639 (1962); C.A. 57, 16513c (1962) [cf. C.A. 55, 27231h (1961)].
459. De Paolo, P. A., Jr., and R. A. Doran, S. African 68 00 278 (1968); C.A. 70, 48265n (1969).
460. De Paolo, P. A., and H. P. Smith, Advan. Chem. Ser., 85, 202 (1968).
461. Parker, E., U.S. 3 245 926 (1966); C.A. 64, 19911b (1966).
462. Pellegrini, J. P., Jr., and H. I. Thayer, U.S. 2 863 743 (1958); C.A. 53, 8614a (1959).
463. Pengilly, B. W., and R. J. Stein, Fr. 1 446 948 (1966); C.A. 66, 38504t (1967).
464. Perry, E., U.S. 3 293 207 (1966); C.A. 66, 38492n (1967).
465. Perkow, W., Ger. 941 972 (1956); C.A. 53, 9153i (1959).
466. Petro-Tex Chemical Corporation, Neth. Appl. 6 507 591 (1966); C.A. 67, 2776q (1967).
467. Petrov, K. A., and E. E. Nifant'ev, Vysokomolekul. Soedin., 4, 242 (1962); C.A. 57, 3665e (1962).
468. Petrov, K. A., E. E. Nifant'ev, and R. G. Gol'tsova, Vysokomolekul. Soedin., Geterotsepnye Vysokomolekul. Soedin., 1964, 170; C.A. 61, 5783a (1964).
469. Petrov, K. A., E. E. Nifant'ev, R. G. Gol'tsova,

and S. M. Korneev, Vysokomolekul. Soedin., Geterotsepnye Vysokomolekul. Soedin., 1964, 68; C.A. 61, 5781e (1964).
470. Petrov, K. A., E. E. Nifant'ev, R. G. Gol'stova, and L. M. Solntseva, Vysokomolekul. Soedin., 4, 1219 (1962); C.A. 59, 4048g (1963).
471. Petrov, K. A., E. E. Nifant'ev, R. G. Gol'tsova, and L. M. Solntseva, U.S.S.R., 152 571 (1964); C.A. 62, 9260d (1965).
472. Petrov, K. A., E. E. Nifant'ev, and T. N. Lysenko, Vysokomolekul. Soedin., Geterotsepnye Vysokomolekul. Soedin., 1964, 240; C.A. 61, 5991h (1964).
473. Petrov, K. A., E. E. Nifant'ev, T. N. Lysenko, and A. I. Suzanskii, Vysokomolekul. Soedin., 5, 712 (1963); C.A. 60, 12120g (1964).
474. Petrov, A. K., E. E. Nifant'ev, I. I. Sopikova, and M. A. Belaventsev, U.S.S.R. 165 697 (1964); C.A. 62, 5441e (1965).
475. Petrov, K. A., E. E. Nifant'ev, I. I. Sopikova, and M. I. Merkulova, Vysokomolekul. Soedin., Tsellyulozy Proizvodnye, Sb. Statei, 1963, 86; C.A. 60, 10913a (1964).
476. Petrov, K. A., and I. I. Sopikova, Zh. Prikl. Khim., 40, 1564 (1967); C.A. 67, 101148t (1967).
477. Pfizer, Chas., and Company, Inc., Brit. 1 117 407 (1968); C.A. 69, 43683s (1968).
478. Phillipson, J. M., Brit. 1 132 701 (1968); C.A. 70, 20218a (1969).
479. Pleshakov, M. G., T. I. Tikhonova, A. M. Kulier., I. I. Namazov, F. N. Mamedov, N. P. Mustafaev, and N. M. Novruzov, Khim. Volokna, 1968 70; C.A. 70, 68955r (1969).
480. Pollock, M. W., Belg. 658 003 (1965); C.A. 64, 3802b (1966).
481. du Pont de Nemours & Co., Brit. 905 824 (1962); C.A. 58, 5870g (1963).
482. Popkin, A. H., U.S. 3 047 372 (1962); C.A. 59, 7288g (1963).
483. Porret, D., and P. Kohler, Swiss 331 847 (1958); C.A. 53, 5196h (1959).
484. Predvoditelev, D. A., E. E. Nifant'ev, and Z. A. Rogovin, Vysokomolekul. Soedin., 7, 1005 (1965); C.A. 63, 18444c (1965).
485. Predvoditelev, D. A., E. E. Nifant'ev, and Z. A. Rogovin, Vysokomolekul. Soedin., 8, 213 (1966); C.A. 64, 19963d (1966).
486. Predvoditelev, D. A., E. E. Nifant'ev, and Z. A. Rogovin, Zh. Prikl. Khim., 40, 413 (1967); C.A. 67, 3577n (1967).
487. Predvoditelev, D. A., M. A. Tyuganova, E. E. Nifant'ev, and Z. A. Rogovin, Zh. Vses. Khim.

Obshchestva D. I. Mendeleeva, 10, 459 (1965); C.A. 63, 15087c (1965).
488. Predvoditelev, D. A., M. A. Tyuganova, E. E. Nifant'ev, and Z. A. Rogovin, Zh. Prikl. Khim., 40, 171 (1967); C.A. 66, 77115m (1966).
489. Progil, S. A., Neth. Appl. 6 607 598 (1966); C.A. 66, 95912y (1967).
490. Pure Chemicals Ltd., Fr. 1 293 479 (1962); C.A. 58, 1600f (1963).
491. Pure Chemicals Ltd., Fr. 1 295 668 (1962); C.A. 58, 1600d (1963).
492. Pure Chemicals Ltd., Brit. 949 914 (1964); C.A. 61, 2037a (1964).
493. Pure Chemicals Ltd., Brit. 1 006 580 (1965); C.A. 64, 3724a (1966) [cf. C.A. 52, 19251i (1958); C.A. 55, 25761i (1961)].
494. Pure Chemicals Ltd., Brit. 1 011 290 (1965); C.A. 64, 11402d (1966).
495. Pure Chemicals Ltd., Brit. 1 037 371 (1966); C.A. 65, 15229a (1966).
496. Pure Chemicals Ltd., Brit. 1 118 559 (1968); C.A. 69, 44386j (1968).
497. Pure Chemicals Ltd., Brit. 1 136 038 (1968); C.A. 70, 46836a (1969).
498. Quesnel, G. L., J. Firard, and A. Thiot, Fr. 1 213 894 (1960); C.A. 55, 17103i (1961).
499. Rabourn, W. J., and E. E. Jones, U.S. 3 322 863 (1967); C.A. 67, 54261n (1967).
500. Raden, D. S., U.S. 3 406 225 (1968); C.A. 70, 11228b (1969).
501. Raeuber, A. E., E. A. Verchot, and S. E. Mileski, U.S. 3 313 650 (1967); C.A. 67, 44950x (1967).
502. Rafikov, S. R., G. N. Chelnokova, and G. M. Dzhilkibaeva, Vysokomolekul. Soedin. Ser. B, 10, 329 (1968); C.A. 69, 28047u (1968).
503. Rapoport, I. A., Dokl. Akad. Nauk SSSR, 134, 1447 (1960); C.A. 55, 9666b (1961).
504. Rath, H., O. De Riz, H. J. Rothe, and R. Kioch, Brit. 1 132 454 (1968); C.A. 70, 38814b (1969).
505. Razumov, A. I., and V. V. Moskva, U.S.S.R. 175 961 (1965); C.A. 64, 9596h (1966).
506. Reichel, I., and D. Purdela, Acad. Rep. Populare Romine, Baza Cercetari Stiint Timisoara, Studii Cercetari Stiinte Chim., 5, 71 (1958); C.A. 55, 1540d (1961).
507. Riley, C. P., Jr., U.S. 3 422 030 (1969); C.A. 70, 58617t (1969).
508. Ritchie, K., and J. W. Addleburg, Brit. 996 254 (1965); C.A. 63, 7202h (1965).
509. Ritchie, K., and J. W. Addleburg, Ger. 1 298 261 (1969); C.A. 71, 82822x (1969).

510. Rode, V. V., T. N. Balykova, and S, R. Rafikov, Dokl. Akad. Nauk, 176 606 (1967); C.A. 68, 3744k (1968).
510a. Rogovin, Z. A., M. A. Tyuganova, G. A. Gabrielyan, and N. F. Konnova, Khim. Volokna, 1966, 27; C.A. 65, 7333f (1966).
511. Romanick, J. F., and A. B. Pruiksma, Brit. 1 137 046 (1968); C.A. 70, 48449a (1969).
512. Romanov, L. M., L. M. Matveeva, and A. N. Shapovalova, U.S.S.R. 205 293 (1967); C.a. 69, 10926v (1968).
513. Rombusch, K., and U. Eichers, Ger. 1 247 007 (1967); C.A. 67, 82803j (1967).
514. Rombusch, K., and U. Eichers, Ger. 1 248 932 (1967); C.A. 67, 91334j (1967).
515. Rosenfelder, W. J., and E. Breznak, U.S. 3 133 043 (1964); C.A. 61, 4552 (1964).
516. Ross, S. D., U.S. 2 934 681 (1960); C.A. 54, 15032e (1960).
517. Roussos, M., and Y. Bourgeois, Fr. 1 500 821 (1967); C.A. 69, 86604j (1968).
518. Roussos, M., Y. Bourgeois, and B. Delmas, Fr. 1 548 478 (1968); C.A. 71, 22598u (1969).
519. Rubtsova, I. K., V. I. Kirihovich, N. V. Andrianov, O. A. Klapovskaya, and G. I. Zhigadlo, U.S.S.R. 178 103 (1966); C.A. 64, 19910c (1966).
520. Rybar, A., J. Holcik, J. Masek, and F. Strand, Fr. 1 461 139 (1966); C.A. 67, 22740b (1967); Czech. 115 828 (1965); C.A. 65, 901f (1966).
521. Sakurai, S., and K. Inouek, Japan. 8625 (1957); C.A. 52, 11727g (1958).
522. Sallmann, R., Swiss 327 715 (1958); C.A. 53, 7012c (1959).
523. Sallmann, R., Ger. 974 569 (1961); C.A. 56, 10640i (1962).
524. Samuel, A. A., R. Bouvet, S. Hittner, and M. de Beaulieu, Fr. 1 157 174 (1958); C.A. 54, 18360d (1960).
525. Sandri, J. M., U.S. 3 318 980 (1967); C.A. 67, 55955d (1967).
526. Sanin, P. I., E. S. Shepeleva, A. V. Ul'yanova, and B. V. Kleimenov, Wear, 3, 200 (1960); C.A. 54, 25731b (1960).
527. Sanin, P. I., E. S. Shepeleva, A. V. Ul'yanova, and B. V. Kleimenov, Tr. Inst. Nefti, Akad. Nauk SSSR, 14, 98 (1960); C.A. 56, 13158f (1962).
528. Sanin, P. I., and A. V. Ul'yanova, Prisadki Maslam Toplivam, Trudy Naucha. Tekhn. Soveshch. 1960, 189; C.A. 56, 13157i (1962).
529. Sanin, P. I., and A. V. Ul'yanova, Neftekhimiya, 3, 775 (1963); C.A. 60, 3921g (1964) [cf. C.A. 56, 13158f (1962)].
530. Schlueter, D., Ger. 1 229 296 (1966); C.A. 66, 19241p (1967).

531. Schmelzer, H. G., and H. Holtschmidt, Fr. 1 373 471 (1964); C.A. 62, 10622b (1965).
532. Schmidt, P., Swiss 314 635 (1956); C.A. 52, 2066g (1958).
533. Schmidt, P., Swiss 316 731 (1956); C.A. 52, 2066h (1958).
534. Schrader, G., U.S. 2 870 191 (1959); C.A. 53, 11223g (1959).
535. Schrader, G., U.S. 2 976 312 (1961); C.A. 62, 446c (1965).
536. Schrader, M., and I. Lerner, U.S. 3 404 023 (1968); C.A. 70, 5027u (1969).
537. Schwenke, K. D., Faserforsch. Textiltech., 15, 297 (1964); C.A. 62, 1560f (1965) [cf. C.A. 54, 18379h (1960)].
538. Schwyzer, R., B. M. Iselin, H. Kappeler, and W. Rittel, Ger. 1 121 059 (1962); C.A. 59, 7646h (1963).
539. Schwyzer, R., B. M. Iselin, and W. Rittel, U.S. 2 938 915 (1960); C.A. 54, 20897g (1960).
540. Sconce, J. S., J. J. Hodan, and W. L. Schall, U.S. 3 415 907 (1968); C.A. 70, 38687n (1969).
541. Scott, L. B., A. C. Nixon, and D. G. Roddick, U.S. 2 763 617 (1956); C.A. 51, 700c (1957).
542. Scullin, J. P., and A. F. Fletcher, Belg. 635 728 (1964); C.A. 62, 1803c (1965).
543. Scullin, J. P., and A. F. Fletcher, U.S. 3 321 423 (1967); C.A. 67, 3318w (1967).
544. Seda de Barcelona, S. A., Span. 333 714 (1967); C.A. 68, 30993x (1968).
545. Shell Internationale Research, Maatschappij N. V., Brit. 869 969 (1961); C.A. 57, 1068h (1962).
546. Shell Internationale Research, Maatschappij N. V., Belg. 613 863 (1962); C.A. 58, 1600e (1963).
547. Shell Internationale Research, Maatschappij N. V., Neth. Appl. 6 505 404 (1965); C.A. 64, 11008d (1966).
548. Shell Internationale Research, Maatschappij N. V., Neth. Appl. 6 612 239 (1967); C.A. 67, 110332a (1967).
549. Shell Internationale Research Maatschappij N. V., Neth. Appl. 6 606 664 (1967); C.A. 68, 50633h (1968).
550. Shepherd, W. T., U.S. 3 458 443 (1969); C.A. 71, 83334b (1969).
551. Shimomura, T., K. Ando, E. Tsuchida, and I. Shinohara, Kogo Kagaku Zasshi, 71, 279 (1968); C.A. 69, 27922p (1968).
552. Shimoi, M., Y. Koda, and H. Hattori, Japan 68 03 012 (1968); C.A. 69, 44396n (1968).
553. Shokal, E. C., U.S. 2 840 617 (1958); C.A. 52,

16795b (1958).
554. Shokal, E. C., U.S. 3 044 996 (1962); C.A. 57, 11386i (1962).
555. Siegle, G., and Co. G.m.b.H., Fr. 1 513 317 (1968); C.A. 70, 58630s (1969).
556. Smith, C. W., G. B. Payne, and E. C. Shokal, U.S. 2 856 369 (1958); C.A. 53, 2686i (1959).
557. Smith, J. C., B. H. Miles, L. Levine, and W. E. Presley, U.S. 3 297 796 (1967); C.A. 67, 82265k (1967).
558. Smith, P. V., Jr., F. K. Knoth, Jr., and W. E. Waddey, U.S. 2 722 517 (1955); C.A. 50, 3751i (1956).
559. Societe Anon. Argus-Chemical N.V., Neth. Appl. 6 612 825 (1967); C.A. 67, 117676z (1967).
560. Spacht. R. B., and C. E. Gleim, Fr. 1 450 685 (1966); C.A. 66, 116286q (1967).
561. Spasskii, S. S., and M. E. Mat'kova, Zh. Obshch. Khim., 29, 3438 (1959); C.A. 54, 15234c (1960).
562. Spiegler, L., U.S. 2 957 903 (1960); C.A. 55, 6438a (1961).
563. Spivack, J. D., U.S. 3 367 870 (1968); C.A. 68, 78420k (1968).
564. Sroog, C. E., U.S. 2 728 790 (1955); C.A. 50, 10759e (1956).
565. Stamicarbon, N. V., Belg. 613 285 (1962); C.A. 57, 16880h (1962).
566. Standard Electrica, S. A., Span. 290 150 (1963); C.A. 61, 13497d (1964).
567. Standard Oil Development Co., Brit. 676 553 (1952); C.A. 47, 7544c (1953).
568. Stanley, J. W., Jr., U.S. 3 296 221 (1967); C.A. 66, 56392j (1967).
569. Starmer, P. H., and A. H. Jorgensen, Jr., Ger. Offen. 1 808 484 (1969); C.A. 71, 31195n (1969).
570. Stauffer Chemical Company, Neth. Appl. 6 613 257 (1967); C.A. 67, 65056a (1967).
571. Strauss, R., and J. Bottomley, U.S. 3 367 996 (1968); C.A. 68, 79142h (1968).
572. Sumi, M., and M. Imoto, Makromol. Chem., 50, 161 (1961); C.A. 56, 14463h (1962).
573. Sumi, M., Japan. 20 986 (1963); C.A. 60, 3123h (1964).
574. Sun Oil Company, Brit. 888 673 (1962); C.A. 56, 10397g (1962).
575. Sun Oil Company, Brit. 936 494 (1963); C.A. 60, 729h (1964).
576. Suzuki, M., K. Watanabe, and S. Yokobori, Japan. 18 985 (1961); C.A. 60, 9387f (1964).
577. Suzuki, M., K. Watanabe, and M. Yokohori, Japan. 16 280 (1961); C.A. 57, 3593g (1962).

578. Szczepanek, A., and G. Koenen, Ger. 1 295 821 (1969); C.A. 71, 22601q (1969).
579. Takigawa, M., T. Takada, and K. Harada, Japan 17 073 (1966); C.A. 66, 76576a (1967).
580. Tarasova, Z. N., G. A. Kirpichnikov, and T. F. Fedorova, Kauchuk Rezina, 22, 14 (1963); C.A. 60, 5719b (1964).
581. Taylor, G. W., and D. H. Wood, Brit. 1 066 404 (1967); C.A. 67, 12147z (1967).
582. Taylor, G. W., and D. H. Wood, U.S. 3 450 802 (1969); C.A. 71, 49542h (1969).
583. Tenneco Chemicals, Inc., Neth. Appl. 6 601 832 (1966); C.A. 66, 3197t (1967).
584. Terekhova, G. M., N. V. Mickhailov, and L. G. Tokareva, Khim. Volokna, 1964, 33; C.A. 61, 13480d (1964).
585. Thie, V., and H. Wurker, Ger. 1 093 361 (1960); C.A. 55, 25142h (1961).
586. Thinius, K., Plaste Kautschuk, 11, 324 (1964); C.A. 63, 10124b (1965).
587. Tholstrup, C. E., and J. C. Ownby, Fr. Addn. 88 842 (1967); C.A. 67, 91374x (1967).
588. Thornley, G. H., U.S. 2 796 400 (1957); C.A. 51, 15113f (1957).
589. Thurston, E. F. W., Ger. 1 187 925 (1965); C.A. 62, 11336h (1965).
590. Tokisawa, M., Japan 5 436 (1967); C.A. 67, 54790j (1967).
591. Toyo Rayon Company, Ltd., Fr. 1 457 783 (1966); C.A. 67, 11930n (1967).
592. Toyo Rayon Company, Ltd., Fr. 1 481 339 (1967); C.A. 68, 2568u (1968).
593. Trescher, V., G. Braun, and H. Nordt, Ger. 1 106 489 (1961); C.A. 55, 26524d (1961).
594. Uchiyama, Y., Y. Arima, and K. Taniguchi, Japan 7891 (1967); C.A. 67, 53664j (1967).
595. Union Carbide Corporation, Fr. 1 298 363 (1962); C.A. 58, 3562g (1963).
596. Union Carbide Corporation, Neth. Appl. 6 408 010 (1965); C.A. 63, 7191f (1965).
597. Union Carbide Corporation, Neth. Appl. 6 609 452 (1967); C.A. 67, 12661n (1967).
598. Uniroyal, Inc., Brit. 1 143 375 (1969); C.A. 70, 96394a (1969).
599. Upjohn Company, Brit. 1 165 994 (1969); C.A. 71, 125256k (1969).
600. Utsunomiya, T., Japan. 14 166 (1967); C.A. 68, 13722r (1968).
601. Verizhnikov, L. V., P. A. Kirpichnikov, and L. G. Angert, Trudy Kazansk. Khim. Tekhnol. Inst. No. 33, 287 (1964); C.A. 65, 2160c (1966).

602. Vinogradov, G. V., A. Ya. Malkin, L. A. Mishina, and G. A. Ermilova, Plast. Massey, 1967, 62; C.A. 68, 60181y (1968).
603. Volgina, N. M., I. M. Seregina, L. A. Vol'f, Yu. K. Kirilenko, S. N. Borisov, E. E. Nifant'ev, N. G. Sviridova, and K. D. Tammik, Zh. Prikl. Khim. (Leningrad), 41, 2563 (1968); C.A. 70, 485342z (1969).
604. Vrany, M., and I. Ruzicka, Czech. 125 577 (1967); C.A. 69, 97450g (1968).
605. Wagner, F. A., Fr. 1 465 976 (1967); C.A. 67, 54737x (1967).
606. Wagner, F. A., Brit. 1 090 249 (1967); C.A. 68, 22463j (1968).
607. Wagner, F. S., Jr., S. African 67 02 846 (1968); C.A. 71, 37904q (1969).
608. Wallis, M., and E. Griebsch, Ger. 1 154 624 (1963); C.A. 59, 14168e (1963).
609. Warren, G. W., U.S. 3 124 551 (1964); C.A. 60, 14689d (1964).
610. Warren, G. W., U.S. 3 127 369 (1964); C.A. 60, 16060f (1964).
611. Warren, G. W., U.S. 3 149 181 (1964); C.A. 62, 1803h (1965).
612. Wasserbach, T. B., and R. W. Walker, U.S. 2 764 866 (1956); C.A. 51, 3986a (1957).
613. Wasson, J. I., and D. L. Bonham, U.S. 2 672 444 (1954); C.A. 48, 9680c (1954).
614. Waters, R. M., and J. C. Smith, U.S. 3 372 208 (1968); C.A. 68, 87856j (1968).
615. Watson, F. J., R. C. Morris, and J. L. Van Winkle, U.S. 2 753 368 (1956); C.A. 51, 1246b (1957).
616. Wehrle, J. J., U.S. 2 638 448 (1953); C.A. 47, 7205h (1953).
617. Wehrle, J. J., U.S. 2 674 603 (1954); C.A. 48, 11777d (1954).
618. Wenborne, R. F. A. E., and D. M. Soul, Fr. 1 456 756 (1966); C.A. 67, 4686j (1967).
619. Williams, D., Brit. 997 952 (1965); C.A. 63, 14759h (1965).
620. Williamson, W. I., Brit. 1 077 390 (1967); C.A. 68, 31196b (1968).
621. Wittenberg, D., Ger. 1 244 172 (1967); C.A. 67, 90456p (1967).
622. Wolff, F., Ger. 1 044 506 (1958); C.A. 55, 9855i (1961).
623. Wright, B., and G. R. Williamson, Brit. 937 630 (1963); C.A. 60, 731h (1964).
624. Wright, B., and G. R. Williamson, Brit. 983 119 (1965); C.A. 62, 11978b (1965).
625. Wu, C., U.S. 3 341 632 (1967); C.A. 68, 86840n

(1968).
626. Yamada, N., K. Shimada, and T. Takemura, Japan. 18 338 (1967); C.A. 68, 22347z (1968).
627. Yamamoto, K., K. Ema, M. Iwasa, and A. Arai, Japan. 69 01 239 (1969); C.A. 70, 107149c (1969).
628. Yoshioka, S., and T. Shiomura, Japan. 22 052 (1967); C.A. 40290e (1968).
629. Yuldashev, A., U. M. Muratova, and M. A. Askarov, Vysokomolekul. Soedin., 7, 1923 (1965); C.A. 64, 3824f (1966).
630. Zaslavskii, Yu. S., G. I. Shor, and R. N. Shneerova, Trudy 3-ei (Tret'ei) Vsesoyuz, Konf Treniyu Iznosu Mashinakh, Akad. Nauk SSSR., Inst. Mashinoved (1958), 3, 348 (1960); C.A. 55, 26421e (1961).
631. Zaslavskii, Yu. S., G. M. Shor, E. V. Evstigneev, F. B. Lebedeva, and I. G. Kirillov, Trudy, Vses. Nauch., Issled. Inst. Pererab. Nefti No. 10, 191 (1967); C.A. 68, 61271q (1968).
632. Zavlin, P. M., M. A. Sokolovskii, and R. S. Tenisheva, Zh. Prikl. Khim., 37, 928 (1964); C.A. 61, 3275f (1964).
633. Zavlin, P. M., M. A. Sokolovskii, and R. S. Tenisheva, U.S.S.R. 163 750 (1964); C.A. 61, 13507b (1964).
634. Zhukov, V. P., and L. A. Pershina, U.S.S.R. 242 892 (1969); C.A. 71, 82825a (1969).
635. Zil'berman, E. N., S. B. Meiman, and A. E. Kulikova, Vysokomolekul. Soedin., Ser. A, 9, 1554 (1967); C.A. 67, 82525v (1967).
636. Zimmermann, H., Faserforsch. Textiltech., 19, 372 (1968); C.A. 69, 87694a (1968).
637. Zeuch, E. A., and R. A. Gray, U.S. 3 310 591 (1967); C.A. 66, 115360x (1967).